Plant Microbiology

Plant Microbiology

Editor: Bill Docherty

www.callistoreference.com

Callisto Reference,
118-35 Queens Blvd., Suite 400,
Forest Hills, NY 11375, USA

Visit us on the World Wide Web at:
www.callistoreference.com

© Callisto Reference, 2018

ISBN: 978-1-63239-980-9 (Hardback)

Trademark Notice: Registered trademark of products or corporate names are used only for explanation and identification without intent to infringe.

Cataloging-in-Publication Data

Plant microbiology / edited by Bill Docherty.
    p. cm.
Includes bibliographical references and index.
ISBN 978-1-63239-980-9
1. Plants--Microbiology. 2. Agriculture. I. Docherty, Bill.
QR351 .P53 2018
579.178--dc23

Table of Contents

Preface...IX

Chapter 1   **Protozoan ALKBH8 Oxygenases Display both DNA Repair and tRNA Modification Activities**..1
Daria Zdżalik, Cathrine B. Vågbø, Finn Kirpekar, Erna Davydova, Alicja Puścian, Agnieszka M. Maciejewska, Hans E. Krokan, Arne Klungland, Barbara Tudek, Erwin van den Born and Pål Ø. Falnes

Chapter 2   **Efficient Transformation of Oil Palm Protoplasts by PEG-Mediated Transfection and DNA Microinjection**..14
Mat Yunus Abdul Masani, Gundula A. Noll, Ghulam Kadir Ahmad Parveez, Ravigadevi Sambanthamurthi and Dirk Prüfer

Chapter 3   **Arbuscular-Mycorrhizal Networks Inhibit *Eucalyptus tetrodonta* Seedlings in Rain Forest Soil Microcosms**..25
David P. Janos, John Scott, Catalina Aristizábal and David M. J. S. Bowman

Chapter 4   **A Rapid, Highly Efficient and Economical Method of *Agrobacterium*-Mediated *In planta* Transient Transformation in Living Onion Epidermis**..............36
Kedong Xu, Xiaohui Huang, Manman Wu, Yan Wang, Yunxia Chang, Kun Liu, Ju Zhang, Yi Zhang, Fuli Zhang, Liming Yi, Tingting Li, Ruiyue Wang, Guangxuan Tan and Chengwei Li

Chapter 5   **Recombinant Plants Provide a New Approach to the Production of Bacterial Polysaccharide for Vaccines**..43
Claire M. Smith, Stephen C. Fry, Kevin C. Gough, Alexandra J. F. Patel, Sarah Glenn, Marie Goldrick, Ian S. Roberts, Garry C. Whitelam and Peter W. Andrew

Chapter 6   **High-Throughput Construction of Intron-Containing Hairpin RNA Vectors for RNAi in Plants**...51
Pu Yan, Wentao Shen, XinZheng Gao, Xiaoying Li, Peng Zhou and Jun Duan

Chapter 7   **6-Hydroxy-3-Succinoylpyridine Hydroxylase Catalyzes a Central Step of Nicotine Degradation in *Agrobacterium tumefaciens* S33**...59
Huili Li, Kebo Xie, Haiyan Huang and Shuning Wang

Chapter 8   **Isolation of a Novel Peroxisomal Catalase Gene from Sugarcane, which is Responsive to Biotic and Abiotic Stresses**...67
Yachun Su, Jinlong Guo, Hui Ling, Shanshan Chen, Shanshan Wang, Liping Xu, Andrew C. Allan and Youxiong Que

Chapter 9   **A Novel Dual Allosteric Activation Mechanism of *Escherichia coli* ADP-Glucose Pyrophosphorylase: The Role of Pyruvate**..78
Matías D. Asención Diez, Mabel C. Aleanzi, Alberto A. Iglesias and Miguel A. Ballicora

Chapter 10   **Carbon Transfer from the Host to *Tuber melanosporum* Mycorrhizas and Ascocarps followed using a ¹³C Pulse-Labeling Technique**..86
François Le Tacon, Bernd Zeller, Caroline Plain, Christian Hossann, Claude Bréchet and Christophe Robin

Chapter 11   **pSiM24 is a Novel Versatile Gene Expression Vector for Transient Assays as well as Stable Expression of Foreign Genes in Plants**..95
Dipak Kumar Sahoo, Nrisingha Dey and Indu Bhushan Maiti

Chapter 12   **Determining the GmRIN4 Requirements of the Soybean Disease Resistance Proteins Rpg1b and Rpg1r using a *Nicotiana glutinosa*-Based Agroinfiltration System**..113
Ryan Kessens, Tom Ashfield, Sang Hee Kim and Roger W. Innes

Chapter 13   **Interaction of *Medicago truncatula* Lysin Motif Receptor-Like Kinases, NFP and LYK3, Produced in *Nicotiana benthamiana* Induces Defence-Like Responses**..120
Anna Pietraszewska-Bogiel, Benoit Lefebvre, Maria A. Koini, Dörte Klaus-Heisen, Frank L. W. Takken, René Geurts, Julie V. Cullimore and Theodorus W. J. Gadella

Chapter 14   **Generation and Characterization of the Western Regional Research Center Brachypodium T-DNA Insertional Mutant Collection**..133
Jennifer N. Bragg, Jiajie Wu., Sean P. Gordon, Mara E. Guttman, Roger Thilmony, Gerard R. Lazo, Yong Q. Gu and John P. Vogel

Chapter 15   **Development of an Agrobacterium-Mediated Stable Transformation Method for the Sensitive Plant *Mimosa pudica***..147
Hiroaki Mano, Tomomi Fujii, Naomi Sumikawa, Yuji Hiwatashi and Mitsuyasu Hasebe

Chapter 16   **Overexpression of the *AtSHI* Gene in Poinsettia, *Euphorbia pulcherrima*, Results in Compact Plants**..158
M. Ashraful Islam, Henrik Lütken, Sissel Haugslien, Dag-Ragnar Blystad, Sissel Torre, Jakub Rolcik, Søren K Rasmussen, Jorunn E Olsen and Jihong Liu Clarke

Chapter 17   **Temperature Effects on Bacterial Phytochrome**..168
Ibrahim Njimona, Rui Yang and Tilman Lamparter

Chapter 18   **Detection of the Virulent Form of AVR3a from *Phytophthora infestans* following Artificial Evolution of Potato Resistance Gene *R3a***..176
Sean Chapman, Laura J. Stevens, Petra C. Boevink, Stefan Engelhardt, Colin J. Alexander, Brian Harrower, Nicolas Champouret, Kara McGeachy, Pauline S. M. Van Weymers, Xinwei Chen, Paul R. J. Birch and Ingo Hein

Chapter 19   **Profound Impact of Hfq on Nutrient Acquisition, Metabolism and Motility in the Plant Pathogen *Agrobacterium tumefaciens***..187
Philip Möller, Aaron Overlöper, Konrad U. Förstner, Tuan-Nan Wen, Cynthia M. Sharma, Erh-Min Lai and Franz Narberhaus

Chapter 20   **Cloning of Insertion Site Flanking Sequence and Construction of Transfer DNA Insert Mutant Library in *Stylosanthes Colletotrichum***..202
Helong Chen, Caiping Hu, Kexian Yi, Guixiu Huang, Jianming Gao, Shiqing Zhang, Jinlong Zheng, Qiaolian Liu and Jingen Xi

Chapter 21   **Enhancement of Lipid Productivity in Oleaginous *Colletotrichum* Fungus through Genetic Transformation using the Yeast *CtDGAT2b* Gene under Model-Optimized Growth Condition**..216
Prabuddha Dey, Nikunj Mall, Atrayee Chattopadhyay Monami Chakraborty and Mrinal K. Maiti

Permissions

List of Contributors

Index

Preface

Every book is initially just a concept; it takes months of research and hard work to give it the final shape in which the readers receive it. In its early stages, this book also went through rigorous reviewing. The notable contributions made by experts from across the globe were first molded into patterned chapters and then arranged in a sensibly sequential manner to bring out the best results.

Plant microbiology is a branch of microbiology that studies the benefits of microorganisms found in the soil. Soil microorganisms provide nutrients, improve soil structure and reduce soil erosion. Some of the microorganisms present in the soil are protozoa, actinomycetes, fungi and bacteria. Fungi like mycorrhiza form a mutualistic relationship with the roots of plants and provide constant access to nutrients like glucose, sucrose, iron, etc. This book discusses the fundamentals as well as modern approaches of plant microbiology. It strives to provide a fair idea about this discipline and to help develop a better understanding of the latest advances within this field. Students, researchers, experts and all associated with plant microbiology will benefit alike from this book.

It has been my immense pleasure to be a part of this project and to contribute my years of learning in such a meaningful form. I would like to take this opportunity to thank all the people who have been associated with the completion of this book at any step.

Editor

Protozoan ALKBH8 Oxygenases Display both DNA Repair and tRNA Modification Activities

Daria Zdżalik[1,2], Cathrine B. Vågbø[3], Finn Kirpekar[4], Erna Davydova[1], Alicja Puścian[2,5], Agnieszka M. Maciejewska[6], Hans E. Krokan[3], Arne Klungland[7,8], Barbara Tudek[2,6], Erwin van den Born[1], Pål Ø. Falnes[1]*

1 Department of Biosciences, University of Oslo, Oslo, Norway, 2 Institute of Genetics and Biotechnology, University of Warsaw, Warsaw, Poland, 3 Department of Cancer Research and Molecular Medicine, Norwegian University of Science and Technology, Trondheim, Norway, 4 Department of Biochemistry and Molecular Biology, University of Southern Denmark, Odense, Denmark, 5 Department of Neurophysiology, Nencki Institute of Experimental Biology, Polish Academy of Sciences, Warsaw, Poland, 6 Institute of Biochemistry and Biophysics, Polish Academy of Sciences, Warsaw, Poland, 7 Clinic for Diagnostics and Intervention and Institute of Medical Microbiology, Oslo University Hospital, Rikshospitalet, Oslo, Norway, 8 Institute of Basic Medical Sciences, University of Oslo, Oslo, Norway

Abstract

The ALKBH family of Fe(II) and 2-oxoglutarate dependent oxygenases comprises enzymes that display sequence homology to AlkB from *E. coli*, a DNA repair enzyme that uses an oxidative mechanism to dealkylate methyl and etheno adducts on the nucleobases. Humans have nine different ALKBH proteins, ALKBH1–8 and FTO. Mammalian and plant ALKBH8 are tRNA hydroxylases targeting 5-methoxycarbonylmethyl-modified uridine (mcm^5U) at the wobble position of $tRNA^{Gly(UCC)}$. In contrast, the genomes of some bacteria encode a protein with strong sequence homology to ALKBH8, and robust DNA repair activity was previously demonstrated for one such protein. To further explore this apparent functional duality of the ALKBH8 proteins, we have here enzymatically characterized a panel of such proteins, originating from bacteria, protozoa and mimivirus. All the enzymes showed DNA repair activity *in vitro*, but, interestingly, two protozoan ALKBH8s also catalyzed wobble uridine modification of tRNA, thus displaying a dual *in vitro* activity. Also, we found the modification status of $tRNA^{Gly(UCC)}$ to be unaltered in an ALKBH8 deficient mutant of *Agrobacterium tumefaciens*, indicating that bacterial ALKBH8s have a function different from that of their eukaryotic counterparts. The present study provides new insights on the function and evolution of the ALKBH8 family of proteins.

Editor: Albert Jeltsch, Universität Stuttgart, Germany

Funding: This work was supported by the Research Council of Norway; the Norwegian Cancer Society; the Polish-Norwegian Research Programme (PNRF-143-A1-1/07) and by Iceland, Liechtenstein and Norway by means of co-financing from the European Economic Area Financial Mechanism and the Norwegian Financial Mechanism as part of the Scholarship and Training Fund. The funders had no role in study design, data collection and analysis, decision to publish, or preparation of the manuscript.

Competing Interests: The authors have declared that no competing interests exist.

* E-mail: pal.falnes@ibv.uio.no

Introduction

Over 30 years ago, inactivation of the *alkB* gene of *Escherichia coli* was shown to result in hypersensitivity towards certain methylation agents that target the DNA bases, such as methyl methane sulfonate (MMS) [1]. Correspondingly, *E. coli* AlkB (EcAlkB) was later shown to be a repair enzyme that was able to demethylate the DNA lesions 1-methyladenine and 3-methylcytosine, as well as their structural analogs 1-methylguanine and 3-methylthymine [2–6]. AlkB belongs to the superfamily of Fe(II) and 2-oxoglutarate (2OG) dependent dioxygenases, a group of enzymes which require ferrous iron and couple the oxidation of a primary substrate to the decarboxylation of the co-substrate 2OG [7,8]. In the AlkB reaction, hydroxylation of the deleterious methyl group is followed by spontaneous release of the resulting hydroxymethyl moiety as formaldehyde (Fig. 1A) [3,6]. EcAlkB is active both on single-stranded (ss) and double-stranded (ds) DNA, and, intriguingly, also on RNA substrates, suggesting a possible role in RNA repair [9–11]. In addition to methyl lesions, EcAlkB can also repair bulkier adducts, such as ethyl, hydroxyethyl, propyl and hydroxypropyl groups, as well as exocyclic etheno, ethano, hydroxyethano, and hydroxypropano adducts [12–18]. Repair of etheno adduct leads to the release of the etheno moiety as glyoxal (Fig. 1B) [12].

Mammals have nine different AlkB homologues (ALKBH): ALKBH1 to ALKBH8, as well as the fat mass- and obesity-associated protein FTO [7,19,20]. *In vitro* and *in vivo* studies indicated that ALKBH2 and ALKBH3 are repair enzymes with a function similar to that of EcAlkB, while non-repair functions have been demonstrated for several other ALKBHs [9,10,13,21,22]. ALKBH8 was shown to be a bifunctional tRNA modification enzyme (see below), while FTO and ALKBH5 both demethylate the mRNA modification N^6-methyladenine [23,24]. Knock-out mice lacking ALKBH1 or ALKBH4 displayed elevated levels of methylation on specific lysine residues in histone 2B or actin, respectively, suggesting that these proteins may be lysine-specific protein demethylases [25,26].

tRNAs from all three kingdoms of life are subject to extensive post-transcriptional modification, and approximately 100 distinct-

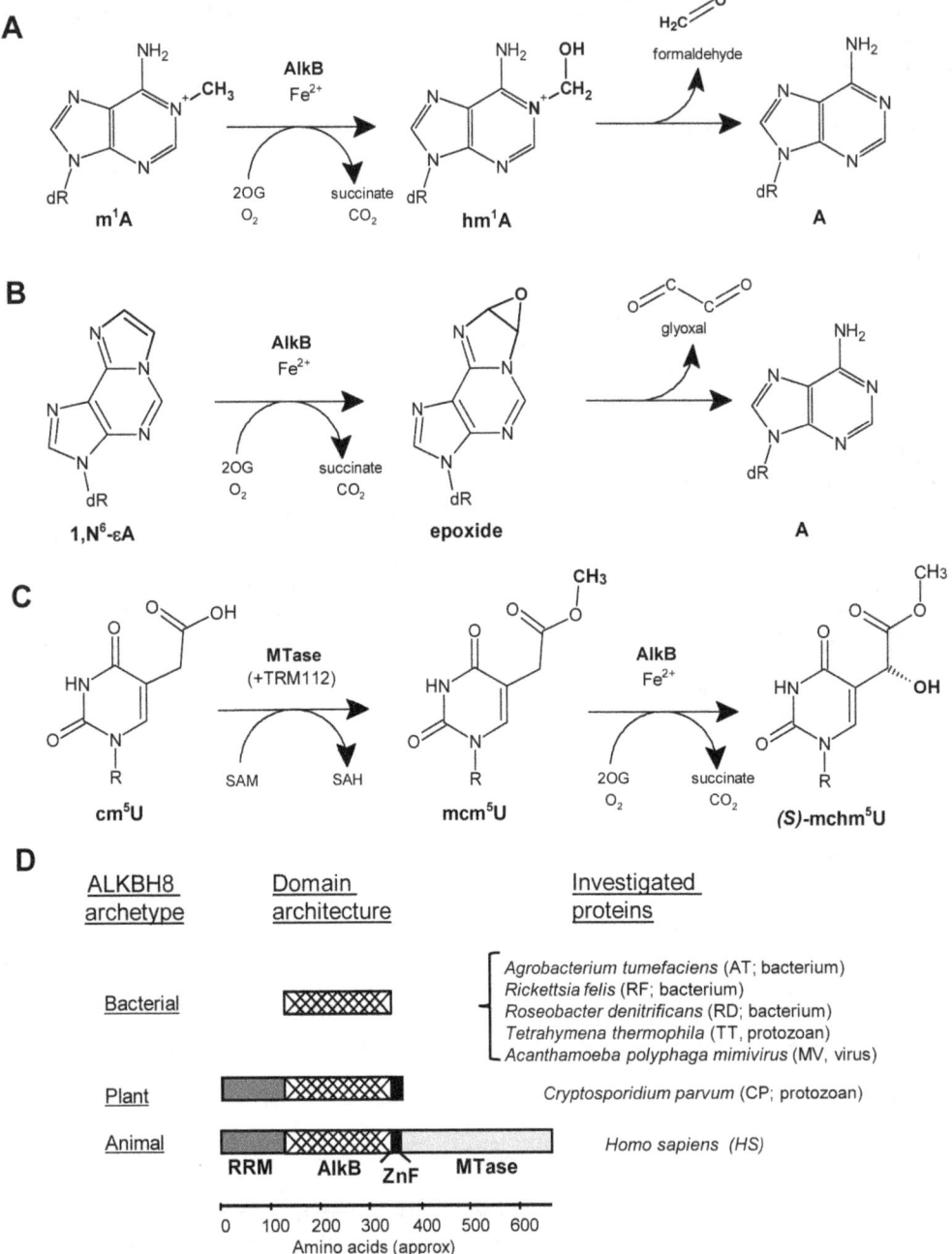

Figure 1. Selected ALKBH8 homologues and modification/repair reactions catalyzed by ALKBH8/AlkB. (A) AlkB-catalyzed DNA repair reaction on the methyl adduct m^1A. (B) AlkB-catalyzed DNA repair reaction on the etheno adduct 1, N^6-εA. (C) ALKBH8-catalyzed modification reactions on tRNA. SAM, S-adenosylmethionine; SAH, S-adenosylhomocysteine. (D) Overview of presently investigated and canonical ALKBH8 proteins and their domain architecture. Gene identification (gi) numbers of the investigated proteins are as follows (size is indicated in parenthesis): AT (195 aa), 159184347; RF (188 aa), 67458989; RD (195 aa), 115345714; TT (199 aa), 118368517; MV (210 aa), 311978317; CP (350 aa), 66362644; HS (664 aa), 195927056.

ly modified tRNA nucleosides have been identified [27]. Uridine, when present at the wobble position of the anticodon (position 34), is usually modified, and such modification substantially affects tRNA decoding properties [28]. In eukaryotes, wobble uridines are modified to either 5-methoxycarbonylmethyluridine (mcm^5U), 5-carbamoylmethyluridine (ncm^5U), or derivatives thereof [29]. In the yeast *Saccharomyces cerevisiae*, the last step in the synthesis of mcm^5U is mediated by Trm9, a methyltransferase (MTase) catalyzing the methyl-esterification of 5-carboxymethyluridine

(cm^5U) into mcm^5U [30]. In mammals, this reaction is catalyzed by the Trm9-like, C-terminal MTase domain of ALKBH8, together with the accessory protein TRM112 (Fig. 1C) [31,32]. Mammalian ALKBH8 also contains an N-proximal AlkB domain flanked by a RNA-recognition motif (RRM) and a cysteine-rich zinc-finger (ZnF) (Fig. 1D). Recently, the RRM/AlkB/ZnF portion of ALKBH8 was shown to hydroxylate mcm^5U, leading to the formation of wobble (S)-5-methoxycarbonylhydroxymethyluridine [(S)-mchm^{5}U] specifically in tRNA$^{Gly(UCC)}$ (Fig. 1C) [33,34].

ALKBHs appear to be present in all multicellular eukaryotes and in a wide range of unicellular eukaryotes and prokaryotes [35,36]. Moreover, AlkB-like proteins are also found in viruses, particularly in RNA viruses that infect plants, and these proteins display repair activities similar to EcAlkB, suggesting that they may be involved in removing methyl lesions from the viral RNA genome [37]. Bacterial AlkB proteins can be subdivided into four groups based on sequence similarity. Three of the groups comprise proteins with similarity to the mammalian repair proteins ALKBH2 and ALKBH3, whereas members of the fourth group are similar to ALKBH8, but lack the RRM, ZnF and MTase domains (Fig. 1D) [38]. Interestingly, when AlkB proteins from the four groups were investigated, the tested proteins all displayed repair activity on DNA. Notably, the tested ALKBH8-like protein, originating from *Rhizobium etli* (*R. etli*), did not show DNA repair activity on methylated bases, but rather on etheno adducts [38]. Plants, such as *Arabidopsis thaliana*, possess putative orthologues of several of the ALKBH proteins found in mammals, including ALKBH8. The plant orthologue of the mammalian ALKBH8 oxygenase contains the RRM/AlkB/ZnF moiety (Fig. 1D), whereas a different gene encodes a Trm9-like MTase. In *Arabidopsis*, these two proteins were shown to represent the functional equivalent of the bifunctional mammalian ALKBH8 [39].

As both repair and modification activities have been demonstrated for ALKBH8 proteins, we have here sought to further illuminate their biological function by investigating such proteins from several different organisms. The proteins were investigated both for DNA repair activity and for the ability to convert mcm^{5}U to (S)-mchm^{5}U in tRNA. We detected *in vitro* DNA repair activity for all tested recombinant proteins, and for two protozoan ALKBH8s we could also detect tRNA modifying activity. To analyse whether bacterial ALKBH8, like its mammalian counterpart, is involved in wobble uridine modification of tRNA$^{Gly(UCC)}$, we also generated an ALKBH8-deficient strain of *Agrobacterium tumefaciens*, and analysed the tRNA$^{Gly(UCC)}$ modification status. We found the wobble uridine modification status of tRNA$^{Gly(UCC)}$ to be unaltered in this ALKBH8-deficient strain, indicating that bacterial ALKBH8s are not involved in modifying this site.

Materials and Methods

Bioinformatics Analysis

Protein sequences of putative ALKBH8 homologues were retrieved from BLAST searches and previous publications [34,38,40]. Multiple sequence alignments were constructed using the MUSCLE algorithm [41]. Alignments were manually edited in the Jalview package [42].

Plasmid Construction

Genes encoding ALKBH8 proteins were amplified by polymerase chain reaction (PCR) from genomic DNA in case of bacterial or protozoan AlkB genes, and a plasmid in case of the mimivirus AlkB protein. For expression purposes, all *Tetrahymena thermophila* TAA and TAG codons were changed to CAA and CAG, respectively, by PCR-mediated mutagenesis, since *T. thermophila* uses an alternative genetic code where UAA and UAG (which normally are stop codons) code for glutamine. Primers used for PCR are listed in Table S1. The PCR products were subsequently cloned into the appropriate restriction sites in plasmid pET-28a(+) (Novagen, Darmstadt, Germany), which was used for expression of recombinant protein in *E. coli*. For bacterial reactivation assays

AlkB-encoding fragments from the pET-28a(+)-derived plasmids were subcloned to pJB658 using appropriate restriction sites [43]. The construction of plasmids encoding EcAlkB, human ALKBH2, human ALKBH8, or its AlkB domain (RRM-AlkB; aa 1–354), and subsequent protein expression and purification were previously described [9,32,38]. To remove the RRM domain from both ALKBH8 and RRM-AlkB, a PCR was performed to obtain the coding regions for AlkB-MTase (aa 129–664) and for the AlkB core (aa 129–338), respectively. To inactivate the Zn-finger, three Zn-coordinating cysteines were mutated to alanine (*i.e.* Cys341Ala, Cys343Ala and Cys349Ala) by fusion PCR. PCR products were placed into the pET28a(+) vector using the NdeI and SalI restriction sites.

Protein Expression and Purification

Plasmids pET-28a(+) containing sequences coding for N-terminally 6xHis-tagged ALKBH8 proteins were transformed into the *E. coli* strain BL21-CodonPlus(DE3)-RIPL (Stratagene, La Jolla, CA, USA). When the bacterial culture reached an optical density of 1 measured at 600 nm (OD$_{600}$), expression of recombinant proteins was induced by adding isopropyl-beta-D-thiogalactopyranoside (IPTG) to a final concentration of 0.5 mM and incubation was continued at 16°C for 16 hours. Cells were harvested by centrifugation at 5000×g for 10 min at 4°C and resuspended in a buffer containing 50 mM sodium phosphate (pH 7.0), 150 mM NaCl, 5 mM imidazole, 0.01% Tween 20, EDTA-free Complete-Protease inhibitor (Roche), and 5 mM β-mercaptoethanol. The cells were lysed by addition of lysozyme to a final concentration of 1 mg/ml, incubation on ice for 30 min and subsequent sonication with three 11 W pulses of 30 sec with 30 sec intervals. Cell debris was removed by centrifugation at 12,000×g for 10 min at 4°C. The obtained supernatant was directly mixed with TALON Metal Affinity Resin (Clontech, Mountain View, CA, USA) and recombinant proteins were obtained by a single affinity purification step according to the manufacturer's instructions. Protein purity and yield were assessed by 15% SDS–PAGE followed by coomassie brilliant blue-staining of the gel.

Bacterial Survival Assay and Phage Reactivation Assay

To test the ability of ALKBH8 proteins for complementation of the repair function of EcAlkB protein, pJB658-derived plasmids encoding these proteins were transformed into the F-pilus-expressing, *alkB*-deficient *E. coli* strain HK82/F'. Protein expression was induced by the addition of 2 mM toluic acid (Fluka/Sigma-Aldrich). To introduce methyl lesions or etheno adducts, ssDNA bacteriophage M13mp18, ssRNA bacteriophage MS2 or bacteria were treated with methyl methanesulphonate (MMS; Sigma-Aldrich) or chloroacetaldehyde (CAA; Sigma-Aldrich), respectively. Phage and bacteria survival was scored by counting the resulting plaques or bacterial colonies, respectively. The experiments were performed essentially as previously described [38].

Assay for AlkB-mediated Decarboxylation of 2-oxoglutarate

To determine whether the recombinant ALKBH8 proteins can catalyze uncouple decarboxylation of 2OG, we used the method described earlier [37], which measures the level of radioactive [1-^{14}C] succinate produced as a result of decarboxylation of [5-^{14}C] 2-oxoglutarate.

In vitro DNA Repair Assay

The oligonucleotide substrates containing m^{1}A, m^{3}C, 1, N^6-ethenoadenine (1, N^6-εA) and 3, N^4-ethenocytosine (3, N^4-εC) were purchased from Chemgenes Corporation, USA. Repair reactions were performed by incubating 100 pmoles of AlkB protein with 1 pmol of ^{32}P-labeled ssDNA or dsDNA oligonucleotides at 37°C for 30 min in a 50 μl reaction mixture containing 50 mM Tris-HCl (pH 8.0), 2 mM ascorbic acid, 1 mM 2-oxoglutarate, and 80 μM (NH$_4$)$_2$Fe(SO$_4$)$_2$•6H$_2$O. Reactions were stopped by incubation at 65°C for 20 minutes. In order to remove AlkB proteins, the reaction mixtures were incubated with 1 μl of 20 mg/ml proteinase K (Sigma-Aldrich) for 15 min at 42°C, and then proteinase K was heat inactivated at 90°C for 10 minutes. When reactions were performed with ssDNA substrates, the complementary DNA oligonucleotide was added prior to the next step. AlkB-treated oligonucleotides with methyl lesions were incubated with 20 units of DpnII for 1 hour at 37°C [22], whereas the oligonucleotides containing etheno adducts were incubated for 30 min at 37°C with human alkyl-N-purine-DNA glycosylase (ANPG) for 1, N^6-εA or E. coli uracil-DNA glycosylase Mug in case of 3, N^4-εC and 1, N^2-εG, followed by abasic site cleavage with human AP endonuclease 1 (HAP1) for 30 minutes at 37°C. Reaction products were resolved by 20% denaturing PAGE in the presence of 7 M urea and visualized by phosphorimaging using FLA-7000 screens (Fujifilm). Quantification was performed by using MultiGauge Software (Fujifilm).

Construction of an *Agrobacterium tumefaciens alkB* Null Mutant

An *Agrobacterium tumefaciens* C58 *alkB* null mutant (*A. tumefaciens* C58 *alkB*$^-$) was made by insertion of a group II intron using the TargeTron Gene Knockout System (Sigma Aldrich) according to the provided manual. An *alkB*-targeted intron was generated by PCR using primers designed by using the TargeTron algorithm (SigmaAldrich) (Table S1) [44]. The generated PCR products were cloned into the HindIII and BsrGI sites of the pBL1 plasmid, which was a kind gift from Dr. Alan Lambowitz [45]. The donor plasmid, pBL1/*alkB*, was transformed into *A. tumefaciens* C58 by electroporation and the bacteria were selected on YEB agar plates containing 2 μg/ml tetracycline and 100 μg/ml ampicillin. For gene targeting, *A. tumefaciens* C58 containing pBL1/*alkB* was grown at 30°C in YEB medium containing 2 μg/ml tetracycline and expression of the TargeTron cassette was induced by adding m-toluic acid to a final concentration of 5 mM when the culture reached early log phase (OD$_{600}$ = 0.3 to 0.4) followed by 3 h growth under the same conditions. After induction, cells were selected on YEB agar plates containing 2 μg/ml tetracycline and 100 μg/ml ampicillin, and incubated at 30°C until colonies were formed (2–3 days). Colonies were screened by colony PCR using ALKBH8-specific primers. Bacteria from a single colony containing the TargeTron cassette inserted into the *alkB* gene were cured of the TargeTron donor plasmid by several passages on YEB agar plates with ampicillin, but without tetracycline. Intron insertion was confirmed by sequencing analysis.

Survival of *Agrobacterium tumefaciens* C58 Wild-type and *alkB* Mutant after MMS Treatment

A. tumefaciens C58 wild-type and *alkB* mutant were grown at 30°C in YEB medium with 50 μg/ml ampicillin. As a negative and positive control HK82/F', E. coli containing pJB658 or pJB658-EcAlkB, respectively, were used. The MMS and CAA treatment was done as described above for the bacterial survival assay with the difference that the induction step with toluic acid was omitted for *A. tumefaciens* C58.

Total tRNA and Isoacceptor Isolation

Total tRNA from S. cerevisiae and E. coli were purchased from Roche. Total tRNA from *A. tumefaciens* C58 wild-type and *alkB* mutant was purified using an RNA/DNA maxi kit (Qiagen) according to the manual provided. tRNA$^{Gly(UCC)}$ was purified from total tRNA using 3′-biotinylated oligonucleotides (Table S1), as previously described [32].

Enzymatic Treatment of tRNA

tRNA from S. cerevisiae (5 to 10 μg) was incubated with 100 pmoles of recombinant proteins for 30 min at 37°C in a 50 μl reaction mixture containing 50 mM HEPES-KOH pH 7.5, 0.5 mM MgCl$_2$, 2 mM ascorbic acid, 100 μM 2-oxoglutarate, 40 μM FeSO$_4$, and 10 U RNasin Plus RNase inhibitor (Promega). Reactions were stopped by incubation for 15 min at 42°C with 1 μl of 20 mg/ml protease K (Sigma-Aldrich). tRNA was extracted with 1 volume of acidic phenol pH 4.0 and chloroform, followed by precipitation with 1 volume of isopropanol in the presence of 1 M NH$_4$Ac and 10 μg of glycogen. Pellets were washed with 70% EtOH, dried and dissolved in H$_2$O. The samples were subjected to nucleoside analysis by LC–MS/MS.

Mass Spectrometry

LC-MS/MS of nucleosides was performed essentially as described previously [32]. Briefly, tRNA was enzymatically digested to nucleosides [46], which were separated by reverse phase high-performance liquid chromatography, followed by mass spectrometry detection. Quantification was performed by comparison with pure nucleoside standards run in between the samples.

For MALDI-TOF mass spectrometry, tRNA isoacceptors were digested with RNase T1 (Ambion) and samples prepared for MALDI mass spectrometry as previously described [32].

Results

Selection and Bioinformatics Analysis of ALKBH8 Proteins from Various Organisms

Previous studies have established a tRNA modification function for ALKBH8 proteins from animals and plants, whereas robust DNA repair activity on etheno lesions was detected for ALKBH8 from the α-proteobacterium R. etli [33,34,38,39]. To further address the function of the ALKBH8 proteins, for the present study we selected such proteins from a wide range of organisms for analysis with respect to both tRNA modification and DNA repair capabilities.

Bacterial ALKBH8 proteins are primarily found in α-proteobacteria, and from this group we chose to examine the ones from *Rickettsia felis* (RF), *Roseobacter denitrificans* (RD), and *Agrobacterium tumefaciens* (AT). In fact, the AT protein was investigated by us previously, but we failed to purify soluble recombinant protein [38]. It is included here because it is possible to generate gene knock-outs of the corresponding bacterium by the so-called TargeTron technology [45]. We also included in our study ALKBH8 proteins from the protozoan *Tetrahymena thermophila* (TT) and from the mimivirus *Acanthamoeba polyphaga mimivirus* (MV), which, like the bacterial ALKBHs, lack annotated domains apart from the defining AlkB domain. In addition, we chose to study the ALKBH8 protein from the protozoan *Cryptosporidium parvum* (CP) which, similarly to plant ALKBH8, has an RRM/AlkB/ZnF domain architecture. The selected proteins and their domain architecture are outlined in Fig. 1D.

To illustrate the degree of sequence similarity between the ALKBH8 proteins, we generated a sequence alignment of the proteins selected for the current study, as well as human and *Arabidopsis* ALKBH8, which have been the focus of previous studies. The alignment shows that these proteins, despite their diverse origins, have a relatively high degree of sequence similarity, both in the core oxygenase region and in the so-called NRL (nucleotide recognition lid) region, which has been implicated in specific binding of the nucleic acid substrate [47]. Moreover, Fe(II)-coordinating residues characteristic of the 2OG/Fe(II)-dependent oxygenases, as well as residues characteristic of the ALKBH subfamily, are conserved between these proteins (Fig. 2).

Uncoupled Enzymatic Activity of Recombinant ALKBH8 Proteins

Many members of the 2OG/Fe(II)-dependent oxygenase superfamily, including several AlkB proteins, are able to convert 2-oxoglutarate to succinate and carbon dioxide at a low rate in the absence of their true substrate, in the so-called uncoupled reaction. As an initial characterization of the ALKBH8 proteins, we expressed and purified hexahistidine (6xHis) tagged recombinant proteins from *E. coli*, and tested their ability to catalyse the uncoupled reaction. The RF protein, which was expressed and purified at very low yield (Fig. 3A), and the AT protein, which we previously have failed to recover in a soluble form [38], were excluded from this analysis. EcAlkB showed robust uncoupled activity, the TT and CP proteins displayed lower, but significant activity, whereas no activity was detected in the case of the MV and RD proteins (Fig. 3B). These results establish the TT and CP proteins as bona fide members of the 2OG/Fe(II)-dependent oxygenase superfamily, but are inconclusive with respect to the

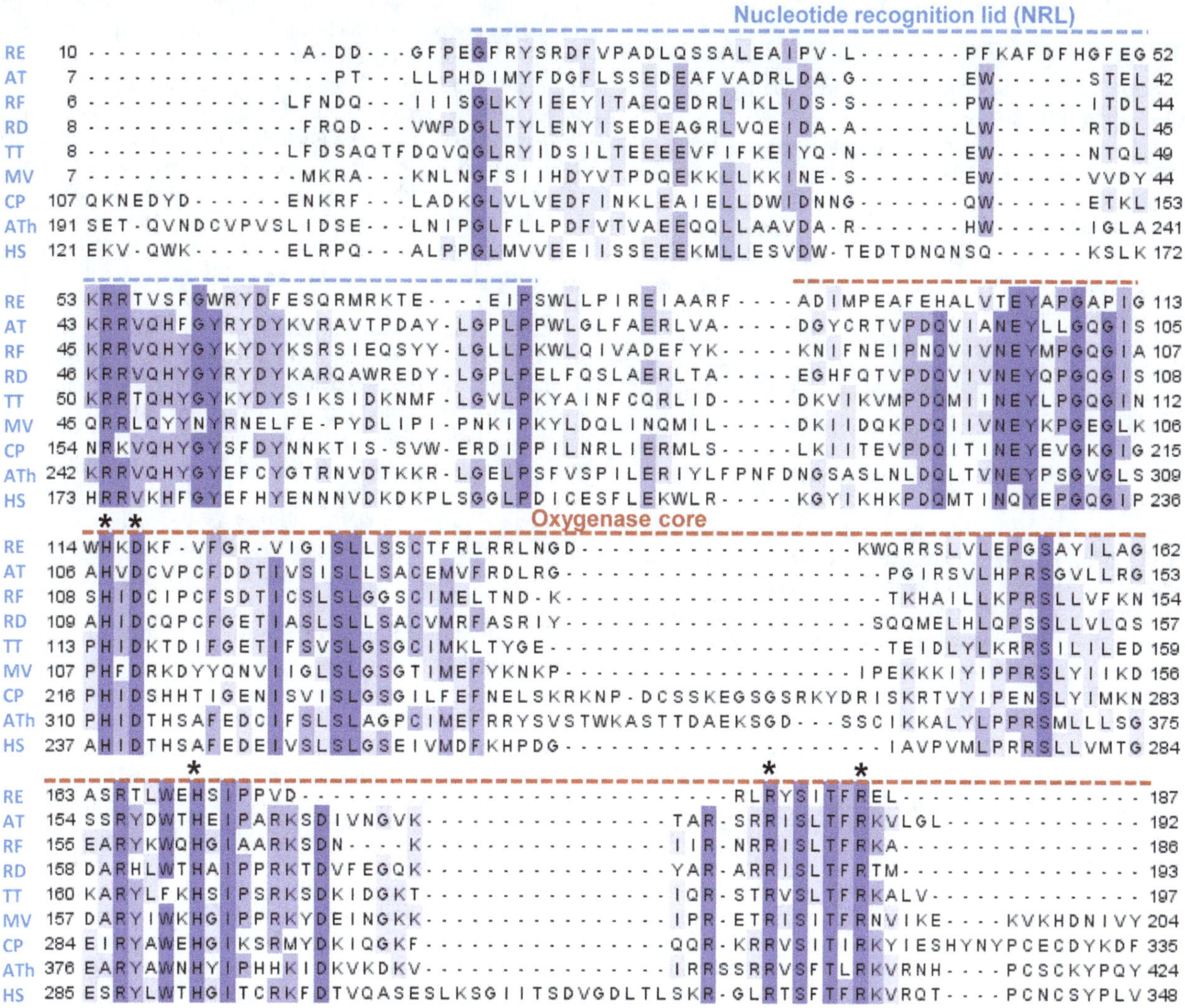

Figure 2. Sequence alignment of ALKBH8 proteins investigated in present study. The ALKBH8 proteins indicated in Fig. 1D were aligned by using the MUSCLE algorithm. For comparison, the previously characterized ALKBH8 proteins from the plant *Arabidopsis thaliana* (Ath; gi|159184347) and the bacterium *Rhizobium etli* (RE; gi|86360251) were also included. The dotted blue line indicates the nucleotide recognition lid, a region implicated in the nucleic acid binding of the EcAlkB protein, whereas the red dotted line indicates the oxygenase core, the region shared between all 2OG/Fe(II) dependent oxygenases. Asterisks indicate the $HXDX_nH$ triad involved in Fe(II) coordination and the RX_5R motif characteristic of the ALKBH family of proteins.

A

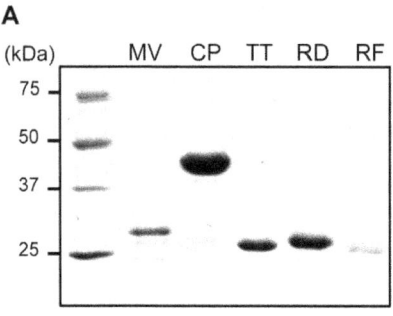

B

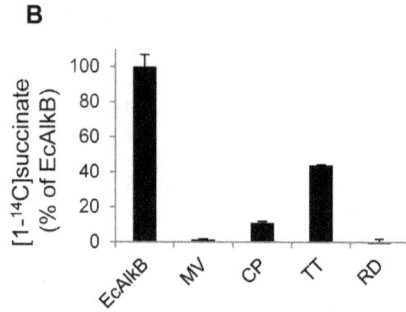

Figure 3. Purification and initial characterization of ALKBH8 proteins. (A) Purification of recombinant His-tagged ALKBH8 proteins used in present study. Proteins were expressed in *E. coli*, purified by metal affinity chromatography and visualized on a Coomassie stained SDS–PAGE gel. (B) Uncoupled activity of recombinant ALKBH8 proteins. [5-^{14}C] 2-oxoglutarate was incubated for 30 min at 37°C with 100 pmoles of examined proteins, and remaining [5-^{14}C] 2-oxoglutarate was precipitated with 2,4-dinitrophenylhydrazine. The generated [1-^{14}C] succinate present in the supernatant was measured by scintillation counting.

MV and RD proteins, since the extent of uncoupled activity varies between 2OG/Fe(II)-dependent oxygenase superfamily members [48].

In vitro Repair of Site Specific Lesions in DNA by Recombinant ALKBH8 Proteins

To test the repair activity of the ALKBH8 proteins towards methyl lesions we used oligonucleotides containing a single m^{1}A or m^{3}C lesion within the recognition sequence (GATC) for the methylation specific restriction enzyme *Dpn*II as substrates [22]. Specifically, these were 25-mer 5′-[^{32}P]-end-labeled ssDNA or dsDNA oligomers (the dsDNA substrates contained a lesion-free, unlabeled complementary strand). After incubation with recombinant ALKBH8 protein, the DNA substrates were treated with *Dpn*II to distinguish repaired from unrepaired oligonucleotides, as *Dpn*II cleavage will only occur if the methyl lesion has been removed (Fig. 4A). When ssDNA oligonucleotides were used in the reaction, they were annealed to the complementary (unlabeled) strand after the repair reaction but prior to *Dpn*II digestion, since *Dpn*II will cleave dsDNA but not ssDNA. Human ALKBH2 was included as a positive control for repair of m^{1}A and m^{3}C. The tested ALKBH8 proteins were unable to repair m^{1}A lesions, with the exception of MV AlkB, which showed a very weak repair activity towards m^{1}A in ssDNA (Fig. 4B,C). Two of the four tested proteins, RD and TT, exhibited repair activity towards m^{3}C in ssDNA and dsDNA (Fig. 4B,C).

Etheno (ε) lesions, such as 1,N^6-ethenoadenine (1,N^6-εA), 3,N^4-ethenocytosine (3,N^4-εC), N^2,3-ethenoguanine (N^2,3-εG), and 1,N^2-ethenoguanine (1,N^2-εG) represent exocyclic adducts resulting from the formation of a new imidazole ring on nucleic acid bases, typically induced by lipid peroxidation products or metabolites of vinyl chloride. These highly mutagenic and cytotoxic lesions interfere with normal Watson-Crick base pairing. Although the etheno lesions are repaired mainly through the base excision repair pathway [49], it has been shown that EcAlkB and its human homologues ALKBH2 and ALKBH3 are also able to repair 1,N^6-εA and 3,N^4-εC *in vitro* [12,18,50]. The simultaneous deletion of ALKBH2, ALKBH3 and the alkyl adenine DNA glycosylase (AAG) in the mouse confers a massively synergistic phenotype after acute inflammation, indicative of overlapping substrate specificities and an *in vivo* role for repairing etheno adducts [51]. It has also been shown that some bacterial AlkB proteins efficiently repair etheno adducts, while having low or no activity on methylated bases [38]. To test the ability of the ALKBH8 proteins to repair etheno adducts *in vitro*, they were incubated with 5′-[^{32}P]-end-

labeled oligonucleotides containing 1,N^6-εA or 3,N^4-εC. To assess whether repair has occurred, the oligonucleotides were treated with a DNA glycosylase which will only cleave the substrate if the lesion is intact, *i.e.* a conceptually opposite approach to that of using *Dpn*II, and one which does not require a particular sequence at the lesion site (Fig. 4D). For cleavage of 1,N^6-εA containing substrates, the DNA glycosylase ANPG (also known as AAG or MPG) was used, while Mug was used for substrates with 3,N^4-εC. After treatment with DNA glycosylase, strand-breaks were introduced at resulting abasic sites through cleavage with the human AP endonuclease 1 (Fig. 4D). When single-stranded oligonucleotides were used in the AlkB reaction, they were annealed to their complementary strand prior to DNA glycosylase treatment. Again, human ALKBH2 was included as a positive control. The four recombinant ALKBH8 proteins (MV, CP, RD, TT) showed strong repair activity towards 3,N^4-εC and somewhat weaker activity towards 1,N^6-εA *in vitro* (Fig. 4E,F). The MV protein was more active on ssDNA compared to dsDNA, whereas the CP protein repaired 3,N^4-εC and 1,N^6-εA only in ssDNA (Fig. 4E,F). In summary, the tested ALKBH8 proteins displayed *in vitro* repair activity on DNA, and they are generally more active on etheno adducts than on methyl lesions, similarly to the *R. etli* ALKBH8 protein investigated previously [38].

ALKBH8-mediated Repair of Chemically Induced Methyl and Etheno Lesions *in vivo*

E. coli alkB mutants are sensitive to the methylating agent methyl methanesulfonate (MMS), due to their inability to repair replication blocking lesions, such as m^{1}A and m^{3}C. Such lesions are introduced at a particularly high frequency in ssDNA, relative to dsDNA. Consequently, when infected by MMS-treated ssDNA bacteriophage, *alkB* mutants show a dramatically reduced ability to generate progeny phage, as they are unable to reactivate the damaged phage DNA through removal of deleterious methyl lesions [52]. To examine if the ALKBH8 proteins were able to complement the MMS-sensitive phenotype, they were expressed in AlkB-deficient bacteria, which were subsequently exposed to MMS and their survival assessed. While the expression of EcAlkB complemented the MMS-sensitive phenotype of the mutant bacteria, none of the ALKBH8 proteins had this effect (Fig. 5A). Similarly, only EcAlkB was able to increase the reactivation of the MMS-treated ssDNA phage M13 (Fig. 5B).

To test the ability of the ALKBH8 proteins to repair etheno adducts, they were expressed in *alkB E. coli*, and their ability to reactivate chloroacetaldehyde (CAA)-treated ssDNA phage M13

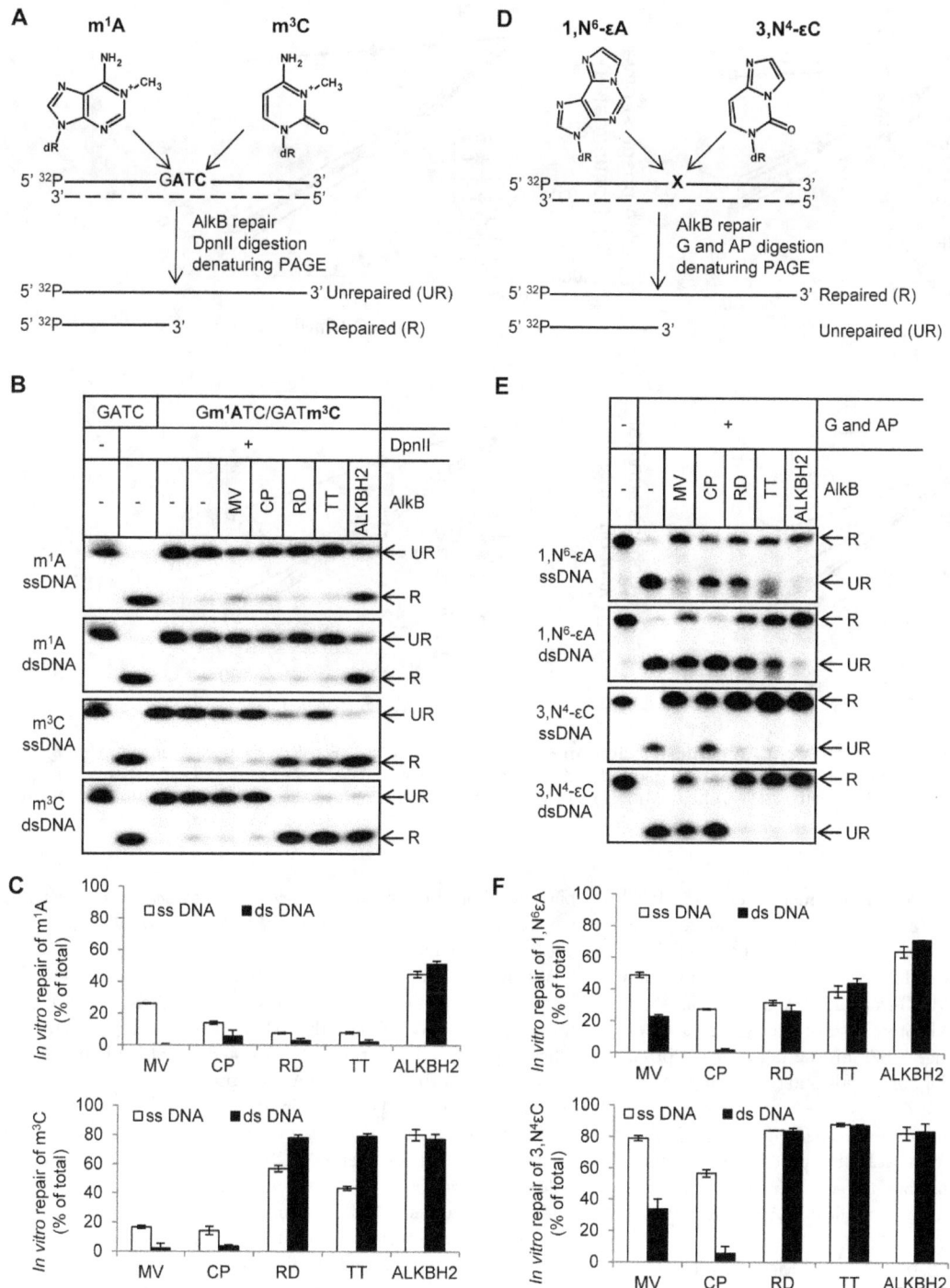

Figure 4. *In vitro* repair activity of ALKBH8 proteins. (A, D) Schematic representation of assay for repair of site specific methyl (A) and etheno lesions (D) in DNA. The dashed line indicates the complementary, lesion-free unlabeled oligonucleotide, which was either present during the repair reaction (dsDNA repair) or added post-repair (ssDNA repair). For repair of methyl lesions, lesion-free oligonucleotide substrates were selectively cleaved by DpnII (A), whereas for etheno adduct repair (D) the lesion-containing base was selectively removed by a glycosylase (G), followed by conversion of the resulting AP site into a single-strand break by an AP endonuclease (AP). (B) Repair activity of purified ALKBH8 proteins on m^1A and m^3C in ssDNA and dsDNA. (C) Quantification of results from experiments exemplified in (B). (E) Repair activity of purified ALKBH8 proteins on 1, N^6-ɛA and 3, N^4-ɛC in ssDNA and dsDNA. The DNA glycosylase ANPG was used on 1, N^6-ɛA containing substrates, while Mug was used for substrates with 3, N^4-ɛC. (F) Quantification of results from experiments exemplified in (E).

was measured. CAA causes the formation of exocyclic DNA adducts with the following relative efficiencies: 1, N^6-ɛA>3, N^4-ɛC>N^2,3-ɛG>1, N^2-ɛG [53,54]. The majority of the ALKBH8 proteins were unable to reactivate CAA-treated ssDNA phage;

only the TT (*T. thermophila*) and RF (*Rickettsia felis*) proteins caused a modest increase in progeny phage formation, but the effect was substantially lower than that observed for EcAlkB (Fig. 5C).

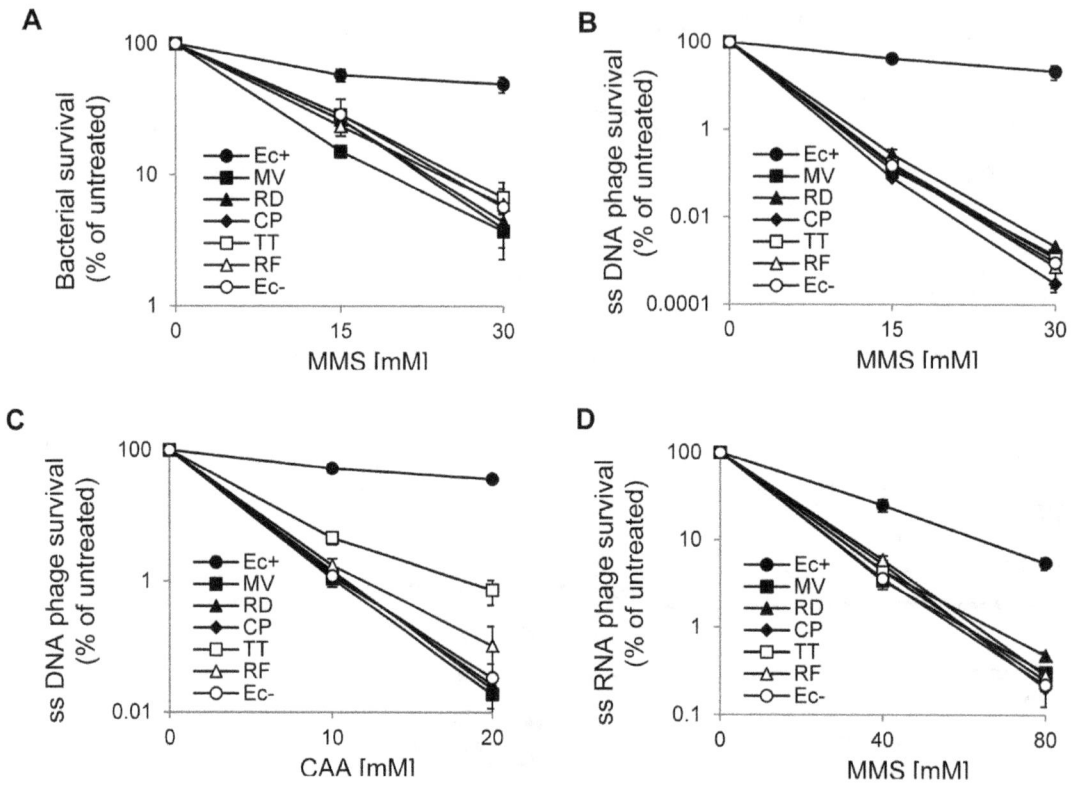

Figure 5. ALKBH8 mediated repair of MMS- or CAA induced lesions in *E. coli*. AlkB-deficient (*alkB*) *E. coli* carrying either an empty expression plasmid (Ec–), or corresponding plasmids for expression of the indicated ALKBH8 proteins or EcAlkB (Ec+) were used in all experiments. (A) MMS-sensitivity of bacteria. (B) survival of MMS-treated ssDNA phage M13. (C) survival of CAA-treated ssDNA phage M13. (D) survival of MMS-treated ssRNA phage MS2.

Certain AlkB proteins are able to repair methyl lesions in RNA *in vitro*, and can reactivate MMS treated RNA phage when expressed in AlkB-deficient *E. coli* [9,11,37]. To test the activity of the ALKBH8 proteins towards RNA, *alkB E. coli* expressing these proteins were infected with MMS-treated RNA phage MS2. While overexpression of the EcAlkB protein substantially increased the survival of MMS-treated phage MS2, this was not the case for any of the ALKBH8 proteins (Fig. 5D), indicating that they are not RNA repair proteins.

These *in vivo* complementation experiments indicate, similarly to the *in vitro* repair assays, that ALKBH8 proteins prefer etheno adducts over methyl lesions, but they also indicate that these enzymes do not efficiently repair canonical EcAlkB substrates.

MMS/CAA Sensitivity and tRNA$^{Gly(UCC)}$ Modification Status of ALKBH8-deficient *Agrobacterium tumefaciens*

In our previous study, the ALKBH8 protein from the bacterium *R. etli* showed robust repair activity on etheno adducts [38]. On the other hand, the bacterial ALKBH8 proteins display a high degree of sequence similarity to human ALKBH8 not only in the core oxygenase domain, but also in the so-called nucleotide recognition lid region (Fig. 2), pointing towards a role in tRNA modification. Human and plant ALKBH8 are both involved in wobble uridine modification of tRNA$^{Gly(UCC)}$, and the sequence of this tRNA is rather well conserved from humans to bacteria, especially in the anticodon loop, which is identical. To address the potential role of bacterial ALKBH8 proteins in tRNA modification, we decided to investigate the wobble uridine modification status of tRNA$^{Gly(UCC)}$ in ALKBH8-deficient versus wild-type

bacteria. For this purpose, we selected *Agrobacterium tumefaciens*, which can be subjected to gene knock-out by the so-called TargeTron technology [44,45].

When using the TargeTron technology, the gene of interest can be disrupted by site specific insertion of a redesigned Group II intron [44,45]. We found the ALKBH8-encoding gene to be efficiently targeted; 3 out of 7 clones tested by colony-PCR carried the inserted intron (Fig. 6A). As the majority of bacterial AlkB proteins appear to be DNA repair enzymes, it was first investigated if disruption of the *A. tumefaciens alkB* gene caused sensitivity towards the genotoxic agents MMS and CAA. AlkB-deficient and wild-type *A. tumefaciens* bacteria were similarly sensitive as to MMS and CAA treatments, whereas AlkB-deficient *E. coli*, as expected and previously reported, were more sensitive to treatment with these DNA damaging agents than bacteria expressing EcAlkB (Fig. 6B,C). These results showed that the AT protein does not protect *A. tumefaciens* against the tested DNA damaging agents, suggesting that the AT protein does not play an important role in repair of methyl and etheno lesions.

To assess the wobble uridine modification status, MALDI-TOF mass spectrometry analysis was performed on RNase T1 digested tRNA$^{Gly(UCC)}$ isolated from *A. tumefaciens*. The data indicated that *A. tumefaciens* tRNA$^{Gly(UCC)}$ contains 5-hydroxyuridine (ho^{5}U) at the wobble position, as an RNase T1 fragment containing the anticodon displayed a mass increase of 16 Da relative to the unmodified sequence (Fig. 6D,E), a result compatible with the action of a hydroxylase such as ALKBH8. However, this uridine modification was also present in the ALKBH8-deficient bacteria, showing that ALKBH8 is not involved in wobble uridine

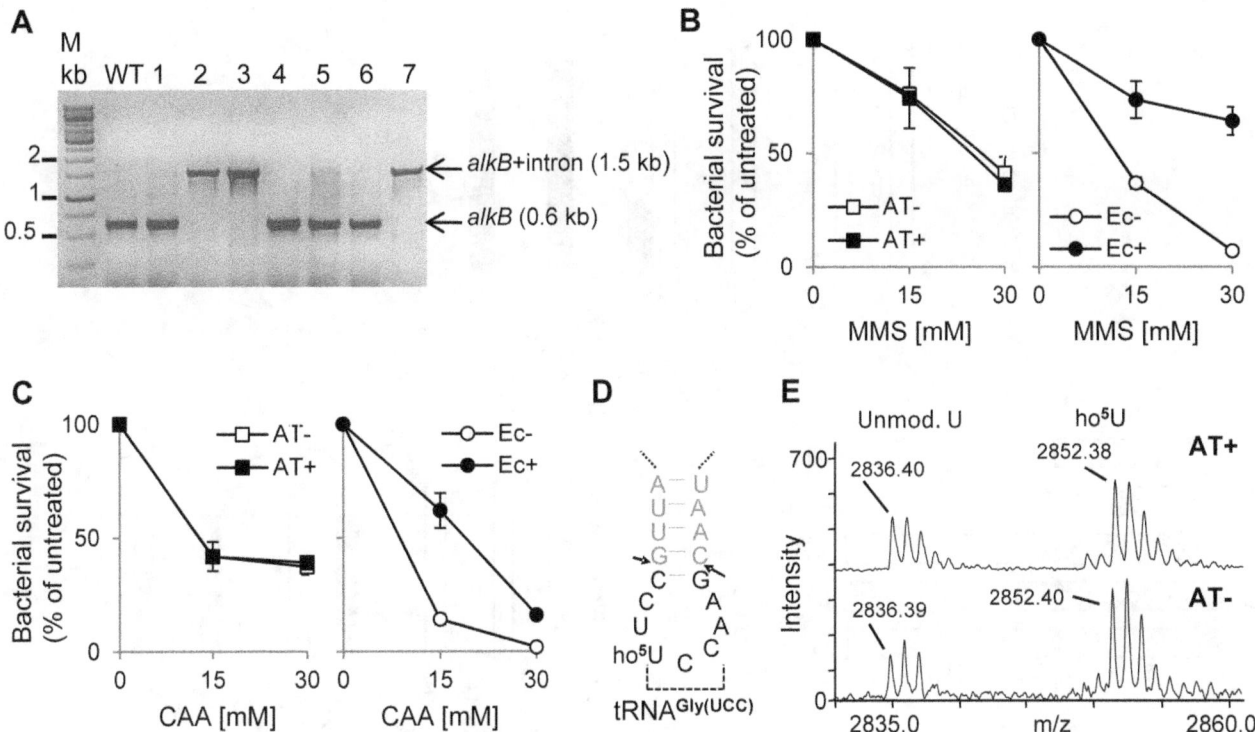

Figure 6. Generation and characterization of AT-deficient *Agrobacterium tumefaciens.* (A) Inactivation of the *A. tumefaciens alkB* (*AT*) gene by site specific intron insertion. After intron induction, bacteria were plated and resulting colonies were subject to colony PCR using *alkB* (*AT*) specific primers. The lower arrow indicates the 0.6 kb fragment resulting from the non-disrupted gene (colonies 1, 4, 5, and 6), while the upper arrow indicates 1.5 kb fragment generated from the *alkB* gene disrupted by intron integration (colonies 2, 3, and 7). (B) MMS sensitivity of AT-deficient (AT–) versus AT-proficient (AT+) *A. tumefacies.* Bacteria were incubated in the presence of the indicated concentrations of MMS, then plated on agar plates, and survival scored by colony counting. *E. coli* served as control. (C), CAA sensitivity of AT-deficient (AT–) versus AT-proficient (AT+) *A. tumefacies.* Same as (B), but CAA was used instead of MMS. (D) Anticodon stem-loop of tRNA$^{\text{Gly(UCC)}}$ from *A. tumefaciens.* Black print indicates the anticodon-containing fragment generated by cleavage with RNase T1 (at arrows). (E) Wobble uridine modification status of tRNA$^{\text{Gly(UCC)}}$ from wild-type and *alkB* (*AT*) mutant *A. tumefaciens.* MALDI-TOF MS spectra of the anticodon-containing RNase T1 fragment illustrated in (D) are shown, and measured masses indicated. Calculated masses for the unmodified and ho^{5}U modified versions of the fragment (CCUUCCAAG) are 2836.37 and 2852.37, respectively (the masses refer to fragments with 2′–3′ cyclic phosphate termini, which represent the major digestion products).

modification of tRNA$^{\text{Gly(UCC)}}$, and suggesting that this may also be the case for other bacterial ALKBH8 proteins.

ALKBH8-mediated Hydroxylation of mcm^{5}U in tRNA

Mammalian and plant ALKBH8 specifically hydroxylate mcm^{5}U into (S)-mchm^{5}U at the wobble position of the anticodon in tRNA$^{\text{Gly(UCC)}}$ (Fig. 1C) [33,34]. As *S. cerevisiae* lacks an ALKBH8 orthologue, yeast tRNA$^{\text{Gly(UCC)}}$ contains wobble mcm^{5}U, and total yeast tRNA is thus a suitable substrate for testing the potential mcm^{5}U hydroxylating ability of ALKBH8 proteins [34]. Yeast tRNA was incubated with various ALKBH8 enzymes in the presence of appropriate cofactors, and then enzymatically digested to nucleosides, which were analyzed by LC–MS/MS. The RRM/AlkB/ZnF portion of human ALKBH8 (RRM-AlkB; aa 1–354) was included as a positive control. Only for the two eukaryotic, protozoan ALKBH8 proteins, CP and TT, was conversion of mcm^{5}U to (S)-mchm^{5}U observed (Fig. 7A). These results suggest that the ALKBH8 proteins from the protozoa *C. parvum* and *T. thermophila* are involved in biosynthesis of wobble (S)-mchm^{5}U.

Enzymatic Activity of Deletion Mutants of Human ALKBH8

The CP protein was the most active out of the two protozoan ALKBH8 proteins that showed tRNA modification activity, and

also contained all three domains found in plant ALKBH8, *i.e.* RRM/AlkB/ZnF. The less active TT protein, in contrast, only consisted of an AlkB moiety, suggesting that the RRM and ZnF domains may contribute positively to ALKBH8 activity. To investigate this, we tested the tRNA modifying activity of various mutants of human ALKBH8, containing deletions or point mutations in these domains, depicted in Fig. 7B. As substrate in these assays, we used tRNA from a gene-targeted mouse (denoted KI(MT$^+$)) expressing the MTase activity, but not the oxygenase activity of ALKBH8, thereby showing an accumulation of mcm^{5}U [34]. We found that changing three of the conserved cysteine residues to alanine in the ZnF domain did not affect ALKBH8 activity, indicating that this structure is not crucial for ALKBH8 activity (Fig. 7C). In contrast, the two mutants lacking the RRM domain ("AlkB core" and "AlkB-MTase") were devoid of enzymatic activity, indicating the importance of this domain.

Discussion

While ALKBH8 proteins from mammals and plants have been established as tRNA modification enzymes, ALKBH8 from the bacterium *R. etli* was shown to possess repair activity towards etheno adducts in DNA [33,34,38,39]. Based on this apparent duality of the ALKBH8 proteins, we have here investigated such proteins from a wide range of species, both with respect to DNA/

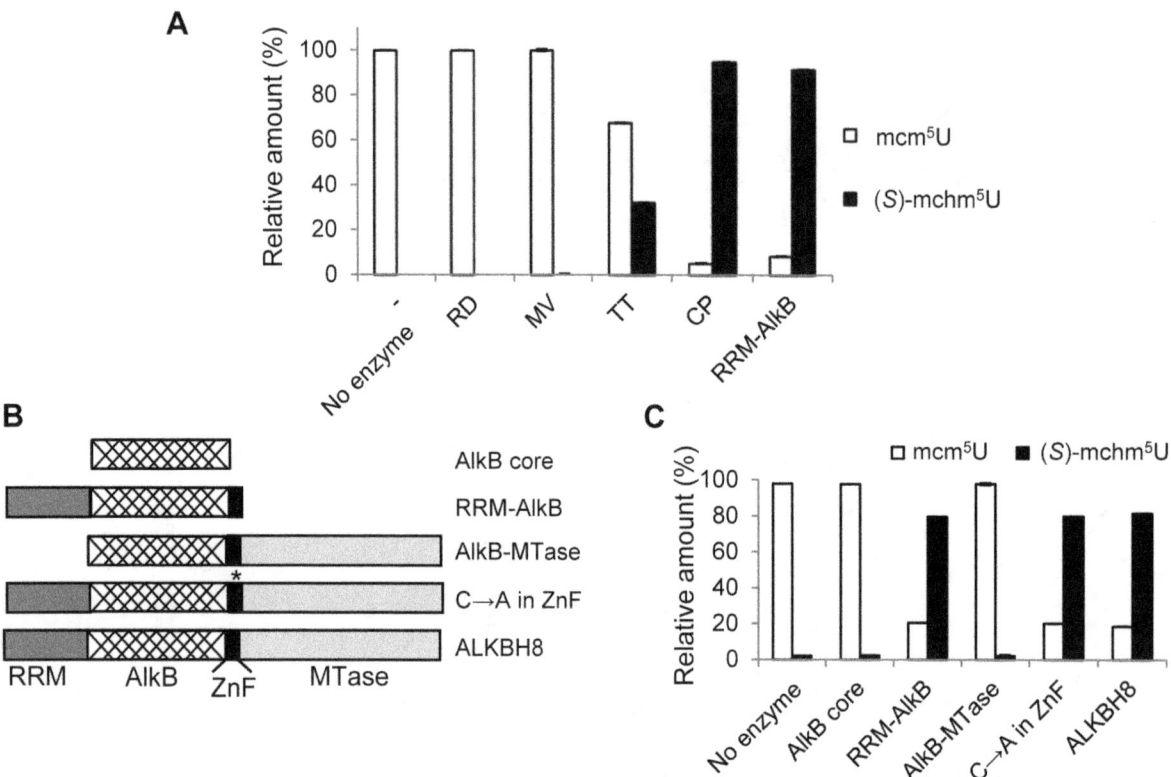

Figure 7. *In vitro* **analysis of tRNA modifying activity of ALKBH8 proteins.** (A) tRNA modifying activity of ALKBH8 proteins from various organisms. The indicated proteins were incubated with *S. cerevisiae* tRNA, and the ability of enzymes to catalyze the conversion of mcm⁵U to (S)-mchm⁵U was investigated by LC-MS/MS analysis of tRNA nucleosides. The RRM-AlkB part of human ALKBH8 was used as positive control. (B, C) Analysis of the tRNA modifying activity of deletion and point mutants of human ALKBH8. (B) Overview of tested proteins. "C→A in ZnF" refers to a mutant where the three conserved Cys residues (Cys341, Cys 343 and Cys 349) of the ZnF moiety have been replaced by alanine. (C) The indicated proteins were incubated with mcm⁵U containing tRNA from the so-called KI(MT⁺) mouse, *i.e.* a gene-targeted mouse expressing the ALKBH8 MTase, but not the oxygenase (AlkB) activity. Enzymatic conversion of mcm⁵U to (S)-mchm⁵U was investigated by LC-MS/MS analysis of tRNA nucleosides.

RNA repair and tRNA modification activities, and the results are summarized in Table 1. The major findings in our study were that DNA repair activity actually could be detected *in vitro* for all tested proteins, and that two ALKBH8 proteins, originating from protozoa, also displayed tRNA modification activity similar to that of ALKBH8 from multicellular eukaryotes.

Many enzymes display promiscuous activities *in vitro*, *i.e.* activities that are different from the one for which the enzyme evolved, and that have no physiological role [55]. It is believed that such promiscuity may play an important part in the evolution of novel enzyme functions through gene duplication and mutation. Gene duplication allows for the retention of the original activity by one gene copy, whereas the other copy can be subjected to optimization of the promiscuous activity through amino acid substitution, until the promiscuous activity actually becomes a physiologically beneficial and selectable trait. Bacteria typically have 1–2 AlkB proteins, and the vast majority (>90%) of these appear to be DNA repair enzymes, whereas multicellular eukaryotes typically have several ALKBH proteins, most of which have other roles, *e.g.* in RNA modification. Thus, it is rather likely that the ALKBH family of enzymes in eukaryotes has evolved from an ancestral DNA repair enzyme, and it is not very surprising that some DNA repair activity can be detected *in vitro* for ALKBH proteins involved in other processes. AlkB proteins that function in DNA repair, such as the founding member *E. coli* AlkB, are in themselves rather promiscuous, as they can repair a wide range of DNA adducts (methyl, etheno, ethano, etc) on several different

nucleobases independent of sequence context. In contrast, the tRNA modification activity of ALKBH8 is much more specific, as it appears to exclusively occur on a single mcm⁵U-containing tRNA species, tRNA$^{Gly(UCC)}$. We therefore believe that the tRNA modification activity observed with the protozoan CP and TT proteins *in vitro* reflect their true, physiologically relevant function, whereas the detected repair activity represents a "ghost" of an evolutionary precursor, now manifested as a promiscuous activity. Indeed, a similar promiscuity has previously been observed for the mammalian ALKBH1 proteins, which show activity on methyl lesions in ssDNA and ssRNA as well as on histone proteins [26,56].

Two of the ALKBH8 proteins studied here, TT and CP, both of which originate from protozoa, catalyzed the hydroxylation of mcm⁵U into (S)-mchm⁵U in tRNA. The observed activity of the CP protein is not very surprising, as this protein has a RRM/AlkB/ZnF architecture also found in the tRNA-modifying ALKBH8s from mammals and plants. The importance of the RRM and ZnF domains in human ALKBH8 was recently demonstrated; the RRM domain provides affinity towards RNA (whereas the AlkB domain does not contribute to RNA binding), and the ZnF moiety increases the overall stability of the protein [57]. Accordingly, we observed that deletion of the RRM moiety abolished the activity of human ALKBH8, whereas mutation of the conserved Cys residues of the ZnF moiety had no effect. Therefore, the observed activity of the TT protein was somewhat unexpected, as this protein solely consists of an AlkB domain.

Table 1. Summary of experiments.

Protein	In vivo experiments			In vitro experiments									
	ssDNA (MMS)	ssRNA (MMS)	Survival (MMS)	ssDNA (CAA)	ssDNA m^{1}A	dsDNA m^{1}A	ssDNA m^{3}C	dsDNA m^{3}C	ssDNA εA	dsDNA εA	ssDNA εC	dsDNA εC	tRNA mcm^{5}U
RF	−			+	*	*	*	*	*	*	*	*	*
RD	−			−	−	−	++	++	+	+	++	++	−
TT	−			+	−	−	+	++	+	+	++	++	+
MV	−			−	+	−	−	−	++	+	++	+	−
CP	−			−	−	−	−	−	+	−	+	−	++

++, activity comparable to positive control (PC); +, substantially lower activity than PC (>30% reduction); −, no detectable activity; *, unable to produce recombinant protein.

However, this result gives important clues regarding the evolution of the tRNA modifying ALKBH8 function: point mutations in an ancestral repair protein may have yielded a beneficial, but suboptimal tRNA modifying activity, followed by the acquisition of RRM and ZnF domains, giving improved substrate affinity and enzyme stability, respectively. Finally, the fusion between the RRM/AlkB/ZnF moiety and a Trm9-like methyltransferase in animals has likely further improved the efficacy of the modification system by providing a direct channeling of mcm^{5}U-modified tRNA$^{Gly(UCC)}$ from the ALKBH8 methyltransferase to the hydroxylase.

We have here generated an *A. tumefaciens* mutant with an inactivated ALKBH8 gene and found the modification status of tRNA$^{Gly(UCC)}$ to be unaltered relative to the wild-type bacterium. This clearly suggests that bacterial ALKBH8s, unlike their eukaryotic counterparts, do not target the wobble uridine of tRNA$^{Gly(UCC)}$. However, the close sequence similarity between bacterial and eukaryotic ALKBH8s in the so-called nucleotide lid region (NRL), which is responsible for interaction with the nucleic acid substrate, may still suggest that bacterial ALKBH8s are RNA modification enzymes. Also, our demonstration of tRNA modification activity for the TT protein, which, like the bacterial ALKBH8s, lacks the RRM- and ZnF-domains, supports this notion. We have previously studied the ALKBH8 protein from the bacterium *R. etli*, and found it to be as efficient as *E. coli* AlkB in reactivating CAA-treated (etheno adduct-containing) ssDNA phage [38]. In contrast, the two bacterial ALKBH8 proteins studied here, RD and RF, showed substantially lower (RF) or negligible (RD) repair activity, and also considerably lower than for the protozoan TT protein, which also displayed tRNA modifying activity. Moreover, it should be noted that the *R. etli* ALKBH8 protein, when compared with the RD and RF proteins, is less similar to eukaryotic ALKBH8s, and lacks several of the conserved residues shared between other ALKBH8s (Fig. 2). Based on the above, we favor the notion that the "canonical" bacterial ALKBH8s such as RF and RD are RNA modification enzymes targeting a (yet unidentified) substrate resembling the mcm^{5}U moiety recognized by the eukaryotic ALKBHs.

The present work provides important insights on the ALKBH8 proteins, but many key questions remain unanswered. It will be of great interest to reveal the physiologically relevant substrate(s) of the bacterial ALKBH8 proteins. Here, the *A. tumefaciens* mutant described in the present work may represent a useful tool. Conceivably, a systematic, global analysis of the RNA modification pattern in the mutant versus wild-type bacteria may uncover this substrate. Furthermore, we have here shown that the ALKBH8 protein (TT) from *Tetrahymena thermophilus*, unlike its mammalian counterpart, is able to catalyse mcm^{5}U hydroxylation even in the absence of an RRM domain. This indicates that its AlkB domain (catalytic moiety) has an intrinsic affinity for the tRNA substrate, suggesting that this protein may be particularly well suited for structural studies aimed at solving the structure of an enzyme/substrate complex, thereby yielding insights on the detailed ALKBH8 mechanism.

Supporting Information

Table S1   Oligonucleotides used in the present study.

Acknowledgments

We would like to thank Chantal Abergel (CNRS, Aix-Marseille Université) for providing the plasmid containing the *Acanthamoeba polyphaga mimivirus* ALKBH8 gene, Barbara Honchak (Washington University in St. Louis) for providing *Roseobacter denitrificans* OCh 114 genomic DNA, Didier Raoult (Université de la Méditerranée) for providing *Rickettsia felis* URRWXCal2 genomic DNA, and Bob Coyne (J. Craig Venter Institute) for providing *Tetrahymena thermophila* SB210 genomic DNA. Human alkyl-N-purine-DNA glycosylase ANPG, *E. coli* uracil-DNA glycosylase Mug and human AP endonuclease 1 HAP-1 were a kind gift from Dr. M. Saparbaev (Institut Gustave Roussy, Villejuif, France). We would like to thank the Proteomics and Metabolomics Core Facility, PROMEC, at NTNU, supported in part by the Faculty of Medicine and the Central Norway Regional Health Authority.

Author Contributions

Conceived and designed the experiments: DZ CBV FK ED AP AMM EvdB PØF. Performed the experiments: DZ CBV FK ED AP AMM EvdB. Analyzed the data: DZ CBV FK ED AP AMM BT EvdB PØF. Contributed reagents/materials/analysis tools: CBV FK HEK AK BT. Contributed to the writing of the manuscript: DZ EvdB PØF.

References

1. Kataoka H, Yamamoto Y, Sekiguchi M (1983) A new gene (alkB) of Escherichia coli that controls sensitivity to methyl methane sulfonate. J Bacteriol 153: 1301–1307.

2. Delaney JC, Essigmann JM (2004) Mutagenesis, genotoxicity, and repair of 1-methyladenine, 3-alkylcytosines, 1-methylguanine, and 3-methylthymine in alkB Escherichia coli. Proc Natl Acad Sci U S A 101: 14051–14056.

3. Falnes PO, Johansen RF, Seeberg E (2002) AlkB-mediated oxidative demethylation reverses DNA damage in Escherichia coli. Nature 419: 178–182.

4. Falnes PO (2004) Repair of 3-methylthymine and 1-methylguanine lesions by bacterial and human AlkB proteins. Nucleic Acids Res 32: 6260–6267.

5. Koivisto P, Robins P, Lindahl T, Sedgwick B (2004) Demethylation of 3-methylthymine in DNA by bacterial and human DNA dioxygenases. J Biol Chem 279: 40470–40474.

6. Trewick SC, Henshaw TF, Hausinger RP, Lindahl T, Sedgwick B (2002) Oxidative demethylation by Escherichia coli AlkB directly reverts DNA base damage. Nature 419: 174–178.

7. Aravind L, Koonin EV (2001) The DNA-repair protein AlkB, EGL-9, and leprecan define new families of 2-oxoglutarate- and iron-dependent dioxygenases. Genome Biol 2: RESEARCH0007.

8. Loenarz C, Schofield CJ (2008) Expanding chemical biology of 2-oxoglutarate oxygenases. Nat Chem Biol 4: 152–156.

9. Aas PA, Otterlei M, Falnes PO, Vagbo CB, Skorpen F, et al. (2003) Human and bacterial oxidative demethylases repair alkylation damage in both RNA and DNA. Nature 421: 859–863.

10. Falnes PO, Bjoras M, Aas PA, Sundheim O, Seeberg E (2004) Substrate specificities of bacterial and human AlkB proteins. Nucleic Acids Res 32: 3456–3461.

11. Ougland R, Zhang CM, Liiv A, Johansen RF, Seeberg E, et al. (2004) AlkB restores the biological function of mRNA and tRNA inactivated by chemical methylation. Mol Cell 16: 107–116.

12. Delaney JC, Smeester L, Wong C, Frick LE, Taghizadeh K, et al. (2005) AlkB reverses etheno DNA lesions caused by lipid oxidation in vitro and in vivo. Nat Struct Mol Biol 12: 855–860.

13. Duncan T, Trewick SC, Koivisto P, Bates PA, Lindahl T, et al. (2002) Reversal of DNA alkylation damage by two human dioxygenases. Proc Natl Acad Sci U S A 99: 16660–16665.

14. Frick LE, Delaney JC, Wong C, Drennan CL, Essigmann JM (2007) Alleviation of 1, N6-ethanoadenine genotoxicity by the Escherichia coli adaptive response protein AlkB. Proc Natl Acad Sci U S A 104: 755–760.

15. Koivisto P, Duncan T, Lindahl T, Sedgwick B (2003) Minimal methylated substrate and extended substrate range of Escherichia coli AlkB protein, a 1-methyladenine-DNA dioxygenase. J Biol Chem 278: 44348–44354.

16. Maciejewska AM, Ruszel KP, Nieminuszczy J, Lewicka J, Sokolowska B, et al. (2010) Chloroacetaldehyde-induced mutagenesis in Escherichia coli: the role of AlkB protein in repair of 3, N(4)-ethenocytosine and 3, N(4)-alpha-hydro-xyethanocytosine. Mutat Res 684: 24–34.

17. Maciejewska AM, Poznanski J, Kaczmarska Z, Krowisz B, Nieminuszczy J, et al. (2013) AlkB dioxygenase preferentially repairs protonated substrates: specificity against exocyclic adducts and molecular mechanism of action. J Biol Chem 288: 432–441.

18. Mishina Y, Yang CG, He C (2005) Direct repair of the exocyclic DNA adduct 1, N6-ethenoadenine by the DNA repair AlkB proteins. J Am Chem Soc 127: 14594–14595.

19. Gerken T, Girard CA, Tung YC, Webby CJ, Saudek V, et al. (2007) The obesity-associated FTO gene encodes a 2-oxoglutarate-dependent nucleic acid demethylase. Science 318: 1469–1472.

20. Kurowski MA, Bhagwat AS, Papaj G, Bujnicki JM (2003) Phylogenomic identification of five new human homologs of the DNA repair enzyme AlkB. BMC Genomics 4: 48.

21. Dango S, Mosammaparast N, Sowa ME, Xiong LJ, Wu F, et al. (2011) DNA unwinding by ASCC3 helicase is coupled to ALKBH3-dependent DNA alkylation repair and cancer cell proliferation. Mol Cell 44: 373–384.

22. Ringvoll J, Nordstrand LM, Vagbo CB, Talstad V, Reite K, et al. (2006) Repair deficient mice reveal mABH2 as the primary oxidative demethylase for repairing 1meA and 3meC lesions in DNA. EMBO J 25: 2189–2198.

23. Zheng G, Dahl JA, Niu Y, Fedorcsak P, Huang CM, et al. (2013) ALKBH5 is a mammalian RNA demethylase that impacts RNA metabolism and mouse fertility. Mol Cell 49: 18–29.

24. Jia G, Fu Y, Zhao X, Dai Q, Zheng G, et al. (2011) N6-methyladenosine in nuclear RNA is a major substrate of the obesity-associated FTO. Nat Chem Biol 7: 885–887.

25. Li MM, Nilsen A, Shi Y, Fusser M, Ding YH, et al. (2013) ALKBH4-dependent demethylation of actin regulates actomyosin dynamics. Nat Commun 4: 1832.

26. Ougland R, Lando D, Jonson I, Dahl JA, Moen MN, et al. (2012) ALKBH1 is a histone H2A dioxygenase involved in neural differentiation. Stem Cells 30: 2672–2682.

27. Cantara WA, Crain PF, Rozenski J, McCloskey JA, Harris KA, et al. (2011) The RNA Modification Database, RNAMDB: 2011 update. Nucleic Acids Res 39: D195–D201.

28. Agris PF, Vendeix FA, Graham WD (2007) tRNA's wobble decoding of the genome: 40 years of modification. J Mol Biol 366: 1–13.

29. Johansson MJ, Esberg A, Huang B, Bjork GR, Bystrom AS (2008) Eukaryotic wobble uridine modifications promote a functionally redundant decoding system. Mol Cell Biol 28: 3301–3312.

30. Kalhor HR, Clarke S (2003) Novel methyltransferase for modified uridine residues at the wobble position of tRNA. Mol Cell Biol 23: 9283–9292.

31. Fu D, Brophy JA, Chan CT, Atmore KA, Begley U, et al. (2010) Human AlkB homolog ABH8 Is a tRNA methyltransferase required for wobble uridine modification and DNA damage survival. Mol Cell Biol 30: 2449–2459.

32. Songe-Moller L, van den Born E, Leihne V, Vagbo CB, Kristoffersen T, et al. (2010) Mammalian ALKBH8 possesses tRNA methyltransferase activity required for the biogenesis of multiple wobble uridine modifications implicated in translational decoding. Mol Cell Biol 30: 1814–1827.

33. Fu Y, Dai Q, Zhang W, Ren J, Pan T, et al. (2010) The AlkB domain of mammalian ABH8 catalyzes hydroxylation of 5-methoxycarbonylmethyluridine at the wobble position of tRNA. Angew Chem Int Ed Engl 49: 8885–8888.

34. van den Born E, Vagbo CB, Songe-Moller L, Leihne V, Lien GF, et al. (2011) ALKBH8-mediated formation of a novel diastereomeric pair of wobble nucleosides in mammalian tRNA. Nat Commun 2: 172.

35. Drablos F, Feyzi E, Aas PA, Vaagbo CB, Kavli B, et al. (2004) Alkylation damage in DNA and RNA–repair mechanisms and medical significance. DNA Repair (Amst) 3: 1389–1407.

36. Falnes PO, Rognes T (2003) DNA repair by bacterial AlkB proteins. Res Microbiol 154: 531–538.

37. van den Born E, Omelchenko MV, Bekkelund A, Leihne V, Koonin EV, et al. (2008) Viral AlkB proteins repair RNA damage by oxidative demethylation. Nucleic Acids Res 36: 5451–5461.

38. van den Born E, Bekkelund A, Moen MN, Omelchenko MV, Klungland A, et al. (2009) Bioinformatics and functional analysis define four distinct groups of AlkB DNA-dioxygenases in bacteria. Nucleic Acids Res 21: 7124–7136.

39. Leihne V, Kirpekar F, Vagbo CB, van den Born E, Krokan HE, et al. (2011) Roles of Trm9- and ALKBH8-like proteins in the formation of modified wobble uridines in Arabidopsis tRNA. Nucleic Acids Res 39: 7688–7701.

40. Altschul SF, Koonin EV (1998) Iterated profile searches with PSI-BLAST–a tool for discovery in protein databases. Trends Biochem Sci 23: 444–447.

41. Edgar RC (2004) MUSCLE: multiple sequence alignment with high accuracy and high throughput. Nucleic Acids Res 32: 1792–1797.

42. Waterhouse AM, Procter JB, Martin DM, Clamp M, Barton GJ (2009) Jalview Version 2–a multiple sequence alignment editor and analysis workbench. Bioinformatics 25: 1189–1191.

43. Blatny JM, Brautaset T, Winther-Larsen HC, Haugan K, Valla S (1997) Construction and use of a versatile set of broad-host-range cloning and expression vectors based on the RK2 replicon. Appl Environ Microbiol 63: 370–379.

44. Perutka J, Wang W, Goerlitz D, Lambowitz AM (2004) Use of computer-designed group II introns to disrupt Escherichia coli DExH/D-box protein and DNA helicase genes. J Mol Biol 336: 421–439.

45. Yao J, Lambowitz AM (2007) Gene targeting in gram-negative bacteria by use of a mobile group II intron ("Targetron") expressed from a broad-host-range vector. Appl Environ Microbiol 73: 2735–2743.

46. Crain PF (1990) Preparation and enzymatic hydrolysis of DNA and RNA for mass spectrometry. Methods Enzymol 193: 782–790.

47. Yu B, Edstrom WC, Benach J, Hamuro Y, Weber PC, et al. (2006) Crystal structures of catalytic complexes of the oxidative DNA/RNA repair enzyme AlkB. Nature 439: 879–884.

48. Mantri M, Zhang Z, McDonough MA, Schofield CJ (2012) Autocatalysed oxidative modifications to 2-oxoglutarate dependent oxygenases. FEBS J 279: 1563–1575.

49. Gros L, Ishchenko AA, Saparbaev M (2003) Enzymology of repair of etheno-adducts. Mutat Res 531: 219–229.

50. Ringvoll J, Moen MN, Nordstrand LM, Meira LB, Pang B, et al. (2008) AlkB homologue 2-mediated repair of ethenoadenine lesions in mammalian DNA. Cancer Res 68: 4142–4149.

51. Calvo JA, Meira LB, Lee CY, Moroski-Erkul CA, Abolhassani N, et al. (2012) DNA repair is indispensable for survival after acute inflammation. J Clin Invest 122: 2680–2689.

52. Dinglay S, Trewick SC, Lindahl T, Sedgwick B (2000) Defective processing of methylated single-stranded DNA by E. coli AlkB mutants. Genes Dev 14: 2097–2105.

53. Dosanjh MK, Chenna A, Kim E, Fraenkel-Conrat H, Samson L, et al. (1994) All four known cyclic adducts formed in DNA by the vinyl chloride metabolite chloroacetaldehyde are released by a human DNA glycosylase. Proc Natl Acad Sci U S A 91: 1024–1028.

54. Kim MY, Zhou X, Delaney JC, Taghizadeh K, Dedon PC, et al. (2007) AlkB influences the chloroacetaldehyde-induced mutation spectra and toxicity in the pSP189 supF shuttle vector. Chem Res Toxicol 20: 1075–1083.

55. Khersonsky O, Tawfik DS (2010) Enzyme promiscuity: a mechanistic and evolutionary perspective. Annu Rev Biochem 79: 471–505.

56. Westbye MP, Feyzi E, Aas PA, Vagbo CB, Talstad VA, et al. (2008) Human AlkB homolog 1 is a mitochondrial protein that demethylates 3-methylcytosine in DNA and RNA. J Biol Chem 283: 25046–25056.

57. Pastore C, Topalidou I, Forouhar F, Yan AC, Levy M, et al. (2012) Crystal structure and RNA binding properties of the RNA recognition motif (RRM) and AlkB domains in human AlkB homolog 8 (ABH8), an enzyme catalyzing tRNA hypermodification. J Biol Chem 287: 2130–2143.

Efficient Transformation of Oil Palm Protoplasts by PEG-Mediated Transfection and DNA Microinjection

Mat Yunus Abdul Masani[1]*, Gundula A. Noll[2], Ghulam Kadir Ahmad Parveez[1],
Ravigadevi Sambanthamurthi[1], Dirk Prüfer[2,3]*

1 Advanced Biotechnology and Breeding Centre, Malaysian Palm Oil Board (MPOB), Kuala Lumpur, Malaysia, 2 Westfälische Wilhelms-Universität Münster, Institut für Biologie und Biotechnologie der Pflanzen, Münster, Germany, 3 Fraunhofer Institut für Molekularbiologie und Angewandte Ökologie, Münster, Germany

Abstract

Background: Genetic engineering remains a major challenge in oil palm (*Elaeis guineensis*) because particle bombardment and *Agrobacterium*-mediated transformation are laborious and/or inefficient in this species, often producing chimeric plants and escapes. Protoplasts are beneficial as a starting material for genetic engineering because they are totipotent, and chimeras are avoided by regenerating transgenic plants from single cells. Novel approaches for the transformation of oil palm protoplasts could therefore offer a new and efficient strategy for the development of transgenic oil palm plants.

Methodology/Principal Findings: We recently achieved the regeneration of healthy and fertile oil palms from protoplasts. Therefore, we focused on the development of a reliable PEG-mediated transformation protocol for oil palm protoplasts by establishing and validating optimal heat shock conditions, concentrations of DNA, PEG and magnesium chloride, and the transfection procedure. We also investigated the transformation of oil palm protoplasts by DNA microinjection and successfully regenerated transgenic microcalli expressing green fluorescent protein as a visible marker to determine the efficiency of transformation.

Conclusions/Significance: We have established the first successful protocols for the transformation of oil palm protoplasts by PEG-mediated transfection and DNA microinjection. These novel protocols allow the rapid and efficient generation of non-chimeric transgenic callus and represent a significant milestone in the use of protoplasts as a starting material for the development of genetically-engineered oil palm plants.

Editor: Meng-xiang Sun, Wuhan University, China

Funding: This research was funded by the MPOB-Fraunhofer collaborative project "Establishment of transgenic oil palm with high added value for commercial exploitation" (T0003050000-RB01-J) and internal resources of the Institute of Plant Biology and Biotechnology of the Westfälische Wilhelms-Universität Münster. The funders had no role in study design, data collection and analysis, decision to publish, or preparation of the manuscript.

Competing Interests: The authors have declared that no competing interests exist.

* E-mail: masani@mpob.gov.my (MYAM); dpruefer@uni-muenster.de (DP)

Introduction

The oil palm genetic engineering program was initiated by the Malaysian Palm Oil Board (MPOB), then known as the Palm Oil Research Institute of Malaysia (PORIM), in the early 1990s [1]. The main objectives of this program are to produce transgenic oil palm (*Elaeis guineensis*) with a higher content of oleic acid, modified oil quality (e.g. a higher content of stearic acid), and the ability to produce value-added oils such as palmitoleic and ricinoleic acid, as well as novel products such a biodegradable plastics. It has been suggested that such targets could be achieved 80% more rapidly by combining genetic engineering and tissue culture techniques [2]. In addition, oil palm is a perennial crop and high-value products could be produced continuously for at least 30 years, making this species an ideal candidate for genetic engineering.

Particle bombardment and *Agrobacterium*-mediated transformation can be used to introduce genes into oil palm, and stable transformation has been achieved using both methods. Successful particle bombardment requires the establishment of optimal physical and biological parameters during transformation and the use of appropriate selectable markers and promoters [3]. A biolistic protocol for the production of glufosinate-resistant transgenic oil palm has been developed [4], and in light of its success thousands of embryogenic calli have been bombarded with genes involved in fatty acid biosynthesis to increase the accumulation of oleic acid [5,6,7], stearic acid [8], polyhydroxybutyrate (PHB) and polyhydroxyvalerate (PHBV) [9,10,11].

One drawback of particle bombardment is that it often promotes the integration of multiple transgene copies [12], whereas *Agrobacterium*-mediated transformation is more likely to introduce either single-copy or low-copy-number transgenes, as shown e.g. in rice [13] and maize [14]. However, *Agrobacterium*-mediated oil palm transformation is inefficient because oil palm is a monocotyledonous species outside the normal *Agrobacterium tumefaciens* host range. Nevertheless, there have been many attempts to improve transformation efficiency, e.g. by using immature oil palm embryos as the target tissue for particle bombardment and *Agrobacterium*-mediated transformation [15,16], and by optimizing other transformation parameters [17]. This led to the development of insect-resistant transgenic oil palms expressing *Bacillus thuringiensis* (Bt) insecticidal proteins [18] and cowpea trypsin inhibitor (CpTI) [19].

The studies described above revealed that 3–5 years are required to generate transgenic oil palm plants by particle bombardment or *Agrobacterium*-mediated transformation and that the process is highly inefficient. The frequency of escapes and chimeric plants is high because the long selection process during callus formation and somatic embryogenesis encourages the growth of non-transformed cells. It is possible that the optimization of DNA delivery and selection could overcome such challenges but an alternative approach is to use protoplasts as transformation targets because they are totipotent, and chimeras can thus be avoided by regenerating transgenic plants from single cells. Novel approaches for the transformation of oil palm protoplasts could therefore offer a new and efficient strategy for the development of transgenic oil palm plants. As well as particle bombardment and *Agrobacterium*-mediated transformation, protoplasts can also be transformed using polyethylene glycol (PEG), electroporation or DNA microinjection.

Recently, we established an efficient protocol for the preparation of oil palm protoplasts and the regeneration of healthy and fertile oil palm plants [20]. Here we developed novel transformation protocols based on PEG-mediated transfection and DNA microinjection showing that protoplasts are suitable as a target for oil palm genetic engineering. We successfully expressed a reporter gene encoding green fluorescent protein (GFP) allowing the rapid and efficient generation of non-chimeric transgnic callus without the use of standard selectable markers. Our results represent a significant milestone in development of genetically-engineered oil palm plants.

Results

PEG-mediated transfection of oil palm protoplasts

Choice of protoplast source. In order to identify the most suitable protoplasts for PEG-mediated transformation, we tested protoplasts from different sources, namely those isolated 7 or 14 days after the subculture of a cell suspension culture that had been cultivated for either 3 or 4 months (Figure 1A–C). For the initial protoplast transfection experiments we used 10 μg of CFDV-hrGFP plasmid DNA mixed with 40% (w/v) PEG dissolved in Rinse solution, and incubated the protoplasts for 10 min. The appearance of fluorescent protoplasts 72 h later indicated that the hrGFP gene was expressed successfully in protoplasts from all the sources we tested (Figure 1D–F). However, the transfection efficiency was low (< 0.1%) because most of the protoplasts were severely damaged and only a small number of fluorescent protoplasts survived. GFP fluorescence was distributed throughout the cytoplasm and nucleus, extending to the plasma membrane in protoplasts from both the 7 and 14 day subcultures (Figures 1D–F). Protoplasts from the 14-day subcultures also showed pale yellow autofluorescence, which was more intense in the protoplasts derived from the 4-month-old cell suspension culture (Figures 1E–F). We selected protoplasts isolated from the 7-day subculture of the 3-month-old suspension culture as the most suitable substrates for PEG-mediated transformation because of the lack of autofluorescence, thus reducing the likelihood of false positive results.

MgCl₂ concentration and DNA incubation period. We investigated the impact of Mg^{2+} ions on transfection efficiency by incubating oil palm protoplasts as above for 10 min in the presence of 10 μg of CFDV-hrGFP plasmid DNA mixed with 40% (w/v) PEG dissolved in Rinse solution, but this time we

varied the concentration of Mg^{2+} ions by preparing solutions containing 10 mM (Figure 2A), 25 mM (Figure 2B), 50 mM (Figure 2C) and 100 mM $MgCl_2$ (Figure 2D). The presence of 10 mM $MgCl_2$ increased the transfection efficiency by four-fold to 0.39% (Figure 2A) compared to a PEG solution lacking magnesium (<0.1%, data not shown) but higher concentrations were even more beneficial, and the greatest efficiency (2.5%) was achieved in the presence of 50 mM $MgCl_2$ (Figure 2B–D). GFP fluorescence was more intense in the protoplasts transfected at higher Mg^{2+} concentrations, indicating the more efficient uptake of exogenous DNA (Figure 2A–D).

Having established the optimal Mg^{2+} concentration for transfection, we next varied the incubation time following the addition of plasmid DNA but prior to the addition of the PEG/$MgCl_2$ solution (Figure 2E–F). Prolonging the incubation period to 15 min (Figure 2E) or 30 min (Figure 2F) reduced the transfection efficiency to 1.42% and 0.65%, respectively. Therefore we reverted to the original incubation period of 10 min.

Concentrations of DNA and PEG, and heat shock treatment. Next we investigated the impact of DNA concentration on transfection efficiency by incubating protoplasts in the presence of 25 μg (Figure 3A) or 50 μg (Figure 3B) of CFDV-hrGFP plasmid DNA using the 40% (w/v) PEG/50 mM $MgCl_2$ solution discussed above. High transfection efficiencies were achieved in both cases, but the lower DNA concentration was less efficient (2.05%, Figure 3A) than the original 50-μg dose (2.73%, Figure 3B). The GFP fluorescence was also more intense in protoplasts transfected with higher concentrations of DNA probably because more was taken up into the cell.

We also investigated the effects of different PEG concentrations, varying the (w/v) concentration of PEG 4000 from 25% (Figure 3C), to 40% (Figure 3D) and also 50% (Figure 3E). In each case, the different PEG concentrations were tested with the optimal DNA and $MgCl_2$ concentrations and 10-min DNA incubation time established above. The corresponding transfection efficiencies were 3.74%, 2.02% and 1.66%, showing that 25% (w/v) PEG is optimal for the transformation of oil palm protoplasts. There was no difference in terms of GFP fluorescence regardless of the PEG concentration, suggesting that PEG does not affect hrGFP gene expression but may instead affect the viability of the oil palm protoplasts at concentrations higher than 25%.

Finally, we investigated the effect of heat shock treatment by incubating the protoplasts at 45°C for 5 min and then cooling on ice for 1 min before adding 50 μg of CFDV-hrGFP plasmid DNA, incubating for 10 min as above and then adding 25% (w/v) PEG in 50 mM $MgCl_2$. This treatment increased the transfection efficiency even further to 4.76% (Figure 4A) indicating that a heat shock significantly improves DNA uptake. Fluorescent protoplasts were observed continuously for 9 days indicating that hrGFP fluorescence remains stable following transfection, although the frequency declined over time from 4.42% on day 6 (Figure 4B) to 4.35% on day 9 (Figure 4C).

Transformation of oil palm protoplasts by DNA microinjection

Choice of protoplast platform and optimal injection time. A novel DNA microinjection protocol for oil palm protoplasts was developed using protoplasts embedded in an alginate layer (Figure 5A) because microinjection is facilitated if the protoplasts are immobilized in a single plane (Figure 5B). Different concentrations of alginate, ranging from 0.5% to 2%, were dissolved in Y3A liquid to prepare the substrate. We found that 1% alginate was ideal for immobilizing the protoplasts, whereas

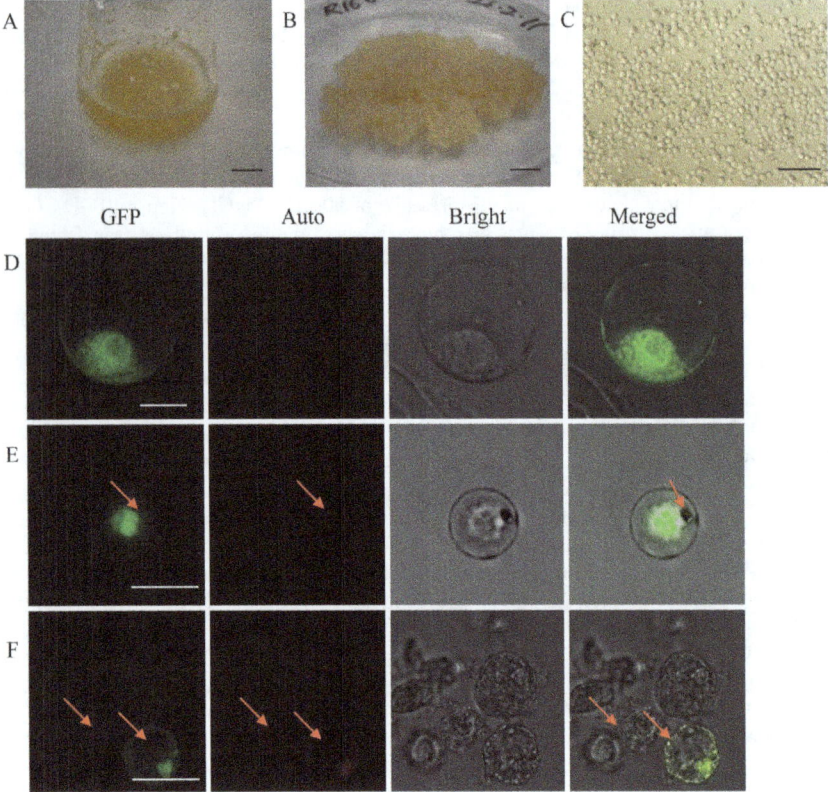

Figure 1. Oil palm protoplasts showing GFP fluorescence. A 3-month-old oil palm cell suspension culture in Y35N5D2iP liquid medium (A) was collected and cultured on Y35N5D2iP solid medium (B) for protoplast isolation (C). Transient GFP fluorescence was observed in protoplasts isolated from the 3-month-old cell suspension culture after subculture for 7 days (D) and 14 days (E), and protoplasts isolated from the 4-month-old cell suspension culture (F). CLSM images are shown representing GFP fluorescence (GFP), autofluorescence (Auto) and bright field (Bright) as well as three-layer images (Merged) of the protoplasts. Red arrows indicate autofluorescence. Scale bar = 1 cm in (A) and (B), 100 µm in (C), 10 µm in (D), 25 µm in (E) and (F).

they remained mobile if lower concentrations were used and higher concentrations promoted the formation of clumps.

The embedded protoplasts were cultured for 3–4 days in a two-compartment dish (Figure 5C) allowing the partial development of the cell wall, which was the ideal time for DNA microinjection. Freshly-embedded protoplasts were damaged by the procedure, demonstrating that the fragile plasma membrane alone cannot withstand penetration by the needle tip. On the other hand, if the protoplasts were left for 5 or more days, the efficiency of microinjection was limited because the cell wall was by this stage fully developed. A single micromanipulator was used to inject all protoplasts because they were immobilized within the alginate layer (Figure 5D).

The injection of DNA and Lucifer yellow into the protoplast cytoplasm. The protoplasts were initially injected with the fluorescent dye Lucifer yellow in order to visualize the cytoplasmic compartment (Figure 5E–F). We then co-injected the dye and the linear CFDV-hrGFP fragment (Figure 5G–H). GFP expression was first detected 72 h after microinjection, and the GFP fluorescence was distributed throughout the cytoplasm and nucleus, extending to the plasma membrane. The emission wavelengths of Lucifer yellow and GFP are distinct, allowing us to clearly distinguish between cells expressing the microinjected DNA and those containing the marker dye alone.

The effect of DNA concentration on transformation efficiency, and the development of microcalli expressing GFP. The optimal DNA fragment concentration was deter-

mined by comparing the transformation efficiencies achieved when injecting 50 embedded protoplasts with ~5 µl of DNA solution at concentrations of 100 ng/µl (Figure 6A–B), 500 ng/µl (Figure 6C–D) and 1000 ng/µl (Figure 6E–F). After one month, we recorded corresponding transfection efficiencies of 74.6% (Figure 6A–B), 39.3% (Figure 6C–D) and 10% (Figure 6E–F), indicating that 100 ng/µl is the optimal concentration of microinjected DNA.

Microcolonies developing from the protoplasts injected with 100 ng/µl DNA were observed for 2–3 months, by which time the proportion of colonies expressing GFP had fallen to 51.3% (Figure 7A–B). GFP expression was maintained for a further 2 months (Figure 7C–D) but the proportion of colonies expressing GFP fell to 14% after 6 months, when microcalli began to develop (Figure 7E–F). The microcalli expressing hrGFP were removed from the alginate layer (Figure 7G) and transferred to Y31N0.1BA solid medium (Figure 7H) for development into embryogenic calli, which is similar to the procedure for regenerating protoplasts into plants using agarose bead cultures [20]. A significant number of wild-type microcalli were also obtained in these experiments reflecting the absence of a selectable marker (data not shown).

GFP	Bright	Merged	Efficiency (%)

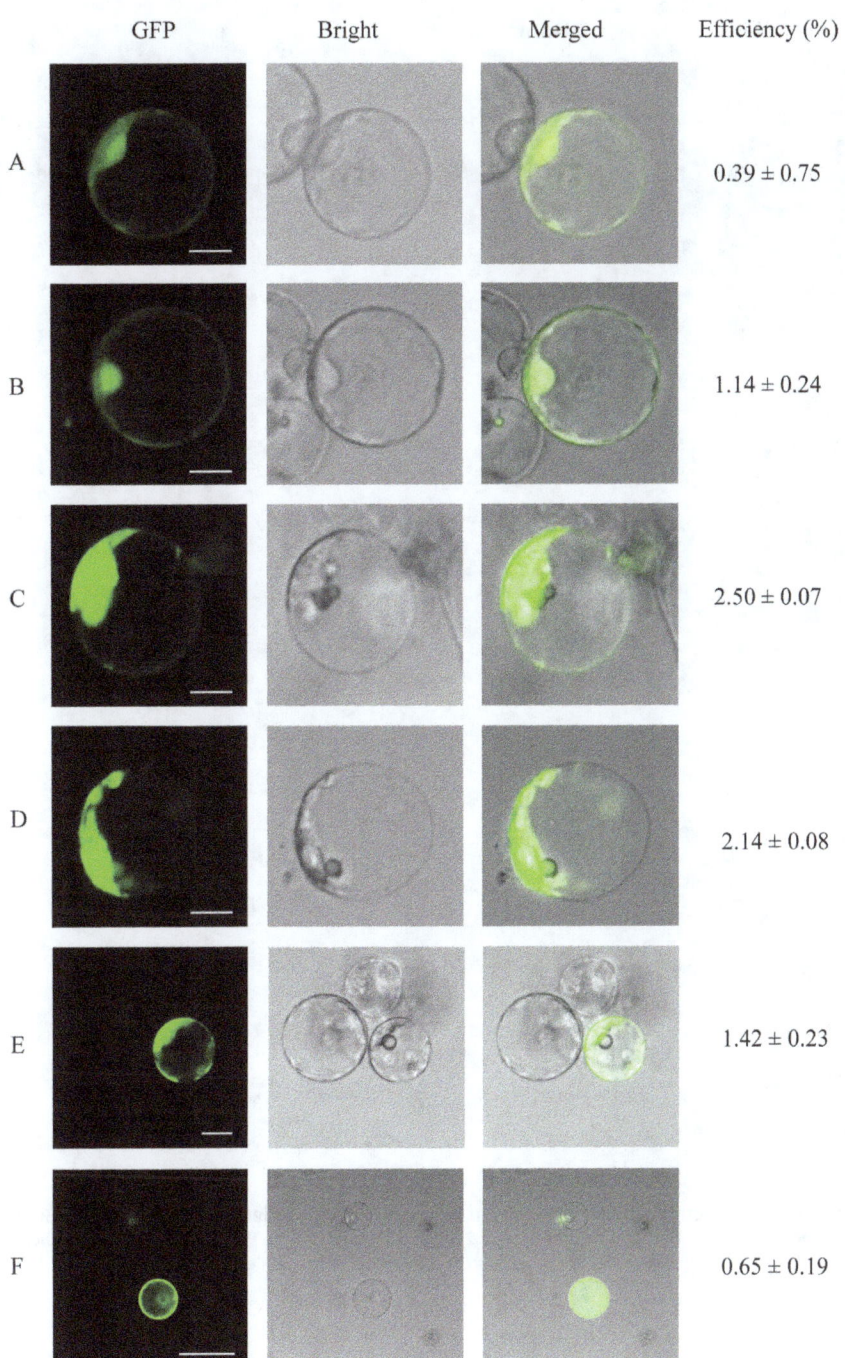

Figure 2. Transfection efficiency is affected by different concentrations of MgCl$_2$ and the DNA incubation time. Oil palm protoplasts were transfected with 10 µg CFDV-hrGFP plasmid using 40% (w/v) PEG solution with MgCl$_2$ at concentrations of 10 mM (A), 25 mM (B), 50 mM (C) and 100 mM (D). Oil palm protoplasts were incubated with 10 µg CFDV-hrGFP plasmid DNA for 15 min (E) or 30 min (F), and then mixed with PEG-MgCl$_2$ solution. Transfection efficiency was calculated as the number of GFP-fluorescent protoplasts divided by the total number of protoplasts in one representative microscope field. The transfection efficiencies represent the mean of three replicates. Scale bar = 10 µm in (A)–(E), 75 µm in (F).

Discussion

PEG-mediated transient expression in oil palm protoplasts

Genetic engineering in oil palm is challenging because the standard transformation approaches based on *Agrobacterium* and particle bombardment are laborious and inefficient, generating a large number of chimeric plants. Following the successful regeneration of oil palm plants from protoplasts derived from cell suspension cultures [20] it is now possible to use protoplasts as the starting material for the development of stable transgenic oil palm lines by genetic engineering. Protoplasts are beneficial as a starting material because they are totipotent, which allows transgenic plants to be regenerated from single cells thus avoiding the issue of chimeras. However, the transformation of oil palm protoplasts using *Agrobacterium* and particle bombardment is limited by the

	GFP	Bright	Merged	Efficiency (%)

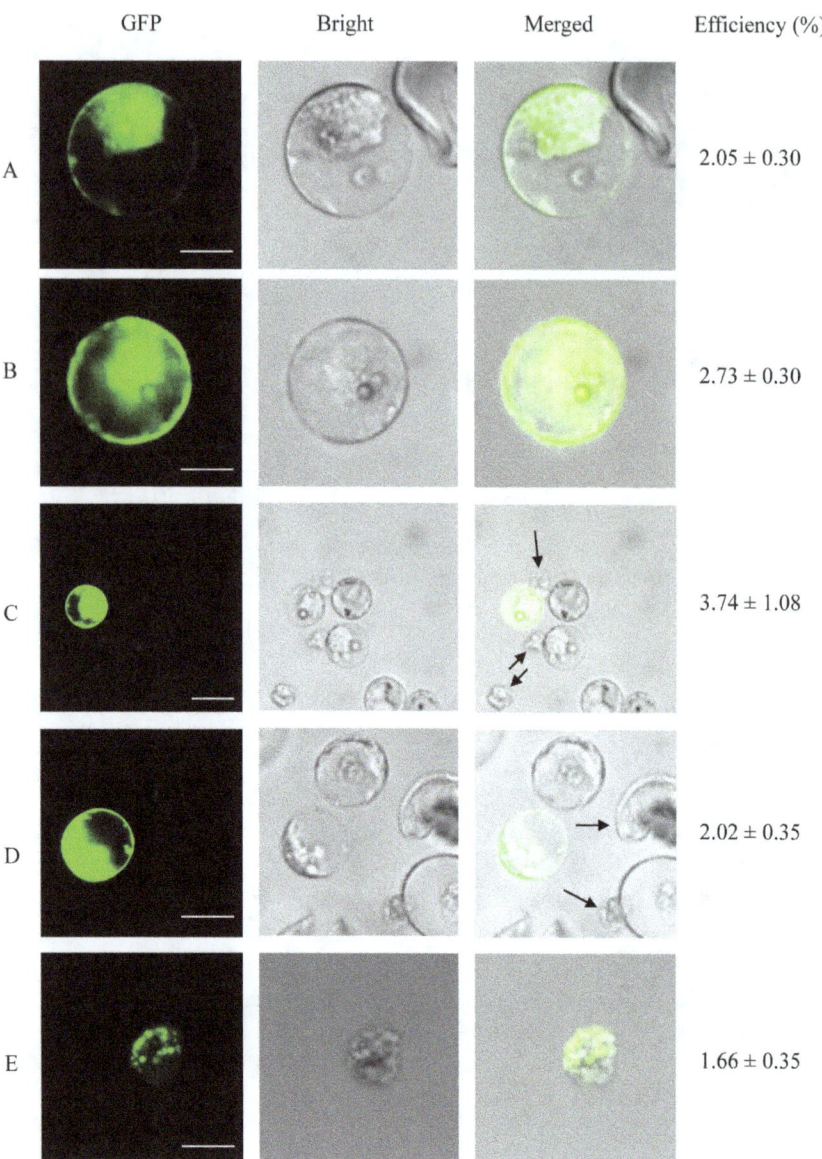

A — 2.05 ± 0.30

B — 2.73 ± 0.30

C — 3.74 ± 1.08

D — 2.02 ± 0.35

E — 1.66 ± 0.35

Figure 3. Effects of DNA and PEG concentrations on transfection efficiency. Oil palm protoplasts were transfected with 25 µg (A) or 50 µg (B) of CFDV-hrGFP plasmid DNA, and with 50 µg CFDV-hrGFP plasmid DNA in the presence of 25% (C), 40% (D) or 50% PEG (E). Black arrows indicate damaged protoplasts caused by PEG toxicity. The transfection efficiencies represent the mean of three replicates. Scale bar = 10 µm in (A) and (B), 25 µm in (C)–(F).

challenges described above, so novel transformation approaches are required in order to develop an efficient strategy for the generation of transgenic oil palm.

The stable integration of exogenous DNA into the genome of oil palm protoplasts following PEG-mediated transfection, electroporation or microinjection, could facilitate the generation of stable transgenic lines because the plants would be regenerated from a single transformed cell. However, these techniques have not been applied in oil palm before and the standard protocols would therefore need to be optimized to maintain the viability of oil palm protoplasts, promote the uptake of DNA and demonstrate the efficiency of transgene expression. Because PEG-mediated transfection is a standard method for gene transfer to protoplasts that allows the rapid analysis of transient reporter gene expression, this method was investigated as a first step in the development of an efficient transformation protocol.

We used GFP as a visual marker because this approach allows the recognition of transient expression just after gene transfer as well as stable expression after transgene integration in plant tissues (and whole plants) derived from transfected cells. In normal transformation approaches, selectable markers are required to allow the propagation of the rare, stably-transformed cells while killing or suppressing the large excess of non-transformed or transiently-transformed cells in the target tissue [21]. When single cells are the target, this selection process is unnecessary. However, the transfection of oil palm protoplasts using PEG is an untested strategy so it was necessary to provide evidence of transformation and regeneration efficiency, and for this purpose a visible marker such as GFP is ideal. Visual selection also allows transgenic plants to be developed without selectable markers, an approach which is considered increasingly attractive in a commercial setting [22].

GFP	Bright	Merged	Efficiency (%)

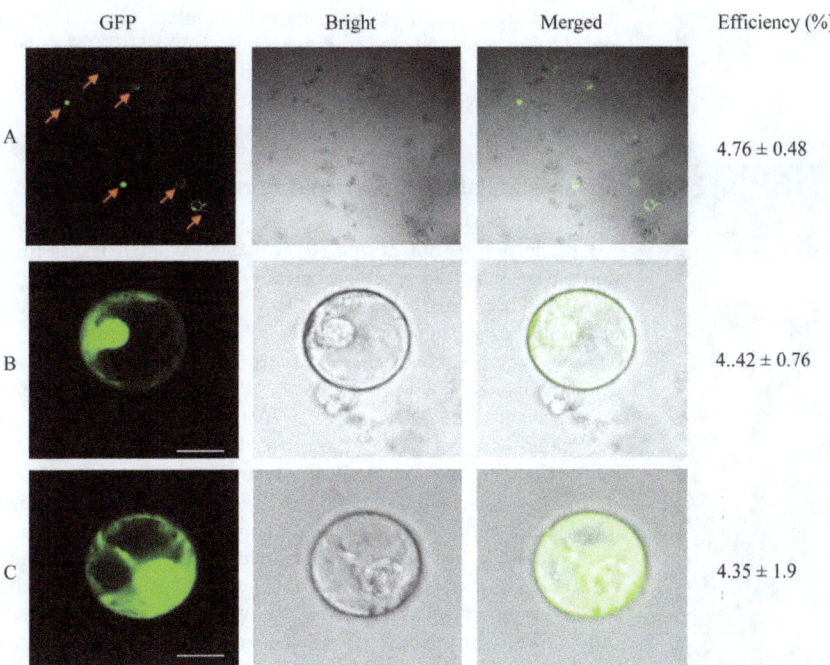

A   4.76 ± 0.48

B   4..42 ± 0.76

C   4.35 ± 1.9

Figure 4. Effect of heat shock treatment on transfection efficiency. Oil palm protoplasts were incubated at 45°C for 5 min and then on ice for 1 min before mixing with 50 μg of CFDV-hrGFP plasmid DNA and then PEG-MgCl₂ solution (25% PEG, 50 mM MgCl₂, 3% KCl and 3.6% mannitol, pH 6.0). The protoplasts were incubated at 26°C for 3 days (A), 6 days (B) and 9 days (C). Red arrows indicate protoplasts showing GFP fluorescence. The transfection efficiencies represent the mean of three replicates. Scale bar = 100 μm in (A), 10 μm in (B) and (C).

Protoplasts isolated from oil palm suspension cultures after 7 days of subculture were identified as the most suitable substrates for PEG-mediated transformation. The protoplasts were uniform in size, and the transfected protoplasts were easily identified due to the absence of autofluorescence. In contrast, autofluorescence was observed in protoplasts from cell suspension cultures that were 3 and 4 months old after 14 days of subculture, which could produce false positive results. Protoplasts isolated from these cell suspension cultures should not have chloroplasts, so the autofluorescence may reflect the presence of small amounts of lipids. Osmotic stress during protoplast isolation can modulate lipid metabolism resulting in the synthesis of up to 27% palmitoleic acid [23].

The concentration of Mg^{2+} is the most important determinant of efficient PEG-mediated transient gene expression in tobacco and maize protoplasts [24,25]. Similarly, we found that the concentration of Mg^{2+} greatly influenced the transfection efficiency of oil palm protoplasts and the intensity of GFP fluorescence. The highest transfection efficiency of 2.50% was achieved in the presence of 50 mM MgCl₂ and the intensity of GFP fluorescence increased as the concentration of MgCl₂ rose from 10 to 100 mM may indicating a more efficient uptake of the exogenous DNA and therefore the stronger expression of GFP, or a better cell survival, or both.

Incubation times with the exogenous DNA exceeding 10 min did not improve the transfection efficiency, and indeed reduced the proportion of cells expressing GFP by approximately four-fold (2.50% to 0.65%; compare Figure 2C to Figure 2E–F) suggesting that longer incubation times prolonged exposure to exogenous and endogenous nucleases, the former released from broken protoplasts. A slight improvement in transfection efficiency was therefore achieved with higher DNA concentrations (2.05% with 25 μg of DNA compared to 2.73% with 50 μg of DNA) because the larger amount of target DNA is likely to saturate the available nuclease pool.

The PEG-mediated transfection of oil palm protoplasts was also inhibited by the potentially toxic effects of excess PEG, resulting in a significant loss of viability in the presence of PEG concentrations exceeding 25% (w/v). High concentrations of PEG also promoted the clumping of protoplasts, making it difficult to identify individual protoplasts producing GFP. In contrast, a heat shock caused a significant improvement in transfection efficiency, probably by transiently permeabilizing the plasma membrane and thus promoting the uptake of DNA.

Stable transformation of oil palm protoplasts by DNA microinjection

Initially, we microinjected DNA into protoplasts embedded in agarose beads, but although this method was successful it was also inefficient, allowing the injection of only 5–10 cells per hour (data not shown). The identification of protoplasts in this environment was often difficult, and the microinjection needle tips often became clogged with agarose debris after only 2–3 injection attempts. Furthermore, the preparation method required protoplasts to be exposed to molten liquid agarose at 45°C which imparted significant heat stress and reduced vitality. In contrast, protoplasts embedded in an alginate layer were ideal for DNA microinjection because alginate is transparent so the protoplasts could be identified easily, there was no heat stress during preparation so the protoplasts were more vital, and the flat surface made it easier to inject the protoplasts more rapidly (50–100 alginate embedded-protoplasts could be successfully injected per hour). Another advantage of the alginate layer was that it could be dissolved in sodium acetate solution to isolate the microcalli, allowing them to be transferred onto the appropriate media for further cultivation.

In previous studies, the nucleus of tobacco protoplasts was identified as the most suitable target compartment for DNA microinjection, with transformation frequencies of 14–20%, mediated transfection (4.76%), particle bombardment (1%) [3] and *Agrobacterium*-mediated transformation (0.7%) [17]. Our novel microinjection approach in oil palm therefore represents a significant increase in the efficiency of transformation in this species.

Advantages of DNA microinjection for the genetic engineering of oil palm

Particle bombardment and *Agrobacterium*-mediated transformation are the most widely used procedures for genetic engineering in plants and both have been applied to oil palm, but in each case the techniques are inefficient and beset by additional disadvantages. The novel and highly efficient transformation approaches we have developed for oil palm protoplasts, i.e. PEG-mediated transfection and particularly DNA microinjection, offer new routes to overcome these barriers and improve the efficiency and applicability of genetic engineering in this erstwhile recalcitrant species.

Both particle bombardment and *Agrobacterium*-mediated transformation require at least 0.5 g of target tissue for each transformation experiment, which involves laborious and time-consuming preparation. In contrast, less than 0.5 g of oil palm tissue is required for protoplast isolation, and each explant yields thousands of protoplasts for subsequent transformation experiments. Another advantages is that 'clean gene' fragments consisting of only the promoter-gene-terminator sequence can be introduced by microinjection into the oil palm protoplast, thereby avoiding the integration of vector backbone sequences that can interfere with transgene expression and raise concerns with the regulatory authorities [22].

The genetic engineering of oil palm protoplasts by microinjection could allow the production of stable and non-chimeric transgenic lines regenerated from a single transformed cell, each carrying a single copy of the integrated transgene (which is an important consideration for the commercialization of transgenic oil palm plants) [27]. Nevertheless, further improvements are required before this approach becomes a standard technique in oil palm transformation programs. For example, the current transformation efficiency of 14% is relatively low compared to the frequencies achieved in other species, such as 26% in alfalfa [28] and 20–53% in tobacco [27,29].

Materials and Methods

Plant material

Oil palm embryogenic cell suspension cultures were cultivated in 100-ml flasks containing 50 ml Y35N5D2iP liquid medium [20] supplemented with 5 μM naphthalene acetic acid (NAA), 5 μM 2,4-dichlorophenoxyacetic acid (2,4-D) and 2 μM 2-γ-dimethylallylaminopurine (2iP). The suspension cultures were incubated in the dark at 28°C on a rotary shaker at 120 rpm. Half of the Y35N5D2iP liquid medium was discarded and replaced with fresh medium every 14 days.

Protoplast isolation and purification

Protoplasts were isolated from 3-month-old and 4-month-old oil palm cell suspension cultures. The cells were collected by filtration through a 300-μm nylon mesh, and 0.5 g fresh weight (fwt) of cells was transferred to a 50-ml centrifuge tube containing 15 ml filter-sterilized enzyme solution (2% (v/v) cellulase (Sigma), 1% (v/v) pectinase (Sigma), 0.5% (w/v) cellulase onuzuka R10 (Duchefa), 0.1% (w/v) pectolyase Y23 (Duchefa), 3% (w/v) KCl, 0.5% (w/v) CaCl$_2$.2H$_2$O and 3.6% (w/v) mannitol, pH 5.6). The cells were resuspended by inverting the tube 6–10 times and then incubated in the dark without shaking at 26°C for 14 h. The mixture was

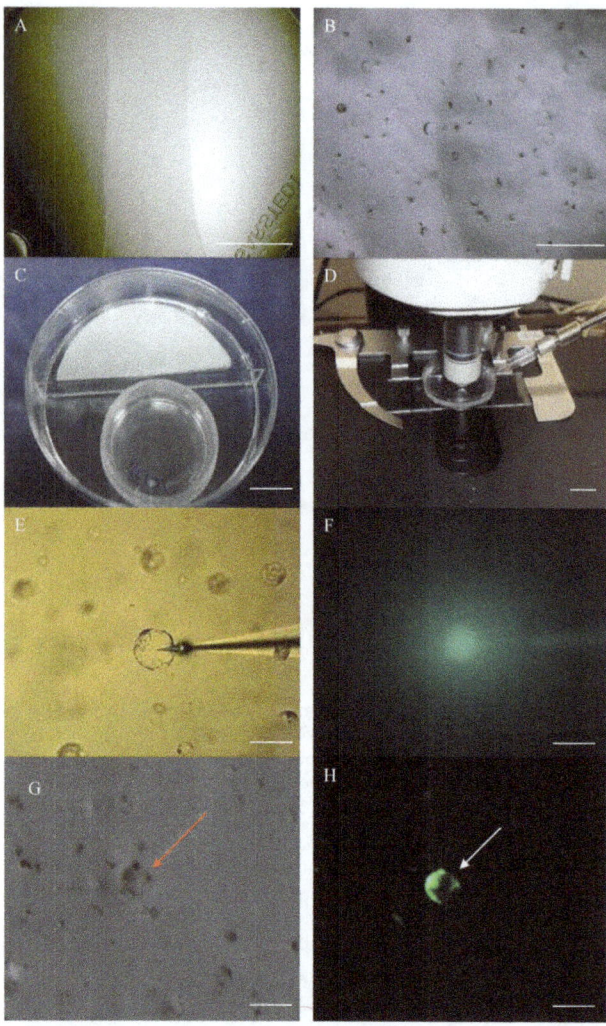

Figure 5. Microinjection of DNA into oil palm protoplasts. Oil palm protoplasts were isolated from a 3-month-old cell suspension culture after subculture for 7 days, mixed with 1% alginate solution in Y3A medium and distributed as a thin layer onto supporting medium (A). The embedded protoplasts were arranged in a single planar layer as confirmed by using the 10× objective (B). The protoplasts were incubated at 28°C in the dark for 3 days (C), and then placed on the microscope stage for DNA microinjection (D). The DNA solution was injected into the protoplast (E) and confirmed by Lucifer yellow fluorescence (F). GFP fluorescence was detected in the cytoplasm after 3 days (G and H). The injected protoplast is indicated by an arrow. Scale bar = 1 cm in (A), (C) and (D), 100 μm in (B), 25 μm in (E)–(H).

compared to 6% when the DNA was injected into the cytoplasm [26,27]. However, we limited DNA microinjection to the cytoplasm because the nucleus was usually difficult to identify, and often became swollen and dislodged after injection resulting in the rapid death of the injected protoplasts (data not shown).

Based on the above experiments, the injection of ~5 μl of DNA (at concentration of 100 ng/μl) into the cytoplasm of protoplasts embedded in an alginate layer was identified as the optimal platform for the transformation of oil palm protoplasts. This resulted in approximately 14% of the injected protoplasts developing into microcalli that continued to express GFP. Although this is the first report of genetic engineering in oil palm by DNA microinjection, the transformation efficiency of 14% is far higher than that achieved using other approaches such as PEG-

Bright   GFP   Efficiency (%)

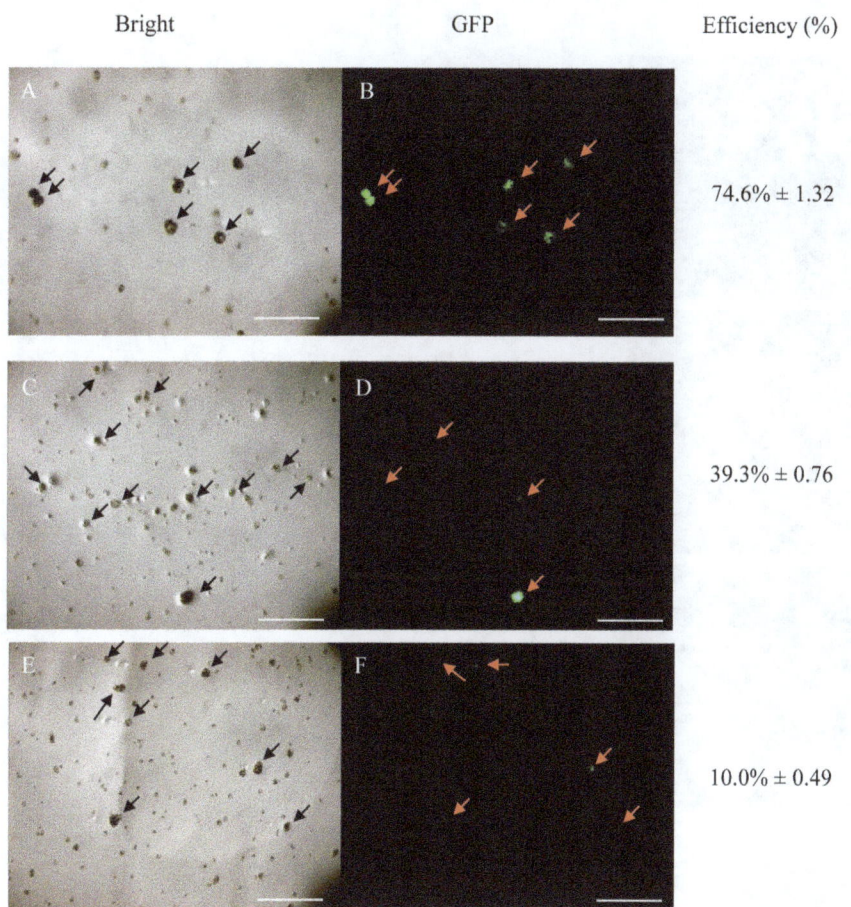

74.6% ± 1.32

39.3% ± 0.76

10.0% ± 0.49

Figure 6. Effect of DNA concentration on transformation efficiency. Protoplasts were injected with 100 ng/µl (A and B), 500 ng/µl (C and D) and 1000 ng/µl (E and F) DNA solutions and monitored for evidence of GFP fluorescence after culture for one month. The transformation efficiencies represent the mean of three replicates. Arrows indicate the surviving injected protoplasts and small dots indicate dead cells. Scale bar = 100 µm. All cells were injected in the visial field shown in this figure, but uninjected cells also developed into (non-fluorescent) microcalli.

diluted with 15ml filter-sterilized washing solution (3% (w/v) KCl, 0.5% (w/v) CaCl$_2$.2H$_2$O, 3.6% (w/v) mannitol, pH 5.6), resuspended by inverting the tube 3–5 times, filtered through a sterilized double layer of miracloth and collected in a 50-ml centrifuge tube. The filtration step was repeated 2–3 times until all undigested tissues, cell clumps and cell debris were removed. The mixture was centrifuged at 60×g for 5 min at 22°C and the supernatant was discarded. The protoplast pellet was resuspended by adding 10 ml washing solution and mixing by inversion, followed by centrifugation as above. The supernatant was removed completely and the protoplast pellet was resuspended in 10 ml filter-sterilized rinse solution (3% (w/v) KCl, 3.6% (w/v) mannitol, pH 5.6) and then centrifuged at 60×g for 5 minutes at 22°C. After three cycles, the supernatant was removed leaving 3 ml in the tube, and was stored at room temperature for further experiments.

PEG-mediated transfection

A 1-ml aliquot of the protoplast suspension was incubated at room temperature for 10 min, or a heat shock was applied by incubation at 45°C for 5 min before immediately cooling on ice for 1 min, and then incubated at room temperature for 10 min. A 500-µl aliquot of the protoplast suspension was then placed as a single droplet in the middle of a 60 mm×15 mm Petri dish (Greiner Bio-One, Germany) and five drops of 100 µl PEG-MgCl$_2$

solution (25–50% (w/v) PEG 4000 (Sigma), 10–100 µM MgCl$_2$ (Sigma), 3% (w/v) KCl, 3.6% (w/v) mannitol, 0.05% (w/v) 2-N-morpholinoethanesulfonic acid (MES), pH 6.0) were added in an adjacent but separate position. We then slowly added 25 or 50 µg of plasmid DNA to the protoplasts and mixed gently by stirring with a 200-µl pipette tip. The mixture was incubated at room temperature in the dark for 10–30 min and then the protoplasts + DNA were mixed with the adjacent PEG/MgCl$_2$ drops by stirring with the 200-µl pipette tip. After a further 30-min incubation, 4 ml of washing solution (3% (w/v) KCl, 0.5% (w/v) CaCl$_2$.2H$_2$O, 3.6% (w/v) mannitol, pH 5.6) was added drop by drop and the mixture was incubated in the dark at 26°C for 9 days. The protoplasts were observed under a Leica TCS 5 SP5 X confocal laser scanning microscope (CLSM) and visualized using a Leica Microsystem LAS AF. GFP fluorescence was observed at an excitation wavelength of 488 nm and an emission range of 500–600 nm, whereas autofluorescence of protoplasts was excited at 543 nm and detected within the emission range 675–741 nm. PEG-mediated transfection efficiency was calculated as the number of GFP-positive protoplasts divided by the total number of protoplasts in one representative microscope field. Each calculation was carried out for a total of three microscope fields containing no fewer than 200 protoplasts.

Bright   GFP   Efficiency (%)

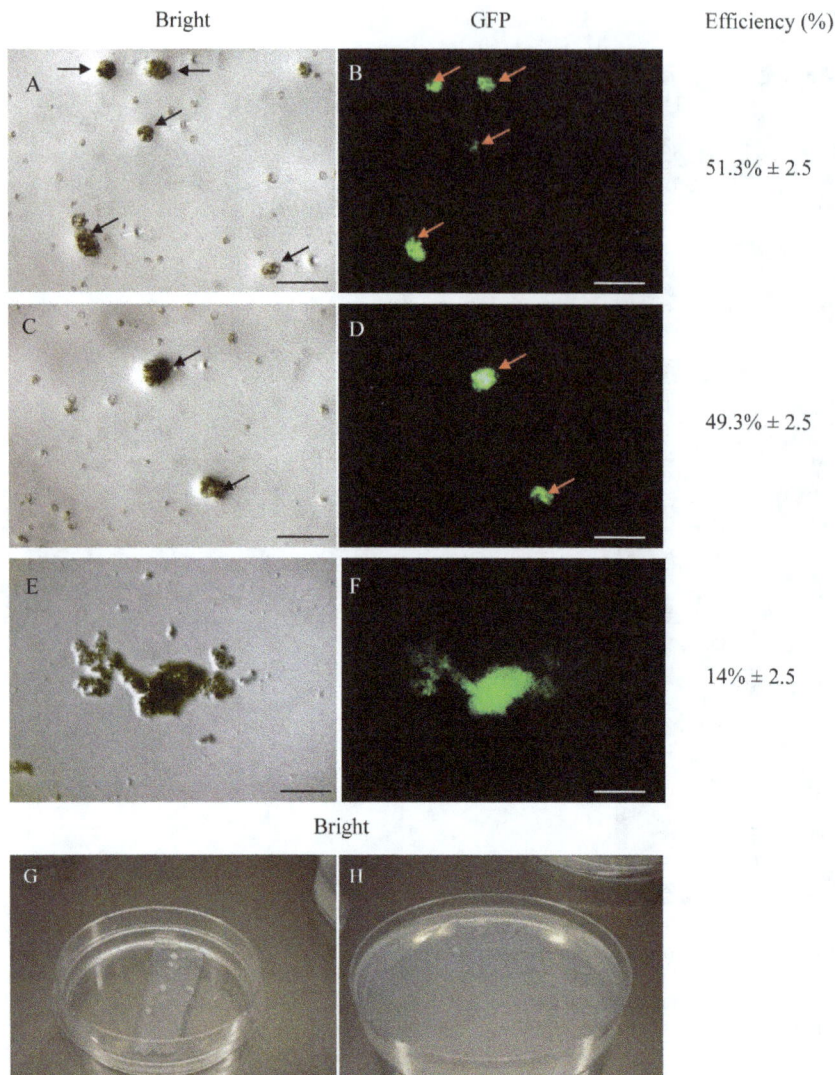

51.3% ± 2.5

49.3% ± 2.5

14% ± 2.5

Bright

Figure 7. Development of microcalli expressing GFP. Five days after DNA microinjection, the alginate layer was transferred to Y3A liquid medium comprising 5.5% (w/v) sucrose and 8.2% (w/v) glucose supplemented with 10 μM NAA, 2 μM 2,4–D, 2 μM IBA, 2 μM GA₃, 2 μM 2iP and 200 mg/l ascorbic acid and cultured at 28°C for 2 weeks. The medium was then replaced with similar Y3A liquid medium comprising 4% (w/v) sucrose and 7.2% (w/v) glucose to allow the development of microcolonies (A and B, after 2 months). The medium was then replaced with Y3A liquid medium comprising 4% (w/v) sucrose to promote the conversion of microcolonies (C and D, after 4 months) into microcalli (E and F, after 6 months). Finally, the alginate layer containing microcalli (G) was transferred onto Y31N0.1BA solid medium (H) for the regeneration of oil palm plants. Arrows indicate the injected protoplasts. The transformation efficiencies represent the mean of three replicates. Scale bar = 100 μm in (A)–(F), 1 cm in (G) and (H).

Preparation of alginate thin layer

After allowing the protoplast suspension to settle for 30 min, the supernatant was removed completely and the pellet was resuspended in 3 ml filter-sterilized alginate solution comprising 1% (w/v) alginic acid sodium salt (A2158, Sigma) dissolved in Y3A liquid medium [20] (5.5% (w/v) sucrose, 11.9% (w/v) glucose, 10 μM NAA, 2 μM 2,4-D, 2 μM IBA, 2 μM GA₃, 2 μM 2iP and 200 mg/l ascorbic acid) adjusted to pH 5.6, and including Y3 macroelements prepared without $CaCl_2$. Alginate-embedded protoplasts were distributed as a thin layer onto a substrate comprising 1.5 ml filter-sterilized Y3A medium (5.5% (w/v) sucrose and 11.9% (w/v) glucose supplemented with 0.1% (w/v) $CaCl_2.2H_2O$ and solidified with 1% (w/v) agarose sea plaque) in a 35×10 mm Petri dish (Greiner Bio-One, Germany). The distribution of alginate-embedded protoplasts was achieved by drop-ping 100 μl alginate-embedded protoplasts at the edge of the Petri dish and tilting to 35° so the drop was distributed as a thin layer. The dishes were placed horizontally into 94×15 mm two-compartment dishes (Greiner Bio-One, Germany) allowing the alginate to solidify within 1–2 min. We added 3 ml of sterile water to the outer compartment to prevent the alginate layer from drying out. The plates were sealed and incubated at 28°C in the dark for 3 days.

Microinjection workstation

The microinjection workstation consisted of a Leica DM LFS upright microscope (Leica Microsystems Wetzlar GmbH, Germany) with a joystick-controlled motorized objective revolver for HCX APOL U-V-I water immersion objectives (10x, 20x, 40x and 63x) mounted on a fixed table and placed under a laminar flow hood. The microscope was equipped with a Luigs and Neumann Manipulator set with a control system SM-5 and SM-6 (Luigs and Neumann, Germany).

Preparation of DNA injection solution

Plasmid DNA was prepared using a midi-scale Plasmid DNA Purification Kit (NucleoBond PC100; MACHEREY-NAGEL, Germany) and was dissolved at concentration of 1 μg/μl in sterile water. The plasmid was digested with HindIII and EcoRI to yield the CFDV-hrGFP-nos cassette as a 1.5-kb fragment. The fragment was separated from the vector sequence (pUC19) by 1% agarose gel electrophoresis, excised using a clean blade and isolated using the PCR clean-up Gel Extraction Kit (NucleoSpin Gel and PCR Clean-up) according to the manufacturer's instructions (Macherey-Nagel, Germany). The DNA cassette was then diluted in sterile water to concentrations of 100 ng/μl, mixed at a 100:1 ratio with Lucifer yellow CH dilithium salt (L0259, Invitrogen) and filter-sterilized using the Ultafree-MC filter (Durapore 0.22 μm, type GV; SK-1M-524-J8; Millipore) at 10,000 rpm, 15 min, 4°C. The eluted DNA was partitioned into 10 μl aliquots and stored at −20°C.

Loading the DNA injection solution into microinjection needle

The DNA injection solution was centrifuged at 14,000 rpm for 30 min at 4°C and a 5-μl aliquot was loaded into the tip of a Femtotip II microinjection needle (no. 5242 957000, Eppendorf) using a microloader (no. 5242 956003, Eppendorf). After 30 min at room temperature, the needle was filled with sterile mineral oil (M8410, Sigma) using the microloader and tightly mounted in the capillary holder of a microinjector CellTram vario (no. 5176 000033, Eppendorf), and then fixed onto the micromanipulator.

Microinjection of oil palm protoplasts

A plate containing protoplasts embedded in an alginate layer was placed on the microscope stage, and the vitality of the protoplasts was confirmed using the 10× objective, which was then raised to its maximum height allowing the needle tip to reach the center of the field view freely with the X/Y-axis controller (Control system SM-5) of the manipulator. The needle was lowered as close as possible to the alginate layer with the Z-axis controller and the cytoplasm or nucleus of the target protoplast

was identified by adjusting the 20× objective to optimal resolution and contrast. The needle tip was then moved to immediately above the protoplast with the X/Y-axis controller and inserted into the alginate layer immediately adjacent to the protoplast using the Z-axis controller before penetration using the X-axis controller. The DNA solution was slowly injected into the protoplast using a microinjector CellTram vario, and confirmed by fluorescence illumination. The needle tip was carefully withdrawn from the protoplast and moved to the next one. The injected protoplasts were monitored periodically using a Leica MZ16F fluorescent stereomicroscope with a GFP3 filter (Leica Microsystems Wetzlar GmbH, Germany).

Alginate layer culture

Following microinjection, the plates containing the alginate layer were incubated in the dark at 28°C for 5 days. The alginate layers were then separated from the substrate and transferred into 60×15 mm Petri dishes containing 3 ml Y3A liquid medium comprising 5.5% (w/v) sucrose and 8.2% (w/v) glucose supplemented with 10 μM NAA, 2 μM 2,4-D, 2 μM IBA, 2 μM GA$_3$, 2 μM 2iP and 200 mg/l ascorbic acid. The dishes were incubated in the dark at 28°C, shaking at 50 rpm. After 2 weeks, the medium was replaced with similar Y3A liquid medium but the concentrations of sucrose and glucose were reduced to 4% (w/v) and 7.2% (w/v), respectively. The alginate layers were cultured in this medium for one month by refreshing the medium at 14-day intervals, then replaced with Y3A liquid medium containing 4% (w/v) sucrose until microcalli had developed.

Acknowledgments

The authors wish to thank the Director General of MPOB for permission to publish this article. The authors would also like to thank Dr. Ahmad Tarmizi Hashim (Clonal Propagation Group Leader, MPOB) for the supply of oil palm cell suspension calli. Finally, the authors would like to acknowledge all members of AG Prof. Dirk Prüfer (IME, Muenster, Germany) and the Transgenic Technology Group (MPOB) for their assistance.

Author Contributions

Conceived and designed the experiments: DP GKAP GN. Performed the experiments: MYAM. Analyzed the data: MYAM GN DP. Wrote the paper: MYAM GKAP RS DP.

References

1. Cheah SC, Sambanthamurthi R, Siti Nor Akmar A, Abrizah O, Manaf MAA, et al. (1995) Towards genetic engineering oil palm (*Elaeis guineensis* Jacq.). In: Plant Lipid Metabolism. Edited by Kader JC, Mazliak P. Netherlands, Kluwer Academic Publishers: 570–572.

2. Majid NA, Parveez GKA (2007) Evaluation of green fluorescence protein (GFP) as a selectable marker for oil palm transformation via transient expression. Asia Pac J Mol Biol Biotech 15: 1–8.

3. Parveez GKA (1998) Optimization of parameters involved in the transformation of oil palm using the biolistics method. PhD Thesis, Universiti Putra Malaysia.

4. Parveez GKA, Christou P (1998) Biolistic-mediated DNA delivery and isolation of transgenic oil palm (*Elaeis guineensis* Jacq) embryogenic callus cultures. J Oil Palm Res 10: 29–38.

5. Parveez GKA, Rasid O, Zainal A, Masri MM, Majid NA, et al. (2000) Transgenic oil palm: production and projection. Biochem Soc T 28: 969–972.

6. Ravigadevi S, Siti Nor Akmar A, Parveez GKA (2002) Genetic manipulation of the oil palm - challenges and prospects. Planter 78: 547–562.

7. Yunus AMM, Kadir APG (2008) Development of transformation vectors for the production of potentially high oleate transgenic oil palm. Electron J Biotech 11: 1–9.

8. Parveez GKA (2003) Novel products from transgenic oil palm. Agri Biotechnol Net 5: 1–8.

9. Parveez GKA, Bohari B, Ayub NH, Yunus AMM, Rasid OA, et al. (2008) Transformation of PHB and PHBV genes driven by maize ubiquitin promoter into oil palm for the production of biodegradable plastics. J Oil Palm Res 2: 76–86.

10. Yunus AMM, Chai-Ling H, Parveez GKA (2008) Construction of PHB and PHBV transformation vectors for bioplastics production in oil palm. J Oil Palm Res 2: 37–55.

11. Masani MYA, Parveez GKA, Izawati DAM, Lan CP, Akmar SNA (2009) Construction of PHB and PHBV multiple-gene vectors driven by an oil palm leaf-specific promoter. Plasmid 3: 191–200.

12. Stanton GB (1998) The introduction and expression of transgenes in plants. Curr Opin Biotech 9: 227–232.

13. Raineri DM, Bottino P, Gordon MP, Nester EW (1990) *Agrobacterium tumefaciens*-mediated transformation of rice (*Oryza sativa* L.). Bio/Technol 8: 33–38.

14. Gould J, Devery M, Hasegawa O, Ulian EC, Peterson G, et al. (1991) Transformation of *Zea mays* L. using *Agrobacterium tumefaciens* and shoot apex. Plant Physiol 95: 426–434.

15. Ruslan A, Zainal A, Heng WY, Li LC, Beng YC, et al. (2005) Immature embryo: A useful tool for oil palm (*Elaeis guineensis* Jacq.) genetic transformation studies. Electron J Biotech 8: 25–34.

16. Bhore SJ, Shah FH (2012) Genetic transformation of the American oil palm (*Elaeis oleifera*) immature zygotic embryos with antisense palmitoyl-acyl carrier protein thioesterase (PATE) gene. World Appl Sci J 16: 362–369.

17. Masli DIA, Kadir APG, Yunus AMM (2009) Transformation of oil palm using *Agrobacterium tumefaciens*. J Oil palm Res 21: 643–652.

18. Lee MP, Yeun LH, Abdullah R (2006) Expression of *Bacillus thuringiensis* insecticidal protein gene in transgenic oil palm. Electron J Biotech 9: 117–126.

19. Ismail I, Lee FS, Abdullah R, Fei CK, Zainal Z, et al. (2010) Molecular and expression analysis of cowpea trypsin inhibitor (CpTI) gene in transgenic *Elaeis guineensis Jacq* leaves. Aust J Crop Sci 4: 37–48.

20. Masani MYA, Noll G, Parveez GKA, Sambanthamurthi R, Prüfer D (2013) Regeneration of viable oil palm plants from protoplasts by optimizing media components, growth regulators and cultivation procedures. Plant Sci 210: 118–127.

21. Twyman RM, Stoger E, Kohli A, Capell T, Christou P (2002) Selectable and screenable markers for rice transformation. In: Jackson JF, Linskens HF (eds) Molecular Methods of Plant Analysis, Volume 22 (Testing for Genetic Manipulation in Plants). Springer-Verlag NY: 1–18.

22. Tuteja N, Verma S, Sahoo RK, Raveendar S, Reddy IN (2013) Recent advances in development of marker-free transgenic plants: regulation and biosafety concern. J Biosci 37: 167–97.

23. Sambanthamurthi R, Parman SH, Mohd Noor MR (1996) Oil palm (*Elaeis quineensis*) protoplast: Isolation, culture and microcallus formation. Plant Cell Tiss Org 46: 35–41.

24. Negrutiu I, Shillito R, Potrykus I, Biasini G, Sala F (1987) Hybrid genes in the analysis of transformation conditions. I. Setting up a simple method for direct gene transfer in plant protoplasts. Plant Mol Biol 8: 363–373.

25. Maas C, Werr W (1989) Mechanism and optimized conditions for PEG mediated DNA transfection into plant protoplasts. Plant Cell Rep 8: 148–151.

26. Crossway A, Oakes JV, Irvine JM, Ward B, Knauf VC, et al. (1986) Integration of foreign DNA following microinjection of tobacco mesophyll protoplasts. Mol Gen Genet 202: 179–185.

27. Schnorf M, Nenhaus-Url G, Galli A, Iida S, Potrykos I, et al. (1991) An improved approach for transformation of plant cells by microinjection: molecular and genetic analysis. Transgenic Res 1: 23–30.

28. Reich TJ, Iyer VN, Miki BL (1986) Efficient transformation of alfalfa protoplasts by the intranuclear microinjection of Ti plasmids. Bio/Technol 4: 1001–1004.

29. Kost B, Galli A, Potrykus I, Neuhaus G (1995) High efficiency transient and stable transformation by optimized DNA microinjection into *Nicotiana tabacum* protoplasts. J Exp Bot 46: 1157–1167.

Arbuscular-Mycorrhizal Networks Inhibit *Eucalyptus tetrodonta* Seedlings in Rain Forest Soil Microcosms

David P. Janos[1]*, **John Scott**[2], **Catalina Aristizábal**[1], **David M. J. S. Bowman**[3]

1 Department of Biology, University of Miami, Coral Gables, Florida, United States of America, **2** Research Institute for the Environment and Livelihoods, Charles Darwin University, Darwin, Northern Territory, Australia, **3** School of Plant Science, The University of Tasmania, Hobart, Tasmania, Australia

Abstract

Eucalyptus tetrodonta, a co-dominant tree species of tropical, northern Australian savannas, does not invade adjacent monsoon rain forest unless the forest is burnt intensely. Such facilitation by fire of seedling establishment is known as the "ashbed effect." Because the ashbed effect might involve disruption of common mycorrhizal networks, we hypothesized that in the absence of fire, intact rain forest arbuscular mycorrhizal (AM) networks inhibit *E. tetrodonta* seedlings. Although arbuscular mycorrhizas predominate in the rain forest, common tree species of the northern Australian savannas (including adult *E. tetrodonta*) host ectomycorrhizas. To test our hypothesis, we grew *E. tetrodonta* and *Ceiba pentandra* (an AM-responsive species used to confirm treatments) separately in microcosms of ambient or methyl-bromide fumigated rain forest soil with or without severing potential mycorrhizal fungus connections to an AM nurse plant, *Litsea glutinosa*. As expected, *C. pentandra* formed mycorrhizas in all treatments but had the most root colonization and grew fastest in ambient soil. *E. tetrodonta* seedlings also formed AM in all treatments, but severing hyphae in fumigated soil produced the least colonization and the best growth. Three of ten *E. tetrodonta* seedlings in ambient soil with intact network hyphae died. Because foliar chlorosis was symptomatic of iron deficiency, after 130 days we began to fertilize half the *E. tetrodonta* seedlings in ambient soil with an iron solution. Iron fertilization completely remedied chlorosis and stimulated leaf growth. Our microcosm results suggest that in intact rain forest, common AM networks mediate belowground competition and AM fungi may exacerbate iron deficiency, thereby enhancing resistance to *E. tetrodonta* invasion. Common AM networks—previously unrecognized as contributors to the ashbed effect—probably help to maintain the rain forest–savanna boundary.

Editor: Minna-Maarit Kytöviita, Jyväskylä University, Finland

Funding: The Cooperative Research Centre (CRC) for Tropical Savannas, and the Parks and Wildlife Commission of the Northern Territory funded this research. The University of Tasmania supported DPJ as a visiting scholar to prepare the manuscript. The funders had no role in study design, data collection and analysis, decision to publish, or preparation of the manuscript.

Competing Interests: The authors have declared that no competing interests exist.

* E-mail: davidjanos@miami.edu

Introduction

Eucalypts predominate across much of monsoon tropical, northern Australia. *Eucalyptus tetrodonta* F.Muell. and *E. miniata* Cunn. ex Schauer are canopy co-dominants of coastal savannas [1] throughout which are scattered patches of rain forest [2,3]. The savannas are highly susceptible to fire, but only after very dry conditions can fires cross the abrupt ecotone between savanna and rain forest [4]. If high-intensity fires penetrate the rain forest, then *E. tetrodonta* seedlings can invade, but otherwise they cannot, even after canopy destruction by tropical cyclones [5]. Reciprocally, in the absence of fire, rain forest plants can colonize savanna, albeit very slowly [6]. Nevertheless, the mechanisms by which fire facilitates rain forest invasion by *E. tetrodonta* are uncertain.

The apparent necessity of fire to facilitate establishment of some species' seedlings, especially those of eucalypts and pines, is known as the "ashbed effect" (a tenet which underpins much Australian forestry practice [7]). Although the ashbed effect has been investigated for more than half a century, its mechanisms remain ambiguous because it likely involves multiple phenomena associated with fire and soil desiccation. Those phenomena may include at different times and places, direct fertilization by ash [7], soil physical and chemical changes that diminish P adsorption [7,8],

release of mineral nutrients from heat-killed soil microorganisms [8] (but see [9]), partial soil sterilization that eliminates pathogenic microbes [10] (but see [11]), or other alterations of the soil microflora, especially ectomycorrhizal fungi [12,13]. Notwithstanding uncertainty about the mechanisms behind the ashbed effect, empirical evidence from across Australia shows that without fire, rain forest resists invasion by savanna plant species, just as fire contributes to savannas' resistance to replacement by rain forest [4].

Bowman and Fensham [5] demonstrated that soil fumigation alone can mimic the ashbed effect in facilitating *E. tetrodonta* seedling growth in rain forest soil. Although the success of fumigation suggests a primary role of soil microflora in the ashbed effect, whether that role mainly is attributable to mineral nutrient release from killed saprotrophs, elimination of pathogens and parasites, or changes in mycorrhizal fungi is uncertain. *E. tetrodonta*, *E. miniata*, and other common tree species of the savanna are ectomycorrhizal (ECM) as adults [14], but Australian rain forest tree species almost exclusively comprise arbuscular mycorrhizal (AM) hosts [15]. So, we hypothesized that rain forest AM fungi are detrimental to *E. tetrodonta* seedlings, and that, much like intense burns, soil fumigation or partial sterilization (as in a pilot experiment that we conducted; Figure 1) relieves the detriment.

Figure 1. *Eucalyptus tetrodonta* seedlings in ambient savanna, ambient rain forest, and microwaved rain forest soil. The seedlings were transplanted 58 d previously to this pilot experiment (data not shown). True leaves of the seedlings in their native savanna soil are dark green, but those of seedlings growing in ambient rain forest soil are chlorotic and extremely stunted. Seedlings in microwave-heated (μ-waved) rain forest soil are not stunted versus those in savanna soil, but their leaves are chlorotic, exhibiting the dark green veins and yellow inter-vein areas characteristic of iron deficiency. All root systems were sparse, and no mycorrhizas of any type were apparent.

Although fire more often negatively affects ECM than AM fungi [16,17], intense fires burning into rain forest kill AM host plants and thereby might disrupt otherwise persistent networks of mycorrhizal interconnections among those plants.

Accumulating evidence supports the inference that plants can be interconnected by mycorrhizal fungus hyphal networks, often called "common mycorrhizal networks" [18–20]. These constellations of hyphal bridges among root systems have been suggested to be pathways for inter-plant movement of fixed-carbon by both ECM [21] and AM [22] fungi, mineral nutrients such as nitrogen [23,24] and phosphorus [25,26] by both types of mycorrhizal fungi, and water, the latter especially by ECM networks [27].

Ectomycorrhizal networks in the savanna might facilitate seedling establishment by redistributing fixed carbon (possibly in the form of nitrogenous compounds [28]), mineral nutrients [20], and water [29]. Additionally, ECM networks may favor seedlings simply by maintaining a high density of fungi that lead to rapid mycorrhiza formation, or by influencing mycorrhizal fungus species composition [20,30]. Arbuscular mycorrhizal networks, however, are not likely to supply fixed carbon to host plants because transported carbon probably remains as storage lipids within the fungi in receiver plant roots [31]. Moreover, because AM fungi generally do not produce rhizomorphs or mycelial strands, they likely do not redistribute water to the same extent as ECM networks. In the savanna, unless coexisting ECM and AM networks include plant species such as eucalypts that can form both types of mycorrhizas [32] and thereby might link their networks, ECM and AM networks probably represent distinct niches [30].

Arbuscular mycorrhizal fungus networks may be as likely to intensify belowground competition as they are to enhance mycorrhiza formation and overall mineral nutrient acquisition. Although Chiariello et al. [25] found in a short-term, pulse-chase, field experiment with P^{32} that it was distributed to plants surrounding a decapitated donor without relationship to distance to recipient, recipient size, or recipient species, several greenhouse experiments have suggested that plants interconnected by AM networks can compete strongly belowground [33–35]. Our hypothesis of AM fungus detriment to *E. tetrodonta* seedlings anticipates that within a rain forest competitive milieu, *E. tetrodonta* seedlings will be disadvantaged by inclusion within a common AM network. Thereby, instead of AM and eventual ECM hosts

coexisting relatively benignly as often tacitly assumed, we propose that in some situations they may interact antagonistically.

We designed a microcosm experiment to examine our hypothesis that failure of *E. tetrodonta* seedlings to grow in rain forest soil is a consequence of inhibition by common AM networks. Because *Eucalyptus* species are susceptible to iron deficiency [36] and we observed symptoms suggestive of iron deficiency in rain forest soil in a pilot experiment with *E. tetrodonta* seedlings (Figure 1), as an additional treatment 130 days after transplant (DAT), we applied soluble iron to one-half of the *E. tetrodonta* seedlings in ambient soil to test an additional hypothesis that iron deficiency limited their growth.

Materials and Methods

Experiment design

We grew 40 *E. tetrodonta* seedlings (10 in each of four treatments) in the end compartments of rectangular microcosms divided into three equal compartments by nylon mesh root barriers through which AM fungus hyphae could pass (Figure 2). In the central compartments, "nurse" plants in ambient (= not fumigated) rain forest soil sustained AM fungi. Hyphae could extend from the nurse plant into the two end compartments potentially to establish common mycorrhizal networks with seedlings at both ends either in ambient soil or in soil fumigated initially to eliminate AM fungi. Hyphae extending into one end compartment of each microcosm could be severed repeatedly to disrupt connections to the nurse plant. Thus, there were four treatments in a split-plot, factorial design: ambient or fumigated soil in both end compartments of individual microcosm "plots" crossed with intact (hereafter called "networked") or severed common mycorrhizal networks.

In order to determine if soil fumigation and repeated hypha severing succeeded in diminishing AM formation as intended, we grew 40 seedlings of an AM-responsive [37] tropical tree species, *Ceiba pentandra* (L.) Gaertn. (Malvaceae) in microcosms to which we applied the same four treatments as for *E. tetrodonta*. We did not grow *C. pentandra* together with *E. tetrodonta* in the same microcosms because we did not wish them to compete with one another (although both could compete with their respective nurse plants).

Figure 2. A disassembled microcosm. The three-compartmented planter box is divided by double layers of nylon silk-screen cloth mesh root barriers (45 μm pore size). A double layer was cemented into a slot on the right side, and single layers were cemented across each cut end of the completely separated left side. When assembled, the compartmented box nested tightly within the fully-intact outer box, but the separated end periodically could be lifted vertically, sliding the layers of cloth mesh against one another and severing hyphae that crossed them.

Establishment of microcosms

We constructed the microcosms from plastic planter boxes (15 cm×45 cm×15 cm deep; Figure 2). Each was divided into thirds, at one side by a slit sawn to, but not through, the rim, and at the opposite side by entirely separating the end one-third. In the slit, we cemented with silicone sealer a double layer of nylon silk-screen cloth mesh (45 μm pore size) that was a barrier to roots but not to AM fungus hyphae or other microorganisms. Similarly, across the fully-cut ends we cemented single pieces of mesh. Thereby, both ends were separated from the central compartment by two layers of mesh. Each modified box was nested within an intact box in which it fit tightly, but which allowed the cut end to be lifted momentarily in order to sever fungus hyphae that crossed the double mesh root barrier. Soil pressed the layers of mesh together tightly so there was no air gap between them.

We filled the microcosm central compartments with rain forest soil (Table 1) collected to no more than 20 cm depth at Gunn Point (12°09.258′ S, 131°02.186′ E) near Darwin, NT, Australia. We filled the two end compartments with either the same ambient soil, or with that soil fumigated with methyl bromide gas at 1 kg m^{-3}. As supplemental inoculum of indigenous AM fungi, we collected fine roots from the Gunn Point rain forest one-day prior to use. The roots were hand cut into 1–2 cm pieces and were layered in central compartments at half their depth. Above this layer of fresh roots, we transplanted a 15–20 cm tall, field-collected root sprout of the Australian rain forest tree species *Litsea glutinosa* (Lour.) C. B. Rob. (Lauraceae). Before transplant, each sprout's source root was treated by dipping it into powdered rooting hormone and micronutrients ("Clonex"). Then, the compartments were filled with ambient soil topped with 2–3 cm of fumigated soil to reduce the chance of spores splashing into end compartments. We filled both end compartments of 20 microcosms with ambient soil, and filled those of an additional 20 microcosms with fumigated soil. The end compartments were neither actively inoculated with AM fungi, nor were they treated with a microbial filtrate [38,39] because the central compartments likely would serve as a source of bacteria and saprotrophic fungi as well as AM fungi.

After transplanting *L. glutinosa* sprouts on 20 June, 1997, we randomized all 40 microcosms with no more than 2 cm separation in two rows on a wire mesh screenhouse bench at the CSIRO Berrimah campus (12°24.800′ S, 130°55.280′ E) in Darwin. The screenhouse provided light shade. The nurse plants grew in the microcosms for six months before we transplanted *E. tetrodonta* and *C. pentandra* seedlings to begin the experiment. This pre-treatment allowed AM fungus hyphae to extend into stationary end compartments, and allowed hyphae from propagules in stationary, ambient-soil end compartments potentially to reach nurse plant roots. In order to disrupt hypha spread, three days after transplanting the *L. glutinosa* sprouts, we began lifting the detached end compartments of each microcosm by 8–10 cm every two or three days. We continued this periodic severing of hyphae until all plants were harvested.

Seedling growth, measurement, and harvest

Seeds of *E. tetrodonta* obtained from Top End Seeds (Nightcliff, NT) and of *C. pentandra* obtained from the George Brown Botanic Gardens (Darwin) were sown in germination flats of fumigated rain forest soil in the screenhouse. On 20 December, 1997, we transplanted two 1–2 week-old *E. tetrodonta* or three just-emerged *C. pentandra* seedlings into both end compartments of separate microcosms. One week later on 28 December, we replaced *C. pentandra* of which all had died in 22 microcosm ends. A week before the first growth measurement, we clipped the weakest

Table 1. Attributes of Australian Northern Territory monsoon rain forest and *Eucalyptus* savanna soils.

Attribute (units)	Rain forest	Savanna	t, P
pH	**6.3** (0.1)	5.0 (0.3)	5.01, 0.038
EC (mS cm^{-1})	**0.26** (0.03)	0.10 (0.02)	4.44, 0.047
Ca (mg kg^{-1})	**3050** (50)	330 (130)	19.53, 0.003
Mg (mg kg^{-1})	**305** (5)	130 (30)	5.75, 0.029
K (mg kg^{-1})	**145** (5)	27 (3)	20.24, 0.002
Na (mg kg^{-1})	<25 (0)	<25 (0)	NA
P (Colwell; mg kg^{-1})	**23** (1)	8 (1)	13.86, 0.005
S (mg kg^{-1})	39 (5)	37 (5)	0.28, 0.804
TKN (%)	**0.31** (0.04)	0.11 (0.02)	4.80, 0.041
Cu (mg kg^{-1})	0.7 (0.0)	0.6 (0.1)	3.01, 0.204
Fe (DTPA; mg kg^{-1})	17 (1)	**43** (2)	−12.85, 0.006
Mn (mg kg^{-1})	101 (12)	156 (60)	−0.91, 0.460
Zn (mg kg^{-1})	**7.8** (0.5)	1.0 (0.7)	8.60, 0.013
OC (%)	**7.20** (0.46)	1.69 (0.07)	11.96, 0.007
Ex Al (me%)	0.01 (0.00)	0.31 (0.29)	−1.03, 0.489

Values are means ±1 standard error in parentheses with significantly highest values in bold. Soils (n = 2 for each) were compared by t-test (approximate t-test for unequal variances for Cu and exchangeable Al); t statistics and probabilities are shown. EC = electrical conductivity; TKN = total Kjeldahl N; OC = organic carbon; Ex Al = exchangeable Al; NA = not tested (below detection limit).

individuals of *C. pentandra* and *E. tetrodonta*, leaving only one seedling in each end compartment. Thus, there were 40 individuals (10 per treatment) in 20 microcosms for each species (40 microcosms in total).

During the rainy season (October–April) all microcosms received natural rainfall and were watered only if necessary, but during the dry season they were watered three times daily by an overhead sprinkler system. Mean monthly rainfall from December, 1997 through April, 1998 ranged from 75 mm to 789 mm, and mean monthly maximum and minimum air temperatures ranged from 31.6 to 33.9 and 24.9 to 25.5 °C, respectively. While we continued to grow *E. tetrodonta* seedlings in ambient soil with or without iron fertilization from May through July, 1998, mean monthly rainfall was less than 0.4 mm, and the maximum and minimum mean monthly temperatures were 33.4 and 21.4 °C, respectively (all weather information for Darwin airport from www.bom.gov.au).

Beginning on 19 January, 1998, one month after transplant, and continuing every two weeks until harvest, we measured height from the soil surface to the shoot apex, length of the longest leaflet or leaf, and counted the number of leaves of every seedling. Additionally, we categorically recorded foliar chlorosis of *E. tetrodonta*. We measured *L. glutinosa* nurse plants only twice, on 22 December, 1997, at the start of the experiment, and on 10 May, 1998. We measured *L. glutinosa* stem diameter, height from the soil surface to the tallest shoot apex, longest leaf length, and counted the number of leaves.

On 10 May, 1998, 141 DAT, we harvested all *C. pentandra* and only the *E. tetrodonta* grown in fumigated soil. We harvested at this time because of declining health and mortality of *E. tetrodonta*, especially in networked, ambient soil. Root systems of both species were extracted from the soil, rinsed gently over a 1 mm sieve, and preserved in 50% ethanol for assessment of mycorrhizal colonization. For *E. tetrodonta* and their accompanying *L. glutinosa* nurse plants, we separated leaf blades from petioles and stems, dried all to constant weight in an oven at 60 °C, and then weighed them.

Iron fertilization

We did not harvest the E. tetrodonta seedlings grown in ambient soil when we harvested the seedlings from fumigated soil. Instead, we continued to grow them for an additional 69 days with or without iron fertilization. We randomly assigned half of each group of hyphae-severed and networked seedlings to receive 81 mg of iron (100 mL of 3.24 g L^{-1} of "Librel Fe–Lo", Allied Colloids, England; water soluble Fe = 13.2% and Fe chelated by EDTA = 12.5%) eight times at irregular intervals (i.e., on 29 April, 13, 21, 28 May, 4, 11, 24 June, and 11 July, 1998). This addition was intended transiently to elevate the available iron concentration of rain forest soil approximately to that of savanna soil (Table 1). We monitored these seedlings' growth, and processed them and their nurse plants at harvest (210 DAT) as described previously.

Mycorrhizal colonization and leaf tissue analyses

Preserved root systems of all E. tetrodonta seedlings were blotted dry and weighed before haphazardly removing a fine-root sample for assessment of mycorrhizal colonization. Afterwards, the remaining roots were weighed again before being dried to constant weight. We used dry-weight to fresh-weight ratios to calculate the dry weights of entire root systems.

For both C. pentandra and E. tetrodonta, we cleared fine-root samples in 10% KOH at room temperature for 48 h before acidifying them in 5% HCl for 15 min, and then stained them in 0.05% trypan blue in lactoglycerol at room temperature for 6 hr. We mounted ten to twenty, 1–2 cm long root segments on microscope slides and scored them for percentage length colonized by AM fungi (typical hyphae and vesicles in the root cortex) with a compound microscope according to the magnified gridline intersection method [40]. We usually examined 160 to more than 500 intersections per plant, but examined fewer than 100 intersections for seven E. tetrodonta with very small root systems. We excluded colonization data of two small individuals from the fumigated soil, networked treatment because they were outliers (below the lower 99.9% confidence limit for the other plants of that treatment).

E. tetrodonta leaf tissues (petioles excluded) were analyzed for element concentrations by the Department of Primary Industries & Fisheries, Berrimah Agricultural Research Centre, Chemistry Section (Berrimah). Leaf tissues of as many individuals as possible were analyzed separately, but because some E. tetrodonta had tiny leaves, we composited tissue of 2–4 plants as necessary. There was one composite of 2 fumigated soil, hyphae-severed plants; three composites (two of 2 and one of 3 plants) of fumigated soil, networked plants; one composite of 4 ambient soil, hyphae-severed plants; and two composites of 2 ambient soil, networked plants. The Chemistry Section also analyzed ambient and fumigated composited samples of the rain forest soil, and for comparison, ambient and fumigated samples of savanna soil collected 0.57 km distant from the rain forest. Soil pH and electrical conductivity were determined in a 1:5 soil:water extract; K, Ca, Mg, and Na in ammonium chloride; Colwell P in sodium bicarbonate; S in calcium dihydrogen phosphate; Zn, Cu, Mn, and Fe in DTPA; organic carbon by modified Walkley-Black digestion; and exchangeable Al in calcium chloride.

Statistical analyses

We used t-tests (approximate t-tests for Cu and exchangeable Al) to compare rain forest and savanna soil (n = 2 for each soil), and we report as significantly different (Table 1) those attributes for which $P \leq 0.05$. All statistical tests except those noted otherwise were performed with Statistix v. 9.0 (Analytical Software, Tallahassee, FL).

We analyzed C. pentandra percentage root length colonized by AM fungi by split-plot, two-factor ANOVA with soil fumigation as the whole-plot factor, but because of different harvest dates, we analyzed E. tetrodonta AM colonization by separate two-sample t-tests for each harvest. Percentage colonized root length was arcsine-square root transformed before analysis. E. tetrodonta dry weight variables were compared by paired t-tests (pairing plants in the same microcosm) for plants in fumigated soil, and by t-tests (because of unequal numbers of surviving plants) for plants in ambient soil after Bonferroni correction for the number of analyzed response variables ($P \leq 0.0125 = 0.05/4$ dry weight response variables).

We analyzed morphological responses of C. pentandra and E. tetrodonta by split-plot, two-factor, repeated-measures ANOVAs after Levene's test of heteroscedasticity. Applying a Bonferroni correction for three morphometric response variables, we report as significant, effects for which $P \leq 0.017$. Mortality of E. tetrodonta in the ambient, networked treatment unbalanced the numbers of seedlings per treatment, so we used JMP Pro v. 10.0.0 (SAS Institute, Inc., Cary, NC) with restricted maximum-likelihood calculations to perform the E. tetrodonta repeated-measures analyses.

We used a one-way MANOVA performed with JMP Pro v. 10.0.0 followed by univariate analyses of the four L. glutinosa response variables to compare the effects on L. glutinosa of being in a microcosm with C. pentandra versus being with E. tetrodonta. Because L. glutinosa individuals were 15–20 cm tall at transplant, we relativized final morphological measurements by using their differences from initial measurements. We examined hypothesized negative associations of C. pentandra and E. tetrodonta with L. glutinosa for effects of severing common mycorrhizal networks with Pearson's correlations. Because we did not determine the dry weights of C. pentandra, we analyzed relativized final morphological measurements for C. pentandra (longest leaflet length change) versus L. glutinosa number of leaves. For E. tetrodonta versus L. glutinosa, however, we analyzed shoot dry weights. One ambient, networked E. tetrodonta seedling with a dry weight (1.67 g) almost twice the upper 99.9% confidence limit for its treatment was excluded.

We analyzed morphological responses of E. tetrodonta to iron fertilization over 69 d following the initial fertilization by repeated-measures ANOVAs performed with JMP Pro v. 10.0.0, followed by Bonferroni correction ($P \leq 0.017$) for having examined three response variables. We used Fisher exact tests to examine the effects of iron fertilization on recovery from chlorosis.

We analyzed element concentrations in E. tetrodonta foliage across both harvests by one-way ANOVA or non-parametric Kruskal-Wallis tests if variances were heteroscedastic. We used a Bonferroni-corrected $P \leq 0.0045 (= 0.05/11$ elements) to assess the significance of these ANOVAs and of Pearson's correlation coefficients between E. tetrodonta leaf dry weights and element concentrations.

Results

Soil attributes

Among the 15 soil attributes shown in Table 1, ten differed significantly, but only one, iron, was greater in savanna soil than in rain forest soil. Those mineral nutrient concentrations that differed significantly were 2.3 (Mg) to 9.2 (Ca) times higher in rain forest than in savanna soil, except for available iron which was 2.6 times more abundant in savanna than in rain forest soil.

C. pentandra mycorrhizas and growth

Fumigation significantly reduced *C. pentandra* AM-colonized root length ($F_{1,9} = 7.13$, $P = 0.026$) but colonization was not affected significantly by severing hyphae ($F_{1,18} = 0.00$, $P = 0.957$) or by the interaction of fumigation and severing hyphae ($F_{1,18} = 1.66$, $P = 0.214$; Figure 3A). At harvest, 141 DAT, *C. pentandra* seedlings averaged 50% AM-colonized root length in ambient soil, but only 39% in fumigated soil.

Fumigation significantly diminished height increase (fumigation×time $F_{9,162} = 4.96$, $P < 0.001$; Figure 3B) and number of leaves increase (fumigation×time $F_{9,162} = 3.02$, $P = 0.002$). At 141 DAT, mean seedling height differed by 3.7 cm and mean number of leaves by 0.7 leaves in fumigated versus ambient soil. Increase of longest leaflet length, which appeared to have reached an asymptote at 8.0 cm, was not significantly affected by soil fumigation (fumigation×time $F_{9,162} = 1.85$, $P = 0.063$). We found no significant effects of severing hyphae on *C. pentandra* growth rates (height: severing×time $F_{9,162} = 1.05$, $P = 0.399$; leaflet length: severing×time $F_{9,162} = 2.30$, $P = 0.018$; number of leaves: severing×time $F_{9,162} = 1.21$, $P = 0.293$), nor did fumigation, severing, and time interact for any morphological response variable (all $P > 0.544$).

E. tetrodonta mycorrhizas and seedling performance

Severing common mycorrhizal network hyphae significantly reduced *E. tetrodonta* AM-colonized root length in fumigated soil 141 DAT (df = 16, $t = 2.15$, $P = 0.047$), but in ambient soil 210 DAT, it did not (df = 15, $t = 0.39$, $P = 0.701$; Figure 4A). In ambient soil, 45% of *E. tetrodonta* root length was colonized by AM fungi, which was similar to the 47% colonized root length of seedlings in fumigated soil with intact networks. Hyphae-severed seedlings in fumigated soil had only 29% colonization. We did not find any ectomycorrhizas.

Three of ten *E. tetrodonta* seedlings in ambient soil with intact common mycorrhizal networks failed to survive until 141 DAT. Among surviving seedlings, severing hyphae significantly elevated height increase (severing×time $F_{9,147.7} = 5.96$, P<0.001), longest leaf length increase (severing×time $F_{9,150.4} = 4.53$, $P < 0.001$; Figure 4B), and number of leaves increase (severing×time $F_{9,146.4} = 3.59$, P<0.001). Hyphae-severed seedlings in fumigated soil exceeded the mean height of seedlings in all other treatments by 3.4 cm, the mean longest leaf length by 2.2 cm, and the mean number of leaves by 3.4. We found no significant effects of fumigation on any morphological variable (height: fumigation×

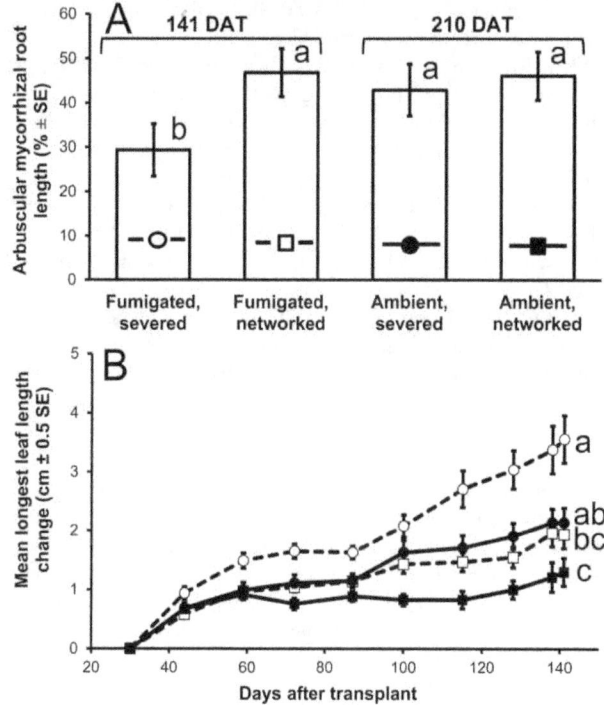

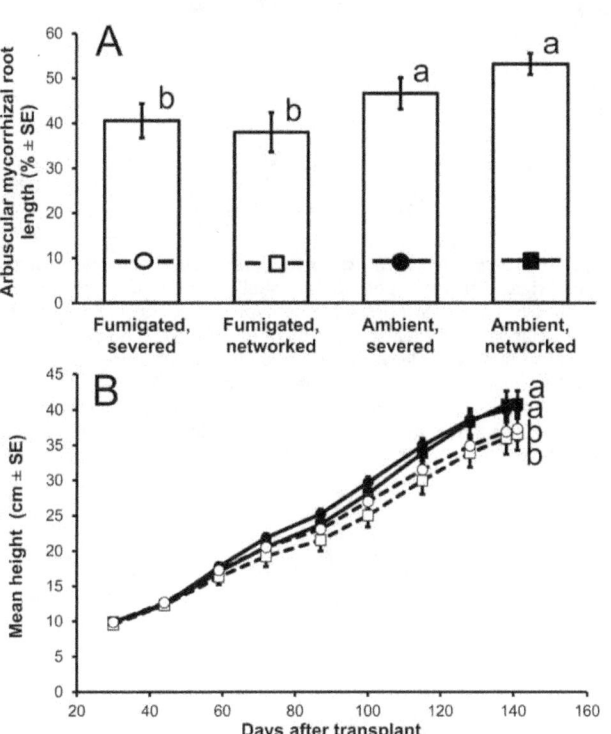

Figure 3. *Ceiba pentandra* **arbuscular mycorrhizas and height growth responses.** (A) arbuscular mycorrhizal root length (%±1 standard error) at 141 days after transplant, and (B) mean height (cm±1 standard error) versus days after transplant to microcosms with *Litsea glutinosa* AM nurse plants containing ambient (filled symbols; solid lines) or methyl-bromide fumigated (open symbols; dashed lines) rain forest soil in which hyphae repeatedly were severed (circles) or not (squares). Symbols on bars in Figure 3A correspond to those in Figure 3B. Bars topped by the same lowercase letter in 3A do not differ significantly; in 3B, letters similarly denote interactions with time. Fumigation reduced colonized root length and diminished mean height increase. (Neither hypha severing nor its interaction with fumigation significantly affected either root colonization or growth.)

Figure 4. *Eucalyptus tetrodonta* **arbuscular mycorrhizas and longest leaf length growth responses.** (A) arbuscular mycorrhizal root length (%±1 standard error) at 141 days after transplant (DAT) to methyl-bromide fumigated rain forest soil and at 210 DAT to ambient soil, and (B) mean longest leaf length change from the initial measurement (cm±0.5 standard error) versus days after transplant to microcosms with *Litsea glutinosa* AM nurse plants containing ambient (filled symbols; solid lines) or methyl-bromide fumigated (open symbols; dashed lines) soil in which hyphae repeatedly were severed (circles) or not (squares). Symbols on bars in Figure 4A correspond to those in Figure 4B. Bars topped by the same lowercase letter in 4A do not differ significantly within a harvest; in 4B, letters similarly denote interactions with time. Severing network hyphae reduced colonized root length in fumigated soil, but not in ambient soil. Severing network hyphae increased mean longest leaf length change, but neither fumigation nor its interaction with severing significantly affected longest leaf length change.

time $F_{9,163.8} = 0.26$, $P = 0.985$; leaf length: fumigation×time $F_{9,158.3} = 1.82$, $P = 0.069$; number of leaves: fumigation×time $F_{9,161.7} = 0.47$, $P = 0.895$), nor did fumigation, hyphae severing, and time interact significantly for any variable (all $P > 0.456$).

Leaf and stem dry weights of *E. tetrodonta* seedlings grown in fumigated soil were significantly increased by severing hyphae, but total root dry weights and fine root-to-leaf dry weight ratio did not differ (Table 2). Hyphae-severed seedlings grown an additional 69 days in ambient soil without iron fertilization consistently exceeded the dry weights of networked seedlings, although not significantly (Table S1).

Common mycorrhizal network effects on plant interactions

C. pentandra seedlings were conspicuously larger than *E. tetrodonta* seedlings at harvest. A one-way MANOVA with all four *L. glutinosa* response variables showed that the *L. glutinosa* nurse plants grew significantly less ($F_{4,35} = 2.95$, $P = 0.033$) when accompanied by *C. pentandra* than when accompanied by *E. tetrodonta*. Univariate analyses of *L. glutinosa* nurse plant height change ($F_{1,38} = 3.19$, $P = 0.082$), largest leaf length change ($F_{1,38} = 4.32$, $P = 0.044$), number of leaves change ($F_{1,38} = 9.96$, $P < 0.001$), and stem diameter change ($F_{1,38} = 6.57$, $P = 0.014$) revealed significant effects on the latter three response variables.

When we examined associations between *C. pentandra* longest leaflet length change (a proxy for growth rate that was not affected significantly by soil fumigation) and *L. glutinosa* number of leaves on 10 May, 1998 (a proxy for plant size), although there was no significant association for hyphae-severed plants (n = 20, $r = -0.05$, $P = 0.832$; Figure 5A), we found a significant negative association (n = 20, $r = -0.66$, $P = 0.002$; Figure 5B) among networked plants. Similarly, for shoot dry weights of *E. tetrodonta* versus *L. glutinosa* harvested with them, although there was no significant association for hyphae-severed plants (n = 20, $r = -0.10$, $P = 0.688$; Figure 6A), there was a significant negative association among networked plants (n = 16, $r = -0.50$, $P = 0.047$; Figure 6B).

Iron fertilization of E. tetrodonta

Iron fertilization of *E. tetrodonta* seedlings in ambient soil stimulated growth and remedied chlorosis. Iron fertilization significantly elevated longest leaf length increase (fertilization× time $F_{6,90} = 4.02$, $P = 0.001$; Figure 7), although neither height increase (fertilization×time $F_{6,90} = 1.87$, $P = 0.095$) nor number of leaves increase (fertilization×time $F_{6,90} = 1.52$, $P = 0.179$) was affected significantly. At harvest, mean longest leaf length of iron-fertilized plants exceeded that of not-fertilized plants by

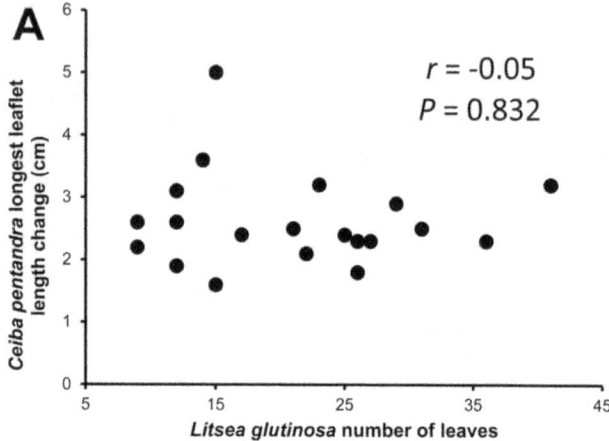

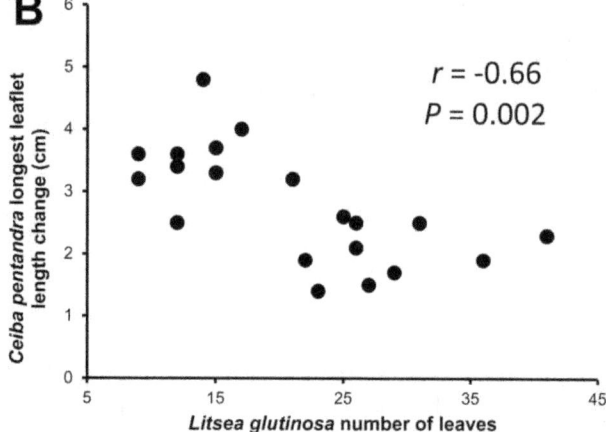

Figure 5. *Ceiba pentandra* **growth versus** *Litsea glutinosa* **nurse plant size.** (A) with potential hyphal network interconnections repeatedly severed, and (B) without hypha severing. Longest leaflet length change from initial measurement (cm) for *C. pentandra* is shown versus the number of leaves on *Litsea glutinosa* nurse plants 141 days after transplant (DAT). With hypha severing, there is no significant association, but in the likely presence of common AM networks a significant negative association suggests belowground competitive interactions.

2.0 cm, and iron fertilization had remedied chlorosis in all 5 chlorotic plants among the 8 plants randomly allocated to be fertilized. In contrast, all 4 chlorotic among 9 not-fertilized plants remained chlorotic. Although the two random groups did not

Table 2. Dry weight responses of *Eucalyptus tetrodonta* seedlings to common mycorrhizal network hypha severing in fumigated rain forest soil and to soluble iron fertilization in non-fumigated, ambient soil.

Response (units)	Fumigated soil (141 DAT)			Ambient soil (210 DAT)		
	Hyphae severed	Networked	DF, *t, P*	No fertilization	Iron fertilization	DF, *t, P*
Leaf weight (g)	**0.480** (0.338)	0.202 (0.053)	9, 3.14, 0.012	0.380 (0.118)	0.504 (0.168)	15, 0.61, 0.549
Stem weight (g)	**0.187** (0.033)	0.085 (0.022)	9, 4.05, 0.003	0.175 (0.044)	0.204 (0.058)	15, 0.40, 0.695
Root weight (g)	0.090 (0.032)	0.037 (0.013)	9, 2.02, 0.074	0.208 (0.088)	0.206 (0.082)	15, 0.02, 0.988
Fine roots:leaf	0.084 (0.025)	0.065 (0.012)	9, 0.91, 0.384	0.319 (0.090)	0.181 (0.033)	10.1, 1.44, 0.181

Values are means ± 1 standard error in parentheses with significantly highest values in bold. Degrees of freedom (DF), *t* statistics, and associated two-tail probabilities (*P*) are shown. Satterthwaite's *t*-test (with fractional degrees of freedom) was used when variances were not homogenous. In methyl-bromide fumigated soil, n = 10 for both groups, but in non-fumigated, ambient soil, 9 plants were not fertilized versus 8 plants fertilized. DAT = days after transplant.

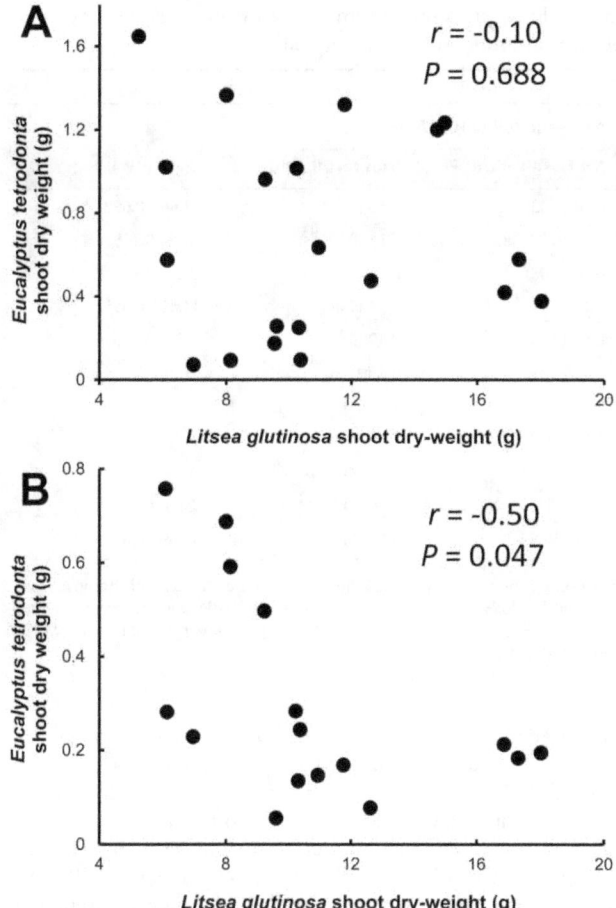

$r = -0.10$
$P = 0.688$

$r = -0.50$
$P = 0.047$

Figure 6. *Eucalyptus tetrodonta* **versus** *Litsea glutinosa* **nurse plant dry weights.** (A) with potential hyphal network interconnections repeatedly severed, and (B) without hypha severing. Shoot dry weights (g) of both *E. tetrodonta* and *L. glutinosa* at harvest are shown. With hypha severing, there is no significant association, but in the likely presence of common AM networks a significant negative association suggests belowground competitive interactions.

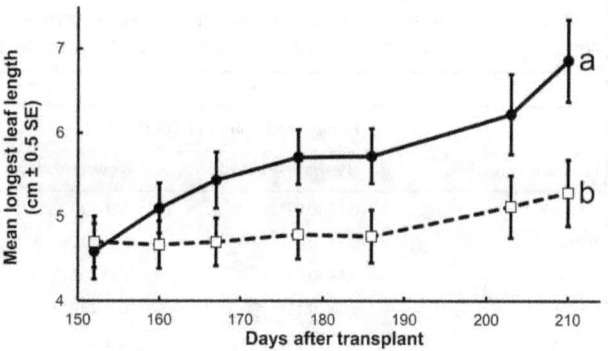

Figure 7. *Eucalyptus tetrodonta* **growth response to iron fertilization.** Mean longest leaf length (cm±0.5 standard error) is shown from 152 to 210 days after transplant (DAT) to microcosms containing ambient rain forest soil either fertilized with soluble iron (filled circles; solid lines) beginning 130 DAT or not fertilized (open squares; dashed lines). Different lower case letters denote significantly different interactions with time. Soluble iron fertilization improved *E. tetrodonta* longest leaf length increase.

significant effects either with or without iron fertilization (Table S2). Among all treatments and eleven elements, the only element that was correlated significantly (negatively) with leaf dry weight was iron (n = 27, $r = -0.62$, $P < 0.001$). There were no significant differences among treatments in the total foliar content of any element.

Discussion

Our results strongly support our hypotheses that AM fungi are detrimental to *E. tetrodonta* seedlings, and that the detriment involves iron limitation. The detriment likely is a consequence of two complementary phenomena: common mycorrhizal network interconnections mediate plant competition [33,41,42] to the disadvantage of *E. tetrodonta*, and active AM fungus mycelia may lead to iron sequestration whether or not those mycelia interconnect plants. Although our study did not trace hyphal interconnections between plants, negative associations between *C. pentandra* or *E. tetrodonta* and *L. glutinosa* only when hyphae were not severed are consistent with common mycorrhizal networks being present, as is the enhancement of *E. tetrodonta* mean root colonization in fumigated soil when networked [41].

The negative effects of potentially networked hyphae on the rates of increase of all *E. tetrodonta* morphological parameters are somewhat surprising because of the extent to which *E. tetrodonta* seedlings formed AM (seedlings in all treatments except fumigated, hyphae-severed, had 45% mean root length colonized, similar to the 45% mean colonization of *C. pentandra* across all treatments). AM improved *C. pentandra* growth in ambient versus fumigated soil, but in distinct contrast, the growth of *E. tetrodonta* seedlings with the lowest mean root colonization (29%) in hyphae-severed, fumigated soil exceeded growth in all other treatments. Consequently, little-colonized seedlings might be the most likely to establish successfully in the stressful milieu of rain forest gaps. Indeed, under our experiment's conditions that simulated non-burnt rain forest gaps, three networked *E. tetrodonta* seedlings in ambient soil died. Although AM previously have been reported to cause growth depressions of plants grown singly in pots, those negative effects usually have reflected AM failing to repay their cost to the host under conditions of low light, extremely low available phosphorus, or high phosphorus fertilization initiated after abundant mycorrhizas already had formed [43]. None of

differ in the proportion of chlorotic plants before fertilization began (Fisher exact two-tail $P = 0.637$), they differed significantly in the proportion of plants that recovered from chlorosis (Fisher exact two-tail $P = 0.008$). Iron fertilization, however, did not significantly elevate *E. tetrodonta* seedling dry weights (Table 2), nor did hyphae-severed plants differ from networked plants after fertilization (Table S1).

E. tetrodonta foliar mineral nutrient concentrations

After Bonferroni correction, only leaf tissue manganese concentration was affected significantly among the four treatments shown in Table 3. Both hyphae-severed and networked *E. tetrodonta* seedlings grown in fumigated soil (at a mean of 204 mg kg^{-1}) had 2.6 times the mean manganese concentration of those in ambient soil. Without Bonferroni correction, boron differed only for fumigated, hyphae-severed versus networked plants (plants in ambient soil are intermediate and cannot be distinguished statistically from either of those groups). Although not significantly different among these four groups ($P = 0.061$), iron had a conspicuously high mean concentration in fumigated soil for networked plants which also had the smallest mean leaf dry weight of all groups (Table 2). In ambient soil, severing hyphae had no

Table 3. Leaf element concentrations of *Eucalyptus tetrodonta* seedlings in response to common mycorrhizal network hypha severing in fumigated rain forest soil and to soluble iron fertilization in non-fumigated, ambient soil.

Element (units)	Fumigated soil (141 DAT)		Ambient soil (210 DAT)		
	Hyphae-severed	Networked	No fertilization	Iron fertilization	$F_{3,23}$ (KW), P
P (%)	0.14 (0.02)	0.26 (0.12)	0.14 (0.03)	0.16 (0.01)	(1.46), 0.692
S (%)	0.14 (0.01)	0.24 (0.08)	0.15 (0.02)	0.14 (0.01)	1.48, 0.246
K (%)	0.54 (0.04)	1.35 (0.62)	0.56 (0.08)	0.55 (0.05)	1.93, 0.153
Ca (%)	1.17 (0.15)	1.26 (0.35)	1.18 (0.14)	0.97 (0.04)	(1.62), 0.655
Mg (%)	0.45 (0.04)	0.48 (0.13)	0.50 (0.05)	0.40 (0.03)	0.34, 0.800
Na (mg kg^{-1})	0.2 (0.02)	0.3 (0.08)	0.4 (0.03)	0.3 (0.02)	(7.28), 0.064
Zn (mg kg^{-1})	57 (7.1)	60 (21.4)	52 (8.9)	43 (2.4)	0.40, 0.755
Cu (mg kg^{-1})	113 (8.6)	116 (32.8)	125 (24.7)	71 (11.1)	1.35, 0.283
Mn (mg kg^{-1})	**204** (17.4)	**204** (62.0)	88 (8.1)	69 (5.4)	5.81, 0.004
Fe (mg kg^{-1})	335 (68.9)	647 (162)	256 (65.0)	317 (94.6)	2.83, 0.061
B (mg kg^{-1})	29 (1.5)	45 (10.6)	39 (2.5)	31 (1.9)	(9.89), 0.020

Values are means±1 standard error in parentheses with significantly highest values after Bonferroni correction ($P \leq 0.0046$) in bold. All elements with homoscedastic variances were compared by one-way analyses of variance (*F* and *P* shown), but P, Ca, Na, and B, for which variances were heteroscedastic were compared by non-parametric Kruskal-Wallis analyses (KW = Kruskal-Wallis statistics shown in parentheses). In methyl-bromide fumigated soil, n = 9 for hyphae-severed and 6 for networked treatments; in non-fumigated, ambient soil, there were 6 samples in each group. DAT = days after transplant.

those effects is likely to explain the suppression of *E. tetrodonta* growth, because under the same conditions in which AM were detrimental to *E. tetrodonta*, AM benefitted *C. pentandra*. Even though mycorrhizas usually are considered to be archetypical mutualisms, our work underscores that mutualism is not constitutive but is context-dependent [44].

It is peculiar that *E. tetrodonta* seedlings sustain AM that might contribute to their deaths in rain forest soil. Natural selection may not favor rejection of AM because AM are beneficial in other contexts or if formed by a different suite of fungus species. For instance, many shrub-layer species in Northern Territory savannas form AM [14], so *E. tetrodonta* seedlings could be connected to common AM networks in savanna. Available iron is two-and-a-half times more abundant there than in rain forest soil, however, perhaps mitigating potential negative effects of AM. Alternatively, if *E. tetrodonta* seedlings quickly form ectomycorrhizas [45–47] amidst dominant ectomycorrhizal adults [14], they may avoid AM networks and diminish selection pressure to reject AM. Nevertheless, a price paid for lacking the capacity to reject AM might be inability to invade rain forest.

E. tetrodonta growth enhancement by hypha severing

Our treatments reduced but did not eliminate AM colonization of either host species. Nevertheless, fumigated, hyphae-severed compartments probably had the least AM inoculum of any treatment while, ambient, networked compartments probably had the most inoculum. Reduction of colonization of *C. pentandra* by fumigation even when hyphae were not severed implies that fumigation diminished viable AM fungus propagules to an extent not entirely compensated by hyphae from the nurse plant. Repeated severing did not significantly affect *C. pentandra* root colonization, however, probably because roots of these relatively large plants came sufficiently close to mesh root barriers to be colonized in the 2–3 d interval between hypha severing. Once established, colonization could spread within a root system to reach an asymptote [48]. In contrast, in fumigated soil, repeated hypha severing did retard root colonization of *E. tetrodonta* which was smaller than *C. pentandra*, thereby probably prolonging the

time needed for roots to closely approach mesh barriers. Not severing hyphae in fumigated soil compartments, however, elevated AM colonization of *E. tetrodonta* at 141 DAT to a level similar to that at 210 DAT, thereby supporting that seedlings and nurse plants indeed may have been connected by common mycorrhizal networks. No effect of hypha severing at 210 DAT, suggests that *E. tetrodonta* seedlings had attained asymptotic colonization.

In fumigated soil, repeated hypha severing doubled *E. tetrodonta* whole plant dry weight versus that of networked plants. Fumigation, however, neither significantly affected *E. tetrodonta* growth rates (assessed with morphological measurements), nor did fumigation interact significantly with hypha severing. Therefore, any fertilization effect of fumigation such as N and P release from killed microbes [8] or an increase in the ratio of ammonium to nitrate that facilitated iron reduction [49] little affected our results. Furthermore, we failed to detect inhibitory effects on *E. tetrodonta* of non-networked AM. That differs from the pot experiments reported by Stocker [50] and Bowman and Fensham [5] in which non-fumigated rain forest soil was inhibitory to singly-grown eucalypt seedlings. Both those studies' ambient soils, however, had not been maintained plant-free for six months prior to eucalypt planting as had ours. In our microcosms, repeated severing of potential network connections to nurse plants maximized *E. tetrodonta* growth rates only when combined with initial elimination of AM inoculum by fumigation, so it is possible that in ambient soil there may have been an inhibitory effect of non-networked AM that we could not detect statistically.

When hyphae were not severed we found significant negative associations between *C. pentandra* and *L. glutinosa*, and between *E. tetrodonta* and *L. glutinosa* which suggest that belowground competition was mediated across common mycorrhizal networks. The belowground competition was not sufficiently strong to produce a significant beneficial effect of hypha severing for *C. pentandra*, however, because *C. pentandra* were the largest plants and probably the strongest competitors overall. Aboveground, because of close spacing of the completely randomized microcosms, tall *C. pentandra* were as likely to shade adjacent *E. tetrodonta* seedlings as they were to shade their accompanying nurse plants, thereby probably distributing aboveground competition relatively evenly across our entire experiment. Belowground, however, hyphal interconnections likely influenced competition within individual microcosms. Even though root system overlap was prevented by mesh barriers, AM fungus hyphae could cross the mesh and might have redistributed mineral nutrients [20,30,51]. Similar to our results, other greenhouse experiments have found that plants interconnected by AM networks can compete strongly belowground [33–35,41,42].

Iron deficiency of *E. tetrodonta* in rain forest soil

The response of *E. tetrodonta* seedlings to iron fertilization of ambient soil unambiguously demonstrated that iron was a growth-limiting mineral nutrient. Fertilization not only stimulated leaf growth but also completely eliminated chlorosis. Iron limitation of *E. tetrodonta* in rain forest soil is consistent with only two-fifths as much iron being available in rain forest as in savanna soil. Moreover, iron fertilization has been reported to remedy chlorosis of several eucalypt species [36]. Although the highest mean foliar iron concentration was found for the smallest plants overall in our experiment, that does not contraindicate iron as a growth-limiting mineral nutrient because iron can remain in leaf veins and be physiologically ineffective [49]. Extreme chlorosis that greatly reduces leaf growth can result in exceptionally high iron concentrations [52–54], as we found.

E. tetrodonta leaf analyses do not suggest that any element other than iron was limiting. Although manganese concentration was highest for all *E. tetrodonta* grown in fumigated soil, iron-fertilized seedlings at 210 DAT attained the highest mean whole plant dry weight with only one-third the manganese concentration of plants in fumigated soil. Furthermore, no analyzed element's total foliar content differed significantly among treatments. Even though we did not analyze nitrogen, it is unlikely that nitrogen limited growth because iron fertilization stimulated growth without supplemental nitrogen.

AM have been reported to improve the iron nutrition of woody plants under some conditions [55–57], but improved iron uptake was unlikely in our experiment because the best *E. tetrodonta* growth was associated with the lowest mean AM colonization. Alternatively, under some conditions AM fungi might exacerbate iron deficiency by producing the glycoprotein glomalin [58] which can contain 8.8% iron by weight [59]. If glomalin sequesters iron, then for such an effect to have operated in our experiment, glomalin in fumigated, hyphae-severed compartments would have had to diminish during the six months before *E. tetrodonta* seedlings were planted. Such a rapid decline of immunoreactive, easily-extractable glomalin is supported by findings of Preger et al. [60].

Fire, mycorrhiza networks, and the rain forest–savanna boundary

If the hypothesized decline of glomalin contributed to *E. tetrodonta* seedling growth in fumigated, hyphae-severed rain forest soil, then intense fires must be capable of leading to similar diminution of glomalin in the field. Knorr et al. [61], however, found no effect of fire on glomalin in Ohio, U.S.A. oak forest soils. Nevertheless, their soils had been stored for 1–7 years at room temperature, and loss of immunoreactivity during storage might have obscured effects of fire. Alternatively, some studies report direct effects of fire on AM fungi, but others report none or only indirect effects through reduced numbers of live host plants [16].

If a severe fire burns into rain forest, microbe death and glomalin decline might elevate iron availability, and AM host death might help *E. tetrodonta* seedlings avoid inclusion in common AM networks that exacerbate belowground competition with rapidly-growing rain forest species. Once ECM fungi invade to associate with establishing seedlings that have survived sufficiently long for the fungi to encounter them [62], ECM fungus siderophores [27,63] could maintain iron availability, a positive feedback.

ECM networks in savanna may enhance the resistance of savanna to invasion by AM rain forest plants. ECM networks probably have far greater potential for hydraulic redistribution [64] than common AM networks because ECM canopy trees have deep roots [65] and some ECM fungi produce rhizomorphs or mycelial strands. Indeed, the open savanna canopy might elevate ground-level temperature [66] and reduce humidity to levels with which rain forest plant species have difficulty. Nevertheless, in the absence of fire, rain forest species slowly may invade savanna, perhaps by sharing AM fungus associates with AM savanna species [14] and thereby coping with low mineral nutrient availability. If such associations extend common AM networks and elevate glomalin, then phosphorus availability might increase because of iron sequestration [67], further favoring invasion by rain forest species.

Fire–mycorrhiza–vegetation feedbacks likely provide resilience to rain forest and savanna alternative stable-state systems, although rain forest resilience is overcome by the ashbed effect. The mechanisms of that effect are unresolved, however, and in our experiments, most suggested ashbed mechanisms played no role. No ECM [12,13] formed, nor were pathogen effects [10] apparent (because iron fertilization resulted in seedlings in ambient soil recovering full health). We detected no fertilization effect of soil fumigation that killed microbes [8], but any release of iron–possibly because of glomalin degradation–likely would have been more important than elevation of nitrogen or phosphorus. The most important feature of our experiment for facilitating *E. tetrodonta* growth was disruption of common AM networks, not previously recognized as part of the ashbed effect, but a potential consequence of fire-caused mortality of AM hosts.

Our results sharply distinguish the possible roles of different mycorrhiza types in influencing plant community composition. ECM and AM should not be viewed simply as alternative plant adaptations that minimize niche overlap and foster coexistence of their hosts. We have shown that common AM networks can be actively antagonistic to an eventual ECM host.

Supporting Information

Table S1 Dry weight responses of *Eucalyptus tetrodonta* seedlings to soluble iron fertilization and to common mycorrhizal network hypha severing in ambient rain forest soil 210 days after transplant.

Table S2 Leaf element concentrations of *Eucalyptus tetrodonta* seedlings in response to soluble iron fertilization and to common mycorrhizal network hypha severing in ambient rain forest soil 210 days after transplant.

Acknowledgments

DPJ and DMJSB are grateful to Harvard University for Bullard Fellowships in Forest Research that led to this collaboration. We thank Tania Wyss for helpful criticism of the manuscript.

Author Contributions

Conceived and designed the experiments: DPJ JS DMJSB. Performed the experiments: JS CA. Analyzed the data: DPJ CA. Wrote the paper: DPJ DMJSB.

References

1. Wilson BA, Brocklehurst PS, Clark MJ, Dickinson KJM (1990) Vegetation of the Northern Territory, Australia. Darwin: Conservation Commission of the Northern Territory, Technical Report No. 49.
2. Russell-Smith J (1991) Classification, species richness, and environmental relations of monsoon rain forest in northern Australia. Journal of Vegetation Science 2: 259–278.
3. Bowman DMJS (1992) Monsoon forests in North-western Australia. II. Forest-savanna transitions. Australina Journal of Botany 40: 89–102.
4. Bowman DMJS (2000) Australian Rainforests: Islands of Green in a Land of Fire. Cambridge: Cambridge University Press.
5. Bowman DMJS, Fensham RJ (1995) Growth of *Eucalyptus tetrodonta* seedlings on savanna and monsoon rainforest soils in the Australian monsoon tropics. Australian Forestry 58: 46–47.
6. Bowman DMJS, Murphy BP, Banfai DS (2010) Has global environmental change caused monsoon rainforests to expand in the Australian monsoon tropics? Landscape Ecology 25: 1247–1260.
7. Humphreys FR, Lambert MJ (1965) An examination of a forest site which has exhibited the ash-bed effect. Australian Journal of Soil Research 3: 81–94.
8. Chambers DP, Attiwill PM (1994) The ash-bed effect in *Eucalyptus regnans* forest: chemical, physical and microbiological changes in soil after heating or partial sterilisation. Australian Journal of Botany 42: 739–749.
9. Ashton DH, Kelliher KJ (1996) Effects of forest soil dessication on the growth of *Eucalyptus regnans* F. Muell. seedlings. Journal of Vegetation Science 7: 487–496.
10. Florence RG, Crocker RL (1962) Analysis of Blackbutt (*Eucalyptus pilularis* Sm.) seedling growth in a Blackbutt forest soil. Ecology 43: 670–679.
11. Iles TM, Ashton DH, Kelliher KJ, Keane PJ (2010) The effect of *Cylindrocarpon destructans* on the growth of *Eucalyptus regnans* seedlings in air-dried and undried forest soil. Australian Journal of Botany 58: 133–140.
12. Launonen TM, Ashton DH, Kelliher KJ, Keane PJ (2004) The growth and P acquisition of *Eucalyptus regnans* F. Muell. seedlings in air-dried and undried forest soil in relation to seedling age and ectomycorrhizal infection. Plant and Soil 267: 179–189.
13. Warcup JH (1991) The fungi forming mycorrhizas on eucalypt seedlings in regeneration coupes in Tasmania. Mycological Research 95: 329–332.
14. Reddell P, Milnes AR (1992) Mycorrhizas and other specialized nutrient-acquisition strategies: their occurrence in woodland plants from Kakadu and their role in rehabilitation of waste rock dumps at a local uranium mine. Australian Journal of Botany 40: 223–242.
15. Hopkins MS, Reddell P, Hewett RK, Graham AW (1996) Comparison of root and mycorrhizal characteristics in primary and secondary rainforest on a metamorphic soil in North Queensland, Australia. Journal of Tropical Ecology 12: 871–885.
16. McMullan-Fisher SJM, May TW, Robinson RM, Bell TL, Lebel T, et al. (2011) Fungi and fire in Australian ecosystems: a review of current knowledge, management implications and future directions. Australian Journal of Botany 59: 70–90.
17. Brundrett MC, Ashwath N, Jasper DA (1996) Mycorrhizas in the Kakadu region of tropical Australia. II. Propagules of mycorrhizal fungi in disturbed habitats. Plant and Soil 184: 173–184.
18. Leake JR, Johnson D, Donnelly DP, Muckle GE, Boddy L, et al. (2004) Networks of power and influence: the role of mycorrhizal mycelium in controlling plant communities and agroecosystem functioning. Canadian Journal of Botany 82: 1016–1045.
19. Newman EI (1988) Mycorrhizal links between plants: their functioning and ecological significance. Advances in Ecological Research 18: 243–270.
20. Simard SW, Durall DM (2004) Mycorrhizal networks: a review of their extent, function, and importance. Botany 82: 1140–1165.
21. Simard SW, Perry DA, Jones MD, Myrold DD, Durall DM, et al. (1997) Net transfer of carbon between ectomycorrhizal tree species in the field. Nature 388: 579–582.
22. Francis R, Read DJ (1984) Direct transfer of carbon between plants connected by vesicular arbuscular mycorrhizal mycelium. Nature 307: 53–56.
23. Arnebrant K, Ek H, Finlay RD, Soderstrom B (1993) Nitrogen translocation between *Alnus glutinosa* (L.) Gaertn. seedlings inoculated with *Frankia* sp. and *Pinus contorta* Dougl. ex Loud. seedlings connected by a common ectomycorrhizal mycelium. New Phytologist 124: 231–242.
24. van Kessel C, Singleton PW, Hoben HJ (1985) Enhanced N-transfer from soybean to maize by vesicular arbuscular mycorrhizal (VAM) fungi. Plant Physiology 79: 562–563.
25. Chiariello N, Hickman JC, Mooney HA (1982) Endomycorrhizal role for interspecific transfer of phosphorus in a community of annual plants. Science 217: 941–943.
26. Woods FW, Brock K (1964) Interspecific transfer of Ca-45 and P-32 by root systems. Ecology 45: 886–889.
27. Courty P-E, Buée M, Diedhiou AG, Frey-Klett P, Le Tacon F, et al. (2010) The role of ectomycorrhizal communities in forest ecosystem processes: new perspectives and emerging concepts. Soil Biology and Biochemistry 42: 679–698.
28. Teste FP, Simard SW, Durall DM, Guy RD, Berch SM (2010) Net carbon transfer between *Pseudotsuga menziesii* var. *glauca* seedlings in the field is influenced by soil disturbance. Journal of Ecology 98: 429–439.
29. Booth MG, Hoeksema JD (2010) Mycorrhizal networks counteract competitive effects of canopy trees on seedling survival. Ecology 91: 2294–2302.
30. Bever JD, Dickie IA, Facelli E, Facelli JM, Klironomos J, et al. (2010) Rooting theories of plant community ecology in microbial interactions. TRENDS in Ecology and Evolution 25: 468–478.
31. Lekberg Y, Hammer EC, Olsson PA (2010) Plants as resource islands and storage units - adopting the mycocentric view of arbuscular mycorrhizal networks. FEMS Microbial Ecology 74: 336–345.
32. Adams F, Reddell P, Webb MJ, Shipton WA (2006) Arbuscular mycorrhizas and ectomycorrhizas on *Eucalyptus grandis* (Myrtaceae) trees and seedlings in native forests of tropical north-eastern Australia. Australian Journal of Botany 54: 271–281.
33. Kytöviita MM, Vestberg M, Tuomi J (2003) A test of mutual aid in common mycorrhizal networks: established vegetation negates benefit in seedlings. Ecology 84: 898–906.
34. Wilson GWT, Hartnett DC, Rice CW (2006) Mycorrhizal-mediated phosphorus transfer between tallgrass prairie plants *Sorghastrum nutans* and *Artemisia ludoviciana*. Functional Ecology 20: 427–435.
35. Moora M, Zobel M (1998) Can arbuscular mycorrhiza change the effect of root competition between conspecific plants of different ages? Canadian Journal of Botany 76: 613–619.
36. Parsons RF, Uren NC (2007) The relationship between lime chlorosis, trace elements and Mundulla Yellows. Australasian Plant Pathology 36: 415–418.
37. Allen EB, Allen MF, Egerton-Warburton L, Corkidi L, Gómez-Pompa A (2003) Impacts of early- and late-seral mycorrhizae during restoration in seasonal tropical forest, Mexico. Ecological Applications 13: 1701–1717.
38. Allen EB, Cannon JP, Allen MF (1993) Controls for rhizosphere microorganisms to study effects of VA mycorrhizae on *Artemisia tridentata*. Mycorrhiza 2: 147–152.
39. Koide RT, Li M (1989) Appropriate controls for vesicular-arbuscular mycorrhiza research. New Phytologist 111: 35–44.
40. McGonigle TP, Miller MH, Evans DG, Fairchild GL, Swan JA (1990) A new method which gives an objective measure of colonization of roots by vesicular-arbuscular mycorrhizal fungi. New Phytologist 115: 495–501.
41. Eissenstat DM, Newman EI (1990) Seedling establishment near large plants: Effects of vesicular-arbuscular mycorrhizas on the intensity of plant competition. Functional Ecology 4: 95–99.
42. Janouskova M, Rydlova J, Pueschel D, Szakova J, Vosatka M (2011) Extraradical mycelium of arbuscular mycorrhizal fungi radiating from large plants depresses the growth of nearby seedlings in a nutrient deficient substrate. Mycorrhiza 21: 641–650.
43. Janos DP (2007) Plant responsiveness to mycorrhizas differs from dependence upon mycorrhizas. Mycorrhiza 17: 75–91.
44. Hoeksema JD, Chaudhary VB, Gehring CA, Johnson NC, Karst J, et al. (2010) A meta-analysis of context-dependency in plant response to inoculation with mycorrhizal fungi. Ecology Letters 13: 394–407.
45. McGuire KL (2007) Common ectomycorrhizal networks may maintain monodominance in a tropical rain forest. Ecology 88: 567–574.
46. Nuñez MA, Horton TR, Simberloff D (2009) Lack of belowground mutualisms hinders pinaceae invasions. Ecology 90: 2352–2359.
47. Collier FA, Bidartondo MI (2009) Waiting for fungi: the ectomycorrhizal invasion of lowland heathlands. Journal of Ecology 97: 950–963.
48. McGonigle TP (2001) On the use of non-linear regression with the logistic equation for changes with time of percentage root length colonized by arbuscular mycorrhizal fungi. Mycorrhiza 10: 249–254.
49. Mengel K (1994) Iron availability in plant tissues - iron chlorosis on calcareous soils. Plant and Soil 165: 275–283.
50. Stocker GC (1969) Fertility differences between the surface soils of monsoon and eucalypt forests in the Northern Territory. Australian Forest Research 4: 31–38.
51. Selosse M-A, Richard F, He X, Simard SW (2006) Mycorrhizal networks: des liaisons dangereuses? TRENDS in Ecology and Evolution 21: 621–628.
52. Bavaresco L, Giachino E, Colla R (1999) Iron chlorosis paradox in grapevine. Journal of Plant Nutrition 22: 1589–1597.
53. Morales F, Grasa R, Abadia A, Abadia J (1998) Iron chlorosis paradox in fruit trees. Journal of Plant Nutrition 21: 815–825.
54. Römheld V (2000) The chlorosis paradox: Fe inactivation as a secondary event in chlorotic leaves of grapevine. Journal of Plant Nutrition 23: 1629–1643.
55. Janos DP, Schroeder MS, Schaffer B, Crane JH (2001) Inoculation with arbuscular mycorrhizal fungi enhances growth of *Litchi chinensis* Sonn. trees after propagation by air-layering. Plant and Soil 233: 85–94.

56. Treeby M (1992) The role of mycorrhizal fungi and non-mycorrhizal micro-organisms in iron nutrition of citrus. Soil Biology & Biochemistry 24: 857–864.

57. Wang M, Christie P, Xiao Z, Qin C, Wang P, et al. (2008) Arbuscular mycorrhizal enhancement of iron concentration by *Poncirus trifoliata* L. Raf and *Citrus reticulata* Blanco grown on sand medium under different pH. Biology and Fertility of Soils 45: 65–72.

58. Janos DP, Garamszegi S, Beltran B (2008) Glomalin extraction and measurement. Soil Biology and Biochemistry 40: 728–739.

59. Wright SF, Upadhyaya A (1998) A survey of soils for aggregate stability and glomalin, a glycoprotein produced by hyphae of arbuscular mycorrhizal fungi. Plant and Soil 198: 97–107.

60. Preger AC, Rillig MC, Johns AR, Du Preez CC, Lobe I, et al. (2007) Losses of glomalin-related soil protein under prolonged arable cropping: a chronosequence study in sandy soils of the South African Highveld. Soil Biology & Biochemistry 39: 445–453.

61. Knorr MA, Boerner REJ, Rillig MC (2003) Glomalin content of forest soils in relation to fire frequency and landscape position. Mycorrhiza 13: 205–210.

62. Janos DP (1980) Mycorrhizae influence tropical succession. Biotropica 12 (supplement): 56–64.

63. Szaniszlo PJ, Powell PE, Reid CPP, Cline GR (1981) Production of hydroxamate siderophore iron chelators by ectomycorrhizal fungi. Mycologia 73: 1158–1174.

64. Egerton-Warburton LM, Querejeta JI, Allen MF (2007) Common mycorrhizal networks provide a potential pathway for the transfer of hydraulically lifted water between plants. Journal of Experimental Botany 58: 1473–1483.

65. Janos DP, Scott J, Bowman DMJS (2008) Temporal and spatial variation of fine roots in a northern Australian *Eucalyptus tetrodonta* savanna. Journal of Tropical Ecology 24: 177–188.

66. Turton SM, Duff GA (1992) Light environments and floristic composition across an open forest-rainforest boundary in northeastern Queensland. Australian Journal of Ecology 17: 415–423.

67. Cardoso IM, Boddington CL, Janssen BH, Oenema O, Kuyper TW (2006) Differential access to phosphorus pools of an Oxisol by mycorrhizal and nonmycorrhizal maize. Communications in Soil Science and Plant Analysis 37: 1537–1551.

A Rapid, Highly Efficient and Economical Method of *Agrobacterium*-Mediated *In planta* Transient Transformation in Living Onion Epidermis

Kedong Xu[1], Xiaohui Huang[2], Manman Wu[2], Yan Wang[1,3], Yunxia Chang[2], Kun Liu[1], Ju Zhang[1], Yi Zhang[1], Fuli Zhang[1], Liming Yi[2], Tingting Li[2], Ruiyue Wang[2], Guangxuan Tan[1], Chengwei Li[1]*

1 Key Laboratory of Plant Genetics and Molecular Breeding, Zhoukou Normal University, Zhoukou, People's Republic of China, **2** Department of Life Science, Zhoukou Normal University, Zhoukou, People's Republic of China, **3** College of Life Science, Henan Agricultural University, Zhengzhou, People's Republic of China

Abstract

Transient transformation is simpler, more efficient and economical in analyzing protein subcellular localization than stable transformation. Fluorescent fusion proteins were often used in transient transformation to follow the *in vivo* behavior of proteins. Onion epidermis, which has large, living and transparent cells in a monolayer, is suitable to visualize fluorescent fusion proteins. The often used transient transformation methods included particle bombardment, protoplast transfection and *Agrobacterium*-mediated transformation. Particle bombardment in onion epidermis was successfully established, however, it was expensive, biolistic equipment dependent and with low transformation efficiency. We developed a highly efficient *in planta* transient transformation method in onion epidermis by using a special agroinfiltration method, which could be fulfilled within 5 days from the pretreatment of onion bulb to the best time-point for analyzing gene expression. The transformation conditions were optimized to achieve 43.87% transformation efficiency in living onion epidermis. The developed method has advantages in cost, time-consuming, equipment dependency and transformation efficiency in contrast with those methods of particle bombardment in onion epidermal cells, protoplast transfection and *Agrobacterium*-mediated transient transformation in leaf epidermal cells of other plants. It will facilitate the analysis of protein subcellular localization on a large scale.

Editor: Keith R. Davis, University of Louisville, United States of America

Funding: The research was supported by National Natural Science Foundation of China (31071807, 31272168) (http://www.nsfc.gov.cn), Plan for Scientific Innovation Talent of Henan Province (124100510021) (http://www.hnkjt.gov.cn), Doctoral Scientific Research Starting Foundation of Zhoukou Normal University (zksybscx201108)(http://www.zknu.edu.cn), Scientific Research and Innovation Fund Projects of Zhoukou Normal University (zksykycx201306) (http://www.zknu.edu.cn) and the Science and Technology Research Major Projects of Department of Education of Henan Province (13B210270)(http://www.haedu.gov.cn). The funders had no role in study design, data collection and analysis, decision to publish, or preparation of the manuscript.

Competing Interests: The authors have declared that no competing interests exist.

* E-mail: lichengweiwau@hotmail.com

Introduction

Onion (*Allium cepa* L.), one kind of biennial herb Liliaceae plant, has been used as classical experimental materials in analyzing structure of plant cells, distribution location of DNA and RNA, reducing sugar of plant tissues [1], plasmolysis and recovery of plant cells [2,3], karyotype [4], protein subcellular localization and interaction [5–7].

Imaging subcellular localization of proteins in living cells has become an important tool for defining protein function. Fluorescent fusion proteins are ideal marker non-enzymatic protein systems for imaging protein subcellular localization in living cells, which have many apparent advantages, such as stable fluorescence properties, easy observation, visualization in living cells, non-toxic to cells, non-specifity for species, without interference to false positives and no substrate etc. In addition, expression of fluorescent fusion proteins had also been used to investigate protein interaction, trafficking, turnover, movement and inheritance in living cells [8].

Transient transformation assays, which were conducted by using particle bombardment [9,10], protoplast transfection [11], and *Agrobacterium*-infiltration [12,13,14], had been used to analyze gene function and protein-protein interactions in *Arabidopsis* [15,16] and rice [11,17], hybrid aspen [18], maize [19,20], potato [21,22], soybean [23–25], tomato [26–28], wheat [23,29,30], white spruce [31], celery and carrot [32].

However, most transient transformation methods have certain disadvantages, such as the lower transformation efficiency, equipment dependency and auxiliary material needed for particle bombardment, complex preparation procedures required for protoplasts transfection. In addition, for transient transformation in plants having complex outline of epidermal cells and the bushy epidermal hairs, laser scanning confocal microscope was usually needed to get ideal micro-images, which increased the reliability on expensive equipments. To avoid these disadvantages, we developed an *in planta* transient transformation method in living onion epidermal cells by using *Agrobacterium*-mediated infiltration. In this protocol, infiltration liquid of *Agrobacterium* carrying constructed vectors were injected into the interface between adaxial epidermis and mesophyll of onion bulb scales, which played an important role in yielding high transformation efficiency, and kept in the living onion bulb for about three days.

With this simple method a higher frequency of transformation was achieved without expensive equipments in comparison with other transient transformation methods like protoplast transfection of *Spinacia oleracea* [33], *A. thaliana* [34] and *Populus euphratica* [35], agroinfiltration of *Arabidopsis* epidermal cells and particle bombardment of onion cells *in vitro*. In addition, following this method it took about 3 days for agroinfiltration to get ideal transient transformation efficiency in living onion epidermis. Therefore, the developed method was rapid, highly efficient, equipment independent and low-cost, which will benefit to analyze protein subcellular localization in a large-scale manner.

Results and Discussion

Special agroinfiltration method could benefit transient transformation

We first modified the agroinfiltration method previously used in *Nicotiana tabacum* [36,37] in order to increase the efficiency of transient transformation. About 200 µl agroinfiltration liquid was slowly injected into the interfaces between adaxial epidermis and mesophyll of onion bulb scales by using a plastic syringe with needle, which resulted in an agroinfiltration bubble at the injection spot filled with infiltration liquid occupying about 1 cm^2 area of epidermis (Figure 1). The formation of agroinfiltration bubble was demonstrated in the schematic diagram (Figure 2). We found that 1 cm^2 agroinfiltration area with forming bubble gave higher transformation efficiency than a larger agroinfiltration area without forming bubble. The reason could be that forming agroinfiltration bubble filled with agroinfiltration liquid gave the infiltrated epidermal cells more chance to be infected by agrobacteria because the ratio of agrobacteria to infiltrated epidermal cells was higher. The special injection method could contribute to the high transformation efficiency of the developed transient transformation method. In contrast with other plant material, the adaxial epidermis and mesophyll of onion bulb scales are more flexible to be separated to form an interface bubble, which can take more agroinfiltration liquid.

Optimization of optical density (OD) of bacteria and agroinfiltration duration

The OD of bacteria and agroinfiltration duration were optimized. It showed that among the different agroinfiltration durations (24 h, 48 h, 72 h and 96 h), three-day agroinfiltration gave the significantly higher transformation efficiency (Table 1). While among three different concentrations of bacteria, agroinfiltration liquid with OD_{600} of 0.10 resulted in significantly higher transformation efficiency (Table 1). Considering both elements, the conditions of agroinfiltration liquid with OD_{600} of 0.10 and three-day infiltration were adopted to achieve the highest transformation efficiency of 43.87% (Table 1).

Agroinfiltration liquid with special combination of components contributed to the high efficiency of transient transformation

The components of infiltration liquid and transfection conditions were referenced to relevant reports [13,38–42] and with special modification. Based on agroinfiltration liquid for floral dip stable transformation and other transient transformation assays, the modified agroinfiltration liquid included 6-benzylaminopurine (BAP), Silwet L-77, D-glucose, acetosyringone (AS), magnesium chloride hexahydrate ($MgCl_2$), calcium chloride dehydrate ($CaCl_2$) and MES-KOH. To our knowledge, adding plant growth regulator BAP into the agroinfiltration liquid for transient

transformation was seldom, although it had previously been applied in floral dip stable transformation. As an analogue of cytokinin, BAP plays roles in promoting cell division and prevent cell senescence, adding of low concentration of BAP in the agroinfiltration liquid could benefit to keep onion epidermal cells vigorous during agroinfiltration. Surfactant, glucose, osmotic buffer, calcium and magnesium ions, and acetosyringone were also supplemented to benefit the transient transformation.

In order to confirm the necessary roles of all the components, the transformation efficiencies of different infiltration liquids without one of the components were investigated (Table 2). The results indicated that all the components of agroinfiltration liquid were necessary and significantly contributed to the high efficiency of *Agrobaterium*-mediated transient transformation in onion epidermal cells.

The developed method is advantageous to particle bombardment and onion is more suitable for the developed method than tobacco and *Arabidopsis*

Through analyzing the side by side experiments of transforming pLPGM413 into onion epidermal cells with the developed method and particle bombardment, it indicated that the developed method resulted in significantly high transformation efficiency and showed advantages on cost and equipment needed (Table 3).

The side by side experiments of *Agrobacterium* mediated transient transformation in onion, tobacco and *Arabidopsis* [14] with the same agroinfiltration liquid showed that the developed transient transformation method of onion epidermis gave significantly high transformation efficiency than tobacco and *Arabidopsis* (Table 4). The developed method of transformation in onion epidermis took only 5 days from the pretreatment of onion bulb to the best time-point for visualizing FFP signals, which is shorter than those in tobacco and *Arabidopsis* if considering the preparation time of plants [14]. In addition, the developed method is easier to get ideal images because onion epidermis has transparent cells arranged in a monolayer. Therefore, the developed method of onion has advantage to those of tobacco and *Arabidopsis*.

Pretreatment of onion bulb benefited the *Agrobacterium* mediated transient transformation in onion epidermis

The effects of different pretreatment times of onion bulb before injection of agroinfiltration liquid were evaluated (Table 5). It indicated that with the increase of pretreatment time the transformation efficiencies were increased (Table 5). However, 48-h and 72-h pretreatments resulted in similar transformation efficiencies, which are significantly higher than those resulted from 0-h and 24-h pretreatments (Table 5). It implied that 48-h pretreatment could be the suitable pretreatment time, which helped to achieve the highest transformation efficiency with the least time (Table 5).

Monitoring of transformation and subcellular localization confirmation

The process of transient transformation was monitored by using four different plasmids carrying report genes mediated by *A. tumefaciens* strain GV3101, which constructed as the following expression binary vectors, pLPGM413 carrying *RcSERK1* (*somatic embryogenesis receptor-like kinase 1* of *Rosa canina*)[43] and *GFP* (green fluorescent proteins), pLPGM413 modified from pSAT6-GFP-N1 [44–47] by adding T-DNA border region of pCAMBIA1303, pLPGM113 modified from pEZS-NL-GFP [48] by adding T-DNA border region of pCAMBIA1303 and pCM1205-RFP (red fluorescent protein) carrying *RFP* [49,50]. The *Agrobacterium* cells

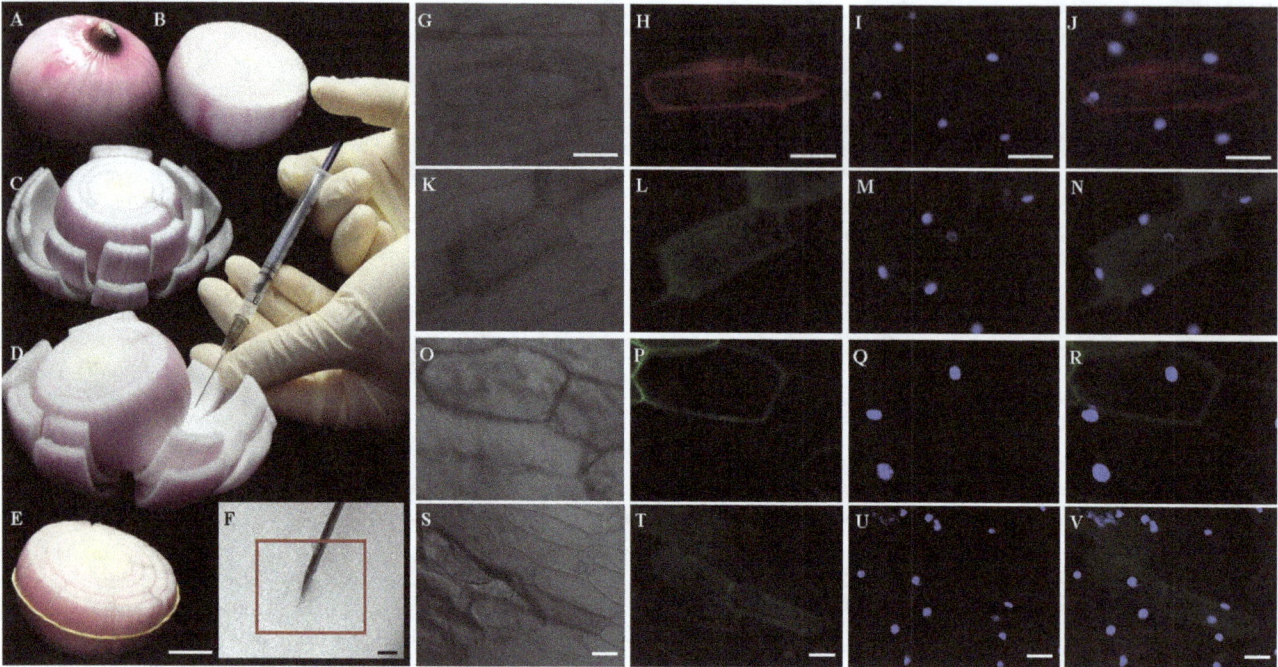

Figure 1. *Agrobacterium* **mediated** *in planta* **transient transoformationin living onion epidermal cells.** (A–F) Operational process of the modified agroinfiltration, (A) Onion bulb without outer scales, (B, C) The cut onion bulb prepared for subsequent injection, (D) The injection of *Agrobacteria*, (E) Bind injected cut scales together with elastic for further incubation, scale bar = 2.5 cm. (F) The magnification of injection location, scale bar = 4 mm. (G–V) Onion epidermal cells were transformed with constructs of pCM1205-RFP (G, H, I and J), pLPGM202 (K, L, M and N), pLPGM413 (O, P, Q and R) and pLPGM113 (S, T, U and V). Bright field images (G, K, O and S), UV excited fluorescence images (L, P, T and H), UV excited DAPI staining images (I, M, Q and U) and the merged images of fluorescence and DAPI (J, N, R and V), scale bar = 10 μm.

harboring above vectors were grown in liquid yeast extract and beef extract medium (YEB) until stationary phase and then re-suspended in infiltration liquid with above special components. Two to three days after injection of infiltration liquid, obvious GFP/RFP signals corresponding to different report genes could be observed in epidermal cells (Figure 1). It indicated that the developed agroinfiltration method can efficiently mediate *in vivo* transient expression of genes in living onion epidermal cells. For pCM1205-RFP, pLPGM202 and pLPGM113, the proteins encoded by their harboring genes were localized throughout the cells (Figure 1), and the fusion protein encoded by *RcSERK1-GFP* in pLPGM413 was localized in cell membrane (Figure 1). The results were the same as those of previous studies [43–50], suggesting the authenticity of the developed method.

The fluorescence signals of the nuclei in the cells transformed by pCM1205-RFP, pLPGM202 and pLPGM113 overlapped with those nuclei stained by 4′,6-diamidino-2-phenylindole (DAPI) (Figure 1), while no fluorescence signals were observed in nuclei of the cells transformed by pLPGM413. It further confirmed the credibility of the developed transient transformation method. It indicated that the developed transient transformation method in onion epidermal cells could be used to determine subcellular localization of the target proteins with the consistence to other transient transformation systems.

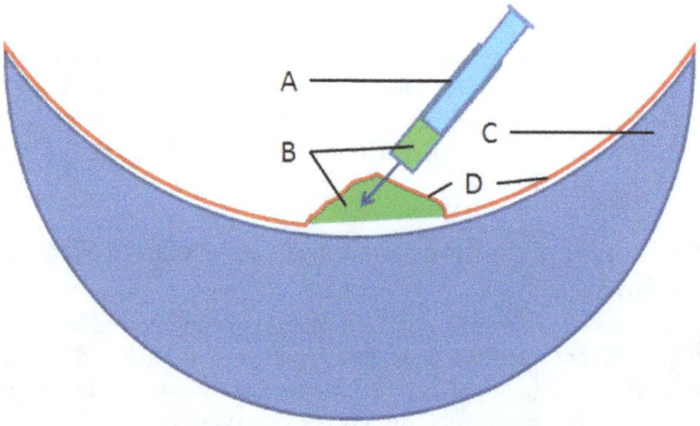

Conclusions

We have developed a rapid and low-cost transient transformation assay by using *in-planta* agroinfiltratoin in living onion epidermal cells, which transformation efficiency is higher than those of the conventional transient assays. It demonstrated that subcellular localization of proteins evaluated by using this method was same as that by other transient transformation assays and DAPI staining. Since it is rapid, efficient and low-cost, this method is suitable for large-scale analyses of protein subcelluar localization.

Materials and Methods

Plant material

Cultivated onion "Hongtaiyang" (*A. cepa* L.), which is a kind of Chinese onion cultivar with red skin, was used as experimental

Figure 2. Schematic diagram of forming agroinfiltration bubble by injecting agroinfiltration liquid into the interface between adaxial epidermis and mesophyll of onion bulb scale. (A) Syringe. (B) Agroinfiltration liquid. (C) Mesophyll of onion bulb scale. (D) Adaxial epidermis of onion bulb scale.

Table 1. Effects of bacterial concentrations and durations of agroinfiltration.

Agroinfiltration durations	Transient transformation efficiencies (%) by using infiltration of *Agrobacterium* with different concentrations		
	0.05 (OD$_{600}$)	0.10 (OD$_{600}$)	0.15 (OD$_{600}$)
24 h	3.30±0.259 cC	12.03±0.282 dD	5.10±0.216 dD
48 h	4.30±0.240 bB	23.77±0.274 cC	12.37±0.256 bB
72 h	13.10±0.326 aA	43.87±0.431 aA	22.53±0.335 aA
96 h	4.77±0.164 bB	17.97±0.282 bB	7.70±0.199 cC

Note: three samples per combination of bacterial concentration and infiltration duration were investigated for ten 2 mm^2 epidermal areas. Each efficiency value represents the mean of transformation efficiencies of thirty replicates of 2 mm^2 epidermal areas from three samples, and the standard errors were calculated by using Excel. Different capital and lowercase letters within the same column exhibit significant difference at the 1% and 5% probability level according to the Duncan test of SPSS 10.0 statistic analysis.

material. Before the injection of agroinfiltration liquid the onion bulbs pre-grew in darkness at 28°C for different days to evaluate the effect of different pretreatment time of onion bulb (Table 5). After injection the onion bulbs were incubated at 28°C in darkness for the agroinfiltration.

Tobacco (*Nicotiana tabacum* L. var. Xanthi nc.) was grown under 25°C with photoperiod of 16 h light (200 μmol m^{-2} s^{-1}) and 8 h darkness. The leaves of fifty-day old plants were injected with agroinfiltration liquid from abaxial surface with syringe.

A. thaliana (Colombia ecotype) was grown under 20°C with photoperiod of 16 h light (120 μmol m^{-2}s^{-1}) and 8 h darkness. The leaves of 15 days old plants, the injection method was same as that of tobacco.

GFP/RFP-based constructs

The binary expression vector pLPGM202 harboring *GFP* genes was modified from pSAT6-GFP-N1 provided by Prof. Tao Wang (State Key Laboratory of Agro-biotechnology, China Agricultural University) by adding T-DNA border region of pCAMBIA1303, which contains the right border, kanamycin (Kan) resistance gene for bacterial resistance selection and the left border. The steps of creating pLPGM202 was as follows: the T-DNA border region (2888-9172) of pCAMBIA1303 was amplified by using PrimeSTAR® GXL DNA Polymerase (TaKaRa, Japan) with PCR primers of Fsat-YR (*Not*I-GTAAACCTAAGAGAAAAGAG) and

Rsat-YR (*PI-Psp*I-TTTGCCTGTTTACACCACAAT) carrying the restriction enzyme cutting sites of *Not*I and *PI-Psp*I, respectively. The PCR products were ligated into pSIMPLE-19 *Eco*RV/BAP vector by T4 DNA Ligase (TaKaRa, Japan) and sequenced to confirm the success of ligation. The constructed recombinant pSIMPLE-19 *Eco*RV/BAP with the T-DNA border region of pCAMBIA1303 was transformed into *Escherichia coli* DH5α for propagation. The pSAT6-GFP-N1 were digested with partial digestion method by using *Not*I and *PI-Psp*I (NEB, USA), the target digested pSAT6-GFP-N1 was obtained with electrophoresis method. The T-DNA border region of pCAMBIA1303 was obtained from the recombinant pSIMPLE-19 *Eco*RV/BAP through complete digestion with *Not*I and *PI-Psp*I, and cloned into the target partially-digested pSAT6-GFP-N1 to form pLPGM202 (Figure S1). The successful construction of pLPGM202 was confirmed by sequencing (Sangon Biotech, China).

The binary expression vector pLPGM113 was modified from pEZS-NL-GFP provided by Dr. Liang Zhang (Department of Life Science, Henan Normal University) by adding T-DNA border region (2888-9172) of pCAMBIA1303 same as that for pLPGM202 described above. The steps of creating pLPGM113 were similar as those of pLPGM202 to be simply described as follows: The T-DNA border region of pCAMBIA1303 was amplified with the primers Fnl-YR (*Pst*I-GTAAACCTAAGA-GAAAAGAG) and Rnl-YR (*Spe*I-TGTTTACACCACAATA-

Table 2. Effects of different components of agroinfiltration liquid (OD$_{600}$ = 0.10).

Infiltration components	Transient transformation efficiencies (%) of different agroinfiltration durations			
	24 h	48 h	72 h	96 h
Complete components	12.93±0.262 aA	22.97±0.376 aA	43.88±0.330 aA	15.68±0.070 aA
D-glucose-	0.15±0.014 fF	0.91±0.013 gG	2.61±0.040 gG	2.40±0.031 hG
CaCl$_2$-	1.21±0.050 eE	2.43±0.047 fF	5.99±0.036 fF	2.55±0.045 gG
MES-KOH-	3.30±0.048 cC	5.24±0.059 dD	13.32±0.089 bB	6.88±0.081 cC
BAP-	2.13±0.062 dD	4.21±0.049 eE	8.28±0.043 dD	5.67±0.043 dD
Silwet L-77-	7.22±0.083 bB	10.07±0.126 bB	12.71±0.082 cC	9.43±0.045 bB
MgCl$_2$-	2.42±0.065 dD	6.34±0.040 cC	8.39±0.040 dD	4.32±0.029 eE
AS-	1.15±0.041 eE	4.39±0.044 eE	7.29±0.253 eE	3.42±0.054 fF

Note: three samples per unit of infiltration components were investigated for ten 2 mm^2 epidermal areas. Each efficiency value represents the mean of transformation efficiencies of thirty replicates of 2 mm^2 epidermal areas from three samples (the percentage of positive cells in total cells per unit area), and the standard errors were calculated by using Excel. Different capital and lowercase letters within the same column exhibit significant difference at the 1% and 5% probability level according to the Duncan test of SPSS 10.0 statistic analysis. The symbol "-", represents that the agroinfiltration liquid included all the components except the referred component.

Table 3. Comparison of agroinfiltration and particle bombardment methods on transformation in onion epidermis.

Comparison items	Transient transformation methods	
	Agroinfiltration	**Particle bombardment**
Transformation time	One day (about 12.93% transformation efficiency)	One day (about 4.67% transformation efficiency)
Cost	Low	High
Special equipment	Not needed	Biolistic equipment
Transformation efficiency (%)	43.73 ± 0.23	4.67 ± 0.11

Note: three samples per unit of infiltration components were investigated for ten 2 mm² epidermal areas. The efficiency value represents the mean of transformation efficiencies of thirty replicates of 2 mm² epidermal areas from three samples (the percentage of positive cells in total cells per unit area) in the part of transformation efficiency, and the standard errors were calculated by using Excel.

TATCC) carrying the restriction enzyme cutting sites of *Pst*I and *Spe*I, respectively. The PCR products were ligated into pSIMPLE-19 *Eco*RV/BAP vector by T4 DNA Ligase (TaKaRa, Japan) and propagated in *E. coli* DH5α. The pEZS-NL-GFP were digested with partial digestion method by using *Pst*I and *Spe*I (NEB, USA), the target digested pEZS-NL-GFP was obtained with electrophoresis method. The T-DNA border region of pCAMBIA1303 was obtained from the recombinant pSIMPLE-19 *Eco*RV/BAP through complete digestion with *Pst*I and *Spe*I, and cloned into the target partially-digested pEZS-NL-GFP to form pLPGM113 (Figure S2). The successful construction of pLPGM113 was confirmed by sequencing (Sangon Biotech, China).

The binary expression vector pCM1205-RFP harboring *RFP* gene was provided by Dr. Wencai Qi (Bioengineering Department, Zhengzhou University).

The cDNA of *RcSERK1* was isolated from cDNA library of *R. canian* PLBs and its sequence was deposited in GenBank as accession number of HM802242. Coding sequence of this gene (without stop codon) was amplified by PCR using Platinum *pfx* DNA polymerase (Invitrogen, USA) with the following primers: *RcSERK1* (F-SKs, forward: 5′-CCCAAGCCTCATGGATAG-CAGGCTT-3′; R-SKs, reverse: 5′-TCCCCCGGGGCCTTG-GACCAGATAAC-3′). The PCR products were digested with *Hind*III and *Sma*I, subsequently ligated into the *Hind*III/*Sma*I sites between the CaMV 35S promoter and *GFP* of pLPGM202 to construct the expression binary vector pLPGM413.

Agrobacterium infiltration

The vectors, pCM1205-RFP, pLPGM113, pLPGM202 and pLPGM413, were transformed into *A. tumefaciens* strain GV3101 for further transient transformation in onion epidermis. Positive

Agrobacterium harboring pCM1205-RFP was selected and cultivated in YEB media supplemented with 100 mg/L rifampicin and 25 mg/L chloramphenicol. Positive *Agrobacterium* harboring pLPGM113, pLPGM202 and pLPGM413 were selected and cultivated in YEB media supplemented with 100 mg/L rifampicin and 100 mg/L Kan. Positive *Agrobacterium* cultivated overnight at 28°C were harvested at OD_{600} of 1.5 to 2.0, centrifuged at 5000 rpm for 10 min and re-suspended in 50 ml of infiltration liquid, and the centrifugation and resuspension procedure was repeated three to five times. Finally, *Agrobacterium* cells were diluted in agroinfiltration liquid to appropriate concentration for agroinfiltration. Different agroinfiltration durations (24 h, 48 h, 72 h and 96 h) and *Agrobacterium* concentrations (OD_{600} 0.05, 0.10 and 0.15) were evaluated to determine conditions to obtain high transformation efficiency (Table 1).

The complete infiltration liquid was made as following: 41.65 mM D-glucose, 100 mM $CaCl_2$, 100 mM MES-KOH (pH 5.6) stock solution, 0.011 μM BAP, 0.01% Silwet L-77, 0.05 mM $MgCl_2$ and 12.5 mM AS (made with DMF, dimethylformamide) stock solution, and suitable amount of ddH_2O to make final volume to 20 ml. About 200 μl infiltration liquid with *Agrobacterium* carrying constructed vectors were injected into the interface of adaxial epidermis and mesophyll of onion scales to make a bubble for agroinfiltration. In order to investigate whether all the components are necessary, different infiltration liquid without one of the components were used to evaluated the effect of different components of the infiltration liquid (Table 2).

Particle bombardment

The adaxial epidermis was obtained from onion bulb and placed on MS medium for 1-day incubation. The binary

Table 4. Comparison of transient transformation efficiencies in different plant materials by using agroinfiltration.

Plant materials	Transient transformation efficiencies (%) by using infiltration of *Agrobacterium* ($OD_{600}=0.10$) with different agroinfiltration durations			
	24 h	**48 h**	**72 h**	**96 h**
Arabidopsis(Col-0)	1.31 ± 0.041 cC	2.56 ± 0.066 cC	1.79 ± 0.043 cC	1.63 ± 0.053 cC
Tobacco(Xanthi)	1.93 ± 0.044 bB	4.97 ± 0.051 bB	6.54 ± 0.074 bB	5.22 ± 0.035 bB
Onion(Hongtaiyang)	12.34 ± 0.098 aA	23.38 ± 0.243 aA	43.35 ± 0.343 aA	14.99 ± 0.082 aA

Note: three samples per material were investigated for ten 2 mm² areas of agroinfiltrated epidermal cells. Each efficiency value represents the mean of transformation efficiencies of thirty replicates of 2 mm² epidermal areas from three samples (the percentage of positive cells in total cells per unit area), and the standard errors were calculated by using Excel. Different capital and lowercase letters within the same column exhibit significant difference at the 1% and 5% probability level according to the Duncan test of SPSS 10.0 statistic analysis. For onion two-day pretreatment was conducted before agroinfiltration.

Table 5. Effects of different pretreatment time of onion before *Agrobacterium* infection.

Pretreatment time	Transient transformation efficiencies (%) by using infiltration of *Agrobacterium* concentration (OD_{600} = 0.10) with different agroinfiltration durations			
	24 h	**48 h**	**72 h**	**96 h**
0 h	2.78±0.052 cC	5.97±0.066 cC	8.74±0.056 dC	5.20±0.048 dD
24 h	8.10±0.068 bB	15.01±0.158 bB	24.10±0.279 cB	13.17±0.272 cC
48 h	12.93±0.203 aA	22.43±0.436 aA	42.40±0.309 bA	15.77±0.298 bB
72 h	13.03±0.286 aA	22.10±0.480 aA	43.57±0.513 aA	17.71±0.276 aA

Note: three repeat samples for each treatment were investigated by observing ten 2 mm^2 epidermal areas of each sample. Each efficiency value represents the mean of transformation efficiencies of thirty replicates of 2 mm^2 epidermal areas from three samples (the percentage of positive cells in total cells per unit area), and the standard errors were calculated by using Excel. Different capital and lowercase letters within the same column exhibit significant difference at the 1% and 5% probability level according to the Duncan test of SPSS 10.0 statistic analysis.

expression vector pLPGM413 was transformed into the onion epidermal cells with particle bombardment method as described by [51].

DAPI staining

To visualize nuclei, the epidermis was stained with DAPI (5 μg/mL, sigma, USA). Materials were soaked in the dye liquid phosphate buffer solution (PBS) (pH 7.0; DAPI: PBS (v/v) = 1:1000) and kept in darkness for 20 min. Pieces of onion epidermis were arranged on slides to make wet mounts, the made slides were observed and photographed in dark-field of fluorescence microscope (Olympus BX 61, Japan).

Microscopic investigation

Transformation efficiency was determined by calculating the proportion of positive epidermal cells with fluorescent signals among the cells in 2 mm^2 epidermis area that was measured with micro ruler under microscope. Images of epidermal cells were taken by using a motorized fluorescence microscope with a mirror unit (U-MNU2), dichroic mirror (DM400), excitation filter (BP360) and barrier filter (BA420) for DAPI (DNA staining). Images of epidermal cells positive for GFP were taken with a mirror unit (U-MSWB2), dichroic mirror (DM500), excitation filter (BP470-490) and barrier filter (BA520IP) for GFP. Images of epidermal cells positive for RFP were taken with a mirror unit (U-MSWG2), dichroic mirror (DM570), excitation filter (BP530-550) and barrier filter (BA590).

Statistical analysis

All statistical analyses were performed with SPSS (version 10.0).

Supporting Information

Figure S1 The sketch map of pLPGM202 originated from pCAMBIA1303 and pSAT6-GFP-N1. Details of creating the vector are given in the Materials and methods.

Figure S2 The sketch map of pLPGM113 originated from pCAMBIA1303 and pEZS-NL-GFP. Details of creating the vector are given in the Materials and methods.

Acknowledgments

The vectors, pSAT6-GFP-N1, pEZS-NL-GFP and pCM1205-RFP, were provided by Prof. Tao Wang (State Key Laboratory of Agro-biotechnology, China Agricultural University), Dr. Liang Zhang (Department of Life Science, Henan Normal University) and Dr. Wen-Cai Qi (Bioengineering Department, Zhengzhou University), respectively.

Author Contributions

Conceived and designed the experiments: KX. Performed the experiments: XH MW. Analyzed the data: KX CL. Contributed reagents/materials/analysis tools: YW YC KL JZ YZ FZ LY TL RW GT. Wrote the paper: KX CL.

References

1. Mitra J, Shrivastava SL, Rao PS (2012) Onion dehydration: a review. Int J Food Sci Tech 49:267–277.
2. Oparka KJ (1994) Plasmolysis: new insights into an old process. New Phytol 126(4):571–591.
3. McLusky SR, Bennett MH, Beale MH, Lewis MJ, Gaskin P, et al. (1999) Cell wall alterations and localized accumulation of feruloyl-3'-methoxytyramine in onion epidermis at sites of attempted penetration by *Botrytis allii* are associated with actin polarisation, peroxidase activity and suppression of flavonoid biosynthesis. Plant J 17:523–534.
4. Keller ERJ, Schubert I, Fuchs J, Meister A (1996) Interspecific crosses of onion with distant *Allium* species and characterization of the presumed hybrids by means of flow cytometry, karyotype analysis and genomic in situ hybridization. Theor Appl Genet 92:417–424.
5. Lee LY, Fang MJ, Kuang LY, Gelvin SB (2008) Vectors for multi-color bimolecular fluorescence complementation to investigate protein-protein interactions in living plant cells. Plant Methods 4:24.
6. Hollender CA, Liu Z (2010) Bimolecular fluorescence complementation (BiFC) assay for protein-protein interaction in onion cells using the helios gene gun. J Vis Exp 40:1963.
7. Eady CC, Lister CE, Suo Y, Schaper D (1996) Transient expression of uidA constructs in *in vitro* onion (*Allium cepa* L.) cultures following particle bombardment and *Agrobacterium*-mediated DNA delivery. Plant Cell Rep 15(12):958–962.
8. Hu CD, Kerppola TK (2003) Simultaneous visualization of multiple protein interactions in living cells using multicolor fluorescence complementation analysis. Nat Biotechnol 21:539–545.
9. Ueki S, Lacroix B, Krichevsky A, Lazarowitz SG, Citovsky V (2009) Functional transient genetic transformation of Arabidopsis leaves by biolistic bombardment. Nat Protoc 4:71–77.
10. Zhang G, Lu S, Chen TA, Funk CR, Meyer WA (2003) Transformation of triploid bermudagrass (*Cynodon dactylon* X *C. transvaalensis* cv. TifEagle) by means of biolistic bombardment. Plant Cell Rep 21:860–864.
11. Chen S, Tao L, Zeng L, Vega-Sanchez ME, Umemura K, et al. (2006) A highly efficient transient protoplast system for analyzing defence gene expression and protein-protein interactions in rice. Mol Plant Pathol 7:417–427.
12. Li JF, Park E, von Arnim AG, Nebenführ A (2009) The FAST technique: a simplified *Agrobacterium*-based transformation method for transient gene expression analysis in seedlings of Arabidopsis and other plant species. Plant Methods 5:6.
13. Marion J, Bach L, Bellec Y, Meyer C, Gissot L, et al. (2008) Systematic analysis of protein subcellular localization and interaction using high-throughput transient transformation of Arabidopsis seedlings. Plant J 56:169–179.

14. Ye GN, Stone D, Pang SZ, Creely W, Gonzalez K, et al. (1999) *Arabidopsis* ovule is the target for *Agrobacterium in planta* vacuum infiltration transformation. Plant J 19:249–257.

15. Abel S, Theologis A (1994) Transient transformation of Arabidopsis leaf protoplasts: a versatile experimental system to study gene expression. Plant J 5:421–427.

16. Blachutzik JO, Demir F, Kreuzer I, Hedrich R, Harms GS (2012) Methods of staining and visualization of sphingolipid enriched and non-enriched plasma membrane regions of *Arabidopsis thaliana* with fluorescent dyes and lipid analogues. Plant Methods 8:28.

17. Zhang Y, Su J, Duan S, Ao Y, Dai J, et al. (2011) A highly efficient rice green tissue protoplast system for transient gene expression and studying light/chloroplast-related processes. Plant Methods 7:30.

18. Takata N, Eriksson ME (2012) A simple and efficient transient transformation for hybrid aspen (*Populus tremula* × *P. tremuloides*). Plant Methods 8:30.

19. Reggiardo MI, Arana JL, Orsaria LM, Permingeat HR, Spitteler MA, et al. (1991) Transient transformation of maize tissues by microparticle bombardment. Plant Sci 75:237–243.

20. Hamilton DA, Roy M, Rueda J, Sindhu RK, Sanford J, et al. (1992) Dissection of a pollen-specific promoter from maize by transient transformation assays. Plant Mol Biol 18:211–218.

21. Sidorov VA, Kasten D, Pang SZ, Hajdukiewicz PT, Staub JM, et al. (1999) Stable chloroplast transformation in potato: use of green fluorescent protein as a plastid marker. Plant J 19:209–216.

22. Bhaskar PB, Venkateshwaran M, Wu L, Ané JM, Jiang J (2009) *Agrobacterium*-mediated transient gene expression and silencing: a rapid tool for functional gene assay in potato. PLoS One 4:e5812.

23. Wang YC, Klein TM, Fromm M, Cao J, Sanford JC, et al. (1988) Transient expression of foreign genes in rice, wheat and soybean cells following particle bombardment. Plant Mol Biol 11:433–439.

24. Santarem ER, Trick HN, Essig JS, Finer JJ (1998) Sonication-assisted *Agrobacterium*-mediated transformation of soybean immature cotyledons: optimization of transient expression. Plant Cell Rep 17:752–759.

25. Trick HN, Finer JJ (1998) Sonication-assisted *Agrobacterium*-mediated transformation of soybean [*Glycine max* (L.) Merrill] embryogenic suspension culture tissue. Plant Cell Rep 17:482–488.

26. Baum K, Gröning B, Meier I (1997) Improved ballistic transient transformation conditions for tomato fruit allow identification of organ-specific contributions of I-box and G-box to the RBCS2 promoter activity. Plant J 12:463–469.

27. Wroblewski T, Tomczak A, Michelmore R (2005) Optimization of *Agrobacterium*-mediated transient assays of gene expression in lettuce, tomato and Arabidopsis. Plant Biotechnol J 3:259–273.

28. Orzaez D, Mirabel S, Wieland WH, Granell A (2006) Agroinjection of tomato fruits. A tool for rapid functional analysis of transgenes directly in fruit. Plant Physiol 140:3–11.

29. Rasco-Gaunt S, Riley A, Barcelo P, Lazzeri PA (1999) Analysis of particle bombardment parameters to optimise DNA delivery into wheat tissues. Plant Cell Rep 19:118–7.

30. Amoah BK, Wu H, Sparks C, Jones HD (2001), Factors influencing *Agrobacterium*-mediated transient expression of uidA in wheat inflorescence tissue. J Exp Bot 52:1135–1142.

31. Li YH, Tremblay FM, Séguin A (1994) Transient transformation of pollen and embryogenic tissues of white spruce (*Picea glauca* (Moench.) Voss) resulting from microprojectile bombardment. Plant Cell Rep 13:661–665.

32. Liu CN, Li XQ, Gelvin SB (1992) Multiple copies of virG enhance the transient transformation of celery, carrot and rice tissues by *Agrobacterium tumefaciens*. Plant Mol Biol 20:1071–1087.

33. Ohlrogge JB, Kuhn DN, Stumpf PK (1979) Subcellular localization of acyl carrier protein in leaf protoplasts of *Spinacia oleracea*. P Natl Acad Sci 76:1194–1198.

34. Pommerrenig B, Popko J, Heilmann M, Schulmeister S, Dietel K, et al. (2013) *SUCROSE TRANSPORTER 5* supplies Arabidopsis embryos with biotin and affects triacylglycerol accumulation. Plant J 73:392–404.

35. Zhang H, Lv F, Han X, Xia X, Yin W (2013) The calcium sensor PeCBL1, interacting with PeCIPK24/25 and PeCIPK26, regulates Na$^+$/K$^+$ homeostasis in *Populus euphratica*. Plant Cell Rep 32:611–621.

36. Yang KY, Liu Y, Zhang S (2001) Activation of a mitogen-activated protein kinase pathway is involved in disease resistance in tobacco. P Natl Acad Sci 98:741–746.

37. Noël LD, Cagna G, Stuttmann J, Wirthmüller L, Betsuyaku S, et al. (2007) Interaction between SGT1 and cytosolic/nuclear HSC70 chaperones regulates Arabidopsis immune responses. Plant Cell 19:4061–4076.

38. Chen X, Equi R, Baxter H, Berk K, Han J, et al. (2010) A high-throughput transient gene expression system for switchgrass (*Panicum virgatum* L.) seedlings. Biotechnol Biofuels 3:9.

39. Manavella PA, Chan RL (2009) Transient transformation of sunflower leaf discs via an *Agrobacterium*-mediated method: applications for gene expression and silencing studies. Nat Protoc 4:1699–1707.

40. Tsuda K, Qi Y, Nguyen LV, Bethke G, Tsuda Y, et al. (2012) An efficient *Agrobacterium*-mediated transient transformation of Arabidopsis. Plant J 69:713–719.

41. Hosein FN, Lennon AM, Umaharan P (2012) Optimization of an *Agrobacterium*-mediated transient assay for gene expression studies in *Anthurium andraeanum*. J Am Soc Hortic Sci 137:263–272.

42. Shah K, Gadella JTW, van EH, Hecht V, de Vries SC (2001) Subcellular localization and oligomerization of the *Arabidopsis thaliana* somatic embryogenesis receptor kinase 1 protein. J Mol Biol 309:641–655.

43. Xu KD, Liu QL, Yang HF, Zeng L, Dong LL, et al. (2011) Isolation and molecular characterization of RcSERK1: A *Rosa canina* gene transcriptionally induced during initiation of protocorm-like bodies. African J of Biotech 10:4011–4017.

44. Liu QL, Xu KD, Ma N, Zeng L, Zhao LJ (2010) Isolation and functional characterization of *DgZFP*: a gene encoding a Cys2/His2-type zinc finger protein in chrysanthemum. Mol Biol Rep 37:1137–1142.

45. Liu QL, Xu KD, Zhao LJ, Pan YZ, Jiang BB, et al. (2011) Overexpression of a novel chrysanthemum *NAC* transcription factor gene enhances salt tolerance in tobacco. Biotechnol Lett 33:2073–2082.

46. Han DG, Yang GH, Xu KD, Shao Q, Yu ZY, et al. (2013) Overexpression of a *Malus xiaojinensis Nas1* gene influences flower development and tolerance to iron stress in transgenic tobacco. Plant Mol Biol Rep 31:802–809.

47. Blanvillain R, Kim JH, Wu S, Lima A, Ow DW (2009) *OXIDATIVE STRESS 3* is a chromatin-associated factor involved in tolerance to heavy metals and oxidative stress. Plant J 57:654–665.

48. Song S, Chen Y, Zhao M, Zhang WH (2012) A novel *Medicago truncatula* HD-Zip gene, *MtHB2*, is involved in abiotic stress responses. Environ Exp Bot 80:1–9.

49. Zhou C, Wang H, Zhu J, Liu Z (2013) Molecular cloning, subcellular localization and functional analysis of *ThCLC-a* from *Thellungiella halophila*. Plant Mol Biol Rep 31:783–790.

50. Tsai AYL, Gazzarrini S (2012) *AKIN10* and *FUSCA3* interact to control lateral organ development and phase transitions in Arabidopsis. Plant J 69:809–821.

51. Wang GL, Fang HY (2002). Gene engineering in plant (the 2nd edition). Beijing: China Science Press, Beijing, pp. 734–736.

Recombinant Plants Provide a New Approach to the Production of Bacterial Polysaccharide for Vaccines

Claire M. Smith[1¤a], Stephen C. Fry[2], Kevin C. Gough[3], Alexandra J. F. Patel[1¤b], Sarah Glenn[1], Marie Goldrick[4], Ian S. Roberts[4], Garry C. Whitelam[5†], Peter W. Andrew[1*]

1 Department of Infection, Immunity and Inflammation, University of Leicester, Leicester, Leicestershire, United Kingdom, 2 The Edinburgh Cell Wall Group, Institute of Molecular Plant Sciences, School of Biological Sciences, University of Edinburgh, Edinburgh, United Kingdom, 3 School of Veterinary Medicine and Science, University of Nottingham, Sutton Bonington Campus, Leicestershire, United Kingdom, 4 Faculty of Life Sciences, University of Manchester, Manchester, United Kingdom, 5 Department of Biology, University of Leicester, Leicester, Leicestershire, United Kingdom

Abstract

Bacterial polysaccharides have numerous clinical or industrial uses. Recombinant plants could offer the possibility of producing bacterial polysaccharides on a large scale and free of contaminating bacterial toxins and antigens. We investigated the feasibility of this proposal by cloning and expressing the gene for the type 3 synthase (cps3S) of *Streptococcus pneumoniae* in *Nicotinia tabacum*, using the pCambia2301 vector and *Agrobacterium tumefaciens*-mediated gene transfer. *In planta* the recombinant synthase polymerised plant-derived UDP-glucose and UDP-glucuronic acid to form type 3 polysaccharide. Expression of the cps3S gene was detected by RT-PCR and production of the pneumococcal polysaccharide was detected in tobacco leaf extracts by double immunodiffusion, Western blotting and high-voltage paper electrophoresis. Because it is used a component of anti-pneumococcal vaccines, the immunogenicity of the plant-derived type 3 polysaccharide was tested. Mice immunised with extracts from recombinant plants were protected from challenge with a lethal dose of pneumococci in a model of pneumonia and the immunised mice had significantly elevated levels of serum anti-pneumococcal polysaccharide antibodies. This study provides the proof of the principle that bacterial polysaccharide can be successfully synthesised in plants and that these recombinant polysaccharides could be used as vaccines to protect against life-threatening infections.

Editor: Eliane Namie Miyaji, Instituto Butantan, Brazil

Funding: This work was supported by a grant from the Welcome Trust (067697) to PWA, GCW and ISR. SCF thanks the UK Biotechnology and Biological Sciences Research Council for support. The funders had no role in study design, data collection and analysis, decision to publish, or preparation of the manuscript.

Competing Interests: The authors have declared that no competing interests exist.

* E-mail: pwa@le.ac.uk

¤a Current address: Department of Respiratory Medicine, Portex Unit, University College London, Institute of Child Health, London, United Kingdom
¤b Current address: Department of Biology, University of Leicester, Leicester, Leicestershire, United Kingdom

†Deceased.

Introduction

Polysaccharide encapsulated bacteria are major causes of disease and death in humans and animals. For example, diseases caused by *Streptococcus pneumoniae* (the pneumococcus), *Neisseria meningitidis* and *Haemophilius influenzae* are responsible for more than two million deaths every year, the majority children under the age of five [1,2][1,2]. *Streptococcus pneumoniae* alone is responsible for more than 50 percent of invasive disease worldwide. Despite the extensive use of pneumococcal vaccines, incidences of disease caused by *S. pneumoniae* remain high, mainly due to serotypes not included in the vaccine [3]. Current anti-pneumococcal vaccines are composed of capsular polysaccharide alone or conjugated to protein. Whatever the formulation, pneumococcal vaccine design has to deal with the facts that there are over 90 different capsular and the serotype distribution varies with time and geography. However, for reasons of economics and biology the current vaccines are limited in coverage (23 in the polysaccharide-only vaccine and 13 in the new version of the conjugate) to the most dominant serotypes in Europe and North America. Ideally

multiple versions of these vaccines are required and they would be regularly reformulated to offer maximum protection. Cost of polysaccharide production then becomes a concern. One of the challenges for pneumococcal vaccine production is to manufacture bacterial polysaccharide on a large-scale, without need for purification procedures to remove contaminating toxins and pyrogens. Currently the preparation of polysaccharides requires expensive fermentation equipment, microbiological containment and high levels of quality control to prevent contamination. Plants offer a solution because they synthesise a large number of high molecular weight polysaccharides, they have many of the sugar precursors of bacterial capsular polysaccharide readily available and plants have compartmentalised metabolic pathways and transport processes that could facilitate polysaccharide extraction [4]. However, until now heterologous antigen production in plants has been limited to the production of proteins [5,6,7]. Here we report that plants can be engineered to synthesise bacterial polysaccharides and these polysaccharides provide protective immunity. We demonstrated this principle using the serotype 3 capsular polysaccharide of *S. pneumoniae*, a serotype that is frequently isolated from disease cases. The type 3 polysaccharide is composed of repeating D-glucose (Glc) and D-glucuronic acid (GlcA) units, as (1→4)-β-D-Glc$p$-(1→3)-β-D-Glc$p$A-(1→4) [8,9] The precursors, UDP-glucose (UDP-Glc) and UDP-glucuronic acid (UDP-GlcA), are polymerised by a type 3 synthase (Cps3S) [8,9].

Results

The Pneumococcal Type 3 Capsule Synthase Gene was Cloned into *Nicotinia tabacum* by *Agrobacterium*-mediated Gene Transfer

The pneumococcal *cps3S* gene {Dillard, 1995 #888} was amplified from genomic DNA of the pneumococcal type 3 strain WU2 using primers CPSFOR and CPSREV and cloned with PR1b signal sequence (which was used to direct secretion of the transgene to the apoplast) into the *Agrobacterium* binary vector pCambia 2301, to give pCMS4. This placed *cps3S* under the control of duplicated constitutive cauliflower mosaic virus promoters, CaMV35S and also enabled selection of transformed plants with kanamycin. Nucleotide sequence analysis of the cloned *cps3S* in pCMS4 showed 100 % identity with the published sequence [8].

Nicotiana tabacum was transformed with pCMS4 by *A. tumefaciens*-mediated gene transfer. A T1 generation was grown from the seeds of six plants and PCR showed that four plants contained the 1.3kb *cps3S* gene (Figure 1A). PCR also confirmed the absence of contaminating *Agrobacterium* DNA (Figure 1B). RT-PCR, with *cps3S*-specific primers, showed that the transgene was expressed in the transgenic plants (Figure 2). No amplicon was generated by direct PCR amplification of RNA extracts, confirming the absence of contaminating *cps3S* DNA (Figure 2A). No amplicon was generated by RT-PCR of untransformed plants (Figure 2A lane 3). A second generation of plants were grown from the seeds of these plants and PCR confirmed stable transgene expression (Figure S1). All subsequent assays were done with second generation (T2) plant material.

Pneumococcal Type 3 Polysaccharide was Detected in the Leaves of Transformed Plants

Double immunodiffusion showed that type 3 antibody-antigen complexes were seen (Figure 2B) between wells which contain purified type 3 pneumococcal polysaccharide, sonicated plant cell extract from transgenic plants (wells 1-4) and type 3 polysaccharide specific antiserum (well A). This was not seen in wells containing extract from a wildtype tobacco plants (wells 5 and 6). Western blotting of transgenic and wildtype plant extracts using type 3 polysaccharide specific antiserum also showed the presence of type 3 polysaccharide in transgenic plant extracts only (Figure 2C). High-voltage paper electrophoresis of hot-acid hydrolysates of cold-acid-extractable tobacco leaf polysaccharides confirmed these findings (Figure 2D). Acid hydrolysis of polysaccharides from transgenic leaves produced a relatively hot-acid-resistant, singly-ionised disaccharide with the same mobility as the β-D-Glc$p$A-(1→4)-D-Glc seen following acid hydrolysis of pneumococcal type 3 polysaccharide. This disaccharide was barely detectable in wild type non-transformed plants (Figure 2D).

Immunisation with Transgenic Plant Extracts Protected Mice from Pneumococcal Disease

To test the immunogenicity of the plant-derived type 3 polysaccharide, mice were immunised with three doses of apoplast extracts from transgenic or wildtype plants. Sera were collected on the day before each immunisation and ten days after the final dose, and anti-type 3 polysaccharide IgG was determined by ELISA. Significantly more (P<0.05) specific anti-type 3 antibody was detected after a single dose of the transgenic leaf extract, with a further increase (P<0.05) after a second dose (Figure 3A), whereas antibody levels remained unchanged in those given wildtype extracts (P>0.05).

Mice were challenged intranasally with the serotype 3 *S. pneumoniae* strain HB565 230 days after the final immunisation. Mice immunised with transgenic plant extract survived significantly longer (P<0.001) than those given wildtype extracts (mean survival: 181±72h and 90±23h for transgenic and wildtype, respectively). Mice immunised with the wildtype extract did not survive longer (P>0.05) than sham-immunised mice (91±38h). None of the fifteen animals given wildtype extract were alive ten

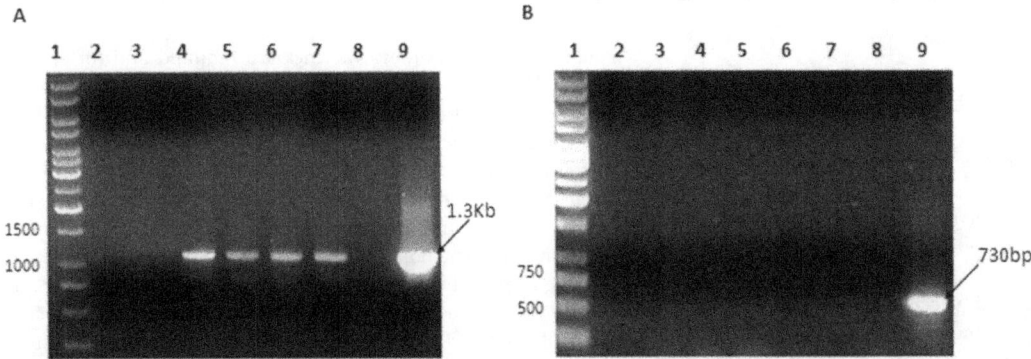

Figure 1. Detection of the *cps3S* gene in transformed tobacco plants. A. DNA was used as a template for PCR (Lanes 2, 3: wild type plants; Lanes 4 – 7: transformed plants.) using *cps3S*-specific primers. PCR products were analysed by agarose gel electrophoresis. The results show the presence of the *cps3S* gene in the transformed plants (Lanes 4 - 7) but not the wild type plants. The PCR reaction in Lane 9 contained purified plasmid DNA containing *cps3S* (pCMS4) as a positive control and Lane 8 contained no template DNA. Molecular sizes are indicated. **B.** PCR showing the absence of *Agrobacterium* DNA contaminating DNA preparations from wild type (Lanes 2, 3) and transformed (Lanes 4 - 7) tobacco plants. PCR was done with *Agrobacterium*-specific primers. The results show that there was no *Agrobacterium* DNA present in the transgenic plant samples. The PCR reaction in Lane 9 contained *Agrobacterium* DNA as a positive control and shows the expected 730bp band and Lane 8 contained no template. Molecular sizes are indicated. DNA was extracted from the same six *N. tabacum* plants for the PCRs shown in A and B.

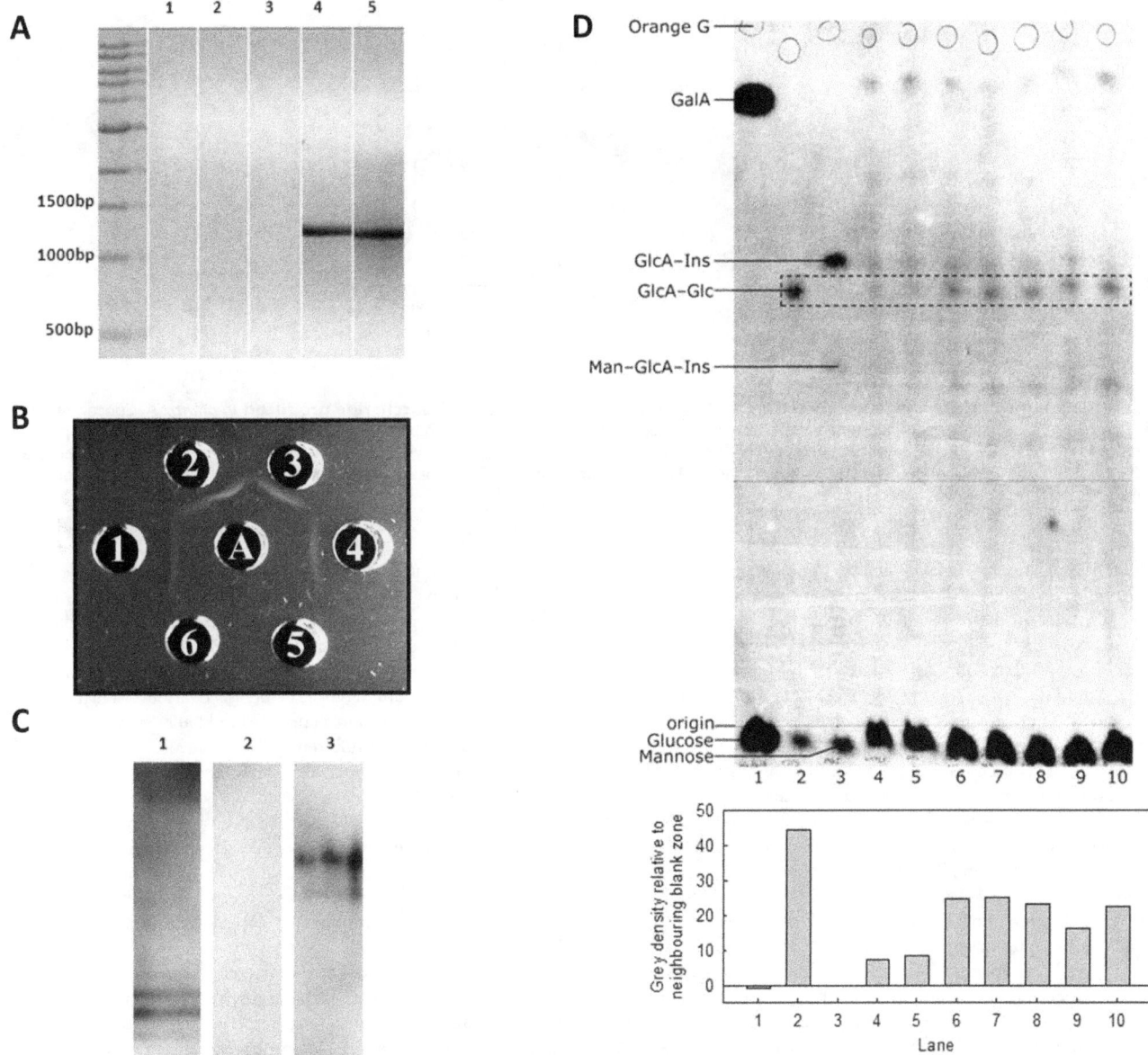

Figure 2. *In planta* **expression of the** *cps3S* **gene and formation of serotype 3 polysaccharide. A.** Reverse transcriptase PCR to detect *cps3S* mRNA in transgenic tobacco plants. RNA was extracted from a wildtype (Lanes 1 and 3) and a transgenic *N. tabacum* containing *cps3S* (Lanes 2 and 4). PCR products, using *cps3S* specific primers were analysed by agarose gel electrophoresis. Lanes 1 and 2 showed the absence of *cps3S* DNA in the RNA. RT-PCR on the same samples showed the presence of *cps3S* mRNA in the transgenic plant (Lane 4) but not in the wildtype (Lane 3). Lane 5 PCR of pCMS4 containing *cps3S*, done as before. The 1.3 kb amplicon in Lanes 4 and 5 shows a full-length transcript of *cps3S* is expressed in the transgenic plant. **B.** Double immunodouble diffusion. Well 1:10 µg purified serotype 3 polysaccharide; Wells 2-4: extract from tobacco plants shown to express *cps3S*: Wells 5 and 6: extract from a wildtype tobacco plant. Well A: type 3 polysaccharide specific antiserum. The preciptin lines identify the presence of type 3 polysaccharide. **C.** Western blotting using type 3 polysaccharide specific antiserum. Lane 1: purified type 3 polysaccharide; Lane 2: wildtype plant extract; Lane 3: transgenic plant extract. **D.** High-voltage paper electrophoresis of tobacco leaf acid hydrolysates. Lanes 1-3: 25 µg of each marker, (Lane 1) galacturonic acid (GalA) and glucose, (Lane 2) glucose and β-D-glucuronosyl-(1→4)-D-glucose (GlcA–Glc) (partial hydrolysate of 10 µg type 3 pneumococcal polysaccharide) and (Lane 3) 10 µg of a mixture of mannose, α-D-glucuronosyl-(1→2)-*myo*-inositol (GlcA–Ins) and a trace of α-D-mannosyl-(1→4)-α-D-glucuronosyl-(1→2)-*myo*-inositol (Man–GlcA–Ins). Lanes 4–10: hydrolysate of polysaccharides cold-acid-extracted from 32 mg fresh weight of wildtype (Lanes 4 and 5) or transgenic (Lanes 6–10) tobacco leaves. Each lane also contains a trace of Orange G (coloured internal marker). All lanes show similar levels of staining for neutral sugars (co-migrating with glucose, near the origin). The samples were electrophoresed at pH 6.5, at 3.0 kV for 60 min (anode at top) and stained with AgNO₃. Spots of the disaccharide, GlcA–Glc, diagnostic of type 3 pneumococcal polysaccharide, are highlighted by the dashed box; these spots were quantified for grey density in PhotoShop (see histogram).

days after the challenge, whereas eight of the fourteen immunised with transgenic extract survived (Figure 3B).

Discussion

This study has shown that bacterial polysaccharide vaccine antigens can be synthesised in plants and that simple extracts of these plants are immunogenic and protect against an otherwise

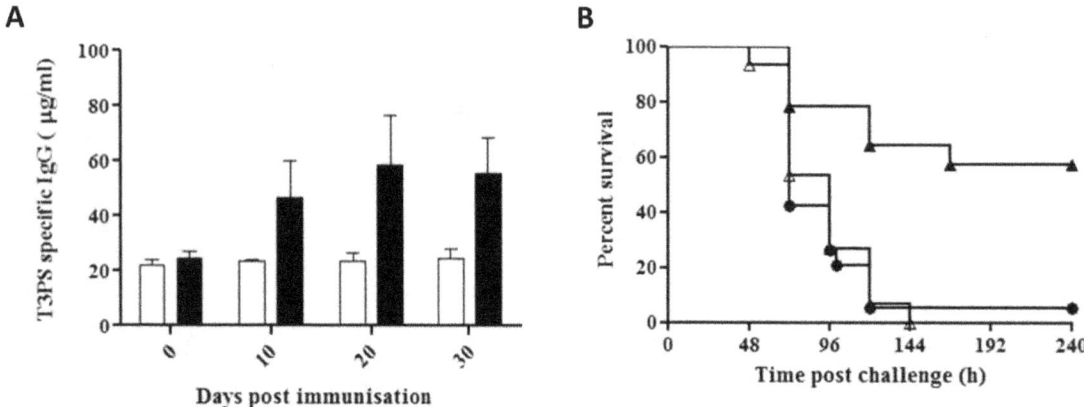

Figure 3. Immunogenicity and protective efficacy of serotype 3 pneumococcal polysaccharide produced *in planta*. A. Concentration of serotype 3 polysaccharide-specific IgG in serum of mice immunised with extracts from tobacco plants expressing *cps3S* (black bars) or wildtype plant (white bars); n = 5. **B.** Survival of mice challenged with virulent type 3 pneumococci 230 days after the final immunisation with transgenic plant extract (closed triangles), wildtype extracts (open triangles) or sham-immunised mice (closed circles). Mice alive at 240h post-infection were considered to have survived the infection.

lethal infection. Transgenic plants are recognised as good expression systems for proteins but the synthesis of bacterial polysaccharide in plants has not been demonstrated before.

We chose the production of the type 3 polysaccharide of *S. pneumoniae* for the reason that it is a relatively simple carbohydrate, being composed of repeating D-glucose (Glc) and D-glucuronic acid (GlcA) organised as (1→4)-β-D-Glc*p*-(1→3)-β-D-Glc*p*A-(1→4) units [8,9]. The precursors, UDP-glucose (UDP-Glc) and UDP-glucuronic acid (UDP-GlcA) are naturally synthesised by plants, which transport them into the endomembrane system as substrates for cell wall polysaccharide synthesis [4,10]. Polymerisation of these substrates into type 3 polysaccharide requires the enzyme, type 3 synthase (Cps3S). Therefore, we cloned the pneumococcal *cps3S* gene into *N. tabacum* using pCambia 2301. This strategy not only placed *cps3S* under the control of duplicated constitutive cauliflower mosaic virus promoters, CaMV35S, but it also enabled selection of transformed plants with kanamycin. Growth of a second generation of kanamycin-resistant plants confirmed stable transgene expression. Although not the primary purpose of the study, we did a limited investigation of how to extract the pneumococcal polysaccharide from plant tissue. The method that yielded the highest concentration of pneumococcal polysaccharide was to grind the plant tissue under liquid nitrogen, suspend the tissue in water and lyse the cells by sonication. Despite cloning the signal sequence PR1b we detected no type 3 polysaccharide in its destination, the apoplastic fluid. This implied that PR1b was not functioning correctly. Previous studies replaced the start codon of the transgene with PR1b [3], however, we maintained the start codon and cloned an in-frame sequence of *cps3S*. This may have led to a reduction in PR1b activity and improving this may increase the yield of type 3 polysaccharide. Another method to increase yield is to use root tissues, since the continuously growing primary cell wall may contain higher concentrations of the UDP-precursors. In this study we focussed on leaf tissue as we were working with parent and F1 generations and removal of the roots may have restricted growth of the plant and seed development. Leaf tissue was also much easier to obtain. For these reasons, the levels of polysaccharide extracted from plant tissue may not have been optimal.

Having shown the principle of *in planta* synthesis using the linear type 3 polysaccharide, the next challenge is the production of more complex, branched, bacterial polysaccharides. All the genes involved in pneumococcal capsular polysaccharide synthesis are closely linked on the bacterial chromosome, arranged within a single locus (a "type specific" cassette). Therefore, it is possible that the introduction of whole cassettes could lead to the synthesis of sugar precursors not naturally occurring in the plant and their assembly into more complex polysaccharides. Furthermore, effective signal or transport peptides should allow easier extraction by compartmentalising different polysaccharides.

Because anti-polysaccharide antibodies are protective against several bacterial pathogens of humans and animals there is great interest in polysaccharides as vaccines. However, some of the problems with these vaccines are illustrated by vaccines against *S. pneumoniae*. The current pneumococcal vaccine contains twenty-three polysaccharide serotypes but protection is serotype-specific and some are not immunogenic in children. The vaccine was formulated on the prevalence of serotypes in North America and Europe, but elsewhere the coverage can be considerably less [11]. In addition to protection being serotype-specific, polysaccharide immunogenicity also varies with serotype and age [12]. Furthermore, temporal variation occurs in the serotypes isolated from adults and children [13]. Thus it has been suggested that different vaccine formulations should be manufactured for differing situations [14], but unless low-cost solutions are found it will not become a reality. Polysaccharide vaccines can be expensive, which restrains their use in developing countries. Production of polysaccharide vaccines in plants can introduce economies of scale that can drive down the production costs. The alternative, of using microorganisms as the vaccine production system, requires expensive fermentation equipment and high levels of quality control to prevent contamination. In contrast, the use of plants for vaccine production offers an achievable solution, opening the possibility of local production, which increases the likelihood of adoption of the vaccine [15].

Efforts to improve the poor immunogenicity of polysaccharide vaccines in the young are focused on the development of polysaccharides covalently linked to protein, but these make difficulties with serotype coverage worse. When the US FDA licensed a 7-valent anti-pneumococcal conjugate vaccine (Prevnar) the serotypes covered by the vaccine caused 90% of disease in North America and Europe, but less than 70% in Asia [15]. This emphasises the desirability of a 'tailor-made' vaccine. However, formulations of conjugates for a particular country or for

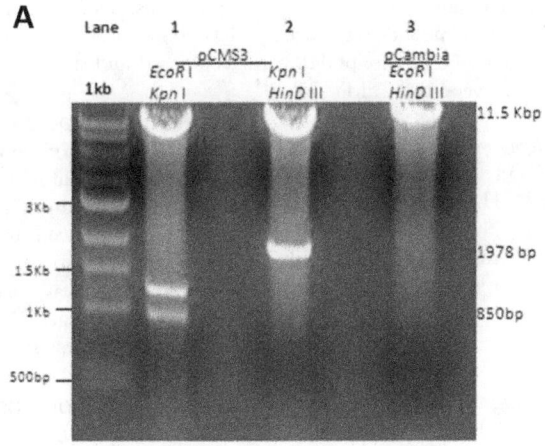

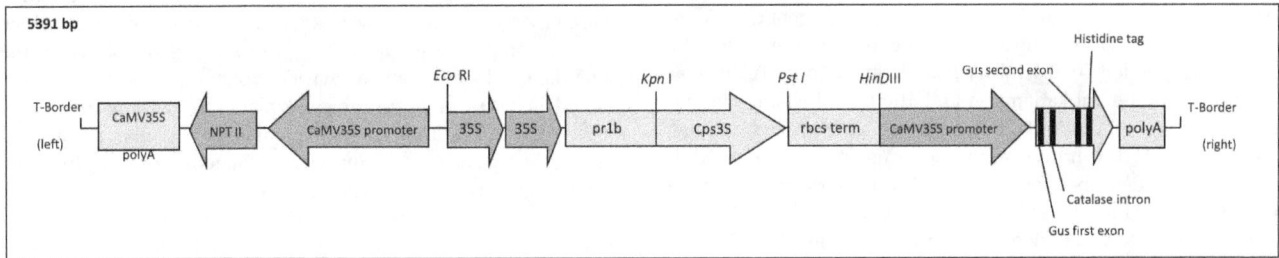

Figure 4. Construction of the pCMS4 vector. A. The digested products of three separate restriction digests of pCMS3 and pCambia 2301. Lane 1 shows the expected 3 fragments of pCMS3 when digested with EcoR I and Kpn I. Lane 2 shows the expected 2 fragments produced from the digestion of pCMS3 with Kpn I and HinD III. Lane 3 shows the expected band of 11.5 Kbp when pCambia 2301 was digested with EcoR I and HinD III. The DNA fragments of 864bp (Lane 1) and 1978bp (Lane 2), indicated by the arrows, were used to reconstruct the pCMS3 T-DNA fragment containing cps3S and these were ligated to the 11.5 Kb pCambia fragment (Lane 3). This plasmid was termed pCMS4. Undigested DNA is not shown. A 1 Kb ladder (New England Biolabs) was used as a molecular size marker. **B.** The expression vector pCHF2. A CaMV35S promoter, rbcS terminator region, and a PR1B signal sequence were cloned into the T-DNA region of the pPZP222 vector [17]. The unique restriction endonuclease sites are also shown.

childhood and for adult vaccination programmes are even less likely than pure polysaccharide formulations, unless cheaper production methods can be found. Conjugate vaccines are very expensive and the pneumococcal conjugates will be more expensive than any other current vaccine. Even with tiered pricing the price of these vaccines is a real concern for their uptake [16]. The ability to produce polysaccharide vaccines in plants, on a large scale, could lead to a ready availability of polysaccharides for protein conjugate vaccine production. There is, however, a more exciting possibility, namely to exploit the plant's glycosylation machinery to glycosylate heterologous proteins with heterologous polysaccharide to make conjugate vaccines *in planta*. Plants making heterologous immunogen represent an innovative technology for the development of childhood vaccines. The longer-term objective of this research is to synthesise polysaccharide-protein conjugates in plants.

In planta synthesis of bacterial polysaccharide and conjugates offers an innovative contribution to vaccinology. Synthesising heterologous polysaccharides in plants represents the first proof of concept step in this process. These experiments have yielded exciting data even though no attempt was made to optimise gene expression, polysaccharide purification or immunising protocol. These are developmental issues for the future now that the concept that bacterial polysaccharide can be synthesised in plants has been proven.

Materials and Methods

Construction of the Plant Expression Vector

The sequence of the type 3 pneumococcal capsular polysaccharide biosynthesis cassette was obtained from GenBank (www.ncbi.nlm.nih.gov), accession number U15171. A 1.3 kb DNA fragment containing the *cps3S* gene was obtained by PCR from the genomic DNA of *S. pneumoniae* serotype 3 (WU2) using the oligonucleotide primers sense (CPSFOR) 5'-CTG GTA ↓ CCC ATG TAT ACA TTT ATT TTA ATG TTG TTG G-3' corresponding to 2227bp - 2254bp with a *KpnI* restriction site inserted at the 5' end; anti-sense (CPSREV) 5'- TCA TCA CTC TGT TAA ATT CCT AGT TCC -3', corresponding to 3454bp - 3479bp of the cassette. The PCR amplification was performed in a total volume of 50 µl. using 2 µg genomic template under the following conditions: 94°C for 4 minutes and then a cycling procedure comprising denaturation at 94°C for 45 seconds, annealing at 58°C for 1 minute, and extension at 72°C for 1 minute 30 seconds, which was repeated either 10 or 30 times, and a final extension at 72°C for 10 minutes. Amplified DNA resulting from PCR were purified from the PCR reagents using the QIAquick PCR purification kit (Qiagen). The amplified fragment was inserted into the multiple cloning site of pCR4-TOPO (Invitrogen, Carlsbad, CA, USA). The *Agrobacterium* binary vector pPZP221 (GenBank U10463) [17] has been previously engineered to produce vector pCHF2 (Figure 4) containing the constitutively expressed cauliflower mosaic virus (CaMV) 35S promoter and a *rbcS* terminator (C. Fankhauser, personal communication). Here, synthetic primers homologous to the PR1b signal peptide sequence [18] were annealed, phosphorylated and inserted into pCHF2 using *Sac*I and *Kpn*I cutting sites situated between the CaMV 35S promoter and the *rbcS* terminator. The *cps3S* gene was removed from the pCR4-TOPO vector using *Kpn*I and *Pst*I and inserted between the PR1b signal sequence and the *rbcS* terminator sequence, to form the clone pCMS3. The entire expression cassette was then excised from pCMS3 with *Eco*RI and *Hin*DIII and ligated into the binary plant vector pCAMBIA2301 (GenBank accession number AF234316) (containing a kanamycin-resistance gene and *gus*) to give the resulting plasmid, pCMS4.

Transformation of Plants

pCMS4 was introduced into *Agrobacterium tumefaciens* strain GV3101 directly by the heat shock method, using 0.5 μl (1 μg) pCMS4 and 0.1 ml of frozen CaCl$_2$-competent *A. tumefaciens* cells. Cells were thawed at 37°C for 5 minutes, re-suspended in 1 ml of YEP broth (10 g/L Yeast Extract, 10 g/L Peptone, 5 g/L NaCl, pH 7.0) and incubated at 28°C for 2 hours with gentle shaking. Cells were harvested by centrifugation at 600 *g* for 10 minutes, the pellet re-suspended in 0.1 ml YEP and spread onto YEP agar containing 100 μg/ml kanamycin and 100 μg/ml rifampicin and incubated overnight at 28°C. Subsequently, *A. tumefaciens* carrying pCMS4 was used to transform tobacco (*Nicotiana tabacum cv* SR1) leaf discs, as described previously [19]. Leaf discs were then transferred to Murashige-Skoog (MS) agar containing 3% (w/v) sucrose, the plate was sealed with parafilm and incubated in the dark at room temperature for 2 days. Discs were then transferred to selective medium (MS agar containing 3% (w/v) sucrose, 250 μg/ml cefotaxime (to kill the *A. tumefaciens* on the surface of explants), 100 μg/ml kanamycin, 100 μg/ml ampicillin, 100 μg/ml naphthalene acetic acid (NAA) and 1 mg/ml 6-benzylaminopurine) to select transgenic progeny. Ten days post-infection, those discs showing shoot formation were transferred to fresh selective medium and re-incubated for approximately 1-2 weeks. Larger shoots were transferred to powder rounds containing the same medium and incubated for a further 3 weeks, or until roots started to form. After eight weeks, kanamycin-resistant shoot regenerants were removed to a rooting medium containing 100 μg/ml NAA and 25 μg/ml kanamycin. Rooted plantlets were transferred to soil, self-pollinated and the seeds stored desiccated at 4°C.

Detection of *cps3S* Gene Expression in Transgenic Plants

Total RNA was isolated from 100 mg of leaf tissue using a RNA isolation kit (RNeasy Plant Mini Kit; Qiagen, Surrey, UK) and used in the production of first strand cDNA using a cDNA synthesis kit (RNase H$^-$ reverse transcriptase kit; Invitrogen) and a random hexa-nucleotide primer. PCR was then performed using the CPSFOR and CPSREV primers as described above, RNA extracts without prior reverse transcriptase treatment were used as a control to indicate the presence of *cps3S* specific DNA. RNA extracted from wild-type tobacco leaves also was used as negative control. To confirm the absence of contaminating *Agrobacterium* DNA PCR was done with the primers VCF (5′-ATC ATT TGT AGC GAC T-3′) and VCR (5′-AGC TCA AAC CTG CTT C-3′), designed to amplify a 730bp region of the *virC* gene of Agrobacterial Ti and Ri plasmids [20].

Preparation of Leaf Extracts

Leaves were collected from tobacco plants, tissue ground under liquid nitrogen and nanopure water added to give 0.5 g/ml plant tissue. The cells were lysed by sonication: 6×30 second sonications at an amplitude of 50 microns with 30 second rests in between sonications. The cell lysate then was centrifuged for 5 minutes at 10,000 *g* and the supernatant divided into 1 ml volumes, lyophilised and stored at 4°C.

Extraction of type 3 Polysaccharide from Apoplastic Fluid

A modification of the method described by Fry and co-workers was used [21]. Leaf material (1 g) was added to 50 ml of 50 mM CaCl$_2$ and vacuum-infiltrated for a period of 30 minutes. The leaves were removed and dried gently on a paper towel before being transferred to the barrel of a 25 ml syringe with the plunger removed. This was placed in a 50 ml Falcon tube and the assembly centrifuged at 800 *g* at 10°C for 10 minutes. The aqueous extract was stored at 4°C until required.

Double Immunodiffusion

A modification Ouchterlony's method was used [22]. 0.2% (w/v) Ouchterlony agarose in barbitone buffer (1.84 g/l diethylbarbituric acid, 10.3 g/l sodium diethylbarbiturate, pH 8.6) was used to coat microscope slides. These were left to dry for 1 hour, and then overlaid with 4.5 ml 1% (w/v) agarose in barbitone buffer. Once set, 4 mm holes were cut and 20 μl of sample was placed in the outer holes. Type 3 polysaccharide from *S. pneumoniae* (ATCC), diluted in 20 μl of untransformed plant extract, was used as a positive control. The central hole contained 20 μl neat rabbit anti-type 3 polysaccharide antiserum (Statens Serum Institute, Copenhagen, Denmark). The slides were incubated at 4°C in a humidity box for 1-2 weeks. Precipitin lines were observed and photographed in indirect light. Recombinant pneumococcal polysaccharides were estimated by comparison with the intensity of the precipitin lines of the positive controls.

High-Voltage Paper Electrophoresis (HVPE)

Leaf material was harvested, washed, cut into pieces and ground into a fine powder under liquid nitrogen. Samples were stored at −20°C until required. To 10 g fresh weight was added 50 ml 5% (v/v) formic acid and the suspension was incubated with gentle shaking for 2 days at room temperature. This procedure is expected to extract the capsular polysaccharide, but only a small proportion of the leaf cell-wall polysaccharides and starch. The homogenate was filtered through Miracloth and rinsed with 25 ml water and then the combined filtrate was adjusted to pH 4.0 with pyridine. Co-extracted proteins were denatured at 100°C for 60 min, then cooled and pelleted by centrifugation at 1700×*g* for 15 min; the supernatant was freeze-dried. The dried material was washed exhaustively at room temperature in several changes of 82.6% (v/v) ethanol, which dissolves low-MW sugars. The remaining insoluble, polysaccharide-rich material was then air-dried, incubated at 90°C for 30 min in 22.6 ml water and cooled; trifluoroacetic acid (TFA) was then added to a final concentration of 0.36 M, which solubilised the polysaccharide.

For partial hydrolysis of the polysaccharide to yield the relatively acid-resistant, diagnostic dimer (aldobiouronic acid; GlcA–Glc) a portion of the solution was hydrolysed in 2 M TFA at 120°C for 30 min [conditions optimised in preliminary runs with authentic type 3 polysaccharide; data not shown], then dried *in vacuo*. The hydrolysis products were redissolved in water containing a trace of Orange G (internal anionic marker), and a volume (equivalent to 32 mg fresh weight of leaf) was spotted on to Whatman 3 MM paper. The samples were subjected to HVPE at pH 6.5, at 3.0 kV for 60 min [23] and then stained with silver nitrate [21] to reveal sugars. External markers, run on the same sheet, included hydrolysates of (i) purified Type 3 polysaccharide (yielding glucose plus GlcA–Glc), and (ii) the trimer α-D-mannosyl-(1→4)-α-D-glucuronosyl-(1→2)-*myo*-inositol [24] (which yields a comparable dimer, α-D-glucuronosyl-$(1\rightarrow2)$-*myo*-inositol, plus mannose). Other markers were commercial glucose and galacturonic acid. After staining with silver nitrate [21], electrophoretograms were scanned and relevant spots were quantified for grey density in PhotoShop, as described in Supplementary Figure 1 of Parsons *et al* [25].

Immunisation and Challenge

Nine-week-old MF1 female mice (HarlanOlac, Bicester, UK) were given three doses of control plant extract or plant extract containing 2 μg plant-derived pneumococcal polysaccharide per mouse (as estimated by the Ouchterlony method) in 67 μl PBS and 33 μl Imject alum adjuvant (Pierce, Rockford, IL, USA). Mice were immunised intraperitoneally on days 0, 10, 20 and 30. Sham-immunised mice received alum adjuvant containing an irrelevant immunogen (KLH) using the same schedule. Serum samples were obtained by tail bleeding the day before each immunisation. Mice were challenged intraperitoneally with 2.8×10^6 cfu serotype 3 pneumococci on Day 260. The health status of animals was monitored, according to the scheme of Morton *et al* [26].These experiments were done under a project licence from the UK Home Office.

ELISA

Maxisorb ELISA wells (Gibco BRL, Nunc products) were coated with 2 μg/ml purified type 3 pneumococcal polysaccharide (ATCC) in coating buffer (50 nM NaHCO$_3$ pH 9.6, 0.02% (w/v) NaN$_3$) for 16 h at 22°C. After rinsing with PBS, the wells were blocked with PBS + 5% (w/v) dried milk at 37°C for 1 h and washed three times with washing buffer (50 mM TrisHCl pH 7.5, 150 mM NaCl, 0.05% (v/v) Tween20). Mouse sera were diluted 1:100 in blocking buffer and 100 μl added to the wells and incubated, shaking, for 2 hours at 37°C. The plates were washed three times as before and bound antibodies were detected using alkaline phosphatase-conjugated goat anti-mouse IgG secondary antibody (Fc specific, Sigma; diluted 1:5000), and 1 mg/ml p-nitrophenyl phosphate (Sigma) dissolved in 1 M diethanolamine pH 9.8, 0.5 M MgCl$_2$. Absorbance was read at 405 nm after 1 hour at 37°C and IgG concentration determined by reference to a standard curve prepared with murine IgG (Statens Serum Institute).

Western blot analysis

Western blotting was performed on leaf extracts as described previously [27].

Statistical analysis

Statistical analysis was performed using GraphPad Prism 5 (GraphPad, San Diego, CA, USA). The differences in antibody titres from mice immunised with transgenic or wildtype plant extracts were analysed using a student's T-test. Survival data were analysed by the Kaplan-Meier survival curve analysis.

Supporting Information

Figure S1   Confirmation of the stable transformation of tobacco plants with the *cps3S* gene. Plant RNA was used as a template for reverse transcriptase PCR (RT-PCR) (Lane 1: wild type plants; Lane 2: transformed plants.) using *cps3S*-specific primers. RT-PCR products were analysed by agarose gel electrophoresis. The results show the presence of the *cps3S* gene in the transformed plants (Lane 2) but not the wild type plants.

Acknowledgments

We would like to acknowledge the contribution of Professor Whitelam who died during the production of this manuscript. We thank William. S. Hawes for his technical contribution and help with plant transformation. SCF thanks the UK Biotechnology and Biological Sciences Research Council for support.

Author Contributions

Conceived and designed the experiments: CMS SCF KCG ISR PWA GCW. Performed the experiments: CMS SCF AJFP SG MG ISR PWA. Analyzed the data: CMS SCF AJFP SG ISR. Contributed reagents/materials/analysis tools: SCF KCG ISR PWA. Wrote the paper: CMS SCF ISR PWA.

References

1. Gilbert C, Robinson K, LePage RWE, Wells J (2000) Heterologous expression of an immunogenic pneumococcal type 3 capsular polysaccharide in *Lactococcus lactis*. Infect Immun 68: 3251–3260.
2. Deuren M, Brandtzaeg P, van der Meer JW (2000) Update on meningococcal disease with emphasis on pathogenesis and clinical management. Clin Microbiol Rev 13: 144–166.
3. CDC (2005) Direct and indirect effects of routine vaccination of children with 7-valent pneumococcal conjugate vaccine on incidence of invasive pneumococcal disease—United States, 1998–2003. 54: 893–897.
4. Kunze R, Frommer WB, Flugge UI (2002) Metabolic engineering of plants: the role of membrane transport. Metab Eng 4: 57–66.
5. Gao Y, Ma Y, Li M, Cheng T, Li SW, et al. (2003) Oral immunization of animals with transgenic cherry tomatillo expressing HBsAg. World J Gastroenterol 9: 996–1002.
6. Mason HS, Arntzen CJ (1995) Transgenic plants as vaccine production systems. Trends Biotechnol 13: 388–392.
7. Rigano MM, Alvarez ML, Pinkhasov J, Jin Y, Sala F, et al. (2004) Production of a fusion protein consisting of the enterotoxigenic Escherichia coli heat-labile toxin B subunit and a tuberculosis antigen in *Arabidopsis thaliana*. Plant Cell Rep. 22: 502–508.
8. Arrecubieta C, Garcia E, Lopez R (1996) Demonstration of UDP-glucose dehydrogenase activity in cell extracts of Escherichia coli expressing the pneumococcal cap3A gene required for the synthesis of type 3 capsular polysaccharide. J Bacteriol 178: 2971–2974.
9. Jennings H (1983) Capsular polysaccharides as human vaccine. Adv Carbo Chem 41: 155–208.
10. Sharples SC, Fry SC (2007) Radioisotope ratios discriminate between competing pathways of cell wall polysaccharide and RNA biosynthesis in living plant cells. Plant J. 52: 252–262.
11. Kalin M (1998) Pneumococcal serotypes and their clinical relevance. Thorax 53: 159–162.
12. Douglas RM, Paton JC, Duncan SJ, Hansman DJ (1983) Antibody response to pneumococcal vaccination in children younger than five years of age. J. Infect. Dis. 148: 131–137.
13. Imöhl M, Reinert RR, van der Linden M (2010) Temporal variations among Invasive pneumococcal disease serotypes in children and adults in Germany (1992–2008). Internat J Microbiol 2010: 1–15.
14. Snaidack DH, Schwartz B, Lipman H, Bogaets J, Butler JC, et al. (1995) Potential interventions for the prevention of childhood pneumonia: geographic and temporal differences in serotype and serogroup distribution of sterile site pneumococcal isolates from children. Pediatr. Infect. Dis. J. 14: 503–509.
15. Hausdorff WP, Bryant J, Paradiso PR, Siber GR (2000) Which pneumococcal serotypes cause the most invasive disease: implications for conjugate vaccine formulation and use. Clin Infect Dis 30: 100–121.
16. Levine MM, Levine OS (1997) Influence of disease burden, public perception and other factors on new vaccine development, implementation and continued use. Lancet 350: 1386–1392.
17. Hajdukiewicz P, Svab Z, Maliga P (1994) The small, versatile pPZP family of Agrobacterium binary vectors for plant transformation. Plant Molec Biol 25: 989–994.
18. Lund P, Dunsmuir P (1992) A plant signal sequence enhances the secretion of bacterial ChiA in transgenic tobacco. Plant Mol Biol 18: 47–53.
19. Draper J, Scott R, Armitage P, Walden R (1988) Plant genetic transformation and gene expression: a laboratory manual. Oxford, Blackwell Scientific Publications.
20. Sawada H, Ieki H, Matsuda I (1995) PCR detection of Ti and Ri plasmids from phytopathogenic Agrobacterium strains. Appl Environ Microbiol 61: 828–831.

21. Fry SC (1988). The Growing Plant Cell Wall: Chemical and Metabolic Analysis. New York, John Wiley and Sons.

22. Ouchterlony O, Nilsson,L (1973) Immunodiffusion and immunoelectrophoresis. Oxford, Blackwell Scientific Publications.

23. Fry SC (2011). High-voltage paper electrophoresis (HVPE) of cell-wall building blocks and their metabolic precursors (Springer, New York, High-voltage paper electrophoresis (HVPE) of cell-wall building blocks and their metabolic precursors. The Plant Cell Wall Methods and Protocols. Z. Popper. New York, Springer: 55–80.

24. Smith CK, Hewage CM, Fry SC, Sadler IH (1999) ±-d-Mannopyranosyl-(1->4)-±-d-glucuronopyranosyl-(1->2)-myo-inositol, a new and unusual oligosaccharide from cultured rose cells. Phytochem 52: 387–396.

25. Parsons HT, Yasmin T, Fry SC (2011) Alternative pathways of dehydroascorbic acid degradation in vitro and in plant cell cultures: novel insights into vitamin C catabolism. Biochem J 440: 375–383.

26. Morton DB, Griffiths PH (1985) Guidelines on the recognition of pain, distress and discomfortin experimental animals and an hypothesis for assessment Vet Record 116: 431–436.

27. McNulty C, Thompson J, Barrett B, Lord L, Andersen C, et al (2006) The cell surface expression of group 2 capsular polysaccharides in *Escherichia coli*: the role of KpsD, RhsA and a multi-protein complex at the pole of the cell. Mol. Microbiol. 59: 907–922.

6

High-Throughput Construction of Intron-Containing Hairpin RNA Vectors for RNAi in Plants

Pu Yan[1,2,4], Wentao Shen[2], XinZheng Gao[3], Xiaoying Li[2], Peng Zhou[2]*, Jun Duan[1]*

1 South China Botanical Garden, Chinese Academy of Sciences, Guangzhou, China, **2** Institute of Tropical Bioscience and Biotechnology, Chinese Academy of Tropical Agricultural Science, Haikou, China, **3** Department of Basic Medical Science, Hainan Medical College, Haikou, China, **4** Graduate University of Chinese Academy of Sciences, Beijing, China

Abstract

With the wide use of double-stranded RNA interference (RNAi) for the analysis of gene function in plants, a high-throughput system for making hairpin RNA (hpRNA) constructs is in great demand. Here, we describe a novel restriction-ligation approach that provides a simple but efficient construction of intron-containing hpRNA (ihpRNA) vectors. The system takes advantage of the type IIs restriction enzyme BsaI and our new plant RNAi vector pRNAi-GG based on the Golden Gate (GG) cloning. This method requires only a single PCR product of the gene of interest flanked with BsaI recognition sequence, which can then be cloned into pRNAi-GG at both sense and antisense orientations simultaneously to form ihpRNA construct. The process, completed in one tube with one restriction-ligation step, produced a recombinant ihpRNA with high efficiency and zero background. We demonstrate the utility of the ihpRNA constructs generated with pRNAi-GG vector for the effective silencing of various individual endogenous and exogenous marker genes as well as two genes simultaneously. This method provides a novel and high-throughput platform for large-scale analysis of plant functional genomics.

Editor: Luis Herrera-Estrella, Centro de Investigación y de Estudios Avanzados del IPN, Mexico

Funding: This work is supported by the National Natural Science Foundation of China (Grant no. 30960218) and the National Nonprofit Institute Research Grant of CATAS-ITBB (ITBB110212). The funders had no role in study design, data collection and analysis, decision to publish, or preparation of the manuscript.

Competing Interests: The authors have declared that no competing interests exist.

* E-mail: zhp6301@126.com (PZ); duanj@scib.ac.cn (JD)

Introduction

After double-stranded RNA was discovered as the trigger of RNA interference (RNAi) [1], RNAi has become one of the most powerful tools for the analysis of gene function [2–4]. Hairpin RNA (hpRNA) constructs are commonly used to induce degradation of target genes through RNAi mechanisms [5]. In plants, intron-containing hairpin RNA (ihpRNA) with an intron as a spacer sequence shows the highest gene silencing efficiency [6]. Therefore, ihpRNA constructs have been widely used for gene silencing in plants. With the explosive release of gene sequences and genomic sequences in plants, a high-throughput and cost-efficient system for making ihpRNA constructs is in great demand.

To facilitate the generation of ihpRNA constructs, several methods have been reported. The traditional ligase-based vectors such as pHANNIBAL and pKANNIBAL were first used to generate ihpRNA constructs [6]. This method requires several rounds of restriction and ligation, and is therefore, tedious and time-consuming. Alternatively, GATEWAY cloning system-based RNAi vectors such as the pHELLSGATE series and the pIPK series have been widely used for generating ihpRNA constructs [7–9]. In this system, a single PCR product, from primers containing the appropriate attB1 and attB2 sites, can be simultaneously recombined into the vector to form the two arms of the hairpin. Thus, it is simpler and more rapid than the traditional ligase-based system. However, this method usually requires two cloning steps, including BP and LR reactions to generate the final ihpRNA. In addition, the reagents used in the GATEWAY system are comparatively expensive, especially for

researchers in developing countries. Besides the traditional ligase-based system and the GATEWAY system, several overlap extension (OE) PCR approach were developed to construct the ihpRNA cassettes. A DA-ihpRNA method, for amplifying an ihpRNA construct directly from genomic DNA, was described by Xiao et al [10]. Our laboratory previously reported a mixed one-step OE-PCR method to generate ihpRNA constructs by assembling two inverted repeat fragments of the target genes and an intron in one tube [11]. Chen et al used a similar strategy to construct ihpRNA cassettes which were immediately inserted into the final plant expression vectors by TA cloning [12]. These PCR-based methods are not only simpler and faster than the conventional ligase-based system, but also more cost-effective than the GATEWAY system. However, they are sometimes suffer from low efficient because of the PCR suppression effect caused by the self-annealing of two inverted repeats, particularly for those target sequences that contain GC-rich regions and/or can form secondary structures. Recently, a one-step method for the generation of hpRNA constructs with 100% efficiency based on ligation independent cloning (LIC) using pRNAi-LIC vector, has been developed [13]. This method is simple and fast, as it can generate ihpRNA constructs for plant RNA silencing by one-step transformation. However, it still needs two rounds of PCR for amplifying the two inverted repeats with different adaptors and a step of restriction enzyme digestion for vector linearization before using the ligation- independent cloning.

To develop a simpler system for making RNAi constructs, we have adopted the Golden Gate (GG) cloning approach. This cloning strategy relies on the type IIs restriction enzymes, which cut outside of their recognition sequence resulting in DNA overhangs that are arbitrarily defined. This property has been used to develop protocols for efficient directional assembly of multiple DNA fragments in a single ligation reaction [14,15]. Furthermore, with proper design of the cleavage sites, two digested fragments can be ligated to generate a product lacking the original restriction site. Thus, the two steps of digestion and ligation can be replaced by a single restriction-ligation step. Based on these properties of Golden Gate cloning, we describe here, a one-step and cost-effective method for generating ihpRNA constructs via our new plant RNAi vector pRNAi-GG. Using this system, a plant ihpRNA expression vector can be made from a single PCR product of the gene of interest by one-tube restriction-ligation reaction and one-step transformation. This method has been successfully applied to generate ihpRNA constructs for the effective silencing of various individual endogenous and exogenous marker genes as well as two genes simultaneously. Our results suggest that this novel method provides a high-throughput and reliable platform for making ihpRNA constructs, thereby, may facilitate a large-scale functional genomics analysis in plants.

Results

Development of a Golden Gate strategy for making ihpRNA constructs

To adopt a Golden Gate strategy for the rapid and high-throughput cloning of the intron-containing inverted repeat inserts into the vectors, we developed a new plant T-DNA RNAi vector, pRNAi-GG. This vector consists of a *ccdB* gene - Pdk intron - *ccdB* gene cassette between the duplicated CaMV 35S promoter and Nos terminator (Fig. 1A). One BsaI recognition site, in the original *ccdB* gene, was eliminated by site-specific mutation without changing the coded amino acid sequence. Two BsaI sites flank each *ccdB* gene and are positioned such that the recognition sites are eliminated from the vector after proper digestion and ligation of products. The cleavage site sequences of BsaI on the left of the first *ccdB* gene are the same to that on the right of the second *ccdB*, but with different orientation, and the same to the other two BsaI sites. Thus, a single PCR product can replace the two *ccdB* genes simultaneously at both sense and antisense orientations to form the two arms of the hairpin (Fig. 1B). To make the ihpRNA construct, the target region of the gene of interest is amplified with BsaI recognition sites flanked at both ends, and the cleavage site sequences (adaptor for pRNAi-GG) are complementary to the appropriate sequences on the vector. The universal form for primer design is 5′- protective bases - BsaI - adaptor for pRNAi-GG - gene specific sequence-3′. Incubation of the purified PCR product and pRNAi-GG in the presence of BsaI enzyme and T4 ligase generates the desired vector, which is stable due to the absence of the original BsaI sites (Fig. 1B). The mixture can then be transformed directly into standard E. coli strains such as DH5α and TOP10 (but not in DB3.1), and only recombinants containing both arms would be recovered. Since pRNAi-GG is a binary vector, the recombinant pRNAi-GG constructs can be transformed directly into Agrobacterium for plant transformation.

Fast and efficient cloning of plant ihpRNA constructs with pRNAi-GG

To test the efficiency of our new plant RNAi vector pRNAi-GG for making ihpRNA constructs, a target region with 337 bp in the marker gene *GFP* (GenBank access number U87973) was amplified, and mixed with pRNAi-GG in the presence of BsaI enzyme and T4 ligase. To the efficiency of cloning at different restriction-ligation time, the mixture was incubated at 37°C for 0.5–2 h or for 20–35 cycles (2 min 37°C+5 min 16°C). After 5 minutes at 80°C for heat inactivation, this mixture was transformed into E. coli DH5α cells and transformants were selected on LB plates containing both kanamycin and chloramphenicol. As shown in Table 1, recombinant colonies increased from 0.5 to 2 h or form 20 to 35 cycles. The efficiency of cloning at 37°C for 2 h is higher than 20 cycles, but a little lower than 35 cycles. Considering the time efficiency, incubated at 37°C for 2 h is commended to be the optimal condition. For each transformation, 12 clones (when less than 12, pick all) were identified by PCR using vector specific primes P21 and P22, and insert reverse primer P12 (Fig. S1, Table S1), which can amplify the two arms simultaneously with a difference of 267 bp in length. All the clones showed the expected bands by agarose gel electrophoresis, as part of the results was shown in Fig. 2A. 12 clones were also identified by restriction enzymes digestion using BamHI and SacI (Fig. S1). As shown in Fig. 2B, all of the colonies contain the correct inserts, and the two different sizes of inserts in digestion are due to the different orientation of the intron. The intron orientation can also be easily identified by colony PCR using a combination of two intron specific primers P24 and P25 and one vector specific primer P21 (Table S1, Fig. 2C). The vector–insert junctions, of six colonies, were sequenced using P22 and P23 primers and all of them contained the correct sequences as expected.

In pRNAi-GG, a chloramphenicol-resistant gene was inserted into the Pdk intron to allow for selecting the presence of the intron (Fig. 1A). To test the importance of the chloramphenicol-resistant gene, in recovering the correct recombination products, we compared the recombinant plasmids in the presence and absence of chloramphenicol. Over 50 chloramphenicol-resistant clones were tested by PCR, and all was found to have the correct structure. While in the absence of chloramphenicol, 10–20% of the clones were found to don't have the intron. Sequences from three illegitimate recombinants showed products of recombinant of the vector itself at the position of two Nos terminators (data not shown). Therefore, we conclude that a selectable marker in the intron is necessary for the generation of ihpRNA constructs when using this system.

In order to demonstrate that the use of vector pRNAi-GG is reproducible and reliable to make ihpRNA constructs with the Golden Gate method, we also successfully generated other recombinant pRNAi-GG constructs, including the target regions from the *N. benthamiana* PDS gene (*NbPDS*) (GenBank access number EU165355) and marker gene *GUS* (GenBank access number AY292368), as well as another target region from *GFP* with an internal BsaI site (Fig. 2D). To bypass the internal BsaI site problem, a two step ligation was employed. After the first step of restriction-ligation for 2 h, the enzymes were heat inactivated, and then fresh ligase was added to the mix for the second ligation for 1 h.

Silencing of exogenous marker genes and endogenous NbPDS by agroinfiltration

Coinfiltration has been widely used to determine the functionality of RNAi constructs for silencing of marker genes. In order to demonstrate that the ihpRNA constructs (described above) can be used for gene silencing, we first tested the ability of pRNAi-GUS and pRNAi-GFP for gene silencing by Agrobacterium-mediated transient expression. Agrobacterium cultures, each containing pBIN19-GUS and pBIN19-GFP were mixed with Agrobacterium cultures, each containing the pRNAi-GUS or pRNAi-GFP (for control), and pRNAi-GFP or pRNAi-GUS (for control), respectively. The mixed Agrobacterium cultures were infiltrated into

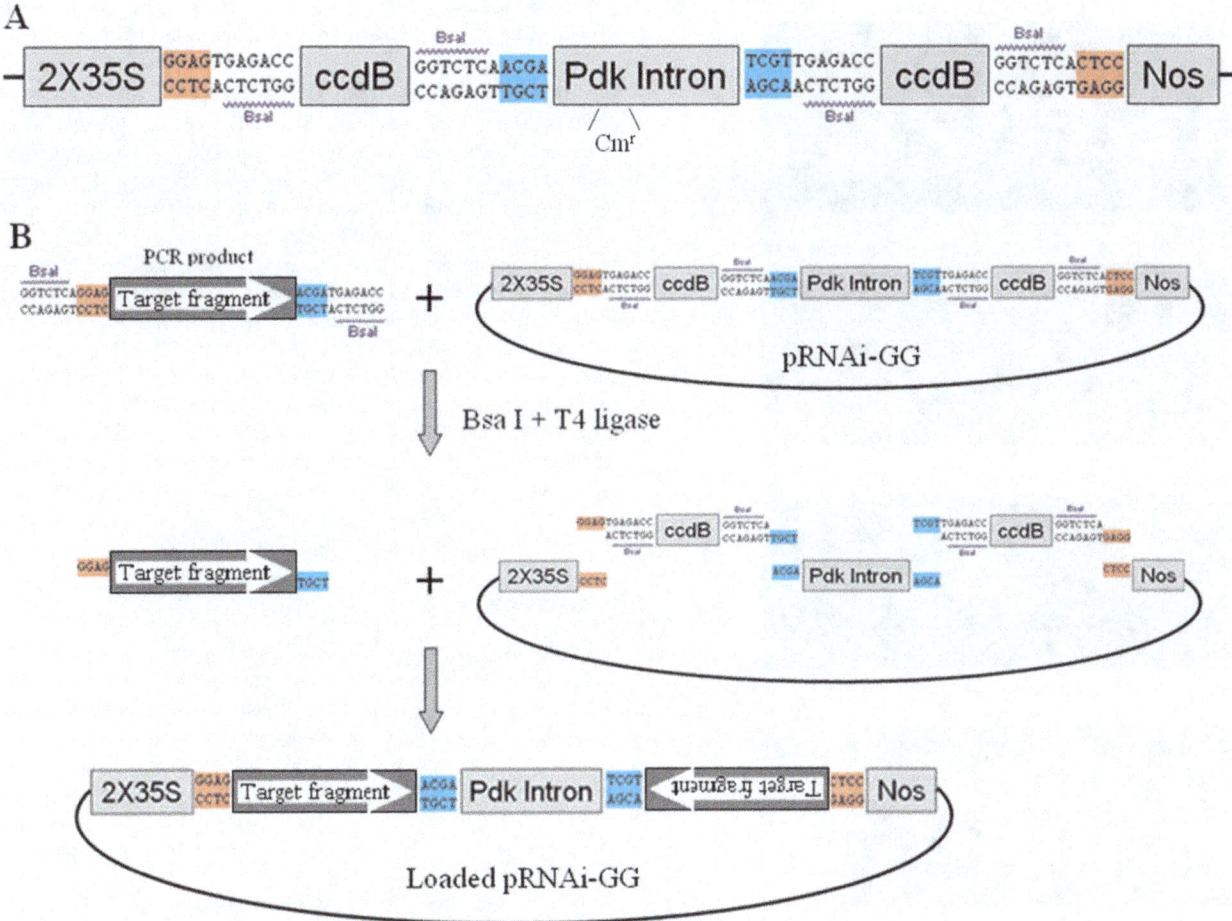

Figure 1. Schematic diagram of ihpRNA construction with pRNAi-GG (Golden Gate) vector. (A) The cassette of pRNAi-GG. The duplicated 35S CaMV promoter, two copies of *ccdB* gene, the Pdk intron, four BsaI recognition sites with designed adaptors (cleavage site sequences of BsaI) are cloned between the HindIII and SacI of T-DNA vector pBI121. Adaptors with the same color have the same sequences but opposite orientation. A chloramphenicol-resistant gene (Cm^r) was contained in the Pdk intron. (B) One-step construction of ihpRNA. The target fragment of the gene of interest is PCR amplified using gene-specific primers carrying BsaI sites and adaptors complementary to the appropriate sequences on the vector. The purified PCR product is mixed, in one tube, with pRNAi-GG vector, BsaI enzyme and T4 ligase for a one-step restriction-ligation reaction.

different parts of the same leaves of *N. benthamiana* plants. The infiltration areas for control gave a strong blue GUS-staining signal (Fig. 3A) or green uorescence (Fig. 3B) at 3 d post-agroinfiltration. However, the infiltration areas for gene silencing gave much weaker signals. To confirm the knock-down of the test marker genes at the molecular level, we performed real time PCR. The results showed that GUS and GFP mRNA levels were reduced to 24% and 36%, respectively, compared to the control samples (Fig. 3C). In addition to the exogenous marker genes, the mRNA level of *NbPDS* was also reduced by the infiltration of Agrobacterium cultures containing pRNAi-PDS. The result of real time PCR showed that the expression of *NbPDS* at 3 d post-agroinfiltration was reduced to 17%, compared to the control samples (Fig. 3C). However, the typical phenotype of *PDS* silencing, such as the photo-bleaching in VIGS (virus-induced gene silencing), was not observed in the infiltration areas even 15 days post infiltration. The reason may be that the RNA interference mediated by agroinfiltration is transient and reacts on small area of the plant, which couldn't reach the level to change the phenotype. Usually, Agrobacterium-mediated transient expression of a transgene achieves highest level in 2–3 days following

argoinfiltration, after which the expression level rapidly decreases [16].

The effect of intron orientation of the recombinant pRNAi-GG on silencing efficiency

Due to the same cleavage sites of BsaI at both sides of Pdk intron in the pRNAi-GG vector, we observed that the recombinant ihpRNA vectors included plasmids in which the intron retained its forward or reversed orientation with respect to the promoter. The intron orientation can be easily identified either by colony PCR using a combination of two intron specific primers P24 and P25 and one vector specific primer P21 (Table S1, Fig. 2D) or by restriction enzymes digestion using BamHI and SacI (Fig. 2C). The position of the primers and the restriction sites on pRNAi-GG were showed in Fig. S1. To test the effect of intron orientation of the recombinant pRNAi-GG on gene silencing, pRNAi-GFP and pRNAi-PDS with introns in either sense or antisense orientations were selected and tested for transient expression. As shown in Fig. 4, no obvious difference in silencing efficiency was observed between the recombinant pRNAi-GG with different intron orientations. Therefore, the selection of

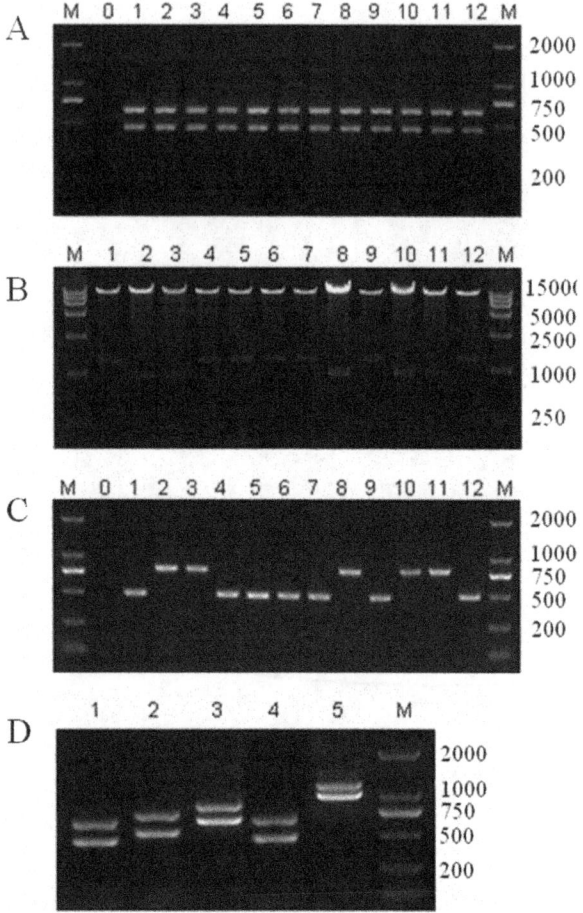

Figure 2. Zero-background cloning of intron-containing hairpin RNA (ihpRNA) constructs. (A) Colony PCR of pRNAi-GFP using primes P21, P22 and P12, which can amplify the two arms simultaneously with a difference of 267 bp in length. (B) Double digestion identification of pRNAi-GFP with BamHI and SacI. The two different sizes of inserts are due to the different orientation of the intron. (C) Identification of the intron orientation of pRNAi-GFP by colony PCR using primes P21, P24 and P25. Lanes 1–12 in (A–C) represent 12 independent colonies, Lane 0 is the negative control and M is DNA marker. (D) Colony PCR results of 5 ihpRNA constructs, lines 1–5, for silencing GFP, GUS, NbPDS, GFP with an internal BsaI site and GFP/NbPDS, respectively. The lengths of the inverted arms are 337 bp, 460 bp, 530 bp, 368 bp and 867 bp, respectively. All constructs gave the predicted bands.

intron orientation is not critical, and if necessary, it can be easily identified by colony PCR or restriction enzyme digestion.

Simultaneous silencing of *GFP* and *NbPDS*

Considering the extremely high efficiency of Golden Gate cloning, it was logical to test whether multiple fragments could be inserted to pRNAi-GG. For this purpose, we developed a strategy to construct an ihpRNA targeting two distinct genes in a single cloning step (Fig. 5A). Both of the inserts are flanked with BsaI sites whose cleavage site sequences are designed such that the two inserts can be ligated after the digestion. The efficiency of cloning at different restriction-ligation time was tested, and the results showed the optimal condition is incubation for 50 cycles of 2 minutes at 37°C and 5 minutes at 16°C (Table 1). By employing this strategy, an ihpRNA, namely pRNAi-GFP/PDS, targeting both *GFP* and *NbPDS* was successfully constructed (Fig. 2D). Agrobacterium cultures containing the GFP and the pRNAi-GFP/PDS were coinfiltrated into the leaves of *N. benthamiana* plants. The simultaneous silencing of *GFP* and *NbPDS* was observed 3 d post-agroinfiltration (Fig. 5B).

Discussion

RNA silencing induced by hpRNA is one of the most powerful and popular tools in reverse genetics. ihpRNA constructs have been widely used for the analysis of gene function in plants. With the explosive release of gene sequences and whole genome sequences in plants, how to rapidly determine the gene function on a large scale is a daunting challenge. In this study, we developed a novel strategy and a new plant RNAi vector, pRNAi-GG, for making ihpRNA constructs rapidly with only one restriction-ligation cloning step. We have demonstrated that the ihpRNA constructs, generated with this RNAi vector based on Golden Gate cloning method, can efficiently induce suppression of target genes. This novel method and our plant RNAi vector provide a high-throughput platform for the large-scale plant functional genomic analyses.

The method we described has many advantages over both the traditional ligase-based and GATEWAY systems. First, ihpRNA constructs can be produced in one tube with one restriction-ligation step. The cloning process of ihpRNA constructs can be completed within several hours. Second, our method requires only a single primer pair and ordinary reagents which are relatively cost-savings. Third, with both the *ccdB* and chloramphenicol-resistant genes, our new vector pRNAi-GG can be used to make ihpRNA constructs with zero-background cloning efficiency. Fourth, pRNAi-GG is a binary vector; thereby, allowing for the recombinant pRNAi-GG constructs to be directly transformed

Table 1. Efficiency of cloning of pRNAi-GG at different restriction-ligation times.

Construct	neg	0.5 h 37°C	1 h 37°C	2 h 37°C	20 cycles	35 cycles	50 cycles
pRNAi-GFP	0	16	24	37	35	48	NT
pRNAi-GFP	0	9	22	45	26	43	NT
pRNAi-GFP/PDS	0	NT	NT	2	2	13	27
pRNAi-GFP/PDS	0	NT	NT	5	3	8	16

The number of recombinant colonies for each transformation was counted. Restriction-ligation was performed with continuous incubation at 37°C or for 20–50 cycles (2 min 37°C+5 min16°C). Cloning for pRNAi-GFP and pRNAi-GFP/PDS were performed in duplicate. The negative control was performed without BsaI enzyme. NT, not tested.

For each transformation, 12 clones (when less than 12, pick all) were identified by PCR using vector specific primes P21 and P22, and insert reverse primer P12 (for pRNAi-GFP) or P16 (for pRNAi-GFP/PDS). All the clones showed the expected bands, as part of the results was shown in Fig. 2A.

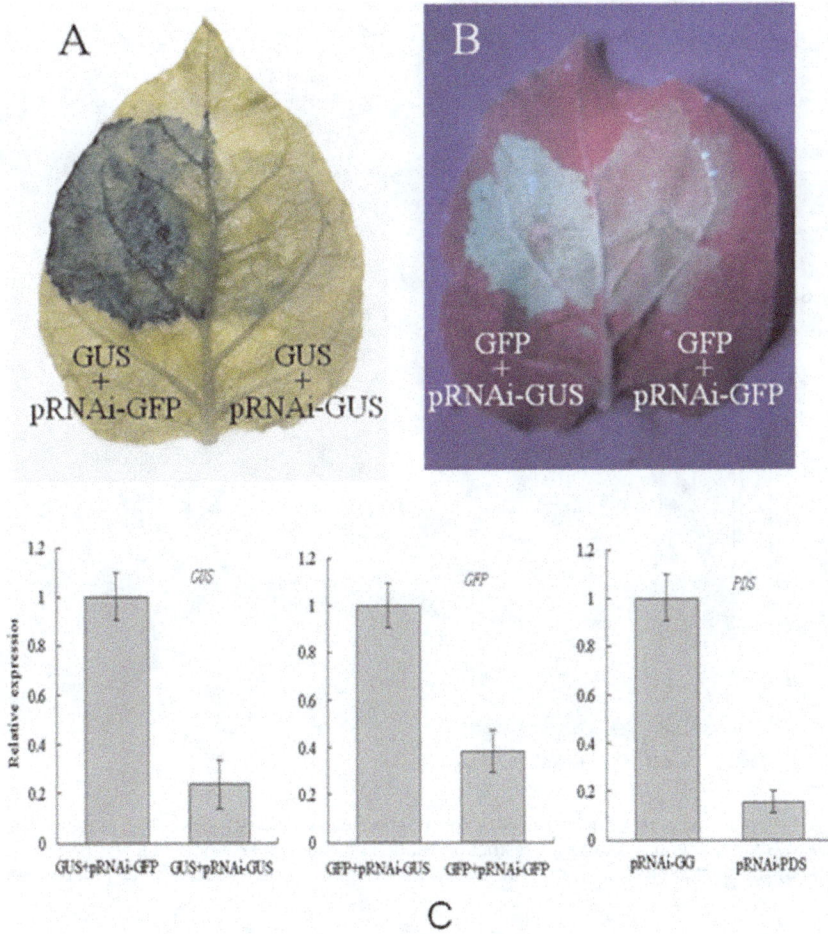

Figure 3. Silencing of two marker genes GUS and GFP and endogenous NbPDS by transient expression of the ihpRNA constructs. (A) Silencing effect of GUS ihpRNA on transient expression of GUS gene. Agrobacterium cultures containing pBIN19-GUS were mixed with Agrobacterium containing pRNAi-GFP (left) or pRNAi-GUS (right), and infiltrated into different parts of the same leaves of *N. benthamiana* plants. (B) Silencing effect of green uorescent protein (GFP) ihpRNA on transient expression of GFP gene. Agrobacterium cultures containing pBIN19-GFP were mixed with Agrobacterium containing pRNAi-GUS (left) or pRNAi-GFP (right), and infiltrated into different parts of the same leaves of *Nicotiana benthamiana* plants. (C) Real time PCR was performed to analyze the silencing effect of GUS, GFP and NbPDS genes. Error bars represent standard deviations (SD) of three independent experiments.

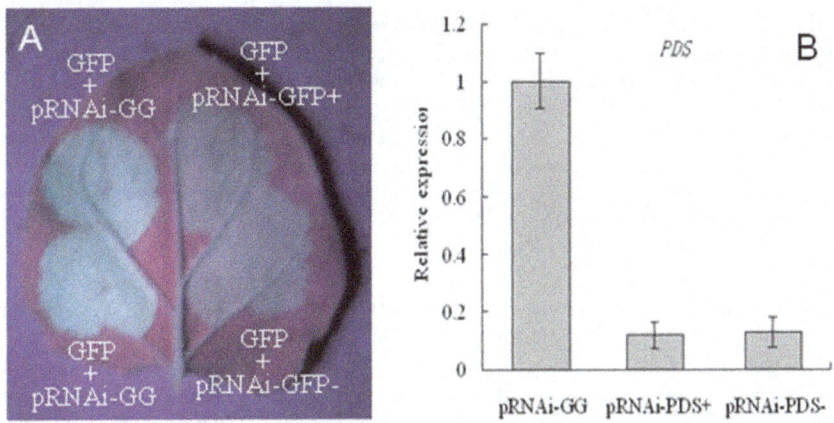

Figure 4. The effect of intron orientation of recombinant pRNAi-GG on silencing efficiency. (A) Silencing of *GFP* by pRNAi-GFP+ and pRNAi-GFP−. (B) Real time PCR was performed to analyze the silencing of *NbPDS* by pRNAi-PDS+ and pRNAi-PDS−. + and − indicate the sense and antisense orientations of intron of recombinant pRNAi-GG, respectively. Empty vector pRNAi-GG was used as a control. Error bars indicate SD of three independent experiments.

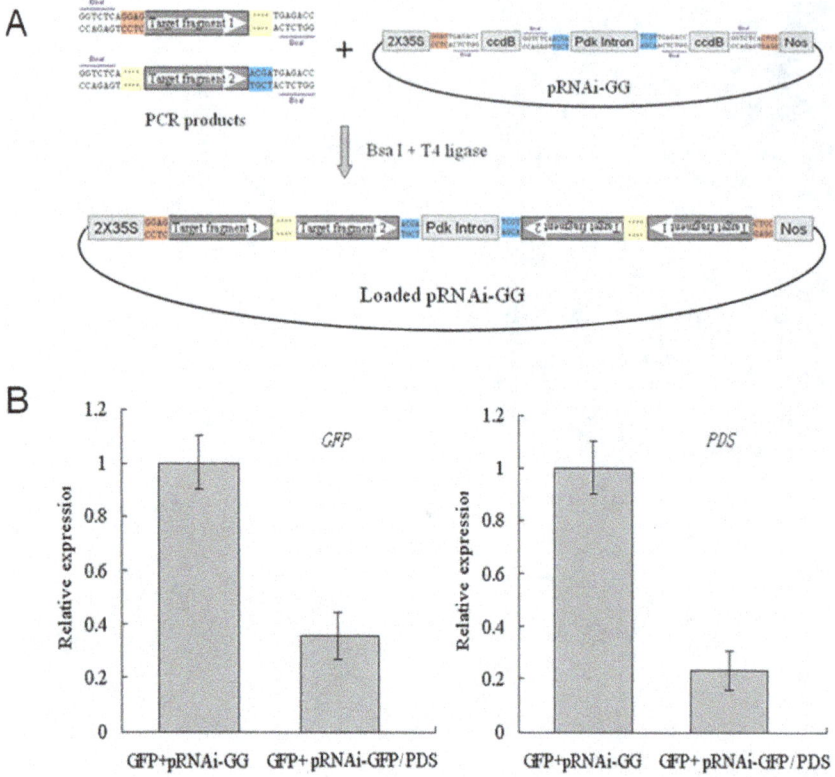

Figure 5. Simultaneous silencing of two genes. (A) Schematic diagram of hybrid ihpRNA construction with pRNAi-GG vector. Two PCR products representing genes of interest can be cloned into pRNAi-GG simultaneously in a single restriction-ligation reaction. (B) Real time PCR was performed to analyze the simultaneous silencing of *GFP* and *NbPDS* by transient expression using pRNAi-GFP/PDS. Error bars represent SD of three independent experiments.

into Agrobacterium for plant transformation. Taken together, these advantages greatly facilitate and enhance the high-throughput applications for large scale gene functional analysis.

Recently, a one-step method for the generation of hpRNA constructs with 100% efficiency based on ligation independent cloning (LIC) using pRNAi-LIC vector, has been developed [13]. This method has almost all of the advantages described above. However, it still requires two rounds of PCR for amplifying the two inverted repeats and a step of restriction enzyme digestion for vector linearization before cloning. On the other hand, our method only calls for a single PCR product, which can be immediately cloned into pRNAi-GG without any other requirements. Therefore, our method is simpler and faster. At least, it is a good alternative system for making RNAi constructs with high-throughput.

Due to the same cleavage sites of BsaI on both sides of Pdk intron in the pRNAi-GG vector, we observed that the recombinant ihpRNA vectors included plasmids in which the intron either was reversed or retained its forward orientation with respect to the promoter. Similar result was also observed in the pHELLSGATE vector based on the GATEWAY system [7,17]. The reverse orientation of intron in the recombinant ihpRNA was usually considered to have negative effect in silencing efficiency, so in some RNAi vectors the spacer fragments were designed to consist of two introns in opposite orientations to ensure one of the introns is in the forward orientation [7]. However, no data was reported for the effect of intron orientation on gene silencing. We compared the silencing efficiency of ihpRNA vectors with both sense and antisense of intron orientations, and found that the orientation of

intron has no obvious effect on silencing efficiency. Therefore, the selection of intron orientation is not critical, and if necessary, it can be easily identified by colony PCR or restriction enzyme digestion.

Our method is based on type IIs restriction enzyme BsaI. Therefore, one potential limitation of this method might come from the occasional presence of one to several internal BsaI site(s) in the gene of interest. In order to solve this problem, we performed the restriction-ligation for 2 h, followed by heat inactivation of the enzymes, and then another 1 h ligation by adding fresh ligase. As a result, an ihpRNA containing an internal BsaI restriction site was successfully constructed (Fig. 2D). We found this to be extremely efficient since the restriction-ligation, in the first step, stably ligates the insert fragments at the ends of the vector, and the second ligation only needs to religate the overhangs at the internal BsaI sites [15]. Since the BsaI restriction site has a 6-base pair recognition sequence, it is expected to be present on average every four kb of the genome. Since the stem length of ihpRNA is usually around 500 bp, the probability of housing more than one BsaI site is very unlikely. Thus, the occasional presence of internal BsaI site will not limit the application of this method.

Golden Gate cloning is based on the use of type IIs restriction enzymes and allows the assembly of multiple DNA fragments in a restriction-ligation reaction. Several protocols based on Golden Gate cloning have been reported for DNA Shuffling [14], assembly of genetic modules [18,19], long-length DNA [20] and transcription activator-like (TAL) effectors [21–24]. In this study, we adopted Golden Gate cloning to make ihpRNA constructs for high-throughput. In our method, the recombinant could be immediately generated between PCR products and the destination plasmid, rather than the entry and destination plasmids in the original Golden Gate cloning. This means that the subcloning of PCR product is unnecessary, which further enhances the high-throughput construction of ihpRNA.

In summary, we have developed a simple, rapid but reliable method for making ihpRNA constructs. This method proves to be promising for a wide application for the large-scale analysis of plant functional genomics.

Methods

Plant and plasmid materials

Wild-type *Nicotiana benthamiana* was used for the analysis of transient silencing. pBIN19-GUS and pBIN19-GFP are pBIN19-based and used for the expression of GUS and GFP, respectively. Both plasmids were kindly provided by Dr. Mestre [25]. pBI121 is a binary vector for plant transformation [26].

Plant RNAi vector pRNAi-GG Construction

pRNAi-GG vector was generated using the T-DNA vector pBI121 as the skeleton. The duplicated CaMV 35S promoter was amplified using pSATN-cEYFP-C1 [27] as template. Two copies of *ccdB* gene and the Pdk intron-containing chloramphenicol resistant gene were PCR amplified using pRNAi-LIC [13] as template. The BsaI site in the original *ccdB* gene was removed by overlap PCR method without changing its amino acid sequence, and both of the modified *ccdB* genes are flanked with BsaI sites. Sequences of these primers are listed in Table S1. All four PCR products were cloned into T-vector pMD-18T and then digested with HindIII and ApaI, ApaI and SalI, SalI and XbaI, XbaI and SacI, respectively. Finally, the four digested fragments were ligated into HindIII- and SacI-digested pBI121, in one tube, to generate pRNAi-GG. pRNAi-GG was used to generate all ihpRNA constructs for the silencing of genes used in this sudy, and maintained in E. coli strain DB3.1, in which the *ccdB* gene is not lethal. The full-length sequence and detailed annotation of pRNAi-GG was deposited in GenBank (access number JQ085427).

Golden Gate cloning with pRNAi-GG for making ihpRNA constructs

Only a single PCR product is required to make the ihpRNA construct for silencing the gene of interest with pRNAi-GG vector by Golden Gate cloning. The product is amplified by the oligo pair: 5′- acca ggtctc aggag - gene specific forward primer-3′ and 5′-acca ggtctc atcgt- gene specific reverse primer-3′ (5′- protective bases - BsaI - adaptor for pRNAi-GG - gene specific sequence-3′). The Golden Gate reaction for making ihpRNA constructs is set up by pipetting into a tube 50 ng purified PCR product, 200 ng pRNAi-GG vector, 5 units BsaI enzyme (NEB) and 10 units T4 DNA ligase (Promega, high concentration ligase - 20 u/μl) in a total volume of 10 μl in 1× ligation buffer (Promega). The restriction-ligation is incubated at 37°C for 2 hours, followed by incubation for 5 minutes at 50°C (final digestion) and then 5 minutes at 80°C (heat inactivation). Then 5–10 μl of the mixture was transformed into E. coli DH5α competent Cells and plated on LB medium containing 25 mg/L kanamycin and 5 mg/L chloramphenicol to select the recombinants. The ihpRNA constructs for silencing GUS, GFP, and NbPDS genes were generated by this restriction-ligation reactions and named as pRNAi-GUS, pRNAi-GFP and pRNAi-NbPDS, respectively. These constructs were confirmed by PCR and DNA sequencing. Sequences of these primers are listed in Table S1.

Transient expression by agroinfiltration

Agroinfiltration was performed following previously reported procedures with a slight modification [28]. *Nicotiana benthamiana* plants were grown in a growth chamber under standard conditions at 25°C under 16-h-light/8-h-dark cycle. All ihpRNA constructs were transformed into Agrobacterium strain GV3101. A single colony of each transformed Agrobacterium was used to inoculate 5 mL YEP medium (Bacto-Trypton, 10 g/l; yeast extract, 10 g/l; NaCl, 5 g/l; pH 7.0) supplemented with 100 mg/l Rifampicinand and 50 mg/l kanamycin. Bacteria were grown overnight to obtain an OD600 of 1.0–1.5 at 28°C, shaking at 200 rpm. The cultures were pelleted by centrifugation at 2,000 g for 5 min and cells were diluted with infiltration buffer (50 mM MES pH 5.6, 10 mM MgCl2, and 100 μM acetosyringone) to a final OD600 of 0.1–0.3, and then incubated for 1–2 h at 25°C in the dark before agroinfiltration of *N. benthamiana* plants using a 1-ml needleless syringe.

GUS and fluorescence assays

GUS assay was performed as per previously described procedures with a slight modification [29]. The infiltrated *Nicotiana benthamiana* leaves were detached 3 d post-agroinfiltration and put into GUS staining buffer (50 mM phosphate buffer (pH 7.0), 10 mM Na$_2$EDTA, 5 mM K3Fe(CN)6, 5 mM K4Fe(CN)6, 0.1% X-Gluc). The leaves were then incubated for 2–10 h at 37°C in the dark, with continuous gentle shaking. When blue color appeared on the leaves, the leaves were washed in ddH$_2$O and then placed into 75% ethanol to fix and remove chlorophyll. Green fluorescence protein was detected using a long-wavelength UV lamp (Black Ray model B 100A; UV products; Upland, CA, USA) and photographs were taken using Cannon EOS 550D digital camera.

RNA isolation and real-time PCR analysis

Real-time PCR was used to examine the level of RNA accumulation. Total RNA was isolated from leaf tissues, by using RNAprep pure plant kit (Tiangen, Beijing, China), and treated with RNase-free DNase I to remove DNA contamination. First-strand cDNA was synthesized from 500 ng of total RNA using SYBR PrimeScript RT-PCR Kit II (Takara, Dalian, China). Real-time PCR was performed on Stratagene MX3000P with three technical and three biological replicates. Sequences of these primers are listed in Table S1. The relative expressions of the genes of interest were calculated using the formula $2^{-\Delta\Delta CT}$ [30], by normalizing to the expression levels of Actin of *N. benthamiana* (GenBank access number AY594294).

Ethics Statement

No specific permits were required for the described field studies. No specific permissions were required for these locations/activities. The location is not privately-owned or protected in any way. The field studies did not involve endangered or protected species.

Supporting Information

Figure S1 The position of the primers and the restriction sites on pRNAi-GG used for the identification of recombinant pRNAi-GG or intron orientation. P21, P22 and insert reverse primer were designed to identify the recombinant pRNAi-GG, by amplifying two arms simultaneously with a difference of 267 bp in length. P21, P24 and P25 were used to identify the intron orientation. The PCR product of the recombinant pRNAi-GG with sense orientation of intron is 309 bp longer than that of the recombinant pRNAi-GG with antisense orientation of intron. BamHI and SacI can also be used to identify the intron orientation. The digested inserts of the recombinant pRNAi-GG with sense orientation of intron is 405 bp shorter than that of the recombinant pRNAi-GG with antisense orientation of intron.

Table S1   Primer sequences used in this study.

Acknowledgments

We are grateful to Professor P. Mestre for kindly providing the Agrobacterium clones containing pBIN19-GUS and pBIN19-GFP, and to Professor Liu for providing pRNAi-LIC.

Author Contributions

Conceived and designed the experiments: PY PZ JD. Performed the experiments: PY. Analyzed the data: WS XG XL. Contributed reagents/materials/analysis tools: WS XL. Wrote the paper: PY.

References

1. Fire A, Xu S, Montgomery MK, Kostas SA, Driver SE, et al. (1998) Potent and specific genetic interference by double-stranded RNA in Caenorhabditis elegans. Nature 391: 806–811.
2. Gunsalus KC, Piano F (2005) RNAi as a tool to study cell biology: building the genome-phenome bridge. Curr Opin Cell Biol 17: 3–8.
3. Kusaba M (2004) RNA interference in crop plants. Curr Opin Biotechnol 15: 139–143.
4. Perrimon N, Ni JQ, Perkins L (2010) In vivo RNAi: today and tomorrow. Cold Spring Harb Perspect Biol 2: a003640.
5. Waterhouse PM, Helliwell CA (2003) Exploring plant genomes by RNA-induced gene silencing. Nat Rev Genet 4: 29–38.
6. Wesley SV, Helliwell CA, Smith NA, Wang MB, Rouse DT, et al. (2001) Construct design for efficient, effective and high-throughput gene silencing in plants. Plant J 27: 581–590.
7. Helliwell C, Waterhouse P (2003) Constructs and methods for high-throughput gene silencing in plants. Methods 30: 289–295.
8. Eamens AL, Waterhouse PM (2011) Vectors and methods for hairpin RNA and artificial microRNA-mediated gene silencing in plants. Methods in molecular biology 701: 179–197.
9. Himmelbach A, Zierold U, Hensel G, Riechen J, Douchkov D, et al. (2007) A set of modular binary vectors for transformation of cereals. Plant Physiol 145: 1192–1200.
10. Xiao YH, Yin MH, Hou L, Pei Y (2006) Direct amplification of intron-containing hairpin RNA construct from genomic DNA. Biotechniques 41: 548, 550, 552.
11. Yan P, Shen W, Gao X, Duan J, Zhou P (2009) Rapid one-step construction of hairpin RNA. Biochem Biophys Res Commun 383: 464–468.
12. Chen S, Songkumarn P, Liu J, Wang GL (2009) A versatile zero background T-vector system for gene cloning and functional genomics. Plant Physiol 150: 1111–1121.
13. Xu G, Sui N, Tang Y, Xie K, Lai Y, et al. (2010) One-step, zero-background ligation-independent cloning intron-containing hairpin RNA constructs for RNAi in plants. New Phytol 187: 240–250.
14. Engler C, Gruetzner R, Kandzia R, Marillonnet S (2009) Golden gate shuffling: a one-pot DNA shuffling method based on type IIs restriction enzymes. PLoS One 4: e5553.
15. Engler C, Kandzia R, Marillonnet S (2008) A one pot, one step, precision cloning method with high throughput capability. PLoS One 3: e3647.
16. Voinnet O, Rivas S, Mestre P, Baulcombe D (2003) An enhanced transient expression system in plants based on suppression of gene silencing by the p19 protein of tomato bushy stunt virus. Plant J 33: 949–956.
17. Helliwell CA, Wesley SV, Wielopolska AJ, Waterhouse PM (2002) High-throughput vectors for efficient gene silencing in plants. Functional plant biology 29: 1217–1225.
18. Weber E, Engler C, Gruetzner R, Werner S, Marillonnet S (2011) A modular cloning system for standardized assembly of multigene constructs. PLoS One 6: e16765.
19. Sarrion-Perdigones A, Falconi EE, Zandalinas SI, Juarez P, Fernandez-del-Carmen A, et al. (2011) GoldenBraid: an iterative cloning system for standardized assembly of reusable genetic modules. PLoS One 6: e21622.
20. Blake WJ, Chapman BA, Zindal A, Lee ME, Lippow SM, et al. (2010) Pairwise selection assembly for sequence-independent construction of long-length DNA. Nucleic Acids Res 38: 2594–2602.
21. Zhang F, Cong L, Lodato S, Kosuri S, Church GM, et al. (2011) Efficient construction of sequence-specific TAL effectors for modulating mammalian transcription. Nat Biotechnol 29: 149–153.
22. Weber E, Gruetzner R, Werner S, Engler C, Marillonnet S (2011) Assembly of designer TAL effectors by Golden Gate cloning. PLoS One 6: e19722.
23. Morbitzer R, Elsaesser J, Hausner J, Lahaye T (2011) Assembly of custom TALE-type DNA binding domains by modular cloning. Nucleic Acids Res 39: 5790–5799.
24. Cermak T, Doyle EL, Christian M, Wang L, Zhang Y, et al. (2011) Efficient design and assembly of custom TALEN and other TAL effector-based constructs for DNA targeting. Nucleic Acids Res 39: e82.
25. Santos-Rosa M, Poutaraud A, Merdinoglu D, Mestre P (2008) Development of a transient expression system in grapevine via agro-infiltration. Plant Cell Rep 27: 1053–1063.
26. Chen PY, Wang CK, Soong SC, To KY (2003) Complete sequence of the binary vector pBI121 and its application in cloning T-DNA insertion from transgenic plants. Molecular Breeding 11: 287–293.
27. Citovsky V, Lee LY, Vyas S, Glick E, Chen MH, et al. (2006) Subcellular localization of interacting proteins by bimolecular fluorescence complementation in planta. J Mol Biol 362: 1120–1131.
28. Sparkes IA, Runions J, Kearns A, Hawes C (2006) Rapid, transient expression of fluorescent fusion proteins in tobacco plants and generation of stably transformed plants. Nat Protoc 1: 2019–2025.
29. Jefferson RA, Kavanagh TA, Bevan MW (1987) GUS fusions: beta-glucuronidase as a sensitive and versatile gene fusion marker in higher plants. EMBO J 6: 3901–3907.
30. Livak KJ, Schmittgen TD (2001) Analysis of relative gene expression data using real-time quantitative PCR and the 2(−Delta Delta C(T)) Method. Methods 25: 402–408.

7

6-Hydroxy-3-Succinoylpyridine Hydroxylase Catalyzes a Central Step of Nicotine Degradation in *Agrobacterium tumefaciens* S33

Huili Li[1][9], Kebo Xie[1][9], Haiyan Huang[2], Shuning Wang[1]*

1 State Key Laboratory of Microbial Technology, Shandong University, Jinan, PR China, **2** Institute of Basic Medicine, Shandong Academy of Medical Science, Jinan, PR China

Abstract

Nicotine is a main alkaloid in tobacco and is also the primary toxic compound in tobacco wastes. It can be degraded by bacteria via either pyridine pathway or pyrrolidine pathway. Previously, a fused pathway of the pyridine pathway and the pyrrolidine pathway was proposed for nicotine degradation by *Agrobacterium tumefaciens* S33, in which 6-hydroxy-3-succinoylpyridine (HSP) is a key intermediate connecting the two pathways. We report here the purification and properties of an NADH-dependent HSP hydroxylase from *A. tumefaciens* S33. The 90-kDa homodimeric flavoprotein catalyzed the oxidative decarboxylation of HSP to 2,5-dihydroxypyridine (2,5-DHP) in the presence of NADH and FAD at pH 8.0 at a specific rate of about 18.8 ± 1.85 μmol min^{-1} mg protein^{-1}. Its gene was identified by searching the N-terminal amino acid residues of the purified protein against the genome draft of the bacterium. It encodes a protein composed of 391 amino acids with 62% identity to HSP hydroxylase (HspB) from *Pseudomonas putida* S16, which degrades nicotine via the pyrrolidine pathway. Considering the application potential of 2,5-DHP in agriculture and medicine, we developed a route to transform HSP into 2,5-DHP with recombinant HSP hydroxylase and an NADH-regenerating system (formate, NAD$^+$ and formate dehydrogenase), via which around 0.53 ± 0.03 mM 2,5-DHP was produced from 0.76 ± 0.01 mM HSP with a molar conversion as 69.7%. This study presents the biochemical properties of the key enzyme HSP hydroxylase which is involved in the fused nicotine degradation pathway of the pyridine and pyrrolidine pathways and a new green route to biochemically synthesize functionalized 2,5-DHP.

Editor: Andrea Motta, National Research Council of Italy, Italy

Funding: This work was supported by the grant (No. 30970027) from National Natural Science Foundation of China (http://www.nsfc.gov.cn/) and the Excellent Middle-Aged and Youth Scientist Award (No. BS2009SW006) from Department of Science and Technology of Shandong province (http://www.sdstc.gov.cn/). The funders had no role in study design, data collection and analysis, decision to publish, or preparation of the manuscript.

Competing Interests: The authors have declared that no competing interests exist.

* Email: shuningwang@sdu.edu.cn

[9] These authors contributed equally to this work.

Introduction

Microbial degradation of nicotine has drawn an increasing interest recently since it has various biochemical and physiological mechanisms and represents a promising biological method to treat the tobacco leaves and wastes [1–6]. Nicotine, the most abundant alkaloid in tobacco plants, makes people addicted to tobacco and leads to a number of diseases such as cancer and pulmonary disease [7,8]. It is also the major toxic compound in tobacco wastes, which are largely produced during tobacco manufacturing process and all activities of tobacco use, and could cause serious environmental problems [9,10]. In addition, as one of the important traditional cash crops, tobacco is planted in a big scale in many countries. However, WHO Framework Convention on Tobacco Control has been adopted by most countries in the world, because of which some new technologies to utilize tobacco are necessary to be developed in the near future. Considering the high content of nicotine in tobacco leaves and wastes ranging from 2% to 8% [9,11], chemists have already tried to use it as a starting material to produce some chemicals of pharmaceutical importance

[12–14]. Nicotine can also be modified into important functionalized pyridines by biocatalytic processes that are difficult to synthesize via chemical methods [15]. For example, it can be transformed by *Arthrobacter* sp. and *Pseudomonas* sp. into valuable chemicals such as 6-hydroxy-nicotine, 6-hydroxy-3-succinoylpyridine (HSP) and 2,5-dihydroxypyridne (2,5-DHP), which are important precursors for synthesis of drugs and insecticides [16–18].

A number of microorganisms able to degrade nicotine have been isolated and characterized till now, including bacteria, actinomycetes and fungi [1,2,19,20], among which only a few were studied for their biochemical pathways to degrade nicotine. The proposed pathways so far can be summarized mainly into three groups: the demethylation pathway used by fungi such as *Aspergillus oryzae*; the pyridine pathway used by Gram-positive bacterium *Arthrobacter* sp.; and the pyrrolidine pathway used by Gram-negative bacterium *Pseudomonas* sp. [6,21,22]. The biochemical and molecular mechanism involved in the pyridine and pyrrolidine pathways have been already well elucidated or are being studied [1,3,6,21,23]. For other microorganisms, however, the biochemical pathway and molecular mechanism in nicotine catabolism are seldom reported.

Previously, a novel pathway different from the reported pathways mentioned above in *Agrobacterium tumefaciens* S33 was characterized and proposed, that is, by way of the identification of its intermediates and measurement of key enzymes activities in cell extracts and partially enriched enzymes [20,24]. In this novel pathway (Figure 1), nicotine is firstly transformed into 6-hydroxy-pseudooxynicotine via the pyridine pathway through 6-hydroxy-L-nicotine and 6-hydroxy-*N*-methyl-myosmine, and then it turns to the pyrrolidine pathway with the formation of HSP and 2,5-DHP. HSP hydroxylase, one of the key enzymes, was partially enriched to 7 folds from *A. tumefaciens* S33, which has also been found as the key enzyme in the pyrrolidine pathway in *Pseudomonas putida* S16 [21], but its biochemical properties and encoding gene in *A. tumefaciens* S33 was still unknown.

In this study, the HSP hydroxylase was purified from *A. tumefaciens* S33, its biochemical properties was characterized, and its encoding gene was identified by determination of the N-terminal amino acids sequence and genome survey. Because 2,5-DHP, the product of the reaction catalyzed by HSP hydroxylase, is a valuable precursor for the chemical synthesis of 5-aminolevulinic acid, which is applied as a plant growth hormone, a herbicide and in cancer therapy, an efficient process to transform HSP into 2,5-DHP was developed here with heterologously expressed HSP hydroxylase and NADH-regenerating system (formate, NAD$^+$ and formate dehydrogenase from *Candida boidinii*).

Materials and Methods

Bacterial strain and growth

Agrobacterium tumefaciens S33, isolated from tobacco plant rhizosphere and deposited at China Center for Type Culture Collection (CCTCC) under accession number CCTCC M 206131 [20], was grown in nicotine medium [24] plus 1.0 g l^{-1} glucose, 0.2 g l^{-1} (NH$_4$)$_2$SO$_4$ and 1.0 g l^{-1} yeast extract at 30°C. Tungstate was omitted from the trace elements solution because it is the specific antagonist of molybdate and would inhibit the molybdenum-containing nicotine dehydrogenase [25]. Nicotine (≥99% purity, Fluka) was added into the media with the final concentration of 1.0 g l^{-1} prior to filter sterilization before inoculation. The pre-culture was obtained by growing the bacteria in nicotine medium with nicotine as the sole source of carbon and nitrogen. Cells were harvested in the late of exponential phase.

Purification of 6-hydroxy-3-succinoylpyridine hydroxylase (HSP hydroxylase)

Ten grams of wet cells obtained from 3 liters of culture were re-suspended in 50-ml 50 mM Tris-HCl (pH 8.0) and disrupted by an ultrasonic liquid processor (Vibra-Cell VCX 500, Sonics and Materials, amplitude 30%, 20 min, pulse on 6 s and pulse off 6 s) in an ice/water bath. Unbroken cells and cell debris were removed by centrifugation at 30,000×g and 4°C for 30 min. The supernatant was used as cell extract for enzyme purification.

All chromatography steps were carried out using an ÄKTA Basic10 chromatography system (GE-Healthcare). All buffers contained 5 μM FAD, which could slow down the activity loss during purification. Firstly, the cell extract was subjected to ammonium sulfate precipitation by slowly adding saturated ammonium sulfate solution to different final concentrations at 4°C. After gently stirred for 30 min, the precipitate was recovered by centrifugation at 30,000×g and 4°C for 20 min. Most HSP

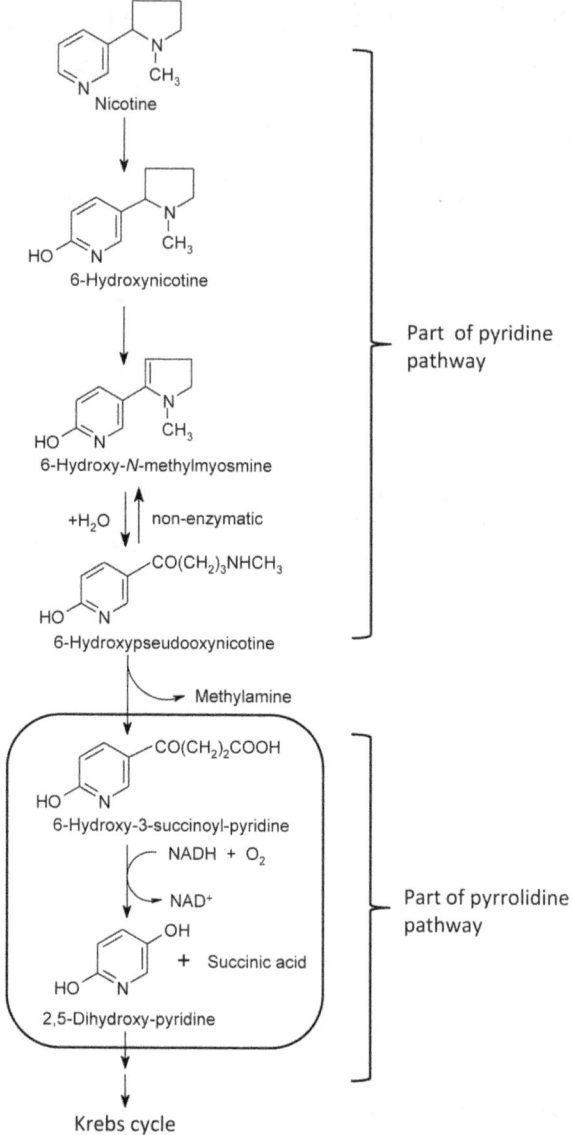

Figure 1. Proposed pathway for nicotine degradation by *A. tumefaciens* S33. The steps from nicotine to 6-hydroxy-pseudoox-ynicotine are same to part of the pyridine pathway, and the step catalyzed by 6-hydroxy-3-succinoylpyridine hydroxylase, which is indicated in the box, is same to the pyrrolidine pathway.

hydroxylase activity was found in the fraction obtained by precipitating between 40% and 60% saturation of ammonium sulfate, which was collected by centrifugation. The precipitate was dissolved in 5 ml 50 mM Tris-HCl (pH 8.0) containing 1.0 M ammonium sulfate. After removing the undissolved proteins by centrifugation, the supernatant was applied on a Butyl Sepharose 6 Fast Flow (high sub, 16 mm×10 cm, 20 ml) equilibrated with 50 mM Tris-HCl (pH 8.0) containing 1.0 M ammonium sulfate. The column was eluted at a 4 ml min^{-1} flow rate with 5 column volumes of each 1.0, 0.8, 0.6, 0.4, 0.2 and 0 M ammonium sulfate step gradients in the same buffer. HSP hydroxylase was eluted at an ammonium sulfate concentration of 0.8 M. The fractions containing the activity were pooled, concentrated, and desalted by the use of an Amicon filter (30-kDa-cutoff). Then the sample was applied on a DEAE Sepharose Fast Flow (16 mm×10 cm, 20 ml) equilibrated with 50 mM Tris-HCl (pH 8.0). The column was

Table 1. Purification of HSP hydroxylase from *A. tumefaciens* S33.

Step	Total protein (mg)	Total activity (unit)	Specific activity (unit/mg)	Yield (%)	Fold
Cell extract	454.6	73.8	0.16	100	1
40–60% (NH4) 2SO4 precipitation	303.8	54.7	0.18	74	1.1
Butyl Sepharose	15.2	30.5	2.0	41.3	12.5
DEAE Sepharose	6.4	26.4	4.1	35.8	25.6
Source 30Q	2.6	24.8	9.4	33.6	58.7
Superdex 200	1.9	19.6	10.3	26.6	64.4

eluted at a 4 ml min^{-1} flow rate with 5 column volumes of each 0.1, 0.2, 0.3, 0.4, and 0.5 M NaCl step gradients in the same buffer. HSP hydroxylase was eluted at a NaCl concentration of 0.2 M. The fractions containing the activity were pooled and desalted, and concentrated. Then the concentrate was applied to a Source 30 Q column (16 mm×10 cm, 20 ml) equilibrated with 50 mM Tris-HCl (pH 8.0). The column was eluted at a 3 ml min^{-1} flow rate with the same procedure as DEAE Sepharose Fast Flow step. HSP hydroxylase was eluted at a NaCl concentration of around 0.2 M. After pooling and concentrating the fractions with the activity, the sample was loaded on a 24-ml Superdex 200 column, which was equilibrated with 50 mM Tris-HCl (pH 8.0) containing 150 mM NaCl prior to loading the sample and eluted with the same buffer. The fraction size collected was set at 1 ml per tube. HSP hydroxylase activity was found in a single peak at 14 ml fraction tube. Fractions with the activity collected during the purification were analyzed by sodium dodecyl sulfate-polyacrylamide gel electrophoresis (SDS-PAGE) using a 12.5% gel according to the standard procedure in a Bio-Rad Mini-Protean III Cell (Laemmli 1970) and stained with Coomassie brilliant blue. The protein concentration was measured by the Bradford method with bovine serum albumin as the standard.

Enzyme assay

The activity of HSP hydroxylase was measured based on the finding by Wang et al. [22,24] and Tang et al. [21,23] that HSP hydroxylase is a NADH-dependent enzyme. The assay was performed at 30°C in a quartz cuvette (1 cm light path) filled with 1-ml reaction mixture. The assay mixture contained 12 μM FAD, 1 mM HSP, 0.5 mM NADH and 50 mM Tris-HCl (pH 8.0). HSP was prepared from the broth of nicotine degrading *P. putida* S16 as described previously [18] with 98% purity verified by HPLC analysis. The reaction was initiated by adding enzyme, and NADH oxidation was spectrophotometrically monitored using an Ultrospec 2100 pro UV/Visible Spectrophotometer (GE Healthcare, USA) by the decrease in absorption at 380 nm (ε = 1.2 mM^{-1} cm^{-1}). The wavelength of 380 nm instead of 340 nm was chosen because neither HSP nor the product 2,5-DHP has absorption at 380 nm. One unit was defined as the oxidation of 1 μmol NADH per min.

The activities of *p*-nitrophenol monooxygenase and 6-hydroxynicotinic acid 3-monooxygenase were measured according to the previous description [16,26].

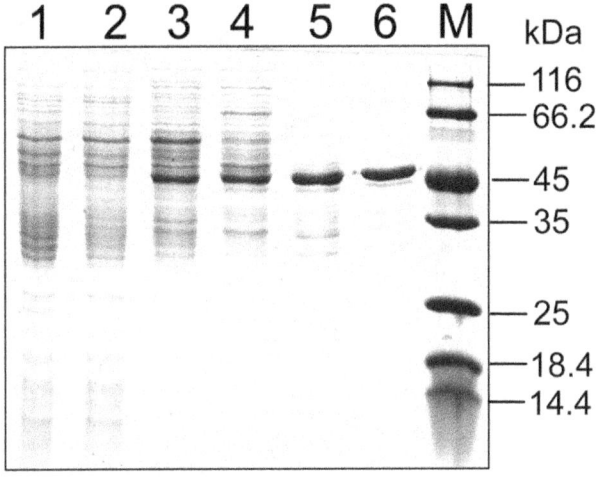

Figure 2. Purification of HSP hydroxylase from *A. tumefaciens* S33. Lane 1, cell extract; lane 2, ammonia sulfate precipitation; lane 3, Butyl Sepharose; lane 4, DEAE Sepharose; lane 5, Source 30Q; lane 6, Superdex 200; and M, protein marker.

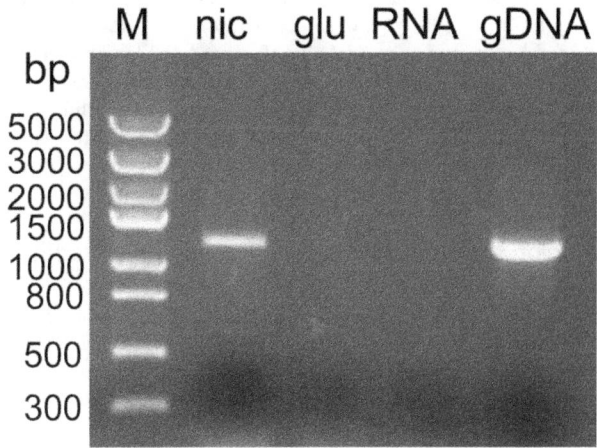

Figure 3. RT-PCR analysis of HSP hydroxylase gene transcription in *A. tumefaciens* S33. The cells were grown in the medium with nicotine as the sole source of carbon and nitrogen (lane nic) and in the glucose and ammonium medium (lane glu), respectively. The gene is 1,176 bp long. Lane RNA, negative control, the total RNA isolated from the culture grown in the medium with nicotine as the sole source of carbon and nitrogen as the template; lane gDNA, positive control, the genomic DNA isolated from the culture grown in the glucose and ammonium medium as the template.

N-terminal amino acid sequence determination of HSP hydroxylase

After analyzed by SDS-PAGE according to the standard procedure, the target protein was transferred to PVDF membrane by electroblotting and visualized with Coomassie brilliant blue solution. Then the corresponding bands were excised and the N-terminal amino acid sequence of the protein was determined by automated Edman degradation using an Applied Biosystems 477A protein/peptide sequencer and the protocol given by the manufacturer.

Genome survey of A. tumefaciens S33

A. tumefaciens S33 genomic DNA was extracted and purified with Wizard Genomic DNA Purification Kit (Promega). The genome sequence of A. tumefaciens S33 was determined by BGI (Shenzhen, China) using the Illumina GA system with a paired-end library. The short reads were assembled with SOAPdenovo, version 1.05 [27]. Gene prediction and annotation were carried out using the RAST annotation server [28].

Isolation of total RNA and RT-PCR

A. tumefaciens S33 was grown in nicotine medium [20] containing 1.0 g l^{-1} nicotine as the sole source of carbon and nitrogen, or in a medium containing 1.0 g l^{-1} glucose, 0.2 g l^{-1} $(NH_4)_2SO_4$ and 1.0 g l^{-1} yeast extract at 30°C. The cells were harvested at the early-exponential phase when the optical density at 620 nm reached around 0.4 in nicotine medium or 0.7 in glucose and ammonium medium. Total RNA was isolated from cell pellets using RNAprep pure Cell/Bacteria Kit (TIANGEN Biotech, China) according to the manufacturer's protocol. Contaminating DNA was removed by treatment with DNase I (RNase-free) provided in the kit at 37°C for 30 min. Total cDNA was synthesized with TransScript First-Strand cDNA Synthesis SuperMix (Beijing TransGen Biotech, China) using 0.7 µg of RNA and random primers provided in the kit according to the manufacturer's manual. The HSP hydroxylase gene was amplified in 20 µl PCR mixture with 95 ng of total cDNA as the template. The amplification condition used for PCR reaction was as follows: 94°C for 5 min; 33 cycles of 94°C for 30 s, 60°C for 30 s and 72°C for 1 min; and then 72°C for 10 min. A full length of HSP hydroxylase gene (1,176 bp) was obtained by using the same primers for heterologous expression (see below). The amplification with genomic DNA and total RNA as the template and the same PCR reaction procedure were used for positive control and negative control, respectively.

Heterologous expression and purification of His-tagged HSP hydroxylase

The HSP hydroxylase gene was amplified by PCR with high-fidelity FastPfu DNA polymerase (Shanghai ShineGene Molecular Bio-tech, China) by the use of A. tumefaciens S33 genomic DNA as a template. The following primers were used: 5'-CGGGATCCTATGAGCGCACATGTCGTTGTCGT (forward primer; the BamHI restriction site is underlined) and 5'-TATCTCGAGTCAATATACTGTCCGCATCTGTTCGTGG (reverse primer; the XhoI restriction site is underlined). The blunt PCR product was ligated into pEASY-Blunt cloning vector (Beijing TransGen Biotech, China), which was subsequently transformed into E. coli Mach 1-T1 cells (Invitrogen). After amplification, the construct was digested by BamHI and XhoI, and the target fragment was ligated into expression vector pETDuet-1, which had been digested by the same restriction endonucleases. The new construct pETDuet-hsph was introduced

into Mach 1-T1 cells again and verified by DNA sequencing. It was then transformed into E. coli BL21(DE3) for expression. The cells were grown in 1 liter of LB medium containing 50 mg of ampicillin $liter^{-1}$ at 16°C at a low stirring speed (100 rpm). IPTG was added to a final concentration of 0.1 mM immediately after inoculation (OD_{620nm}, between 0.2 and 0.3). After incubation for 24 h, the cells were harvested by centrifugation. The induced cells were washed twice with 50 mM Tris-HCl buffer (pH 8.0) and stored at $-20°C$ until use. Via this method, around 60% recombinant protein product were found to be soluble in the supernatant, while most products were in the inclusion bodies when it was induced in classic method with 0.5 mM IPTG at 37°C.

For purification of the His-tagged HSP hydroxylase, the E. coli cells (6 g, wet weight) were re-suspended in 30 ml 50 mM Tris-HCl buffer (pH 8.0), and disrupted by sonication in an ice/water bath. Cell debris was removed by centrifugation at 30,000×g and 4°C for 30 min. The clarified supernatant was applied on a 5-ml His Trap HP column (GE Healthcare). The recombinant protein was eluted with 50 mM Tris-HCl buffer (pH 8.0) containing imidazole in a linear gradient ranging from 10 mM to 100 mM in a total gradient volume of 100 ml at a flow rate of 4 ml min^{-1}. The fractions with the activity were pooled and applied on a Source 30Q column (16 mm×10 cm, 20 ml). The column was eluted with 50 mM Tris-HCl buffer (pH 8.0) containing 0.5 M NaCl in a linear gradient for 20 column volumes at a flow rate of 4 ml min^{-1}. The HSP hydroxylase activity was recovered in the fractions eluted with 0.15 M NaCl. The protein was concentrated by ultrafiltration with an Amicon filter (30-kDa-cutoff) and stored at $-20°C$. All the buffers used were supplemented with 5 µM FAD in order to avoid its disassociation from the enzyme during purification.

Heterologous expression of formate dehydrogenase (Fdh) in E. coli and purification of His-tagged Fdh

The recombinant pETDuet-1 harboring fdh from Candida boidinii was a gift of Professor Cuiqing Ma from Shandong University. The fdh gene was cleaved with NdeI and XhoI and ligated into pACYCDuet-1. And the new construct pACYCDuet-fdh was then transformed into E. coli BL21(DE3) for expression. Induction and purification of His-tagged Fdh was performed with the same procedure as His-tagged HSP hydroxylase. The activity was measured according to what had been previously described with modification [29]. Briefly, the 0.8 ml-mixture contained 1.67 M sodium formate, 1.67 mM NAD^+ and 50 mM sodium phosphate buffer (pH 8.0). The reaction was carried out at 25°C, started by adding formate dehydrogenase, and monitored by determining NAD^+ reduction at 340 nm ($\varepsilon = 6.2$ mM^{-1} cm^{-1}). One unit (U) was defined as the reduction 1 µmol NAD^+ per min.

Transformation of HSP into 2,5-DHP

Two methods were tried to convert HSP to 2,5-DHP. (i) Transformation of HSP into 2,5-DHP with whole cells of E. coli BL21(DE3) harboring pETDuet-hsph and pACYCDuet-fdh. Firstly the constructs pETDuet-hsph and pACYCDuet-fdh were co-transformed into the competent E. coli BL21(DE3) cells. The recombinant E. coli BL21(DE3) was then grown and induced with the same procedure as expressing His-tagged HSP hydroxylase. After incubation for 20 h, the cells were harvested by centrifugation and washed twice with 50 mM Tris-HCl buffer (pH 8.0), then were used as catalysts to perform the transformation reaction. The reaction mixture contained 50 mM sodium phosphate buffer (pH 8.0), 0.75 mM HSP, 1.67 mM NAD^+, 83.5 mM sodium formate and 1.5 g $liter^{-1}$ wet cells of recombinant E. coli

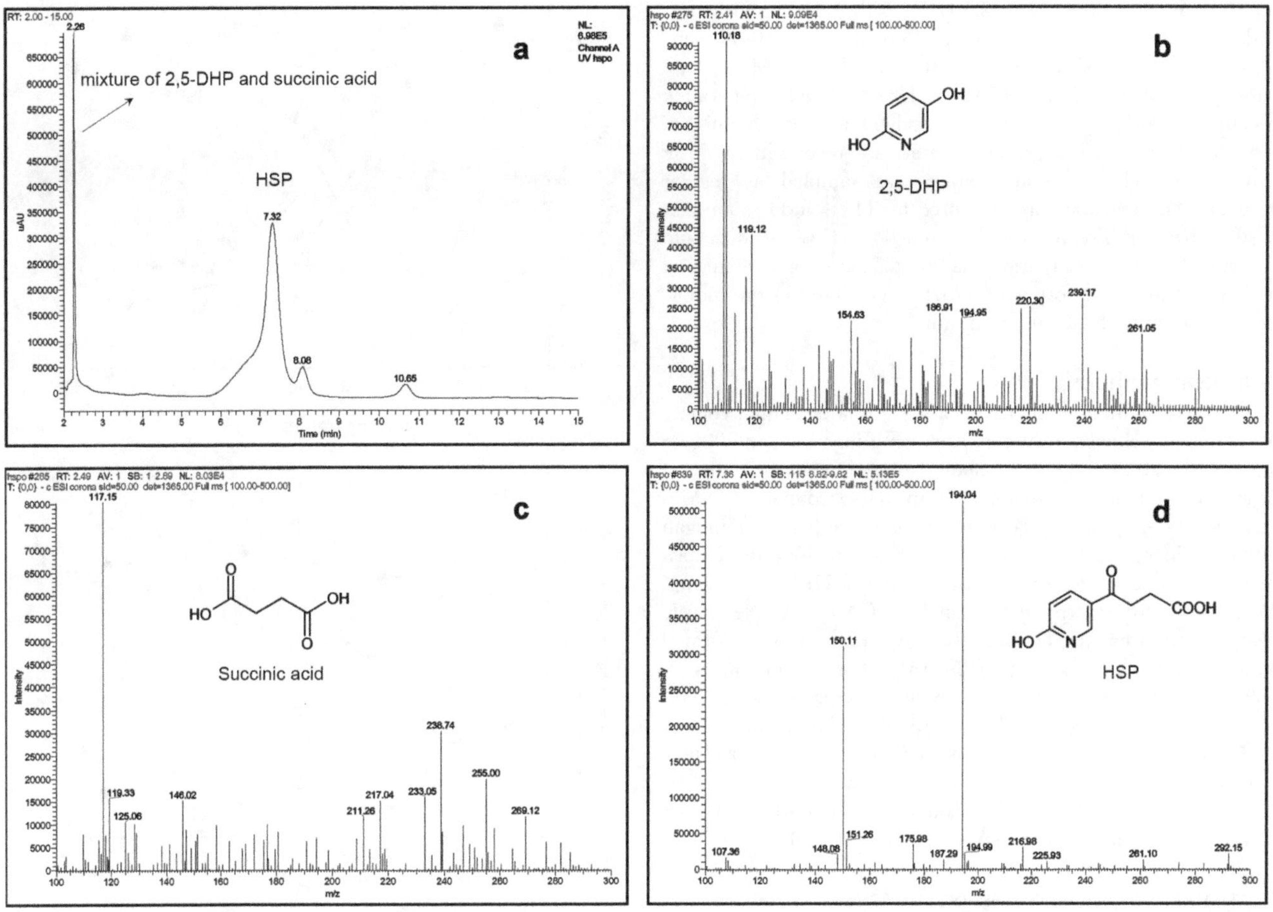

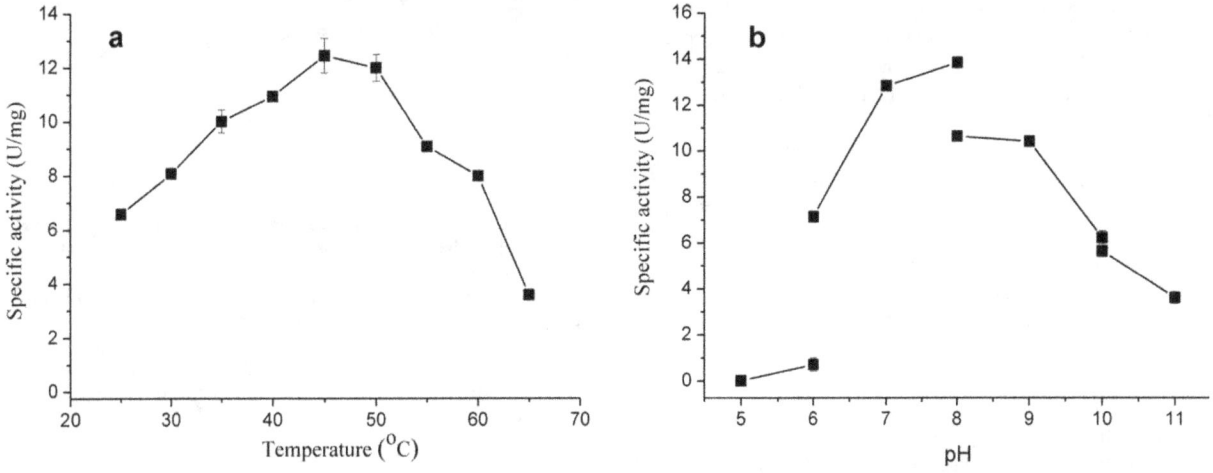

Figure 4. LC-MS profiles of the reaction catalyzed by purified HSP hydroxylase from *A. tumefaciens* S33. (a) HPLC profile monitored with PDA detector; (b–d) mass spectra of products 2,5-DHP (*m/z* 110.18) and succinic acid (*m/z* 117.15) and substrate HSP (*m/z* 194.04), respectively. Negatively charged ions were detected.

BL21(DE3) prepared as described above. The reaction was carried out at 25°C with gentle shaking at 60 rpm. Samples were withdrawn at regular intervals and analyzed by TLC and HPLC after those cells were removed by centrifugation. (ii) Enzymatic

Figure 5. Effects of temperature and pH on the activity of HSP hydroxylase from A. *tumefaciens* S33. (a) The reactions were carried out in 50 mM Tris-HCl (pH 8.0) at the temperature indicated. (b) The reactions were performed at 30°C in different buffers: citric acid-sodium hydrogen phosphate (pH 5.0 and pH 6.0), sodium phosphate (pH 6.0, pH 7.0 and pH 8.0), Tris-HCl (pH 8.0, pH 9.0 and pH 10.0) and sodium bicarbonate-sodium hydroxide (pH 10.0 and pH 11.0). The values are means of three replicates, and the error bars indicate the standard deviations.

transformation of HSP into 2,5-DHP with recombinant HSP hydroxylase. The reaction was performed in a 10-ml tube containing 5 ml of mixture composed of 50 mM sodium phosphate buffer (pH 8.0), 0.75 mM HSP, NADH-regenerating system (1.67 mM NAD$^+$, 83.5 mM sodium formate, 5 units of formate dehydrogenase), and 8 units of recombinant HSP hydroxylase. The reaction mixture was sampled at regular intervals. The reaction was monitored by TLC, and the amount of substrates and products was determined by HPLC according to previous description [24] with small modification of a mixture of methanol and 1 mM formic acid (87:13, v/v) used as the mobile phase and the flow rate at 0.5 ml min^{-1}.

Analytical methods

The reaction mixture of enzyme assay was added equal volume of ethanol and shook for 10 min. After centrifugation at 30,000×g and 4°C for 20 min, the supernatant was used for identification of products with liquid chromatography-mass spectrometry (LC-MS) analysis. LC-ESI-MS data were obtained using a Finnigan Surveyor MSQ single quadrupole electrospray ionization mass spectrometer coupled with a Finnigan Surveyor HPLC (Finnigan/Thermo Electron Corporation, San Jose, CA, USA). Negatively charged ions were detected. The HPLC system was performed with a 20RBAX Eclips XDB-C18 column (column size, 150×4.6 mm; partical size, 5 μm; Agilent, USA) and a photodiode array (PDA) detector. A mixture of methanol and 1 mM formic acid (85:15, v/v) was used as the mobile phase and the flow rate was set at 0.5 ml min^{-1}. The sample was also analyzed by GC-MS after lyophilization and silylation with N,O-Bis(trimethylsilyl) trifluoroacetamide (BSTFA, Supelco) as described [24].

Nucleotide sequence accession number

The nucleotide sequence of HSP hydroxylase gene reported in this study has been deposited in the GenBank database under the accession number KJ129609.

Results and Discussion

Purification and characterization of HSP hydroxylase from A. tumefaciens S33

It was previously found that cell extracts of A. tumefaciens S33 grown on nicotine as the sole source of carbon and nitrogen catalyzed NADH-dependent hydroxylation of HSP with a specific activity of 0.15 U/mg to 0.19 U/mg [24]. The activity was enriched for 7 folds after ammonia sulfate fractionation and DEAE Sepharose treatment. In this study, the bacterium was grown in nicotine medium supplemented with glucose and yeast extract in order to obtain more cells. Based on the previous study [20], the cells were harvested in the late exponential phase to ensure that the enzymes involved in nicotine degradation were induced. Cell extracts prepared in this way presented a NADH-dependent HSP hydroxylase activity of 0.16 U/mg and a relatively low NADPH-dependent HSP hydroxylase activity of 0.04 U/mg. Then HSP hydroxylase was purified from such cell extracts (Table 1).

Purification was achieved by ammonia sulfate fractionation, hydrophobic chromatography (Butyl Sepharose), anion-exchange chromatography (DEAE Sepharose and Source 30Q) and gel filtration (Superdex 200). During the purification, the activities were determined by spectrophotometrically monitoring HSP-dependent NADH oxidation at 380 nm. The enzyme was purified 64-fold, with a yield of 27% and a NADH-dependent specific activity of 10.3 U/mg (Table 1).

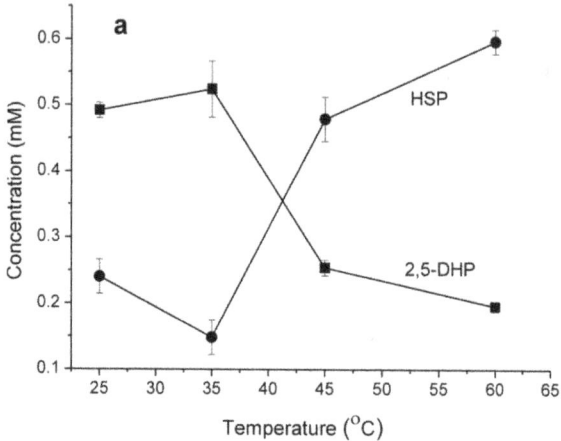
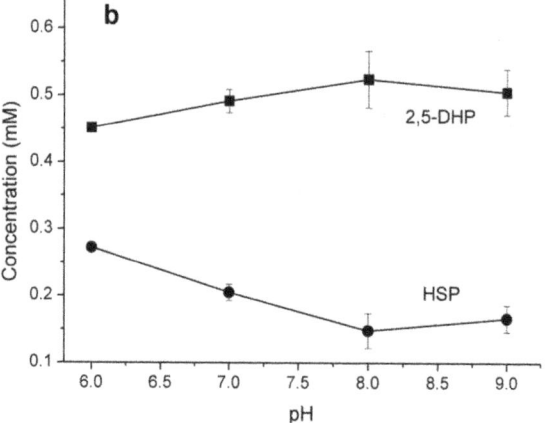
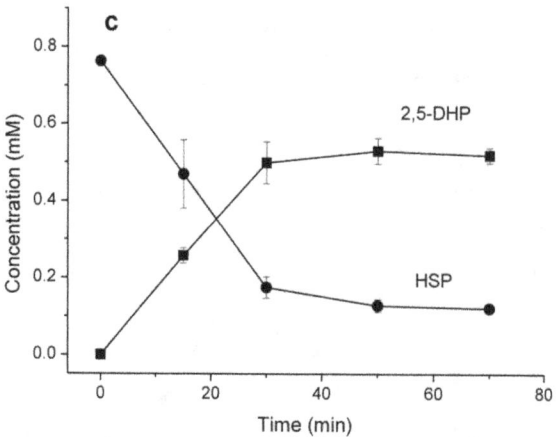

Figure 6. Biotransformation of HSP into 2,5-DHP by HSP hydroxylase from A. tumefaciens S33. Effects of temperature (a) and pH (b) on the enzymatic formation of 2,5-DHP (squares) from HSP (circles). (a) The reactions were carried out in 50 mM sodium phosphate (pH 8.0) at the temperature indicated. (b) The reactions were performed at 35°C in sodium phosphate buffer at pH indicated. (c) The reaction was performed in 50 mM sodium phosphate (pH 8.0) at 35°C. The values are means of three replicates, and the error bars indicate the standard deviations.

During purification, all the buffers were supplemented with 5 μM FAD. When FAD was omitted, the purification yield was very low. FAD could not be substituted by FMN. Because FAD was added to all the buffers during purification, the amount of it in the purified enzyme was not determined.

SDS-PAGE of the purified enzyme showed one protein with a molecular mass of around 45 kDa (Figure 2). The apparent molecular mass of the enzyme determined by gel exclusion chromatography was near 90 kDa, and this indicates that the native enzyme is a homodimer.

Identification of HSP hydroxylase gene in the genome of A. tumefaciens S33

The N-terminal amino acid sequence of the purified HSP hydroxylase was determined by automated Edman degradation to be SAHVVVVGGGPTGLLTTLGL. Its encoding gene was identified by searching the N-terminal amino acid sequence against the genome draft sequence of A. tumefaciens S33, which was automatically annotated on RAST online server. HSP hydroxylase gene (1,176 bp) was found to encode a 44-kDa protein composed of 391 amino acids, which has 62% identity to HspB [21] and no significant similarity with HspA [23]. HspA and HspB are two HSP hydroxylase isozymes involved in nicotine degradation in P. putida S16. In addition, HSP hydroxylase from A. tumefaciens S33 also shows high identity in protein sequence to p-nitrophenol monooxygenase from Pseudomonas sp. WBC-3 (38%) [26], Pseudomonas sp. NyZ402 (37%) [30] and P. putida DLL-E4 (37%) [31]. The three p-nitrophenol monooxygenases are almost identical in protein sequence (more than 85% identity) and convert p-nitrophenol to p-benzoquinone in the presence of NAD(P)H and FAD in p-nitrophenol degradation. And HSP hydroxylase has relatively low identity to 6-hydroxynicotinic acid 3-monooxygenase from P. putida KT2440 (17%) [32] and P. fluorescens TN5 (16%) [16], which catalyzes oxidative decarboxylation of 6-hydroxynicotinic acid to render 2,5-DHP in nicotinic acid degradation. Conserved domains analysis in NCBI shows that N-terminal domain of HSP hydroxylase from A. tumefaciens S33 (amino acids 22–169) belongs to PRK07608 superfamily (ubiquinone biosynthesis hydroxylase family protein), where a GADGA motif was found (amino acids 156–160), and HSP hydroxylase contains a FAD-binding domain (pfam01494, amino acids 22–333). The feature is also found in HspB from P. putida S16 [21]. Alignment of HSP hydroxylase from A. tumefaciens S33 with the proteins mentioned above and other FAD-dependent monooxygenases indicates that HSP hydroxylase contains the conserved motifs for FAD and NAD(P)H binding, including GxGxxG, DGxcSxhR, and GxhhLhGDAAHxxxPxxGxGxNxxxxDxxxL (x = all residues; c = charged residues; h = hydrophobic residues), that are associated with the hydroxylase activity [33]. However, HSP hydroxylase from A. tumefaciens S33 did not show any detectable activity with p-nitrophenol or 6-hydroxynicotinate as substrate in the presence of NAD(P)H and FAD.

Moreover, in order to confirm whether the gene is really involved in the nicotine degradation in A. tumefaciens S33, RT-PCR analysis was performed. Total RNA was isolated from cultures grown in the medium with nicotine as the sole source of carbon and nitrogen and in the medium with glucose and ammonium sulfate as the sources of carbon and nitrogen. After reverse transcription with random primers, the full length of HSP hydroxylase gene (1,176 bp) was amplified with specific primers. As shown in Figure 3, RT-PCR resulted in the specific full fragment of HSP hydroxylase gene when the strain was grown in the medium with nicotine as the sole source of carbon and nitrogen. In contrast, PCR product was not detected when the strain was grown in the glucose and ammonium medium, and this indicates that expression of HSP hydroxylase gene in A. tumefaciens S33 is induced in the presence of nicotine.

Catalytic properties of HSP hydroxylase

In order to identify the reaction catalyzed by the purified enzyme, the products of the NADH-dependent HSP oxidation reaction were determined by LC-MS (Figure 4). Three compounds were found from two peaks in the chromatography (retention time as 2.26 min and 7.32 min, respectively, Figure 4a) monitored with PDA detector by detecting negatively charged ions. The ratios of mass to charge (m/z) of their main fragments in the mass spectra are 110.18 (Figure 4b), 117.15 (Figure 4c) and 194.04 (Figure 4d), which fit the calculated molecular mass of 2,5-DHP ($C_5H_5NO_2$, 111.10), succinic acid ($C_4H_6O_4$, 118.09) and HSP ($C_9H_9NO_4$, 195.17), respectively. Thus, the peak with retention time of 7.32 min was the substrate HSP, the peak (2.26 min) was the mixture of 2,5-DHP and succinic acid, which were difficult to separate from each other even under the best conditions we had in the liquid chromatography. However, succinic semialdehyde found in the previous report [21] was not detected even by scanning all the negatively charged ions in the spectra. One of the possible reasons may be that succinic semialdehyde produced in the reaction was oxidized into succinic acid under the reaction conditions used. In addition, the products were also analyzed with GC-MS after silylation the sample with BSTFA. The trimethylsilyl (TMS) derivatives of HSP and 2,5-DHP were able to be detected and identified, the mass of spectra of whose TMS derivative (GC-MS data not shown) were identical to the previous description [24]. Succinic acid could not be detected because of its poor separation with GC and low abundance based on the LC-MS analysis.

The purified enzyme catalyzed HSP-dependent NADH oxidation with temperature and pH optima near 45°C and pH 8.0, respectively (Figure 5). Compared with Tris-HCl buffer, the enzyme showed higher activity in phosphate buffer. Under the condition of 30°C and 50 mM phosphate buffer (pH 8.0), the enzyme presented an V_{max} of 18.8±1.85 U/mg (the value is the mean ± SD of triplicates, the same hereinafter), an K_m for HSP of 0.15±0.02 mM and an K_m for NADH of 0.1±0.01 mM. NADPH could also be oxidized by HSP hydroxylase with a specific activity of 12.9±0.6 U/mg and K_m of 0.35±0.04 mM. The results indicate that NADH is the preferred physiological electron donor in vivo.

Transformation of HSP into 2,5-DHP

HSP hydroxylase from A. tumefaciens S33 can catalyze HSP to produce 2,5-DHP in the presence of NADH, which is a valuable precursor for synthesis of drugs and insecticides [16]. Therefore, an enzymatic route to produce 2,5-DHP from HSP was tried, which could provide an alternative for utilizing tobacco and its main alkaloid nicotine. Formate dehydrogenase has been used as an effective NADH-regenerating tool for many biocatalytic reactions [34]. In this study, it was used to supply NADH for the biotransformation of HSP into 2,5-DHP. Firstly, induced whole cells of recombinant E. coli BL21(DE3) harboring pETDuet-hsph and pACYCDuet-fdh was used as catalyst to convert HSP to 2,5-DHP. During the reaction, HSP was found to be consumed, however, no 2,5-DHP could be detected based on either TLC or HPLC analysis. This suggested that the product 2,5-DHP might be oxidized in the cells of E. coli. Further, an enzymatic transformation method was tested by using recombinant HSP hydroxylase and formate dehydrogenase. Both His-tagged enzymes were purified by His Trap HP and Source 30Q columns. The purified recombinant HSP hydroxylase showed same properties and specific activity as the wild-type enzyme purified from A. tumefaciens S33 (data not shown). As shown in Figures 6a and 6b, the formation of 2,5-DHP from HSP was in the condition of temperature and pH optima around 35°C and pH 8.0, respectively. Under this condition, around 0.53±0.03 mM 2,5-DHP was produced from 0.76±0.01 mM HSP after 50 min, and the molar conversion was 69.7% (Figure 6c). The amount of 2,5-DHP did not increase when incubated for longer time. On the contrary, the amount of 2,5-DHP began to decrease (data not shown) because of its oxidation in the air as indicated by the appearance of blue gray color [22,35], which suggests that the biotransformation process should be finished as soon as possible in order to avoid product oxidation.

Conclusion

A. tumefaciens S33 was previously proposed to degrade nicotine via a fused pathway of the pyridine and pyrollidine pathways. Here, HSP hydroxylase, one key enzyme involved in the pathway, was purified and characterized. The NADH-dependent flavoprotein catalyzes the oxidative decarboxylation of HSP to produce 2,5-DHP. The finding provides a new insight into the mechanism and diversity of nicotine catabolism in nature.

An enzymatic route to produce a functionalized pyridine, 2,5-DHP, from HSP was developed, which is a precursor of important chemicals and difficult to synthesize by chemical methods. This provides a new method to utilize the tobacco and its wastes.

Acknowledgments

We thank Professor Ping Xu from Shanghai Jiao Tong University and Professor Luying Xun from Washington State University for their valuable discussion and suggestion. We also thank Professor Cuiqing Ma from Shandong University for providing recombinant pETDuet-1 harboring *fdh* from *Candida boidinii*.

Author Contributions

Conceived and designed the experiments: HL KX HH SW. Performed the experiments: HL KX SW. Analyzed the data: HL KX HH SW. Contributed reagents/materials/analysis tools: HH SW. Contributed to the writing of the manuscript: HL HH SW.

References

1. Brandsch R (2006) Microbiology and biochemistry of nicotine degradation. Appl Microbiol Biotechnol 69: 493–498.
2. Li H, Li X, Duan Y, Zhang KQ, Yang J (2010) Biotransformation of nicotine by microorganism: the case of *Pseudomonas* spp. Appl Microbiol Biotechnol 86: 11–17.
3. Tang H, Wang L, Wang W, Yu H, Zhang K, et al. (2013) Systematic unraveling of the unsolved pathway of nicotine degradation in *Pseudomonas*. PLoS Genet 9: e1003923.
4. Wang JH, He HZ, Wang MZ, Wang S, Zhang J, et al. (2013) Bioaugmentation of activated sludge with *Acinetobacter* sp. TW enhances nicotine degradation in a synthetic tobacco wastewater treatment system. Bioresour Technol 142: 445–453.
5. Zhong W, Zhu C, Shu M, Sun K, Zhao L, et al. (2010) Degradation of nicotine in tobacco waste extract by newly isolated *Pseudomonas* sp. ZUTSKD. Bioresour Technol 101: 6935–6941.
6. Qiu J, Ma Y, Wen Y, Chen L, Wu L, et al. (2012) Functional identification of two novel genes from *Pseudomonas* sp. strain HZN6 involved in the catabolism of nicotine. Appl Environ Microbiol 78: 2154–2160.
7. Benowitz NL (2010) Nicotine addiction. N Engl J Med 362: 2295–2303.
8. Hecht SS (1999) Tobacco smoke carcinogens and lung cancer. J Natl Cancer Inst 91: 1194–1210.
9. Civilini M, Domenis C, Sebastianutto N, de Berfoldi M (1997) Nicotine decontamination of tobacco agro-industrial waste and its degradation by microorganisms. Waste Manag Res 15: 349–358.
10. Novotny TE, Zhao F (1999) Consumption and production waste: another externality of tobacco use. Tob Control 8: 75–80.
11. Bush L, Hempfling WP, Burton H (1999) Biosynthesis of nicotine and related compounds. In: Gorrod JW, Jacob PI, editors. Analytical determination of nicotine and related compounds and their metabolites. Amsterdam: Elsevier.
12. Enamorado MF, Ondachi PW, Comins DL (2010) A five-step synthesis of (S)-macrostomine from (S)-nicotine. Org Lett 12: 4513–4515.
13. Ondachi PW, Comins DL (2010) Synthesis of fused-ring nicotine derivatives from (S)-nicotine. J Org Chem 75: 1706–1716.
14. Kagarlitskii AD IM, Turmukhambetov AZh (2002) Catalytic conversion of nicotine into nicotinonitrile–a pharmaceutical intermediate product. Pharma Chem J 36: 26–27.
15. Schmid A, Dordick JS, Hauer B, Kiener A, Wubbolts M, et al. (2001) Industrial biocatalysis today and tomorrow. Nature 409: 258–268.
16. Nakano H, Wieser M, Hurh B, Kawai T, Yoshida T, et al. (1999) Purification, characterization and gene cloning of 6-hydroxynicotinate 3-monooxygenase from *Pseudomonas fluorescens* TN5. Eur J Biochem 260: 120–126.
17. Roduit JP, Wellig A, Kiener A (1997) Renewable functionalized pyridines derived from microbial metabolites of the alkaloid (S)-nicotine. Heterocycles 45: 1687–1702.
18. Wang SN, Xu P, Tang HZ, Meng J, Liu XL, et al. (2005) "Green" route to 6-hydroxy-3-succinoyl-pyridine from (S)-nicotine of tobacco waste by whole cells of a *Pseudomonas* sp. Environ Sci Technol 39: 6877–6880.
19. Meng XJ, Lu LL, Gu GF, Xiao M (2010) A novel pathway for nicotine degradation by *Aspergillus oryzae* 112822 isolated from tobacco leaves. Res Microbiol 161: 626–633.
20. Wang SN, Liu Z, Xu P (2009) Biodegradation of nicotine by a newly isolated *Agrobacterium* sp. strain S33. J Appl Microbiol 107: 838–847.
21. Tang H, Yao Y, Zhang D, Meng X, Wang L, et al. (2011) A novel NADH-dependent and FAD-containing hydroxylase is crucial for nicotine degradation by *Pseudomonas putida*. J Biol Chem 286: 39179–39187.
22. Wang SN, Liu Z, Tang HZ, Meng J, Xu P (2007) Characterization of environmentally friendly nicotine degradation by *Pseudomonas putida* biotype A strain S16. Microbiology 153: 1556–1565.
23. Tang H, Wang S, Ma L, Meng X, Deng Z, et al. (2008) A novel gene, encoding 6-hydroxy-3-succinoylpyridine hydroxylase, involved in nicotine degradation by *Pseudomonas putida* strain S16. Appl Environ Microbiol 74: 1567–1574.
24. Wang S, Huang H, Xie K, Xu P (2012) Identification of nicotine biotransformation intermediates by *Agrobacterium tumefaciens* strain S33 suggests a novel nicotine degradation pathway. Appl Microbiol Biotechnol 95: 1567–1578.
25. Grether-Beck S, Igloi GL, Pust S, Schilz E, Decker K, et al. (1994) Structural analysis and molybdenum-dependent expression of the pAO1-encoded nicotine dehydrogenase genes of *Arthrobacter nicotinovorans*. Mol Microbiol 13: 929–936.
26. Zhang JJ, Liu H, Xiao Y, Zhang XE, Zhou NY (2009) Identification and characterization of catabolic *para*-nitrophenol 4-monooxygenase and *para*-benzoquinone reductase from *Pseudomonas* sp. strain WBC-3. J Bacteriol 191: 2703–2710.
27. Li R, Zhu H, Ruan J, Qian W, Fang X, et al. (2010) De novo assembly of human genomes with massively parallel short read sequencing. Genome Res 20: 265–272.
28. Aziz R, Bartels D, Best A, DeJongh M, Disz T, et al. (2008) The RAST Server: Rapid Annotations using Subsystems Technology. BMC Genomics 9: 75.
29. Schüte H, Flossdorf J, Sahm H, Kula MR (1976) Purification and properties of formaldehyde dehydrogenase and formate dehydrogenase from *Candida boidinii*. Eur J Biochem 62: 151–160.
30. Wei Q, Liu H, Zhang JJ, Wang SH, Xiao Y, et al. (2010) Characterization of a *para*-nitrophenol catabolic cluster in *Pseudomonas* sp. strain NyZ402 and construction of an engineered strain capable of simultaneously mineralizing both *para*-and *ortho*-nitrophenols. Biodegradation 21: 575–584.
31. Shen W, Liu W, Zhang J, Tao J, Deng H, et al. (2010) Cloning and characterization of a gene cluster involved in the catabolism of *p*-nitrophenol from *Pseudomonas putida* DLL-E4. Bioresour Technol 101: 7516–7522.
32. Jiménez JI, Canales A, Jiménez-Barbero J, Ginalski K, Rychlewski L, et al. (2008) Deciphering the genetic determinants for aerobic nicotinic acid degradation: the *nic* cluster from *Pseudomonas putida* KT2440. Proc Natl Acad Sci U S A 105: 11329–11334.
33. Eppink MH, Schreuder HA, Van Berkel WJ (1997) Identification of a novel conserved sequence motif in flavoprotein hydroxylases with a putative dual function in FAD/NAD(P)H binding. Protein Sci 6: 2454–2458.
34. Kratzer R, Pukl M, Egger S, Nidetzky B (2008) Whole-cell bioreduction of aromatic alpha-keto esters using *Candida tenuis* xylose reductase and *Candida boidinii* formate dehydrogenase co-expressed in *Escherichia coli*. Microb Cell Fact 7: 37.
35. Yao Y, Tang H, Ren H, Yu H, Wang L, et al. (2013) Iron(II)-dependent dioxygenase and N-formylamide deformylase catalyze the reactions from 5-hydroxy-2-pyridone to maleamate. Sci Rep 3: 3235.

8

Isolation of a Novel Peroxisomal Catalase Gene from Sugarcane, Which Is Responsive to Biotic and Abiotic Stresses

Yachun Su[1,9], Jinlong Guo[1,9], Hui Ling[1], Shanshan Chen[1], Shanshan Wang[1], Liping Xu[1]*, Andrew C. Allan[2], Youxiong Que[1]*

1 Key Laboratory of Sugarcane Biology and Genetic Breeding, Ministry of Agriculture, Fujian Agriculture and Forestry University, Fuzhou, Fujian, China, 2 The New Zealand Institute for Plant and Food Research, Sandringham, Auckland, New Zealand

Abstract

Catalase is an iron porphyrin enzyme, which serves as an efficient scavenger of reactive oxygen species (ROS) to avoid oxidative damage. In sugarcane, the enzymatic activity of catalase in a variety (Yacheng05–179) resistant to the smut pathogen *Sporisorium scitamineum* was always higher than that of the susceptible variety (Liucheng03–182), suggesting that catalase activity may have a positive correlation with smut resistance in sugarcane. To understand the function of catalase at the molecular level, a cDNA sequence of *ScCAT1* (GenBank Accession No. KF664183), was isolated from sugarcane infected by *S. scitamineum*. *ScCAT1* was predicted to encode 492 amino acid residues, and its deduced amino acid sequence shared a high degree of homology with other plant catalases. Enhanced growth of ScCAT1 in recombinant *Escherichia coli* Rosetta cells under the stresses of $CuCl_2$, $CdCl_2$ and NaCl indicated its high tolerance. Q-PCR results showed that *ScCAT1* was expressed at relatively high levels in the bud, whereas expression was moderate in stem epidermis and stem pith. Different kinds of stresses, including *S. scitamineum* challenge, plant hormones (SA, MeJA and ABA) treatments, oxidative (H_2O_2) stress, heavy metal ($CuCl_2$) and hyper-osmotic (PEG and NaCl) stresses, triggered a significant induction of *ScCAT1*. The ScCAT1 protein appeared to localize in plasma membrane and cytoplasm. Furthermore, histochemical assays using DAB and trypan blue staining, as well as conductivity measurement, indicated that *ScCAT1* may confer the sugarcane immunity. In conclusion, the positive response of *ScCAT1* to biotic and abiotic stresses suggests that *ScCAT1* is involved in protection of sugarcane against reactive oxidant-related environmental stimuli.

Editor: Ji-Hong Liu, Key Laboratory of Horticultural Plant Biology (MOE), China

Funding: This research was supported by National Natural Science Foundation of China (No.31101196), the earmarked fund for the Modern Agriculture Technology of China (CARS-20), Research Funds for Distinguished Young Scientists in Fujian Agriculture and Forestry University (xjq201202), National High Technology Research and Development Program of China (863 Program) Project (2013AA102604). The funders had no role in study design, data collection and analysis, decision to publish, or preparation of the manuscript.

Competing Interests: The authors have declared that no competing interests exist.

* E-mail: xlpmail@126.com (LX); queyouxiong@hotmail.com (YQ)

9 These authors contributed equally to this work.

Introduction

Sugarcane smut, a prevalent and worldwide disease of sugarcane, is caused by the basidiomycete *Sporisorium scitamineum* (*S. scitamineum*). The characteristic symptom of this disease is the emergence of black whips after three months of exposure to infection with smut [1]. The tainted buds may either produce symptoms, or exist as a latent infection and produce black whips in the following season [2]. The enormous quantity of teliospores as well as the quick spread within the sugarcane-producing area makes it almost impossible to completely eliminate this disease. Smut usually results in poor cane growth with profuse tillering, spindly shoots, and narrow leaves, therefore causing considerable loss in yield and sugar content [3]. The release of smut resistant sugarcane varieties, correct quarantine and integrated field management are three main pathways for the control this disease [1]. It is reported that the rates and patterns of colonization of *S. scitamineum* differ in resistant and susceptible sugarcane tissues [4]. Solas et al. found that buds of the resistant sugarcane cultivar were not subjected to intracellular penetration by *S. scitamineum* compared to that of the susceptible cultivar [5]. Susceptible cultivars produce a large number of sori which develop earlier than that in resistant cultivars [6]. Therefore, breeding for smut resistant sugarcane varieties has proved to be the most effective method [7].

Due to the complicated genetic background (a polyploid-aneuploid genome) and pressures of breeding selection (the interaction among sugarcane, smut pathogen and environmental factors), many years and multipoint resistance evaluation tests are needed to obtain relatively high smut resistant sugarcane variety [8]. Alternatively, genetic modification, with directional improvement and molecular assisted breeding technology linked to a target trait, is an alternative way to obtain a resistant variety more quickly and efficiently [9]. By introducing disease-resistance genes to improve gene expression, or by silencing disease-susceptible genes to increase resistance, genetic engineering has made it practical to generate smut resistant sugarcane cultivars [10].

Catalase (E.C.1.11.1.6; H_2O_2:H_2O_2 oxidoreductase; CAT) is an iron porphyrin enzyme, mostly localized in peroxisomes [11]. It serves as an efficient scavenger of reactive oxygen species (ROS). The main function of catalase is to remove excessive H_2O_2 (hydrogen peroxide) during developmental process or biotic/abiotic stress, to avoid oxidative damage [12]. Plant catalases are composed of a multi-gene family and have been reported in many plant species [13]. There are three members identified in *Arabidopsis thaliana* [14], *Nicotiana tabacum* and *Zea mays* [15,16], two in *Hordeum vulgare* [17], one in *Solanum lycopersicum* [18]. In the catalase gene family, different members encode distinct catalase proteins that exhibit different patterns of subcellular localization and expression regulation [19].

The expression of various plant catalase genes is regulated temporally and spatially and responds to developmental and environmental oxidative stimuli [11,13,20,21]. In *Panax ginseng*, *PgCat1* gene was expressed at different levels in leaves, stems, roots of *P. ginseng* seedlings and was induced by different stresses including heavy metals, osmotic agents, plant hormones and high light irradiances [11]. Kwon and An cloned a *Capsicum annuum* catalase cDNA, and northern hybridization showed its transcript was more abundant in stems than in leaves and roots, and more in the early stages than that in the mature stage of fruit development [19]. They also found that aluminum, sodium chloride (NaCl) and light treatment could induce its transcript. Previous research also revealed that the expression of three different maize catalase genes was regulated differentially in response to developmental phase or the fungal toxin cercosporin [22], abscisic acid (ABA) and salicylic acid (SA) [16,22]. Wang et al. found increased transcription of a catalase gene (*MmeCAT*) in resistant clam *Meretrix meretrix* which indicated that *MmeCAT* could most probably benefit the immune system of clams to defend against pathogen infection [23]. The positive response of catalase genes to various stimuli suggested that catalase may help to protect the plant against reactive oxidant related environmental stresses. It is therefore interesting to determine the role of sugarcane catalases and their encoding genes in response to biotic and abiotic stresses.

To date, a partial cDNA sequence (GenBank Accession No. CF572408.1) similar to catalase has been cloned from *Saccharum* hybrid cultivar Q117 [24], while its function remained unclear. In the present research, we analyzed the differences of sugarcane catalase enzyme activity between Yacheng05−179 (resistant) and Liucheng03−182 (susceptible) inoculated with *S. scitamineum*. A novel full-length catalase gene *ScCAT1* (GenBank Accession No. KF664183) from sugarcane bud infected by *S. scitamineum* pathogen was cloned and characterized. Its response in *Escherichia coli* (*E. coli*) Rosetta strains, subcellular localization and expression patterns in sugarcane tissues under various biotic/abiotic stresses were reported. *Agrobacterium*-mediated transient expression of this gene in *Nicotiana benthamiana* (*N. benthamiana*) was used to functionally test ScCAT1.

Materials and Methods

Plant Materials and Treatments

Smut whips were collected in the most popular cultivar "ROC"22 in the Key Laboratory of Sugarcane Biology and Genetic Breeding, Ministry of Agriculture (Fuzhou, China), and stored at 4°C. Sugarcane varieties of Yacheng05−179 (smut resistant) and Liucheng03−182 (smut susceptible) were cultivated in the Key Laboratory of Sugarcane Biology and Genetic Breeding, Ministry of Agriculture (Fuzhou, China). All of the treatments were repeated independently three times.

For tissue distribution studies, six healthy 10 month old plants were selected. For each plant, the youngest fully expanded leaf viz +1 leaf with a visible dewlap (the collar between the leaf blade and sheath), all the buds, stem epidermis and the stem pith were taken for RNA extraction.

During biotic treatments, two-bud sets of both sugarcane genotypes, Yacheng05−179 and Liucheng03−182, were inoculated with 0.5 μL suspension containing 5×10^6 spores·mL^{-1} in 0.01% (v/v) Tween-20, while controls were mock inoculated with 0.01% (v/v) Tween-20 in sterile distilled water instead of spores [25,26]. All the inoculated sets were grown at 28°C in condition of 12 h light/12 h dark. Five buds from each of both genotypes were collected at each of the time point of 0 h, 6 h, 12 h, 24 h, 48 h, 72 h and 96 h. Samples were frozen in liquid nitrogen, and stored at −80°C.

During abiotic treatments, uniform four-month-old sugarcane tissue cultured plantlets of Yacheng05−179 were grown in water for one week and then transferred to the following seven different treatments in conical tubes at 28°C with 16 h light/8 h dark. The plantlets were treated with 5 mM SA solution, 25 μM MeJA (methyl jasmonate) in 0.1% (v/v) ethanol and 0.05% (v/v) Tween-20, 100 μM ABA, and 25% PEG (polyethylene glycol), and the plantlets were set to different periods of time (0 h, 6 h, 12 h and 24 h), respectively. In addition, plantlets were separately treated with 250 mM NaCl and 100 μM CuCl$_2$ (copper chloride) for 0 h, 12 h, 24 h and 48 h [27,28]. For H_2O_2 stress, the leaves were sprayed with 10 mM H_2O_2, and the sampling time points were 0 h, 6 h, 12 h and 24 h, respectively. After treatments, three sugarcane plantlets per time point were collected and immediately fixed in liquid nitrogen, and then kept at −80°C until used for analysis.

Enzyme Extraction and Activity Assay

To analyze quantitative change in catalase activity in Yacheng05−179 and Liucheng03−182 after inoculation with smut pathogen, 0 h, 6 h, 12 h, 24 h, 48 h, 72 h and 96 h buds were sampled as above. Controls were mock inoculated with sterile distilled water. The frozen buds of 0.5 g were homogenized in a mortar and pestle with 3.0 mL of ice-cold phosphoric acid buffer (pH7.8) and a small amount of quartz sand. The supernatant was centrifuged at 4,000 ×g for 15 min at 4°C. The supernatant was used as a crude enzyme solution. After incubation at 25°C (for blank control, incubated in a boiling water for 10 min), 0.2 mL supernatant was mixed with 1.5 mL phosphoric acid buffer (pH7.8) added with polyvinylpyrrolidone (PVP) and 0.3 mL 0.1 mol/L H_2O_2 in 10 mL tube, which initiated the reaction. The decrease in absorbance was recorded by Lambda 35 UV WinLab software (Perkin Elmer, China) followed by the decomposition of H_2O_2 at 240 nm, and measured for a total of 3 min [29]. One unit of enzyme activity (U) was defined as A_{240} reduced 0.1 enzyme quantity per min per g. The enzyme activity was calculated as follows:

$$U = \frac{\Delta A_{240} \times V_T}{0.1 \times V_1 \times t \times FW}$$

$$\Delta A_{240} = A_{S0} - \frac{A_{S1} + A_{S2}}{2}$$

Among them, A_{S0} means the absorbance of the blank control, A_{S1} and A_{S2} stand for the absorbance of the samples, V_T means the total volume of the crude enzyme solution (mL), V_1 represents the volume of the detected crude enzyme solution (mL), *FW* means the fresh weight of sample (g) and t means the time from adding H_2O_2 to the last time (min). The activity of catalase was calculated by the activity level of inoculation minus the level of the mock at each corresponding time point.

RNA Extraction

Total RNA of Yacheng05–179 and Liucheng03–182 was extracted with Trizol reagent (Invitrogen, China) according to the manufacturer's protocol. The quality of the RNA was monitored by measuring the absorbance at 260 nm, 280 nm (NanoVue plus, GE, USA), and the 28 S and 18 S were examined by electrophoresis. DNase I (Promega, China) was used to remove DNA contamination. The first-strand cDNA synthesis was completed by the Prime-Script™ RT Reagent Kit (TaKaLa, China).

Isolation of Sugarcane Catalase Gene

Eighty-six sugarcane expressed sequence tags (ESTs), which share homology to the mRNA sequence of the sugarcane catalase gene (GenBank Accession No. CF572408.1), were obtained from sugarcane sequence database (cultivated sugarcanes (taxid:286192); wild sugarcane (taxid:62335); sugarcane (taxid:128810); sugarcane (taxid:4547)) in GenBank. The CAP3 sequence assembly program (http://pbil.univ-lyon1.fr/cap3.php) was used to construct a putative cDNA sequence of sugarcane catalase gene (*ScCAT1*). Then the cDNA of *ScCAT1* was amplified with primers designed on the basis of this assembled sequence.

Amplification of *ScCAT1* gene was performed with primers ScCAT1-cDNAF and ScCAT1-cDNAR (Table 1) on first-strand cDNA template of Yacheng05–179 post 48 h *S. scitamineum* inoculation in a Mastercycler (Eppendorf, Hamburg, Germany). The reaction was performed at 94°C for 5 min, and then subjected to 35 cycles of 94°C for 30 s, 57°C for 45 s and 72°C for 1 min, followed by a final extension at 72°C for 10 min. The expected length of the amplified fragments was 1,658 bp. PCR products were gel-purified, cloned into the pMD18-T vector (TaKaLa, China) and sequenced (Shenggong, China).

Protein Structural Analysis and Phylogenetic Tree Construction

Sequence data were analyzed by ORF (open reading frame) Finder (http://www.ncbi.nlm.nih.gov/gorf/gorf.html), ProtParam (http://web.expasy.org/protparam/), SignalP 4.0 Server (http://www.cbs.dtu.dk/services/SignalP/), TargetP 1.1 server (http://www.cbs.dtu.dk/services/TargetP/), SMART (http://smart.embl-heidelberg.de/), PSORT Prediction (http://psort.hgc.jp/form.html). After blast comparison, the amino acid sequence of *ScCAT1* was aligned with published plant catalases, including *Sorghum bicolor* catalase (XP_002437631.1), *Z. mays* catalase (NP_001241808.1), *Oryza sativa* catalase (A2YH64.2), *Brachypodium distachyon* catalase (XP_003563243.1), *Puccinellia chinampoensis* catalase (ADN94253.1), *H. vulgare* catalase (P55307.1), *Triticum aestivum* catalase (P55313.1) and *Setaria italica* catalase (XP_004966515.1). Multiple alignment of the amino acid sequences was carried out using the Clustal W software. The phylogenetic tree was constructed following the neighbor-joining (NJ) method (1,000 bootstrap replicates) by using the MEGA 5.05 software [13].

Agrobacterium-mediated Transient Expression and Subcellular Localization Assay

For the studying of subcellular location constructs of pCAMBIA 2300-GFP were generated, *ScCAT1* gene was PCR amplified from pMD18-T-*ScCAT1* using primers ScCAT1-SublocF and ScCAT1-SublocR (*Xba* I and *Spe* I sites) as indicated in Table 1. The fragment was inserted into the vector of pCAMBIA 2300-GFP to construct the fusion protein expression vector of *35S::ScGluA1::GFP* (Fig. S1). The recombinant plasmid, was verified by PCR, double digestion and sequencing followed by transfection of the competent cells of *Agrobacterium tumefaciens* strain EHA105.

The assay for *Agrobacterium*-mediated transformation referred to the method as previously described [30]. *Agrobacterium* strain EHA105 carrying the indicated construct was grown overnight in LB liquid medium containing 35 $\mu g \cdot mL^{-1}$ rifampicin and 50 $\mu g \cdot mL^{-1}$ kanamycin. The suspension at $OD_{600} = 0.8$ (containing 200 μM acetosyringone) was infiltrated into 4–5 weeks old *N. benthamiana* leaves and cultured at 24°C for 2 days (16 h light/8 h darkness). The subcellular localization of the fusion protein was

Table 1. Primers used in this study.

Primer	Sequence	Strategy
ScCAT1-cDNAF	GCGGCTTCCTACTCCTCGTCCTT	RT-PCR
ScCAT1-cDNAR	CGCCTGCTTTCTTCCTTGTCAATC	RT-PCR
ScCAT1-SublocF	TGCTCTAGAATGGATCCGTACAAGCAC	Subcellular localization vector construction
ScCAT1-SublocR	GGACTAGTCATGTTTGGCTTCAGGTTCAG	Subcellular localization vector construction
ScCAT1-32aF	CGGAATTCATGGATCCGTACAAGCAC	prokaryotic expression vector construction
ScCAT1-32aR	CCCTCGAGTTACATGTTTGGCTTCAGGT	prokaryotic expression vector construction
GAPDH-QF	CACGGCCACTGGAAGCA	Q-PCR
GAPDH-QR	TCCTCAGGGTTCCTGATGCC	Q-PCR
ScCAT1-QF	CTCTGCTCCTCCAATCCC	Q-PCR
ScCAT1-QR	GAGTGACCTCAAAGAAACCCT	Q-PCR
ScCAT1-1301F	GCTCTAGAATGGATCCGTACAAGCACCG	Over expression vector construction
ScCAT1-1301R	TCCCCCGGGTTACATGTTTGGCTTCA	Over expression vector construction

visualized using fluorescence microscopy (Axio Scope A1, Germany).

Expression in *Escherichia coli* Rosetta Cells

To study the function of *ScCAT1* in prokaryotes, the *ScCAT1* ORF was amplified by PCR from the identified cDNA clone using the primers ScCAT1-32aF and ScCAT1-32aR (Table 1) followed by 94°C for 4 min; 94°C for 30 s, 58°C for 30 s, 72°C for 1.5 min, 35 cycles; and 72°C for 10 min. The *ScCAT1* ORF with *Eco*R I and *Xho* I sites was subcloned into pET 32a (+) vector with *Eco*RI-*Xho*I sites in the *E. coli* Rosetta strains to generate the putative recombinant (pET 32a-*ScCAT1*). The desired recombinant plasmid was identified by PCR amplification, double digestion and sequencing. The prokaryotic expression product was induced in 1.0 mM isopropyl β-D-thiogalactoside (IPTG) for 8 h at 37°C and analyzed by sodium dodecyl sulfate-polyacrylamide gel electrophoresis (SDS-PAGE). Meanwhile, LB medium with blank *E. coli* Rosetta strains (blank) or Rosetta+pET 32a (control) was each induced in IPTG for 8 h and also analyzed by SDS-PAGE [26,31].

During the response of *E. coli* cells to various abiotic stresses, the growth of *E. coli* Rosetta strains transformed with pET 32a and pET 32a-*ScCAT1* was analyzed using spot assay in different treatments of $CuCl_2$, $CdCl_2$ or NaCl. When OD_{600} of the LB medium (plus 170 μg·mL^{-1} chloramphenicol and 80 μg·mL^{-1} ampicillin) with *E. coli* cells reached 0.6, IPTG was added to a final concentration of 1.0 mM, and then continued growth for 12 h at 37°C. Thereafter, the cultures were diluted to 0.6 (OD_{600}), and then to two levels (10^{-3} and 10^{-4}). Ten microlitres from each dilutions was spotted on LB plates (plus 170 μg·mL^{-1} chloramphenicol and 80 μg·mL^{-1} ampicillin) containing $CuCl_2$ (250, 500 and 750 μM), $CdCl_2$ (250, 500 and 750 μM) or NaCl (250, 500 and 750 mM) [26,31]. All these plates were cultured in 37°C overnight and photographed.

The effect of 750 μM $CuCl_2$, 750 μM $CdCl_2$ and 250 mM NaCl on the growth of *E. coli* strains with pET 32a-*ScCAT1* or pET 32a was studied in LB medium followed with Su et al. [26]. As above, when cells were grown as earlier described and then diluted to 0.6 (OD_{600}), 400 μL of cells were transferred into 10 mL of LB medium containing 170 μg·mL^{-1} chloramphenicol and 80 μg·mL^{-1} ampicillin, 750 μM $CuCl_2$ or 750 μM $CdCl_2$ or 250 mM NaCl [32]. Cultures were shaken at 200 rpm at 37°C and growth of the cells was measured at every 2 h by Lambda 35 UV WinLab software (Perkin Elmer, USA).

Real-time Quantitative PCR Analysis

The each time point of 0 h, 6 h, 12 h, 48 h and 72 h during Yacheng05–179-smut incompatible interaction and Liucheng03–182-smut compatible interaction, as well as mock plants inoculated with sterile distilled water at each corresponding time point, were used to analyze the expression patterns of the *ScCAT1*. The relative expression of the target gene under certain biotic stress was calculated by the expression level of the inoculated sample minus the level of the mock at each corresponding time point. For tissue-specific expression of *ScCAT1*, the leaf, bud, stem epidermis and stem pith of sugarcane variety Yacheng05–179 were used as experimental materials. The expression of *ScCAT1* under the stresses of SA, MeJA, ABA, H_2O_2, PEG, $CuCl_2$ and NaCl were also performed by real-time quantitative PCR (Q-PCR).

The method of Q-PCR followed the instruction of the SYBR Green Master (ROX) (Roche, China) on a 7500 Q-PCR system (Applied Biosystems, USA). The *GAPDH* gene (GAPDH-QF/GAPDH-QR) (Table 1) was chosen as the internal control of the Q-PCR [28]. According to the sequence of *ScCAT1*, a pair of specific primers ScCAT1QF/ScCAT1-QR was designed using the Primer Premier 5.0 software. Q-PCR was carried out with FastStart Universal SYBR Green Master (ROX) in a 20 μL volume containing 10 μL FastStart Universal SYBR Green PCR Master (ROX), 0.5 μM of each primer and 2.0 μL template (100 × diluted cDNA). PCR with distilled water as template was performed as control. The Q-PCR reaction condition was held at 50°C for 2 min, 95°C for 10 min, 40 cycles of 95°C for 15 s and 60°C for 1 min. When the reaction was complete, the melting curve was analyzed. Each Q-PCR was repeated three times. The $2^{-\triangle\triangle Ct}$ method was adopted to analyze the Q-PCR results [33].

Histochemical Assay

For analysis of defense response caused by *ScCAT1* over-expression, primers of ScCAT1-1301F/ScCAT1-1301R in Table 1 (*Xba* I -*Sma* I sites) were used to construct binary vector expressing *ScCAT1* (pCAMBIA 1301-*ScCAT1*). *Agrobacterium* strain EHA105 containing recombinant vector and pCAMBIA 1301 vector alone were grown overnight in LB liquid medium (plus 35 μg·mL^{-1} rifampicin and 50 μg·mL^{-1} kanamycin) at 28°C. Then cultures were pelleted and resuspended in MS liquid medium (plus 200 μM acetosyringone) at $OD_{600} = 0.8$ and infiltrated into *N. benthamiana* leaves at eight-leaf stage [34]. Plants were incubated at 24°C for 1–2 days (16 h light/8 h darkness), which were employed to following different tests.

DAB (3,3′-diaminobenzidinesolution) staining. Agroinfiltrated leaves were incubated in 1.0 mg·mL^{-1} DAB-HCl solution in the dark overnight. Then the leaves were destained by boiling in 95% ethanol for 5 min. The bronzing color of the leaves for H_2O_2 detection which generated in leaves after treatments was photographed [35].

Trypan blue staining. The infiltrated leaves were boiled for 5 min in lactophenol-ethanol trypan blue solution (10 mL glycerol, 10 mL lactic acid, 10 g phenol, 10 mg trypan blue, 30 mL absolute ethanol and 10 mL distilled water). Then the leaves were destained in 2.5 g·mL^{-1} choloral hydrate in distilled water and the blue color indicates the cell death [30].

Measurement of ion conductivity. It was performed as previously described with some modifications [36]. Six leaf discs (11 mm in diameter per leaf) were cut and washed in distilled water and then incubated in 20 mL of distilled water and shaken slowly at room temperature for 60 min. After that, electrolyte leakage was measured using a conductivity meter (SevenEasy, METTLER TOLEDO, Switzerland).

Results

Enzyme Activity of Catalase

To analyze the correlation between catalase activity and smut resistance, the changes in enzyme activity in smut challenged Yacheng05–179 (resistant) and Liucheng03–182 (susceptible) cultivars were studied and different patterns of enzyme activity change were found. As shown in Fig. 1, activity of catalase in Yacheng05–179 increased at 6 h (118.19 U) and reached the peak value of 254.14 U at 24 h compared to its mock. It should be noted that the catalase activity in the resistant sugarcane variety (Yacheng05–179) was always higher than that of the mock at all the sampling time points, but the tendency was an increased at 12 h and 120 h, decreased at 6 h and 24 h and almost unchanged at 0 h, 48 h, 72 h in the susceptible one (Liucheng03–182). In addition, when compared with the mock, the catalase activity was much higher in the resistant variety (from 40.00 U to 254.14 U) than that in the susceptible one (from −76.94 U to 52.97 U) at all the time points. These results suggest that there are positive

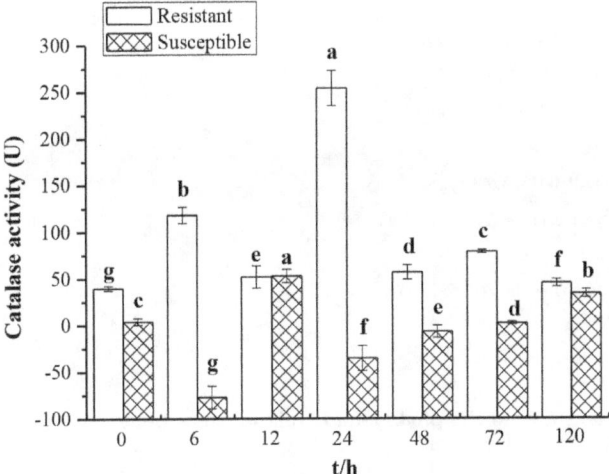

Figure 1. The catalase activity in smut resistant (Yacheng05–179) and smut susceptible (Liucheng03–182) sugarcane varieties inoculated with *Sporisorium scitamineum*. All data points (deduction its mock) are means±SE (n = 3).

correlations between catalase activity and smut resistance in sugarcane.

Cloning and Sequence Analysis of *ScCAT1*

To study the sugarcane catalase at the molecular level, a 1,658 bp full-length catalase gene *ScCAT1* (GenBank Accession No. KF664183) was cloned using RT-PCR method combined with *in silico* cloning technique. There were 1,479 nucleotides in its ORF (Fig. S2). *ScCAT1* encoded a predicted polypeptide of 492 amino acids with no signal peptide. The predicted protein had a molecular mass of 56.81 kDa with a pI of 6.72. A search at the NCBI for conserved protein domains indicated that ScCAT1 belonged to a member of catalase-like superfamily. The catalytic active site and heme binding motifs of ScCAT1 were detected by Motif Scan Online program. 17 amino acids at the position of 54–70 (FDRERIPERVVHARGAS) were reported to be a catalase active site signature, and the heme-ligand signature was detected at the position of 344–352 (RIFSYADTQ). These data suggested clearly that sugarcane *ScCAT1* encoded a putative peroxisomal catalase. Furthermore, it also predicted that ScCAT1 contains no transmembrane helix domain, implying that ScCAT1 is not a membrane located or secretory protein.

A GenBank Blastp comparison showed that ScCAT1 exhibited high identity with other plant catalases, including *S. bicolor* catalase (XP_002437631.1) (97.97% identity), *Z. mays* catalase (NP_001241808.1) (97.56% identity), *O. sativa* catalase (A2YH64.2) (94.92% identity), *B. distachyon* catalase (XP_003563243.1) (93.29% identity), *P. chinampoensis* catalase (ADN94253.1) (92.07% identity), *H. vulgare* catalase (P55307.1) (91.67% identity), *T. aestivum* catalase (P55313.1) (91.26% identity) and *S. italica* catalase (XP_004966515.1) (87.02% identity). Phylogenetic analysis (Fig. 2) revealed that ScCAT1 was closely related to *S. bicolor* catalase (XP_002437631.1), *Z. mays* catalase (NP_001241808.1) and *S. italica* catalase (XP_004966515.1), with the homologies of 97.97%, 97.56% and 87.02%, respectively.

Subcellular Localization of ScCAT1

To further understand the function of *ScCAT1* gene, its subcellular localization was conducted. *ScCAT1* was recombined into plant expression vector pCAMBIA 2300 between the sites of the *35S* promoter and *GFP* (Figs. S1 and S3), and its location was characterized by transient expression of the target gene and GFP in *N. benthamiana* leaves with *Agrobacterium*-mediated transformation. After 2 days of cultivation, the infiltrated leaves were harvested and the reporter protein GFP was observed under a fluorescence microscope. The results revealed that 35S::ScCAT1::GFP was located in plasma membrane and cytoplasm (Fig. 3). In contrast, GFP was shown in the nucleus, cytoplasm and plasma membrane cells transiently transfected with 35S::GFP.

Expressions of ScCAT1 in *E. coli*

As reported before, different stresses, such as copper (Cu), cadmium (Cd), high temperature, wounding, ethylene (ET), H_2O_2, SA, jasmonic acid (JA), ABA and other inducers, could trigger an induction of plant catalases [11,13,21,37,38]. To study the function of ScCAT1 in response to different kinds of adverse environments *in vivo*, pET 32a-*ScCAT1* (Fig. S4A) was transformed into *E. coli* Rosetta cell. The recombinant protein of 62 kDa was specifically induced and accumulated approximately after 8 h IPTG induction on the SDS-PAGE (Fig. S4B).

The growth of gene-expressed cells (Rosetta+pET 32a-*ScCAT1*) and mock (Rosetta+pET 32a) was analyzed on LB plates with different supplements (Figs. 4A, B, C, and D). After one day culture, Rosetta+pET 32a-*ScCAT1* showed an increased number of colonies as compared to the control cells on LB plates containing $CuCl_2$, $CdCl_2$ and NaCl. The growth was also analyzed in the LB liquid medium containing 750 μM $CuCl_2$, 750 μM $CdCl_2$ and 250 mM NaCl (Figs. 4E, F, G and H). All the Rosetta+pET 32a-*ScCAT1* cells showed faster growth as compared to that of the mock which revealed that ScCAT1 had an effect on increasing the tolerance to $CuCl_2$, $CdCl_2$ and NaCl. These results demonstrate that the recombinant protein of ScCAT1 enhanced growth ability of prokaryotic *E. coli* Rosetta strains in stress conditions.

Tissue-specific Expression Analysis of *ScCAT1*

The relative expression of *ScCAT1* was detected in four kinds of sugarcane tissues, including leaf, bud, stem epidermis and stem pith. As showed in Fig. 5, the bud exhibited the highest mRNA expression, while the mRNA expression of stem epidermis and stem pith was at a moderate level. The leaf showed a relatively low level in comparison with the other three kinds of tissues.

ScCAT1 Expression in Response to Different Stress Treatments

Smut challenged sugarcane (Yacheng05–179 and Liucheng03–182) buds were detected by Q-PCR for examination whether the expression of *ScCAT1* was induced or inhibited (Fig. 6). In order to eliminate the influence of wounding, the relative expression of the target gene was calculated by the expression level of the inoculated sample minus the level of the mock at each corresponding time point. As indicated in Fig. 6, after the inoculation of smut pathogen, the mRNA expression of *ScCAT1* in resistant variety Yacheng05–179 was higher than that in susceptible variety Liucheng03–182. During the sugarcane-smut incompatible interaction, the transcript of *ScCAT1* in Yacheng05–179 began was elevated as early as 6 h post-inoculation (6 hpi), while that of *ScCAT1* in Liucheng03–182 appeared delayed (12 hpi). Furthermore, the transcript of *ScCAT1* in Yacheng05–179 and Liucheng03–182 reached the maximum at 48 h, but the expression in incompatible interaction was 1.55 times that of the compatible one, and then decreased in both. During the whole process of

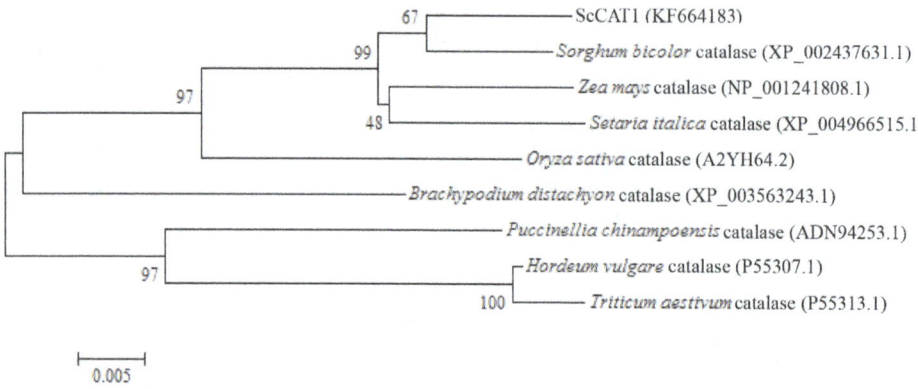

Figure 2. Phylogenetic trees based on catalase amino acid sequences, showing the phylogenetic relationships between ScCAT1 (KF664183) and the catalases from other plant species. Neighbor-joining method was used.

interaction, the transcript of *ScCAT1* in the incompatible cultivar almost always higher than that of the compatible, except at 12 h. These data reveal that the up-regulation of *ScCAT1* expression was most probably associated with smut resistance in sugarcane.

Expression of *ScCAT1* in response to various abiotic stimuli in Yacheng05–179 plantlets was checked after treatment with 5 mM SA, 10 mM H_2O_2, 25 μM MeJA, 100 μM ABA, 25% PEG, 250 mM NaCl and 100 μM $CuCl_2$, and the results shown in Fig. 7. Interestingly, *ScCAT1* showed a positive response to exogenous stresses, including plant hormones stresses of SA, MeJA and ABA, oxidative stress of H_2O_2, hyper-osmotic stresses of PEG and NaCl, as well as mental stress of $CuCl_2$. *ScCAT1* transcription was always up-regulated and the expression level usually increased steadily from 0 h to 24 h or 48 h after post-treatment with these seven exogenous inducers. These results suggest that *ScCAT1* may be a positive responsive component of abiotic stress in sugarcane.

Transient Over-expression of *ScCAT1* in *N. benthamiana* Leaves Induces Hypersensitive Reaction Response

To test whether *ScCAT1* can induce HR and immunity in plant, *ScCAT1* was transient over-expressed in *N. benthamiana* leaves by infiltration with *Agrobacterium* EHA105 carrying pCAMBIA 1301 (mock) and pCAMBIA 1301-*ScCAT1*. The results showed that at the time point of 48 h after infiltration, a typical HR symptom, darker DAB staining and enhanced electrolyte leakage, was found in the leaves expressing the target gene (Figs. 8A and C). Furthermore, injected leaves 5 d after agroinfiltrated by *35S::ScCAT1* presented yellow symptoms (Fig. 8B). What is more, cell death measured by qualitative trypan blue staining showed a darker color than that in mock (Fig. 8B). These results indicate the involvement of *ScCAT1* in cell death responses.

Discussion

Fungal disease is a major concern worldwide for sugarcane production and most other crops. During plant-pathogen interac-

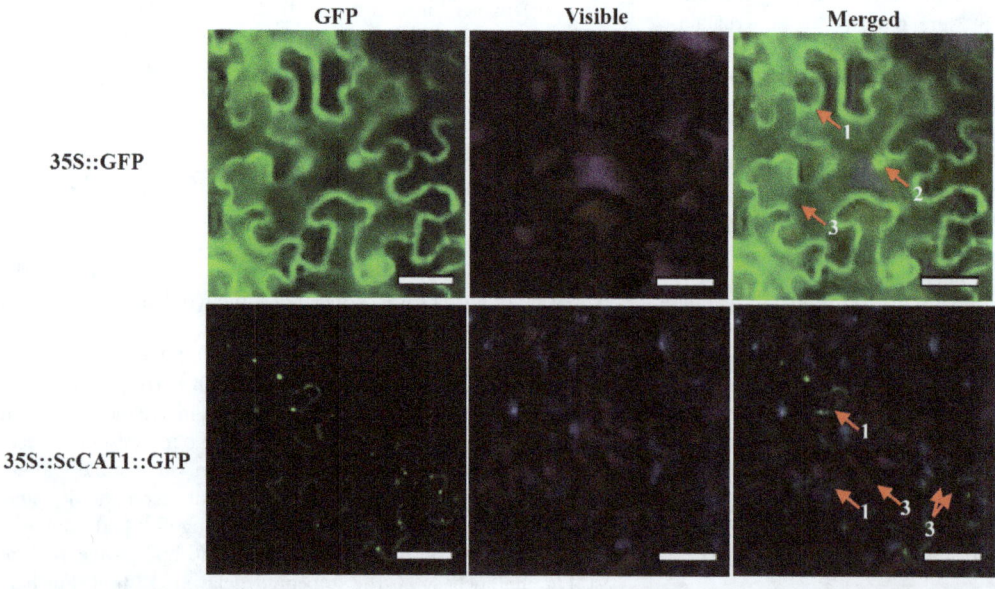

Figure 3. Subcellular localizations of ScCAT1 and empty vector in *Nicotiana benthamiana* leaves 48 h after infiltration. The epidermal cells were used for taking images of green fluorescence, visible light and merged light. Read arrows 1, 2 and 3 indicated plasma membrane, nucleus and cytoplasm, respectively. Bar = 50 μm.

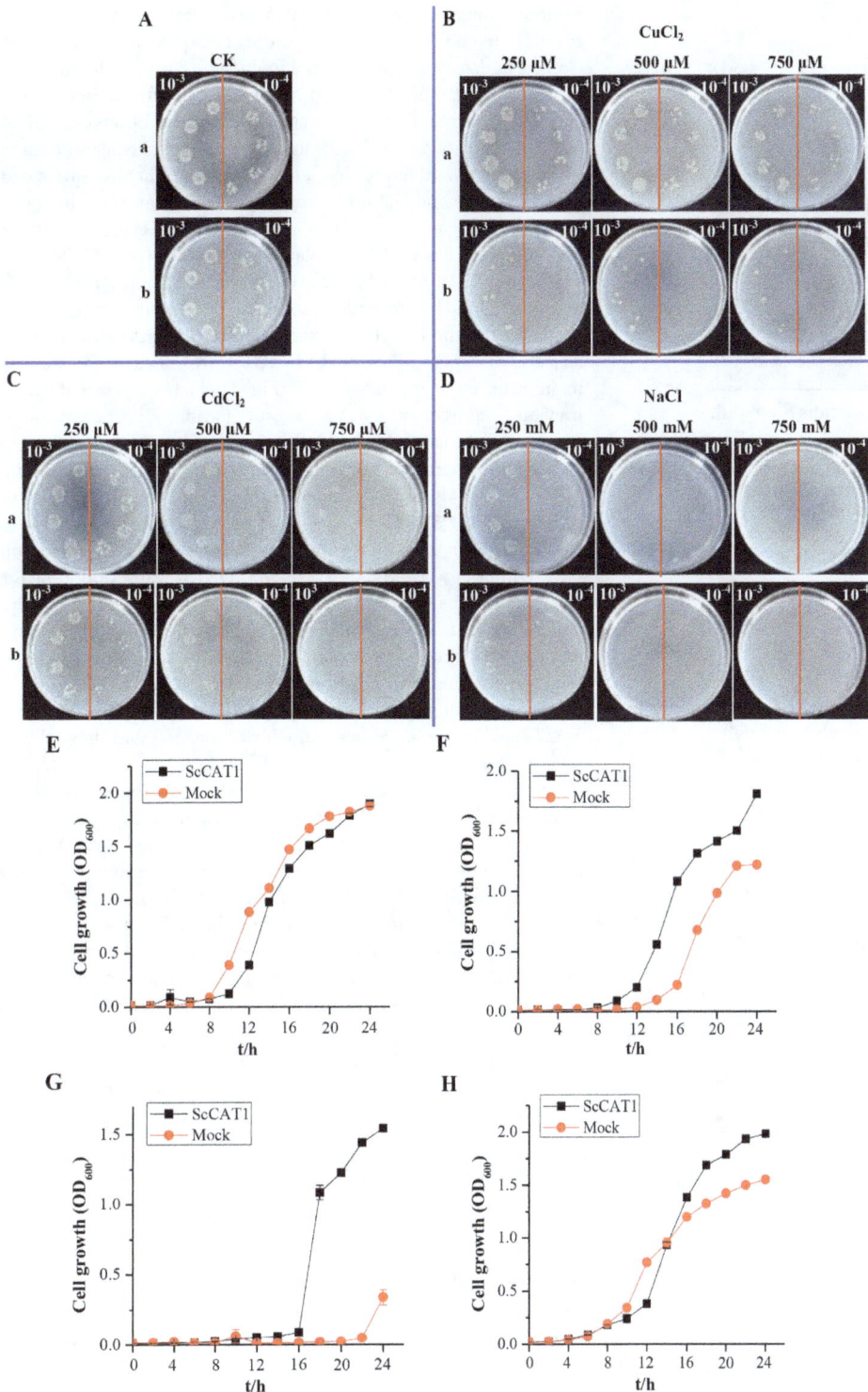

Figure 4. Spot assays of Rosetta+pET 32a-*ScCAT1* (a) and Rosetta+pET 32a (mock) (b) on LB plates with CuCl₂, CdCl₂ and NaCl (A–D). And liquid culture assay in LB liquid medium with 750 μM CuCl₂, 750 μM CdCl₂ and 250 mM NaCl (E–H). IPTG (isopropyl β-D-thiogalactoside) was added to the cultures of Rosetta+pET 32a-*ScCAT1* and Rosetta+pET 32a to induce the expression of recombinant protein. The cultures were adjusted to $OD_{600} = 0.6$. Ten microliters from 10^{-3} (left side of red line on plate) to 10^{-4} (right side of red line on plate) dilutions were spotted onto LB basal (A) plates or with CuCl₂ (250, 500 and 750 μM) (B), CdCl₂ (250, 500 and 750 μM) (C), NaCl (250, 500 and 750 mM) (D). For studying the growth analysis of *ScCAT1*, Rosetta+pET 32a-*ScCAT1* and Rosetta+pET 32a were grown in LB liquid medium with LB basal medium (E) or with 750 μM CuCl₂ (F), 750 μM CdCl₂ (G), and 250 mM NaCl (H). All data points are means ± SE (n = 3). CuCl₂: copper chloride; CdCl₂: cadmium chloride; NaCl: sodium chloride.

tions, many antifungal components have been identified [39]. Peroxidase (POD) activity increased in resistant sugarcane varieties

(and not in susceptible) implies that it may be related to smut resistance [40]. Our previous report showed that β-1,3-glucanase activity in the

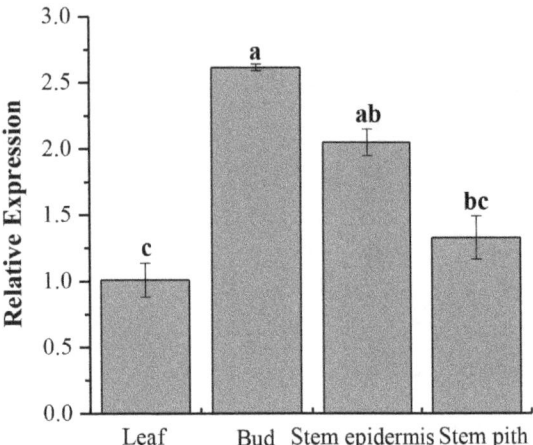

Figure 5. Tissue-specific expression analysis of the *ScCAT1* in sugarcane variety Yacheng05–179. Data are normalized to the *GAPDH* expression level. All data points are means±SE (n = 3).

resistant variety increased faster and lasted longer than that of the susceptible one after challenged by *S. scitamineum*, which showed a positive relationship between the activity of the sugarcane β-1,3-glucanase and smut resistance [26]. Plant catalase, one of the scavenger enzymes, has been also shown to be involved in plant defense and development [41]. Wang et al. found that catalase could be induced by pathogen infection in resistant clam *Meretrix meretrix* [23]. As showed in Fig. 1, the activity of catalase increased in the resistant genotype (Yacheng05–179) after challenged by *S. scitamineum* in comparison to the susceptible cultivar (Liucheng03–182). There appears to be a positive correlation between catalase activity and sugarcane smut resistance. This observation should be repeated in different resistant genotypes.

The capacity of a plant to scavenge H_2O_2 may result from increased activities of scavenger enzymes or up-regulated expression of genes increasing of the levels of the corresponding proteins [42].

Multiple catalase isozymes in plants have been observed. Previous research demonstrated that there were at least six catalase isozymes existing in *A. thaliana* encoded by a multi-gene family including three genes (*cat1*, *cat2* and *cat3*) [14]. In *Z. mays*, three catalase isoenzymes encoded by three different structural genes were observed [38]. A sweet potato catalase *SPCAT1* was cloned from mature leaves treated with ethephon and found that it could alleviate ethephon-mediated leaf senescence and H_2O_2 elevation [13]. Until now, there has been no report on sugarcane catalase genes involved in the sugarcane-smut interaction. In this study, we isolated and characterized a full-length sugarcane catalase gene *ScCAT1* which encoded a polypeptide of 492 amino acids and had high identities with several other plant catalases. Using the method of *Agrobacterium*-mediated transformation in *N. benthamiana* leaves, 35S::ScCAT1::GFP was located in plasma membrane and cytoplasm in cells (Fig. 3) which is consistent with a previous report that catalase mostly localized in peroxisomes, glyoxysome and cytoplasm [11,43].

Recent publications have reported that *E. coli* cells can be enhanced or inhibited under stress expressing recombinant proteins [27,31,32,44]. Some of the protective mechanisms were similar in both eukaryotes and prokaryotes under stress stimuli [45]. Gupta et al. studied an A-2 type DREB transcription factor from extreme halophyte *Salicornia* brachiata and found it conferred abiotic stress tolerance in *E. coli* cells under NaCl, PEG and mannitol treatments, which may be due to the stress regulated function by this transcription factor [32]. Guo et al. tested a sugarcane dirigent protein gene *ScDir* and a metallothionein gene *ScMT2-1-3* in the *E. coli* system, which indicated that they offered different tolerance against PEG, NaCl and mental stresses [27,31]. Chaurasia et al. studied that phytochelatin synthase gene *PCS*, when expressed in *E. coli*, provided better protection against the stresses of heat, salt, carbofuron, cadmium, copper and UV [44]. In the present study, the ScCAT1 recombinant protein expressed in *E. coli* Rosetta cells leads to a better growth under the stresses $CuCl_2$, $CdCl_2$ and NaCl. In eukaryote, the previous studies found that the increased tolerance to stress maybe due to the activity and expression of scavenging enzymes which increased in plants placed in different conditions [12]. It has been proposed that catalase, one

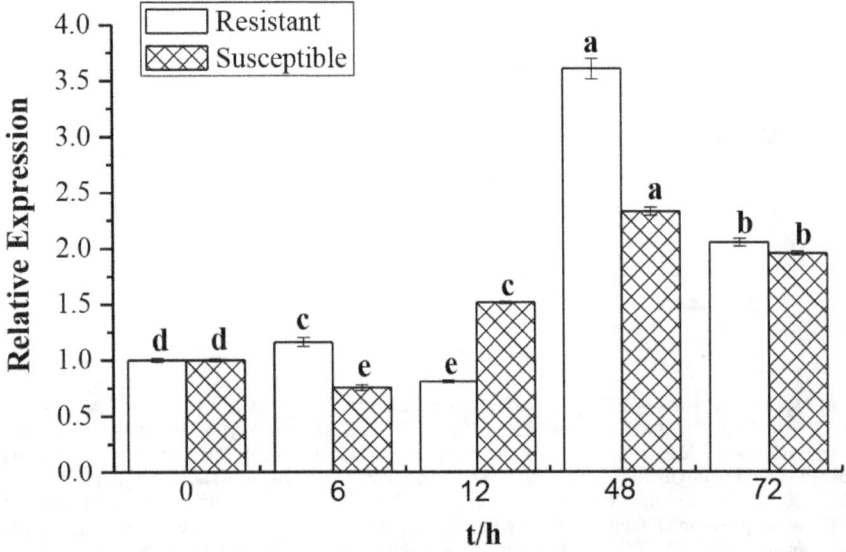

Figure 6. Q-PCR analysis of the *ScCAT1* expression patterns in biosystem of sugarcane-smut (*Sporisorium scitamineum*) interaction. Data was normalized to the *GAPDH* expression level. All data points (deduction its mock) are means±SE (n = 3). Resistant: Yacheng05–179 variety; Susceptible: Liucheng03–182 variety.

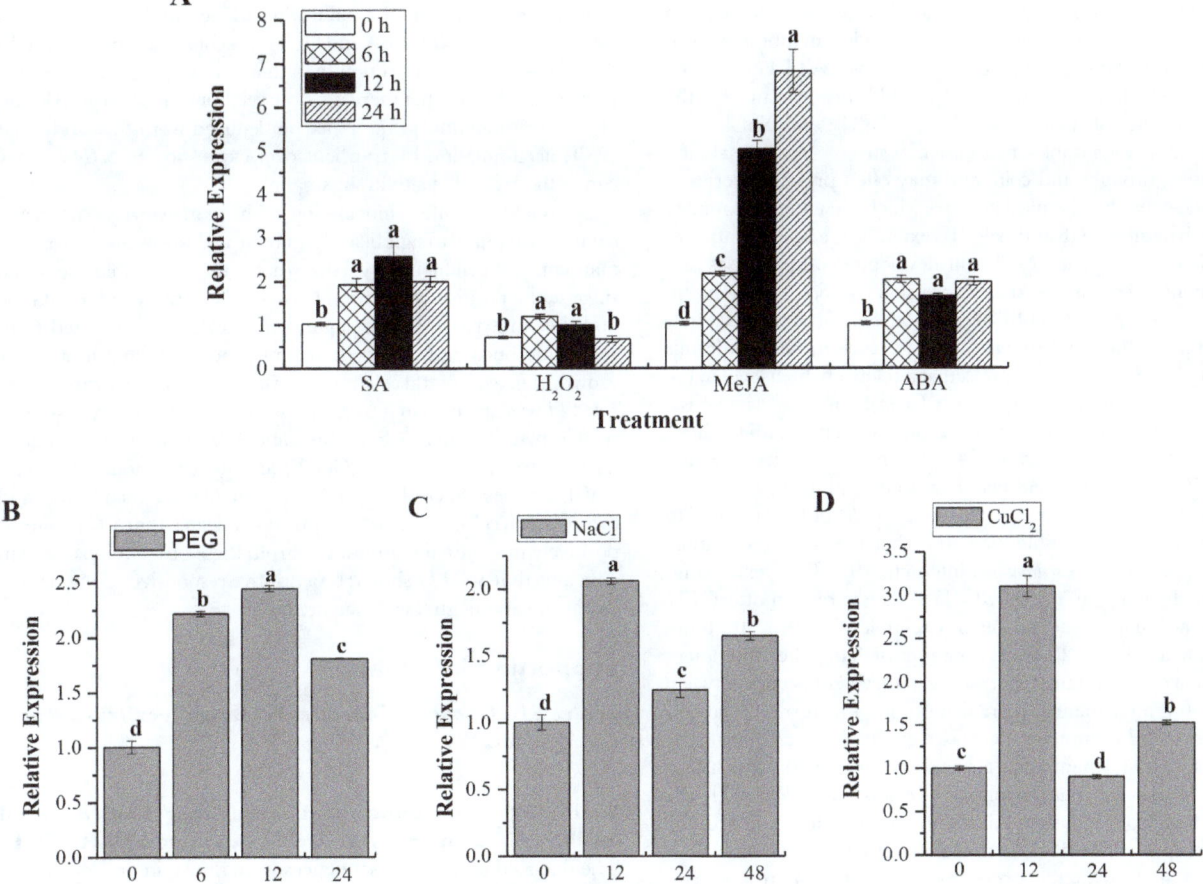

Figure 7. Q-PCR analysis of the *ScCAT1* expression patterns in Yacheng05–179 plantlets with abiotic elicitors. Data are normalized to the *GAPDH* expression level. (**A**) The relative expression of *ScCAT1* under the stresses of 5 mM SA, 10 mM H_2O_2, 25 μM MeJA and 100 μM ABA. (**B**) The relative expression of *ScCAT1* under 25% PEG stress. (**C**) The relative expression of *ScCAT1* under 250 mM NaCl stress. (**D**) The relative expression of *ScCAT1* under 100 μM $CuCl_2$ stress. All data points are means±SE (n = 3). SA: salicylic acid; H_2O_2: hydrogen peroxide; MeJA: methyl jasmonate; ABA: abscisic acid; PEG: polyethylene glycol; NaCl: sodium chloride; $CuCl_2$: copper chloride.

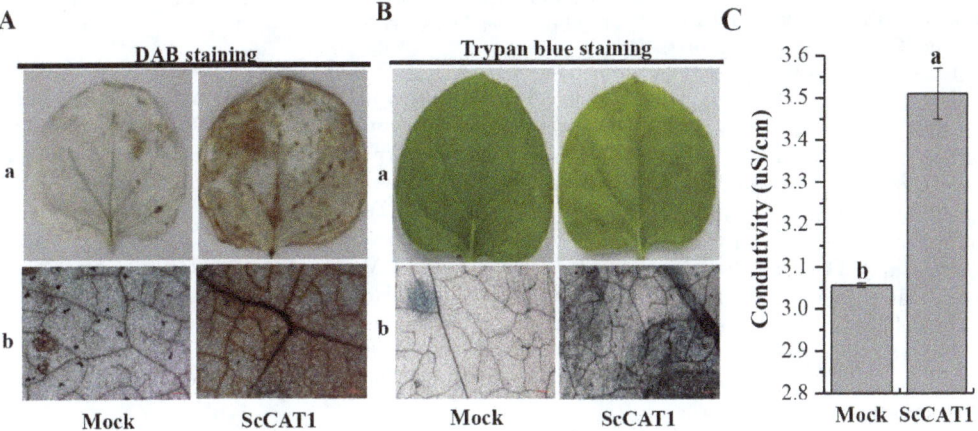

Figure 8. The effect of transient over-expression of *ScCAT1* on immunity induction in *Nicotiana Benthamiana* leaves. (**A**) DAB staining with *N. benthamiana* leaves 48 h after *35S::ScCAT1*-containing *Agrobacterium* strain infiltration to assess the H_2O_2 production; a: images taken by SONY camera; b: images taken by microscope. (**B**) Cell death measured by trypan blue staining of transient expression leaves 5 d after agroinfiltration; a: phenotypes of *N. Benthamiana* 5 d after infiltration taken by SONY camera; b: images of trypan blue staining taken by microscope. (**C**) Conductivity measurement of *N. Benthamiana* leaves infiltrated with *35S::ScCAT1*-containing *Agrobacterium* strain after 48 h. Mock: *Agrobacterium* strain carrying *35S::00*. Bar = 10 μm. All data points are means±SE (n = 3).

of the antioxidant enzymes, can be modulated and controlled in response to excessive iron stress, due to alterations in the electron transport chain and damages to the thylakoidal membranes [46]. Therefore, it is plausible to predict that the ScCAT1 encoded by *ScCAT1* gene cloned in this study could be helpful for the tolerance/stresses of sugarcane to $CuCl_2$, $CdCl_2$ and NaCl.

The plant faces variable environmental stresses like soil salinity, temperature, drought and cold, and may often present a series of physiological and biochemical changes which are a highly complex and disturb plant growth and yield. To examine the accumulation of sugarcane catalase gene in different developmental processes and environmental conditions, the expression of *ScCAT1* gene in sugarcane was analyzed by Q-PCR method (Figs. 5, 6 and 7). Results indicated that while expressed at moderate levels in stem epidermis and stem pith, *ScCAT1* was expressed at a relatively high level in the bud (Fig. 5). Similar to other species of *Ustilago*, the *S. scitamineum* is a parasite of young meristematic tissues and gains entry into the host, exclusively through the bud scales [47]. From above, high expression of *ScCAT1* in sugarcane bud may help to defend against the smut pathogen. In our study, the target transcript of *ScCAT1* was found to be higher in the incompatible interaction than that in the compatible one during sugarcane-*S. scitamineum* interaction (Fig. 6). After the smut pathogen challenge in Yacheng05–179, the expression of *ScCAT1* increased at 6 h and reached the maximum level at 48 h (1.5 times that in Liucheng03–182). As previous reported, the phenomenon of smut hypha entry into the sugarcane bud meristem occurs between 6 h and 36 h after the teliospore deposition [48]. It should be also noted that *ScCAT1* expression decreased gradually after 48 h, but the expression level still maintained at a higher level than that at 0 h, and the gene expression pattern of *ScCAT1* was coincident with the activity change of catalase in this study. So we assume that *ScCAT1* may have a protective effect on smut penetration in sugarcane.

Q-PCR analysis of the expression of *ScCAT1* in response to hydrogen peroxide and plant hormones showed that from 0 h to 24 h its levels increased under the stresses of 10 mM H_2O_2, 5 mM SA, 25 μM MeJA and 100 μM ABA (Fig. 7A). In *Panax ginseng*, *PgCat1* transcript accumulated during 1–12 h of 10 mM H_2O_2 treatment [11]. Maize *Cat1* gene transcript increased in developing embryos by the treatments of 1.5 mM SA, 50 mM H_2O_2, 100 μM JA and 1 mM ABA [38,49]. In the present study, for hyper-osmotic stress, *ScCAT1* mRNA levels increased until 12 h then slightly decreased at 24 h and induced at 48 h under 250 mM NaCl treatment. *ScCAT1* transcript was also stimulated till 24 h after 25% PEG stress (Figs. 7B and C). 500 mM NaCl stress induced the expression of *Cat1* in *Avicennia marina* seedlings till 12 h then subsequently decreased [50]. In *Panax ginseng*, *PgCat1* transcripts accumulated till 24 h then decreased till 72 h after 100 mM NaCl treatment [11]. Plants suffering from NaCl stress not only because of increased osmolarity but also oxidative stress caused by ionic character [51]. In our study, the *ScCAT1* transcript increased 1.5 fold until 48 h under the stress of 100 μM $CuCl_2$. The maximum expression was observed to be 3.0 fold at 12 h after treatment. Previous study revealed that copper toxicity caused ultra structural damage which resulting in the increasing production of ROS [46]. The *Prunus cerasifera Cat1* gene expression and enzyme activity were high for 10 days under 100 mM copper stress [52]. These results lead us to conclude that *ScCAT1* may be a positive responsive component of abiotic stresses in sugarcane.

N. benthamiana has been widely employed in functional character-ization of the target genes by over-expression [30]. Cell death presented at the infected site is the most efficient strategy to restrict pathogen growth and development [30]. The induction of R gene expression, ion fluxes, stimulation of ROS and defense-related hormones, are the common response of cell death [53,54]. Here,

DAB staining showed deep brown in the presence of H_2O_2 in *N. benthamiana* leaves after 48 h infiltration and resulted in an increase of electrolyte leakage (Figs. 8A and C). Trypan blue staining exhibited a darker color post 5 d injection than that in mock (Fig. 8B). Previous studies have shown that there is a close relationship between HR and H_2O_2 accumulation [55]. It can be deduced from this study that H_2O_2 accumulation by transient over-expression of *ScCAT1* may confer the HR cell death in sugarcane.

In conclusion, after inoculation with *S. scitamineum*, sugarcane catalase was found to significantly increase in the resistant variety and maintain at much higher level than that of the susceptible one which suggested a positive correlation between the activity of the catalase and the smut resistance in sugarcane. *ScCAT1* was isolated from sugarcane buds and the recombinant protein resulted in a better growth of *E. coli* Rosetta cells under certain stresses. The expression of *ScCAT1* was up-regulated by smut infection and by different stresses such as plant hormones (SA, MeJA and ABA) treatments, oxidative (H_2O_2) stress, heavy metal ($CuCl_2$) and hyper-osmotic (PEG and NaCl) stresses. ScCAT1 was located in plasma membrane and cytoplasm in cells. Histochemical assays indicated that *ScCAT1* acted positively in sugarcane immunity. From these observations, we can conclude that *ScCAT1* should be a positive responsive component of biotic and abiotic stresses in sugarcane.

Supporting Information

Figure S1　Construction of subcellular localization vector *35S::ScCAT1::GFP*.

Figure S2　Nucleotide acid sequences and deduced amino acid sequences of *ScCAT1* obtained by RT-PCR. The deduced amino acid sequences were shown in one-letter code under the cDNA sequences. The underlines showed the catalase active site signature (FARERIPERVVHARGAS) and the heme-ligand signature (RVFAYADTQ) of ScCAT1.

Figure S3　The enzyme digestion to identify the insert-integrated subcellular localization expression vector *35S::ScCAT1::GFP*. 1, 15,000+2,000 bp DNA marker; 2, *35S::GFP/Xba* I; 3, *ScCAT1* ORF PCR product; 4, *35S::ScCAT1::GFP/Xba* I; 5, *35S::ScCAT1::GFP/Xba* I+*Spe* I; 6, 100 bp ladder DNA marker.

Figure S4　The enzyme digesting identification of insert-integrated prokaryotic expression vector pET 32a-*ScCAT1* (A) and corresponding protein expressions in *Escherichia coli* Rosetta strains (B). (A) 1, 100 bp ladder DNA marker; 2, pET 32a/*Eco*R I; 3, *ScCAT1* ORF PCR product; 4, pET 32a-*ScCAT1/Eco*R I; 5, pET 32a-*ScCAT1/Eco*R I+*Xho* I; 6, 15,000+2,000 bp DNA Marker. **(B)** 1, Protein marker; 2, blank without induction; 3, blank induction for 8 h; 4, control without induction; 5, control induction for 8 h; 6, pET 32a-*ScCAT1* without induction; 7 and 8, pET 32a-*ScCAT1* induction for 4 h and 8 h, respectively. The induced protein was shown by arrow.

Author Contributions

Conceived and designed the experiments: YS JG LX YQ. Performed the experiments: YS JG HL SC SW. Analyzed the data: YS JG LX YQ. Contributed reagents/materials/analysis tools: YS LX YQ. Wrote the paper: YS JG LX YQ AA. Revised and approved the final version of the paper: LX YQ.

References

1. Sundar AR, Barnabas EL, Malathi P, Viswanathan R (2012) A mini-review on smut disease of sugarcane caused by *Sporisorium scitamineum*. In: Mworia J, editor. Botany. Croatia: InTech. 109–128.

2. Agnihotri VP (1990) Diseases of sugarcane and sugarbeet Oxford & IBH Publishing Co Pvt Ltd. New Delhi: 72–103.

3. Hoy JW, Hollier CA, Fontenot DB, Grelen LB (1986) Incidence of sugarcane smut in Louisiana and its effects on yield. Plant Dis 70: 59–60.

4. Singh N, Somai BM, Pillay D (2004) Smut disease assessment by PCR and microscopy in inoculated tissue cultured sugarcane cultivars. Plant Sci 167: 987–994.

5. Solas MT, Pinon D, Vicent C, Legaz ME (1999) Ultrastructural aspects of sugarcane bud infection by *Ustilago scitaminea* teliospores. Sugar Cane 2: 14–18.

6. Waller JM (1970) Sugarcane smut (*Ustilago scitaminea*) in Kenya: II. Infection and resistance. T British Mycol Soc 54: 405–414.

7. Xu LP, Chen RK, Chen PH (2004) Analysis on infection index of smut caused by *Ustilago scitaminea* in sugarcane segregated population. Chin J Trop Crop 25: 33–36.

8. Lin YQ, Chen RK, Gong DM (1996) Analysis of quantitative inheritance for smut resistance in sugarcane. J Fujian Agr U China 25: 271–275.

9. Dussle CM, Quint M, Melchinger AE, Xu ML, Lubberstedt T (2003) Saturation of two chromosome regions conferring resistance to SCMV with SSR and AFLP markers by targeted BSA. Theor Appl Genet 106: 485–493.

10. Lakshmanan P, Geijskes RJ, Aitken KS, Grof CLP, Bonnett GD, et al. (2005) Sugarcane biotechnology: the challenges and opportunities. In Vitro Cell Dev-Pl 41: 345–363.

11. Purev M, Kim YJ, Kim MK, Pulla RK, Yang DC (2010) Isolation of a novel catalase (*Cat1*) gene from *Panax ginseng* and analysis of the response of this gene to various stresses. Plant Physiol Bioch 48: 451–460.

12. Mhamdi A, Queval G, Chaouch S, Vanderauwera S, Van-Breusegem F, et al. (2010) Catalase function in plants: a focus on *Arabidopsis* mutants as stress-mimic models. J Exp Bot 61: 4197–4220.

13. Chen HJ, Wu SD, Huang GJ, Shen CY, Afiyanti M, et al. (2012) Expression of a cloned sweet potato catalase *SPCAT1* alleviates ethephon-mediated leaf senescence and H$_2$O$_2$ elevation. J Plant Physiol 169: 86–97.

14. Frugoli JA, Zhong HH, Nuccio ML, McCourt P, McPeek MA, et al. (1996) Catalase is encoded by a multigene family in *Arabidopsis thaliana* (L.) Heynh. Plant physiol 112: 327–336.

15. Willekens H, Villarroel R, Van-Montagu M, Inze D, Van-Camp W (1994) Molecular identification of catalases from *Nicotiana plumbaginifolia* (L.). FEBS Lett 352: 79–83.

16. Guan L, Scandalios JG (1995) Developmentally related responses of maize catalase genes to salicylic acid. P Natl A Sci 92: 5930–5934.

17. Skadsen RW, Schulze-Lefert P, Herbst JM (1995) Molecular cloning, characterization and expression analysis of two catalase isozyme genes in barley. Plant Mol Biol 29: 1005–1014.

18. Drory A, Woodson WR (1992) Molecular cloning and nucleotide sequence of a cDNA encoding catalase from tomato. Plant Physiol 100: 1605.

19. Kwon SI, An CS (2001) Molecular cloning, characterization and expression analysis of a catalase cDNA from hot pepper (*Capsicum annuum* L.). Plant Sci 160: 961–969.

20. Du YY, Wang PC, Chen J, Song CP (2008) Comprehensive functional analysis of the catalase gene family in *Arabidopsis thaliana*. J Integr Plant Biol 50: 1318–1326.

21. Guan ZQ, Chai TY, Zhang YX, Xu J, Wei W (2009) Enhancement of Cd tolerance in transgenic tobacco plants overexpressing a Cd-induced catalase cDNA. Chemosphere 76: 623–630.

22. Williamson JD, Scandalios JG (1992) Differential response of maize catalases and superoxide dismutases to the photoactivated fungal toxin cercosporin. The Plant J 2: 351–358.

23. Wang C, Yue X, Lu X, Liu BZ (2013) The role of catalase in the immune response to oxidative stress and pathogen challenge in the clam *Meretrix meretrix*. Fish Shellfish Immun 34: 91–99.

24. Casu RE, Dimmock CM, Chapman SC, Grof CP, McIntyre CL, et al. (2004) Identification of differentially expressed transcripts from maturing stem of sugarcane by in silico analysis of stem expressed sequence tags and gene expression profiling. Plant Mol Biol 54: 503–517.

25. Moosawi-Jorf SA, Mahin BI (2007) In vitro detection of yeast-like and mycelial colonies of *Ustilago scitaminea* in tissue-cultured plantlets of sugarcane using polymerase chain reaction. J Appl Sci 7: 3768–3773.

26. Su YC, Xu LP, Xue BT, Wu QB, Guo JL, et al. (2013) Molecular cloning and characterization of two pathogenesis-related β-1,3-glucanase genes *ScGluA1* and *ScGluD1* from sugarcane infected by *Sporisorium scitamineum*. Plant Cell Rep 32: 1503–1519.

27. Guo JL, Xu LP, Su YC, Wang HB, Gao SW, et al. (2013) *ScMT2-1-3*, a metallothionein gene of sugarcane, plays an important role in the regulation of heavy metal tolerance/accumulation. BioMed Res Int: doi.10.1155/2013/904769.

28. Que YX, Xu LP, Xu JS, Zhang JS, Zhang MQ, et al. (2009) Selection of control genes in real-time qPCR analysis of gene expression in sugarcane. Chin J Trop Crop 30: 274–278.

29. Hao JJ, Kang ZL, Yu Y (2001) Plant physiology experiment technology. In: Hao JJ, editor. China: Liaoning Sci Technol Press. 53–55.

30. Hwang IS, Hwang BK (2012) Requirement of the cytosolic interaction between pathogenesis-related protein10 and leucine-rich repeat protein1 for cell death and defense signaling in pepper. Plant Cell 24: 1675–1690.

31. Guo JL, Xu LP, Fang JP, Su YC, Fu HY, et al. (2012) A novel dirigent protein gene with highly stem-specific expression from sugarcane, response to drought, salt and oxidative stresses. Plant Cell Rep 31: 1801–1812.

32. Gupta K, Agarwal PK, Reddy MK, Jha B (2010) SbDREB2A, an A-2 type DREB transcription factor from extreme halophyte *Salicornia brachiata* confers abiotic stress tolerance in *Escherichia coli*. Plant Cell Rep 29: 1131–1137.

33. Livak KJ, Schmittgen TD (2001) Analysis of relative gene expression data using real-time quantitative PCR and the $2^{-\triangle\triangle Ct}$ method. Methods 25: 402–408.

34. Choi HW, Kim YJ, Hwang BK (2011) The hypersensitive induced reaction and leucine-rich repeat proteins regulate plant cell death associated with disease and plant immunity. Mol Plant Microbe In 24: 68–78.

35. Huckelhoven R, Fodor J, Trujillo M, Kogel KH (2000) Barley Mla and Rar mutants compromised in the hypersensitive cell death response against *Blumeria graminis* f. sp. *hordei* are modified in their ability to accumulate reactive oxygen intermediates at sites of fungal invasion. Planta 212: 16–24.

36. Hwang IS, Hwang BK (2011) The pepper mannose-binding lectin gene *CaMBL1* is required to regulate cell death and defense responses to microbial pathogens. Plant Physiol 155: 447–463.

37. Scandalios JG, Acevedo A, Ruzsa S (2000) Catalase gene expression in response to chronic high temperature stress in maize. Plant Sci 156: 103–110.

38. Guan LM, Zhao J, Scandalios JG (2000) Cis-elements and trans-factors that regulate expression of the maize *Cat1* antioxidant gene in response to ABA and osmotic stress: H$_2$O$_2$ is the likely intermediary signaling molecule for the response. The Plant J 22: 87–95.

39. Takken FLW, Joosten MHAJ (2000) Plant resistance genes: their structure, function and evolution. Eur J Plant Pathol 106: 699–713.

40. Xu LP, Wang JN, Chen RK (1994) Biochemical reaction of sugarcane to smut and its relation to resistance. Sugarcane 1: 13–16.

41. Wan LL, Zha WJ, Cheng XY, Liu C, Lv L, et al. (2011) A rice β-1,3-glucanase gene *Osg1* is required for callose degradation in pollen development. Planta 233: 309–323.

42. Rezaee F, Ghanati F, Behmanesh M (2013) Antioxidant activity and expression of catalase gene of (*Eustoma grandiflorum* L.) in response to boron and aluminum. S Afr J Bot 84: 13–18.

43. Song XX, Zhao FY (2007) Research progress on catalase in plants. J Anhui Agric Sci, 35: 9824–9827.

44. Chaurasia N, Mishra Y, Ai LC (2008) Cloning expression and analysis of phytochelatin synthase (*pcs*) gene from *Anabaena* sp. PCC 7120 offering multiple stress tolerance in *Escherichia coli*. Biochem Bioph Res Co 376: 225–230.

45. Liu Y, Zheng Y (2005) PM2, a group 3 LEA protein from soybean, and its 22-mer repeating region confer salt tolerance in *Escherichia coli*. Biochem Bioph Res Co 331: 325–332.

46. Sandmann G, Boger P (1980) Copper-mediated lipid peroxidation processes in photosynthetic membranes. Plant Physiol 66: 797–800.

47. Fawcett GL (1942) Circular, Estacion Experimental Agricola. In: Fawcett GL, editor. Tucuman. 114.

48. Alexander KC, Ramakrishnan K (1980) Infection of the bud, establishment in the host and production of whips in sugarcane smut (*Ustilago scitaminea*) of sugarcane. Proc Int Soc Sugar Cane Technol 17: 1452–1455.

49. Guan LM, Scandalios JG (2000) Hydrogen peroxide-mediated catalase gene expression in response to wounding. Free Radical Bio Med 28: 1182–1190.

50. Jithesh MN, Prashanth SR, Sivaprakash KR, Parida A (2006) Monitoring expression profiles of antioxidant genes to salinity, iron, oxidative, light and hyperosmotic stresses in the highly salt tolerant grey mangrove, *Avicennia marina* (Forsk.) Vierh. by mRNA analysis. Plant Cell Rep 25: 865–876.

51. Munns R (2002) Comparative physiology of salt and water stress. Plant Cell Environ 25: 239–250.

52. Lombardi L, Sebastiani L (2005) Copper toxicity in *Prunus cerasifera*: growth and antioxidant enzymes responses of in vitro grown plants. Plant Sci 168: 797–802.

53. Melech-Bonfil S, Sessa G (2010) Tomato MAPKKKε is a positive regulator of cell-death signaling networks associated with plant immunity. The Plant J 64: 379–391.

54. Li Y, Tessaro MJ, Li X, Zhang Y (2010) Regulation of the expression of plant resistance gene *SNC1* by a protein with a conserved BAT2 domain. Plant physiol 153: 1425–1434.

55. Levine A, Tenhaken R, Dixon R, Lamb C (1994) H$_2$O$_2$ from the oxidative burst orchestrates the plant hypersensitive disease resistance response. Cell 79: 583–593.

A Novel Dual Allosteric Activation Mechanism of *Escherichia coli* ADP-Glucose Pyrophosphorylase: The Role of Pyruvate

Matías D. Asención Diez[1,2], Mabel C. Aleanzi[2], Alberto A. Iglesias[2], Miguel A. Ballicora[1]*

1 Department of Chemistry, Loyola University Chicago, Chicago, Illinois, United States of America, **2** Laboratorio de Enzimología Molecular, Instituto de Agrobiotecnología del Litoral (UNL-CONICET), FBCB Ciudad Universitaria, Santa Fe, Argentina

Abstract

Fructose-1,6-bisphosphate activates ADP-glucose pyrophosphorylase and the synthesis of glycogen in *Escherichia coli*. Here, we show that although pyruvate is a weak activator by itself, it synergically enhances the fructose-1,6-bisphosphate activation. They increase the enzyme affinity for each other, and the combination increases V_{max}, substrate apparent affinity, and decreases AMP inhibition. Our results indicate that there are two distinct interacting allosteric sites for activation. Hence, pyruvate modulates *E. coli* glycogen metabolism by orchestrating a functional network of allosteric regulators. We postulate that this novel dual activator mechanism increases the evolvability of ADP-glucose pyrophosphorylase and its related metabolic control.

Editor: Eric Cascales, Centre National de la Recherche Scientifique, Aix-Marseille Université, France

Funding: This work was supported by grants to AAI from CONICET [PIP 112-201101-00438 and external collaboration], UNL [CAI+D'11 and Orientado], and ANPCyT [PICT'12 2439]; and to MAB from the National Science Foundation [MCB 1024945]. MDAD is Post-Doctoral Fellow from CONICET. AAI is Principal Investigator from the same Institution. The funders had no role in study design, data collection and analysis, decision to publish, or preparation of the manuscript.

Competing Interests: The authors have declared that no competing interests exist.

* Email: mballic@luc.edu

Introduction

Both glycogen synthesis in bacteria and starch synthesis in plants share a key metabolic step: synthesis of the glucosyl-donor ADP-glucose (ADP-Glc). The reaction is catalyzed by ADP-Glc pyrophosphorylase (EC: 2.7.7.27; ADP-Glc PPase), which is allosterically regulated by metabolites from the main carbon assimilation route in the respective organism [1,2]. It belongs to an enzyme family with kinetic properties adapted to different metabolic environments. This is evidenced by a certain degree of promiscuity observed for the substrate and/or activator in some of the groups [3,4]. Pyr was previously reported as weak activator for the enzyme from enterobacteria [5]. However, almost no kinetic data regarding Pyr activation was collected and no physiological relevance was inferred. In this work we found an important role for Pyr in the *E. coli* ADP-Glc PPase.

ADP-Glc PPase catalyzes the reaction ATP+Glc-1P $\rightleftharpoons$ ADP-Glc+PP$_i$ in the presence of Mg^{2+} [1,2]. The enzyme activators are small molecules that indicate high energy within the cell, whereas the inhibitors indicate starvation [1,2,6]. The whole regulatory scenario is compatible with an enzyme involved in synthesis of cellular reserves of carbon and energy, which uses ATP as a substrate.

The crystal structures of the enzyme from *A. tumefaciens* and the small subunit from potato tuber have been solved, but the regulatory mechanism remains largely unknown [7,8]. Several studies have established structure-function-regulation relationships between ADP-Glc PPases from different organisms, and those studies showed enzymes with different specificity for different regulators [1,2,9,10]. Despite the broad diversity in regulator specificity in different species and metabolic environments, we recently identified key common regulatory loops conserved throughout the ADP-Glc PPase family. They are involved in propagating the allosteric signal both in *E. coli* [11] and potato tuber [12]. This indicated that the same allosteric mechanism, but with different effectors, could be shared among very distant species from bacteria and plants.

The allosteric regulatory properties of the *E. coli* ADP-Glc PPase has been extensively characterized, where Fru-1,6-P$_2$ and AMP are the main activator and inhibitor, respectively [1,11,13–19]. Early studies acknowledged that the *E. coli* enzyme has a series of other minor allosteric regulators [17,20]. One of those is Pyr, which produces a weak activation of the enzyme with an $A_{0.5}$ higher than 10 mM [5]. This value suggested that this keto acid was not of physiological relevance for glycogen synthesis regulation in enterobacteria [1,2]. However, in the last decade, it has been found that Pyr is at relatively high levels in *E. coli* and it is critical to control metabolic fluxes [21]. For that reason, the role of the keto acid in enterobacteria as regulator of the polysaccharide metabolism required re-examination.

Herein, we report a detailed study on the kinetic effects exerted by Pyr on the *E. coli* ADP-Glc PPase. These results lead us to reconsider the relevance of the metabolite as a modulator of the enzyme activity and allow us to propose a novel dual allosteric mechanism. Pyr interacts with the established main effectors and reciprocally strengthens the allosteric control.

Materials and Methods

Chemicals, enzymes and bacterial strains

Protein standards, antibiotics, isopropyl-β-thiogalactoside (IPTG), substrates, and inorganic pyrophosphatase were from Sigma-Aldrich (Saint Louis, MO, USA). Stocks solutions of Pyr were prepared and taken to pH 8.0 before adding it to the reaction mixture to avoid pH changes at the highest concentrations. All the other reagents were of the highest quality available. The ADP-Glc PPase was expressed from pETEC (pET24a plasmid derivative) using *E. coli* BL21 (DE3) cells, and then purified to apparent electrophoretic homogeneity as described elsewhere [5,22].

Protein methods

Denaturing protein electrophoresis (SDS-PAGE) was conducted as described by Laemmli [23]. Protein concentration was determined by absorbance at 280 nm with a NanoDrop 1000 (Thermo Scientific, Wilmington, DE) using an extinction coefficient of 1.273 ml mg^{-1} cm^{-1}, determined from the amino acid sequence by using the ProtParam server (http://web.expasy.org/protparam/) [24].

Enzymatic Assays

Synthesis of ADP-Glc was assayed by following the formation of Pi (after hydrolysis of pyrophosphate by inorganic pyrophosphatase) as previously described [25]. Unless otherwise stated, a 50 μl reaction mixture contained 100 mM HEPES (pH 8.0), 10 mM MgCl$_2$, 1.5 mM ATP, 0.2 mg/ml bovine serum albumin, and 0.5 mU/μl yeast inorganic pyrophosphatase, and the given amount of enzyme. Assays were initiated with 1 mM Glc-1P. If mentioned, a given amount of Pyr, Fru-1,6-P$_2$, and/or AMP was also included. After incubation for 10 min at 37°C, addition of Malachite Green reagent terminated the reaction. The complex formed with the released P$_i$ was measured at 630 nm with an ELISA EMax detector (Molecular Devices). Sodium pyrophosphate was used as standard. One unit of enzyme activity was defined as the amount producing of 1 μmol of product in 1 min under the specified conditions.

Kinetic characterization

Data of enzyme activity were plotted versus effector concentration. Kinetic parameters from the Hill equation such as Hill coefficient (n_H), maximal velocity (V_{max}), as well as the activator substrate or inhibitor concentrations giving 50% of the maximal activation ($A_{0.5}$), velocity ($S_{0.5}$), or inhibition ($I_{0.5}$), were acquired by fitting the data with a non-linear least-squares algorithm using the program Origin 7.0 (OriginLab). Parameters are the mean of at least three independent sets of data, reproducible within ±10%. Sample standard deviations of the data were calculated from the Hill equation fitting by using the Levenberg–Marquardt method.

Results

Pyr activates *E. coli* ADP-Glc PPase

Pyr has been reported as a very weak activator of the *E. coli* ADP-Glc PPase [5], but its regulatory effect has not been thoroughly investigated. For that reason, we decided to first examine the effect of Pyr in the absence of any other regulatory effector. A detailed study of the enzyme activity (assayed in the physiological direction of ADP-Glc synthesis) in presence of increasing Pyr concentrations is depicted in Figure 1. Pyr increased the enzyme activity 3.3-fold at the highest concentrations tested (100 mM), following hyperbolic (n_H value of 1.03)

kinetics with an $A_{0.5}$ of 25 mM. In addition, Pyr increased the apparent affinity of the enzyme for the substrates (Table 1, assay condition A, B and C). The $S_{0.5}$ for ATP was reduced 3.9- or 6.2-fold by the presence of Pyr at 25 mM or 50 mM, respectively. Similar, although slightly lower, was the effect of Pyr on the other substrate, Glc-1P. In that case, the $S_{0.5}$ for Glc-1P was lowered 1.7- and 2.6-fold (Table 1).

As a whole, the effect of Pyr on the *E. coli* ADP-Glc PPase is not significant if compared to Fru-1,6-P$_2$, which increases the enzyme activity by near ~50-fold with an $A_{0.5}$ in the sub-millimolar range [11,14,17]. The apparent affinity for Pyr seemed to be poor. Nonetheless, this keto acid not only increased the enzyme maximal activity, but also the affinity for substrates. In fact, the ~4-fold increase in V_{max} of the *E. coli* ADP-Glc PPase resembles the 5-fold Pyr activation of the *A. tumefaciens* enzyme. The main difference is that the *A. tumefaciens* enzyme has an apparent affinity for the keto acid more than two orders of magnitude higher ($A_{0.5}$ of 0.13 mM) and that Fru-6P is the other major activator with comparable kinetic parameters [26].

Interplay between Pyr and Fru-1,6-P$_2$

To determine whether Pyr has a synergistic effect or another type of interaction with other regulators we tested it in presence of the main allosteric effector of the enzyme (Fru-1,6-P$_2$) at concentrations below the $A_{0.5}$. As shown in Figure 1, the effect of Pyr as allosteric activator of the *E. coli* ADP-Glc PPase was enhanced when analyzed in presence Fru-1,6-P$_2$. Even at sub-saturating concentrations of 10 μM and 30 μM, Fru-1,6-P$_2$ improved the apparent affinity of the enzyme for Pyr 4.9- and 8.4-fold, respectively. In those conditions, the keto acid produced a maximal activation of ~6-fold (Figure 1). These results prompted us to further analyze the combination of both activators and its effect on the enzyme behavior. The increment in the *E. coli* ADP-Glc PPase apparent affinity for Pyr was clearly observed at increasing concentrations of Fru-1,6-P$_2$ (Figure 2). The $A_{0.5}$ for Pyr went down from 34 mM to a low mM range. Remarkably, a Fru-1,6-P$_2$ concentration as low as 10 μM (~10 times lower than the $A_{0.5}$) more than doubled the Pyr apparent affinity. This clearly indicates a synergistic effect between Pyr and Fru-1,6-P$_2$.

To establish whether this synergy affected the interaction with the substrates, we determined the ATP and Glc-1P apparent affinities in presence of Pyr (25 or 50 mM) and sub-saturating, slightly above half-saturation, or saturating concentrations of Fru-1,6-P$_2$ (10, 150 or 2000 μM, respectively). As illustrated in Table 1 (D), the combination of activators enhanced the affinity toward the enzyme substrates. This effect was higher than the ones produced by each of them separately (Table 1, C & E). Thus, 10 μM Fru-1,6-P$_2$ alone only slightly (1.2- to 1.3-fold) modified the affinity for either of the substrates. However, when combined with 25 mM Pyr, the affinity for Glc-1P increased 4.1-fold (compared to values in absence of any effector). On the other hand, the increase caused by 25 mM Pyr alone was only 1.7-fold. As well, the presence of both activators lowered the $S_{0.5}$ for ATP 6.6-fold, whereas 25 mM Pyr in absence of Fru-1,6-P$_2$ decreased it 3.9-fold. In addition, saturating concentrations of both activators (Table 1, assay condition F) produced the highest apparent affinities for substrates in the *E. coli* ADP-Glc PPase, indicating that Pyr enhanced the activation exerted by Fru-1,6-P$_2$ alone (Table 1, condition G).

The Pyr to Fru-1,6-P$_2$ synergistic interaction is also reinforced by results showed in Figure 3. We analyzed the *E. coli* ADP-Glc PPase response to Fru-1,6-P$_2$ when the enzyme was in absence or presence of sub-saturating concentrations of Pyr (Figure 3A). The concentration of Pyr, 2.5 mM, was 10-fold lower than its $A_{0.5}$. Notably, this Pyr concentration, which has no prominent intrinsic

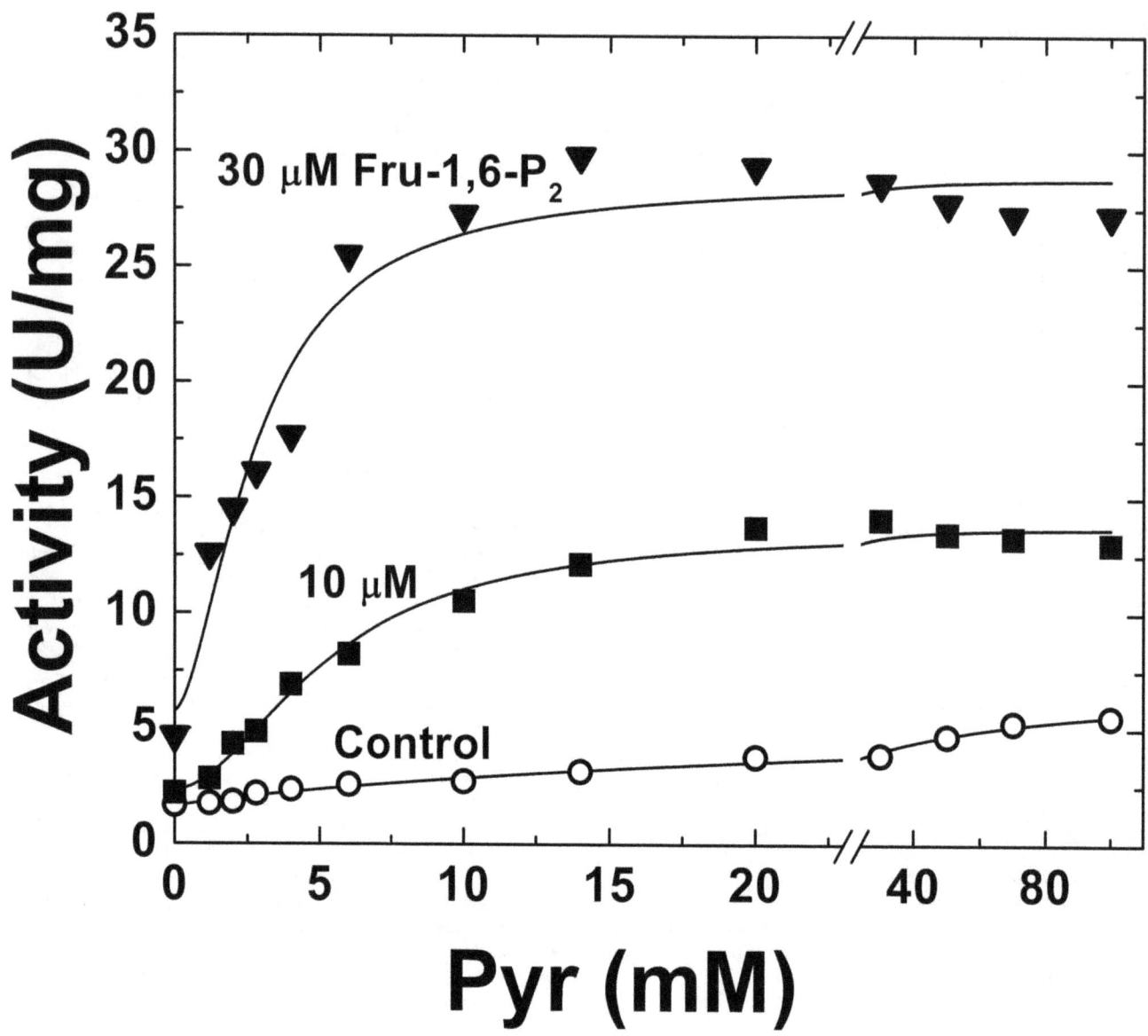

Figure 1. *E. coli* **ADP-Glc PPase Pyr saturation curves in the absence (control, ○) or presence of Fru-1,6-P$_2$ (10 μM, ■; or 30 μM, ▼).** Other conditions for the assays were as described under Materials and Methods. ●□▲.

Table 1. Kinetic parameters for *E. coli* ADP-Glc PPase.

Assay Condition	Glc-1P		ATP-Mg		V_{max} (U/mg)
	$S_{0.5}$ (mM)	n_H	$S_{0.5}$ (mM)	n_H	
A. No effector	0.95±0.07	1.1	6.49±0.21	1.5	1.67±0.15
B. 25 mM Pyr	0.57±0.05	1.0	1.67±0.08	2.5	3.93±0.21
C. 50 mM Pyr	0.37±0.02	0.9	1.05±0.06	2.2	5.31±0.19
D. 25 mM Pyr +10 μM Fru-1,6-P$_2$	0.23±0.02	1.3	0.99±0.07	3.6	31±2
E. 10 μM Fru-1,6-P$_2$	0.77±0.04	1.0	5.02±0.25	2.1	6.21±0.19
F. 50 mM Pyr +150 μM Fru-1,6-P$_2$	0.09±0.01	1.1	0.37±0.03	2.4	104±5
G. 150 μM Fru-1,6-P$_2$	0.12±0.01	1.2	1.31±0.09	3.2	63±4
I. 2 mM Fru-1,6-P$_2$	0.09±0.01	1.3	0.43±0.02	2.1	108±6
J. 50 mM Pyr +2 mM Fru-1,6-P$_2$	0.11±0.01	1.0	0.19±0.02	1.4	110±5

Assays were carried out as described in Material and Methods. Values are average numbers from three independent experiments, using regression analysis.

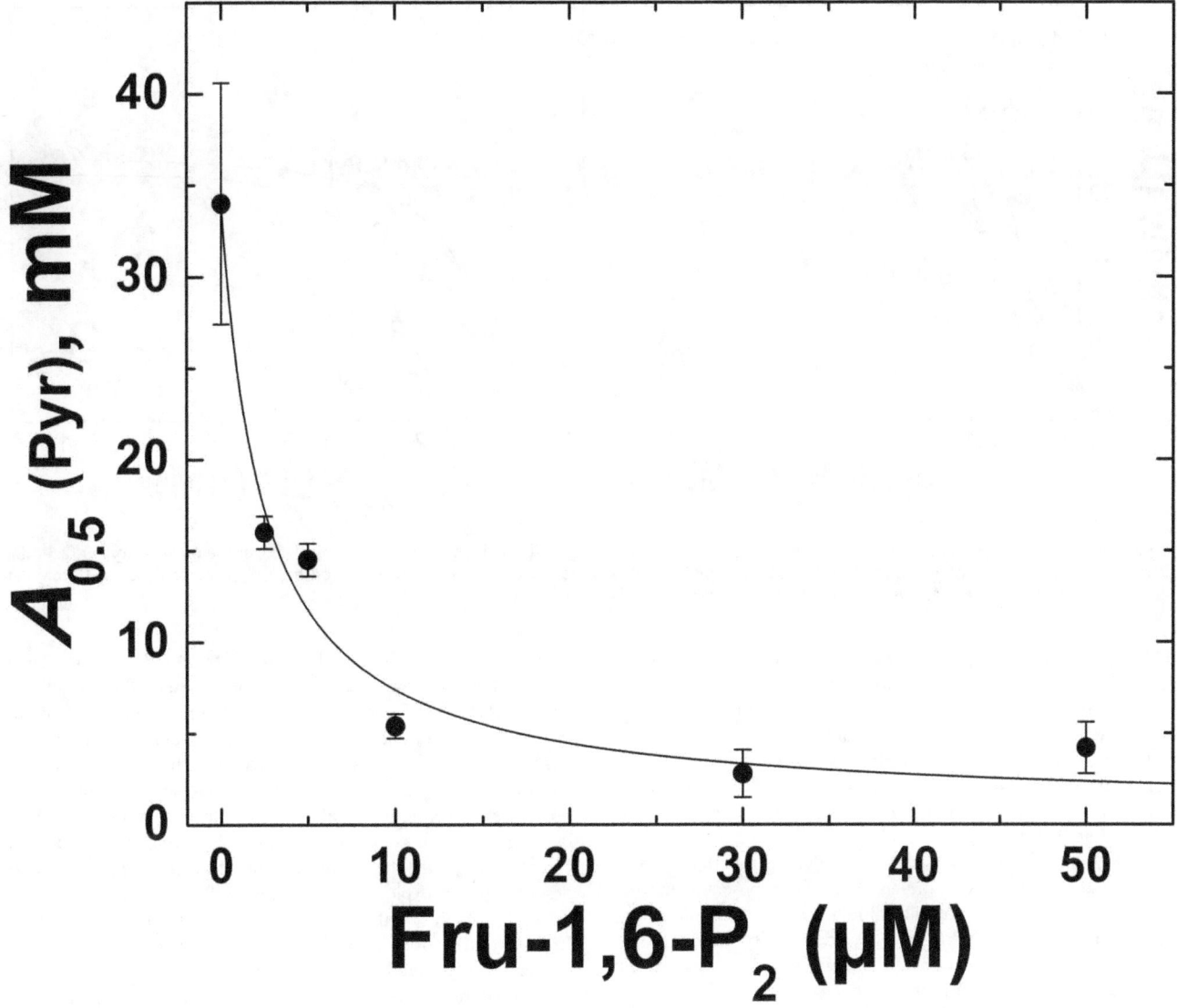

Figure 2. Effect of Fru-1,6-P$_2$ on the apparent affinity of ADP-Glc PPase for Pyr. Other conditions for the assays were as described under Materials and Methods. Values are the mean of three independent measurements ± standard deviation.

effect on the *E. coli* ADP-Glc PPase, doubled the Fru-1,6-P$_2$ apparent affinity (the Fru-1,6-P$_2$ $A_{0.5}$ value changed from 0.12 mM to 0.07 mM). In addition, Pyr clearly increased the enzyme sensitivity to pyridoxal-5′-phosphate (PLP) (Figure 3B). In previous works, it was demonstrated that PLP is incorporated to the *E. coli* ADP-Glc PPase emulating Fru-1,6-P$_2$ activation by binding to Lys[39] [27–29]. In that respect, PLP has been extensively used as an analog of activators, not only in enzymes activated by Fru-1,6-P$_2$, but also by 3-phosphoglycerate. Indeed, the enzyme became more susceptible to PLP in presence of 10 mM Pyr; not only with a decrease in the $A_{0.5}$ value (from 2.7 µM to 1.0 µM), but also with a dramatic change (from sigmoidal to hyperbolic) in the activation pattern (Figure 3B). In addition, we evaluated the Pyr activation of the *E. coli* ADP-Glc PPase in presence of 1 µM PLP, which has no direct effect on the enzyme. In this condition, the $A_{0.5}$ for the keto acid was as low as 3.7 mM (data not shown). This indicated that a sub functional concentration of PLP lowered the $A_{0.5}$ for Pyr one order of magnitude. This result agrees with the observed increase in the affinity for Pyr caused by Fru-1,6-P$_2$

(Figure 1 and Table 1). As a whole, it could be inferred that Pyr at lower concentrations makes the enzyme more sensitive to Fru-1,6-P$_2$, and *vice versa*.

Pyr diminishes the inhibition by AMP

It is well known that AMP is the major inhibitor of the *E. coli* ADP-Glc PPase, mainly acting in a cross-talk with the activator Fru-1,6-P$_2$ [1,14,18,19]. To advance in the characterization of the role of Pyr, we analyzed how this metabolite affected the kinetic behavior of AMP inhibition of the *E. coli* ADP-Glc PPase. Curves of AMP were obtained in the absence or in the presence of 20 mM Pyr. In all cases reaction mixtures contained 100 µM Fru-1,6-P$_2$ because the activator is needed to observe AMP inhibition [20]. It is important to note that this value is near the $A_{0.5}$ for Fru-1,6-P$_2$ (Figure 3), which is high enough to have a significant activation, but low enough to allow a reversion by AMP. Results depicted in Figure 4 clearly indicate that when Pyr was present the enzyme was less sensitive to AMP inhibition. Pyr diminished the apparent affinity for AMP 3-fold. The $I_{0.5}$ for AMP changed from

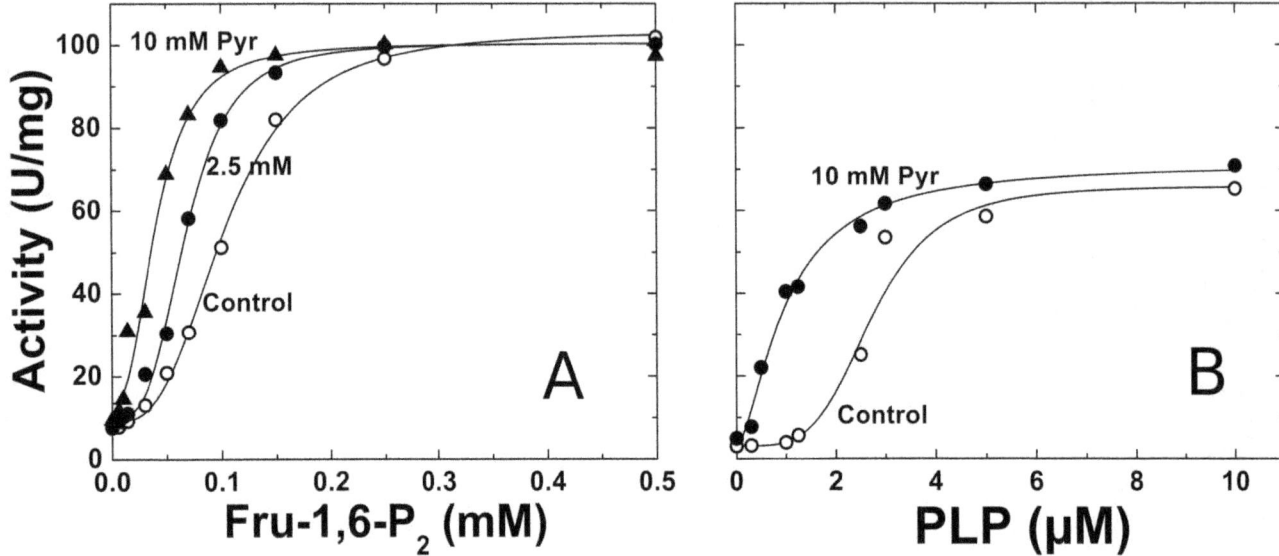

Figure 3. A) Effect of Pyr on the saturation curves for Fru-1,6-P$_2$. Assays were performed in the absence (control, ○), or the presence of Pyr (2.5 mM, ●; or 10 mM, ▲). B) Effect of Pyr on the saturation curves for PLP. Assays were performed in the absence (control, ○), or the presence of 10 mM Pyr (▲). Other conditions for the assays were as described under Materials and Methods.

0.03 mM, a typical value for this condition [1,14], to 0.11 mM. Also, the maximum inhibition was very similar in both cases (around 10% of remaining activity). However, in presence of Pyr, the enzyme depicted a more sigmoidal behavior, where n_H changed from 1.7 to 2.4. When the relative activities from both inhibition curves were compared it was observed that *E. coli* ADP-Glc PPase was up to 3-fold more active in the 0.05–0.15 mM AMP range if Pyr was present.

We evaluated the enzyme sensitivity to Pyr when both the major activator and inhibitor were present. For that purpose, the enzyme was assayed when it was partially inhibited with 50 μM AMP, and in presence of 100 μM Fru-1,6-P$_2$. Under this condition, which maximizes the sensitivity towards those effectors, Pyr activated the enzyme 4-fold, with saturation kinetics slightly deviated from a hyperbolic behavior (n_H 1.2) (Figure 5). Here, the $A_{0.5}$ was 2.76 mM, indicating that in presence of AMP and Fru-1,6-P$_2$ the apparent affinity for Pyr was one order of magnitude higher in absence of other effectors. Remarkably, Pyr has a very significant effect on the partially inhibited enzyme. Noteworthy, the affinity for Pyr in these conditions drops to levels that are lower than the reported intracellular concentration [21], highlighting its putative physiological role [21].

Taken together, results presented *ut supra* show that Pyr has the merit to be considered an indirect or ancillary, yet important activator of *E. coli* ADP-Glc PPase. The keto acid is remarkably involved in augmenting the sensitivity of the enzyme to its main activator (Fru-1,6-P$_2$) and decreasing the inhibition by AMP. Consequently, it further increases V_{max} and the affinity for substrates. Reciprocally, the affinity of the enzyme toward Pyr is greatly increased at very low concentrations of Fru-1,6-P$_2$. A scenario can be proposed where Pyr behaves as a chief modulator of the allosteric regulation of the enzyme, mainly exerting its action by finely orchestrating the effect of the main allosteric effectors.

Discussion

ADP-Glc PPases from diverse sources are classified according to their allosteric effectors [1,2]. The *E. coli* enzyme is mainly

activated by Fru-1,6-P$_2$ and inhibited by AMP. Pyr has been identified as a major allosteric activator of other ADP-Glc PPases where AMP, ADP and/or P$_i$ are inhibitors [1,2]. There is an important amount of work that has been done regarding the characterization of Pyr as effector, mainly with the *A. tumefaciens* enzyme. Despite the fact that the crystal structure of the *A. tumefaciens* enzyme is available [7], it is not known where the Pyr binding site is. Therefore, it is not possible to do proper structural comparisons with the activator bound. This is an area that deserves further exploration. In other ADP-Glc PPases, Pyr shares its activation effect with six-carbon molecules such as Fru-6P [1,2]. In this work, we found that Pyr has an important effect on the *E. coli* enzyme, which is in a class that traditionally has not been considered to have Pyr as an effector. Important elements to evaluate the significance of the Pyr effect are the 5-fold increase in V_{max} and the enhancement of apparent substrate affinities, mainly ATP. The $A_{0.5}$ for Pyr was 25 mM; however, the most relevant effect is based on its interaction with the other regulators.

As demonstrated with different mutant strains of *E. coli* and *S. typhimurium*, there is a clear correlation between the apparent affinity of the enzyme for sub-millimolar concentrations of Fru-1,6-P$_2$ and the ability to accumulate glycogen when compared to the wild type strain [30]. In this work, sub-saturating concentrations of Fru-1,6-P$_2$ significantly increased the apparent affinity for Pyr, which highlights the existence of a cross-talk between these two effectors. This synergy would enhance the physiological effect. For instance, at 50 μM Fru-1,6-P$_2$, Pyr $A_{0.5}$ decreased 5-fold to reach a 5 mM value, which is within physiological range [21]. The reciprocal situation was also observed where 2.5 mM Pyr reduced in half the $A_{0.5}$ for Fru-1,6-P$_2$. Moreover, the apparent substrates affinities also increased by a synergistic effect between Fru-1,6-P$_2$ and Pyr. We are unable to compare this phenomenon in the *E. coli* enzyme with other ADP-Glc PPases since this is the first time the effect of two synergistic allosteric activators in this family is reported. Nevertheless, our findings suggest that the regulatory fine-tuning of other ADP-Glc PPases should be revisited. Studies combining two activators, even some that may have been considered weak or non-physiological, will be helpful to verify

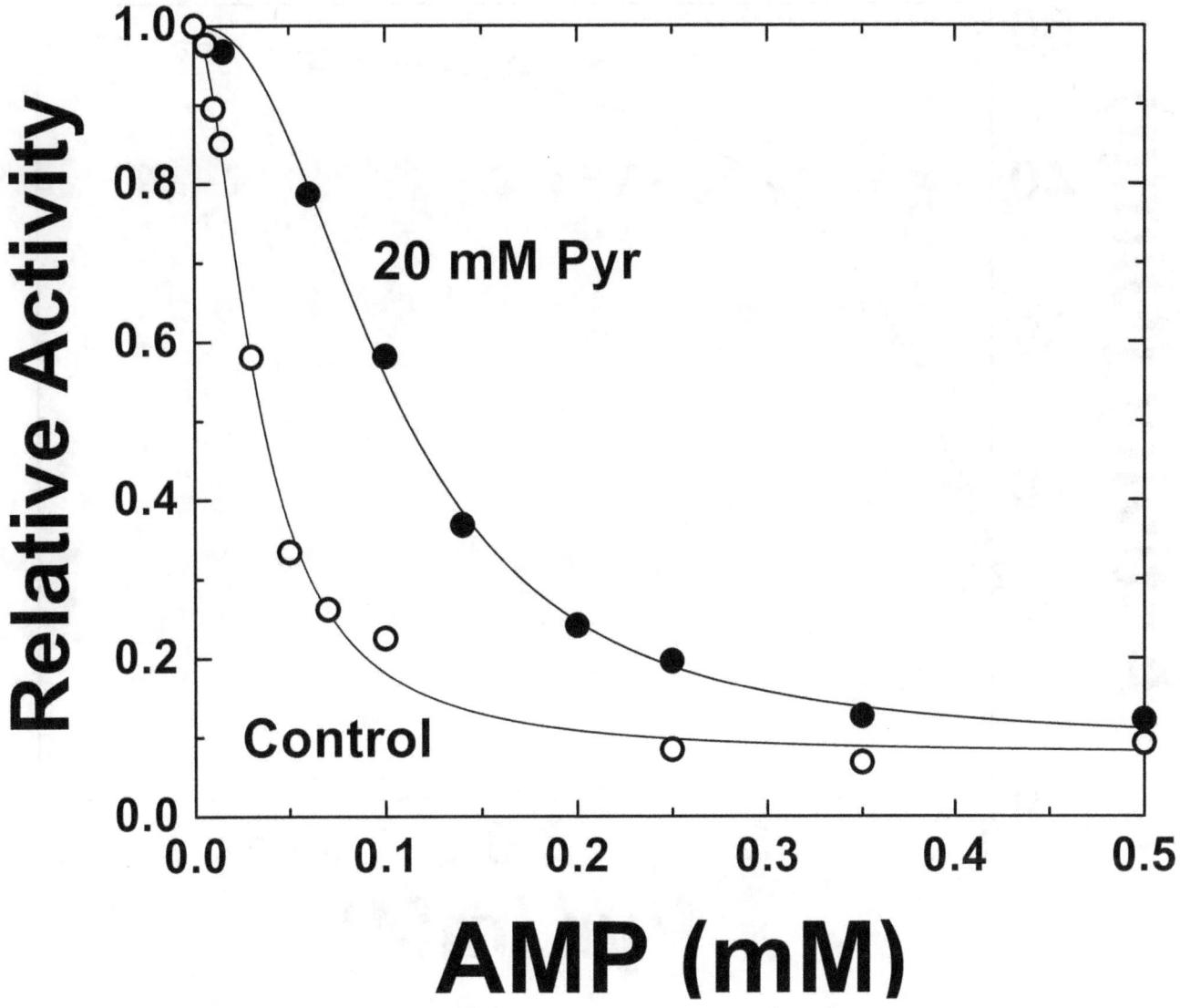

Figure 4. Effect of Pyr on the AMP inhibition. Assays were performed in the absence (control, ○), or the presence of 20 mM Pyr (●). Other conditions for the assays were as described under Materials and Methods. Activity is relative to experiments conducted in absence or presence of Pyr, whose absolute values were 31.2 and 61.8 U/mg, respectively.

whether this synergistic behavior is common to the enzyme from other diverse species.

We postulate that the interplay between Pyr and Fru-1,6-P_2 has key implications at the physiological level. The "weak effect", *per se*, that Pyr has on the enzyme becomes much more relevant in the presence of Fru-1,6-P_2. Intracellular Pyr concentration was reported to be 7–14 mM [21], which is in fact in the range of the $A_{0.5}$ for this metabolite in presence of very low Fru-1,6-P_2 concentrations.

Another important feature is that Pyr diminished AMP inhibition. Given that this inhibition occurs only in presence of Fru-1,6-P_2 [5], and considering that the latter interacts synergistically with Pyr, most likely the Pyr effect is indirect. That is, Pyr enhances the Fru-1,6-P_2 activation, making the *E. coli* ADP-Glc PPase less sensitive to the inhibitor. In addition, Pyr completely reverses the (partial) AMP inhibition with an $A_{0.5}$ of 2.76 mM (the smallest value observed in this work for Pyr).

From a mechanistic point of view, it is important to highlight that the synergistic effect between Pyr and Fru-1,6-P_2 implies that

they do not compete. Hence, they must be binding to two non-overlapping, but interacting allosteric sites. A traditional view of the promiscuity of this enzyme was that the allosteric site could accommodate different negatively charged molecules [1]. However, our results indicate that Pyr is not a promiscuous ligand that binds to the Fru-1,6-P_2 allosteric site, but that it binds to a distinct site in the enzyme. This opens a new perspective to understand the evolvability of the ADP-Glc PPase family.

The presence of a putative Pyr site explains the results observed with chimeric constructs between *E. coli* and *A. tumefaciens* [5]. The C-terminus had a major role in determining the affinity for Pyr, whereas both N- and C-domains shared the effect on Fru-1,6-P_2 [5]. In addition, it is known that in the N-domain Lys[39] interacts with the activator Fru-1,6-P_2 [31]. This agrees with previous works on other ADP-Glc PPases where important residues for the interaction with the hexose-phosphate activator were found in both the N- and C-domain [7,32]. On the other hand, those residues were not critical for Pyr activation [7,32]. If we accept previous evidence that both domains interact with Fru-

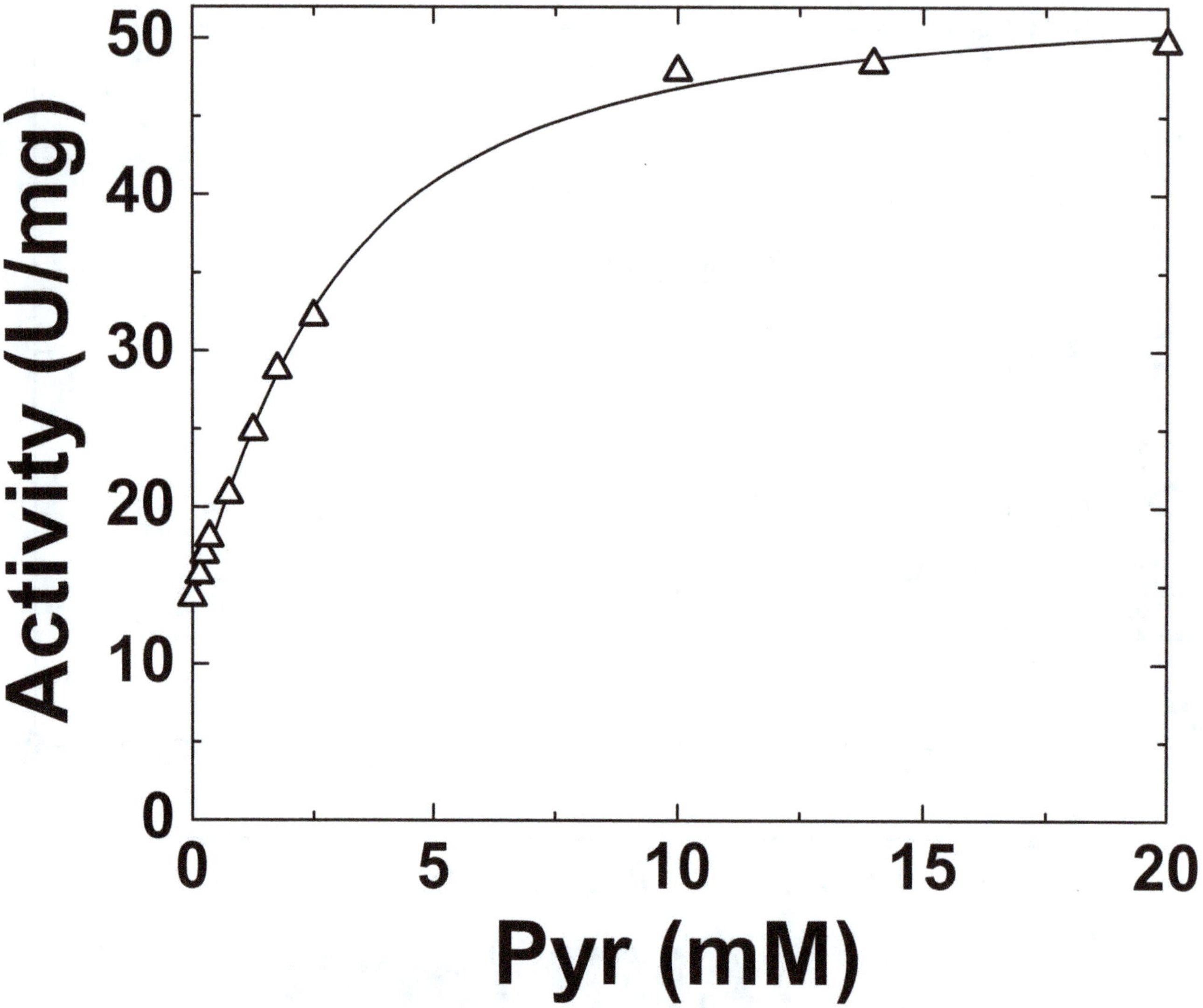

Figure 5. Partially inhibited *E. coli* ADP-Glc PPase is activated by Pyr. The enzyme activity was determined in presence of 100 μM Fru-1,6-P$_2$ and 50 μM AMP. These concentrations were the $A_{0.5}$ and the $I_{0.5}$ values, respectively, as this range maximized the sensitivity of the enzyme for these effectors. Other conditions for the assays were as described under Materials and Methods.

1,6-P$_2$ [11,33] and Pyr mainly with the C-terminus [5,34], we can postulate the activation scheme in Figure 6 for the *E. coli* ADP-Glc PPase. In that scheme, Pyr binds to the C-domain and as a consequence, it activates the enzyme by a direct interaction with the catalytic site present in the N-domain. In addition, Pyr allosterically enhances the Fru-1,6-P$_2$ binding to a site located in the interface between both domains. Finally, Fru-1,6-P$_2$ transmits the allosteric signal to the active site. The increase of Fru-1,6-P$_2$ apparent affinity implies an indirect regulatory effect from Pyr.

If we consider Pyr as a "new" effector, the *E. coli* ADP-Glc PPase regulatory scenario will become more similar to the enzyme from *A. tumefaciens,* which is activated by Fru-6P and Pyr [26] and from anoxygenic bacteria (e.g. *Rhodobacter* spp.) that are regulated by Pyr and an hexose-P (Fru-6P and/or Fru-1,6-P$_2$) [1]. In those examples, the main difference from *E. coli* ADP-Glc PPase is that kinetic parameters for Pyr and the other effector are more similar [1,35,36]. It is very possible that a Pyr site may be present in many different types of bacteria rather than having a promiscuous binding to the hexose-phosphate site. The sensitivity

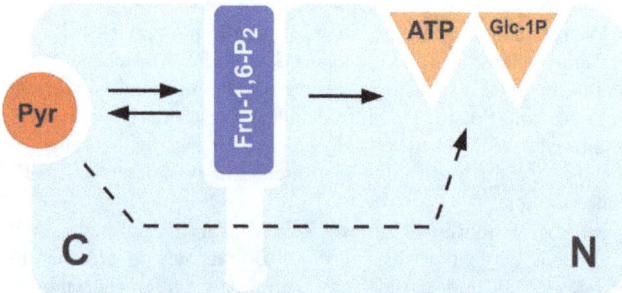

Figure 6. Proposed dual allosteric model for *E. coli* ADP-Glc PPase. The two main domains of the enzyme are illustrated by C and N. Catalysis occurs in the N-domain where the active site for ATP and Glc-1P is located. Full and dashed arrows indicate a strong and weak positive interaction, respectively.

for Pyr may have been "tuned" up or down by evolution according to the metabolic scenario.

As a whole, our results support the hypothesis that the *E. coli* ADP-Glc PPase is concurrently activated by both Pyr and Fru-1,6-P_2 and also regulated by AMP levels. This multi-regulated mechanism reflects how the enzyme should operate in the actual metabolic environment in *E. coli* [11,17]. There are two important aspects of Pyr regulation: (i) the enzyme in presence of activator and inhibitor (Fru-1,6-P_2 and AMP, respectively) has a higher affinity for the keto acid; and (ii) even at sub-saturating concentrations, Pyr enhances the sensitivity of the enzyme for Fru-1,6-P_2 activation. It could be proposed that Pyr orchestrates the activation of the main effector Fru-1,6-P_2, working as an "activator of an activator" and playing consequently as a fine

tuning modulator. Results obtained in this work will also help to understand the interplay between activators for other ADP-Glc PPases where more than one effector was reported. This poses an interesting case of allosterism where one metabolite facilitates the action of another allosteric activator and consequently organizes a global fine-tuned modulatory network.

Author Contributions

Conceived and designed the experiments: AAI MAB. Performed the experiments: MDAD. Analyzed the data: MDAD MCA AAI MAB. Contributed reagents/materials/analysis tools: MCA AAI MAB. Wrote the paper: MDAD AAI MAB.

References

1. Ballicora MA, Iglesias AA, Preiss J (2003) ADP-glucose pyrophosphorylase, a regulatory enzyme for bacterial glycogen synthesis. Microbiol Mol Biol Rev 67: 213–225, table of contents.
2. Ballicora MA, Iglesias AA, Preiss J (2004) ADP-Glucose Pyrophosphorylase: A Regulatory Enzyme for Plant Starch Synthesis. Photosynth Res 79: 1–24.
3. Kuhn ML, Figueroa CM, Iglesias AA, Ballicora MA (2013) The ancestral activation promiscuity of ADP-glucose pyrophosphorylases from oxygenic photosynthetic organisms. BMC evolutionary biology 13: 51.
4. Machtey M, Kuhn ML, Flasch DA, Aleanzi M, Ballicora MA, et al. (2012) Insights into glycogen metabolism in chemolithoautotrophic bacteria from distinctive kinetic and regulatory properties of ADP-glucose pyrophosphorylase from Nitrosomonas europaea. Journal of bacteriology 194: 6056–6065.
5. Ballicora MA, Sesma JI, Iglesias AA, Preiss J (2002) Characterization of chimeric ADPglucose pyrophosphorylases of Escherichia coli and Agrobacterium tumefaciens. Importance of the C-terminus on the selectivity for allosteric regulators. Biochemistry 41: 9431–9437.
6. Iglesias AA, Preiss J (1992) Bacterial glycogen and plant starch biosynthesis. Biochem Educ 20: 196–203.
7. Cupp-Vickery JR, Igarashi RY, Perez M, Poland M, Meyer CR (2008) Structural analysis of ADP-glucose pyrophosphorylase from the bacterium Agrobacterium tumefaciens. Biochemistry 47: 4439–4451.
8. Jin X, Ballicora MA, Preiss J, Geiger JH (2005) Crystal structure of potato tuber ADP-glucose pyrophosphorylase. The EMBO journal 24: 694–704.
9. Asencion Diez MD, Demonte AM, Guerrero SA, Ballicora MA, Iglesias AA (2013) The ADP-glucose pyrophosphorylase from Streptococcus mutans provides evidence for the regulation of polysaccharide biosynthesis in Firmicutes. Molecular Microbiology 90: 1011–1027.
10. Asencion Diez MD, Peiru S, Demonte AM, Gramajo H, Iglesias AA (2012) Characterization of recombinant UDP- and ADP-glucose pyrophosphorylases and glycogen synthase to elucidate glucose-1-phosphate partitioning into oligo- and polysaccharides in Streptomyces coelicolor. Journal of bacteriology 194: 1485–1493.
11. Figueroa CM, Esper MC, Bertolo A, Demonte AM, Aleanzi M, et al. (2011) Understanding the allosteric trigger for the fructose-1,6-bisphosphate regulation of the ADP-glucose pyrophosphorylase from Escherichia coli. Biochimie 93: 1816–1823.
12. Figueroa CM, Kuhn ML, Falaschetti CA, Solamen L, Olsen KW, et al. (2013) Unraveling the activation mechanism of the potato tuber ADP-glucose pyrophosphorylase. PloS one 8: e66824.
13. Ballicora MA, Erben ED, Yazaki T, Bertolo AL, Demonte AM, et al. (2007) Identification of regions critically affecting kinetics and allosteric regulation of the Escherichia coli ADP-glucose pyrophosphorylase by modeling and pentapeptide-scanning mutagenesis. Journal of Bacteriology 189: 5325–5333.
14. Bejar CM, Ballicora MA, Gomez-Casati DF, Iglesias AA, Preiss J (2004) The ADP-glucose pyrophosphorylase from Escherichia coli comprises two tightly bound distinct domains. FEBS Lett 573: 99–104.
15. Hill MA, Preiss J (1998) Functional analysis of conserved histidines in ADP-glucose pyrophosphorylase from Escherichia coli. Biochemical and biophysical research communications 244: 573–577.
16. Preiss J (1996) ADPglucose pyrophosphorylase: basic science and applications in biotechnology. Biotechnology annual review 2: 259–279.
17. Preiss J, Shen L, Partridge M (1965) The Activation of Escherichia Coli Adp-Glucose Pyrophosphorylase. Biochemical and biophysical research communications 18: 180–185.
18. Wu MX, Preiss J (1998) The N-terminal region is important for the allosteric activation and inhibition of the Escherichia coli ADP-glucose pyrophosphorylase. Archives of biochemistry and biophysics 358: 182–188.
19. Wu MX, Preiss J (2001) Truncated forms of the recombinant Escherichia coli ADP-glucose pyrophosphorylase: the importance of the N-terminal region for

allosteric activation and inhibition. Archives of biochemistry and biophysics 389: 159–165.
20. Preiss J, Shen L, Greenberg E, Gentner N (1966) Biosynthesis of bacterial glycogen. IV. Activation and inhibition of the adenosine diphosphate glucose pyrophosphorylase of Escherichia coli B. Biochemistry 5: 1833–1845.
21. Yang YT, Bennett GN, San KY (2001) The effects of feed and intracellular pyruvate levels on the redistribution of metabolic fluxes in Escherichia coli. Metabolic engineering 3: 115–123.
22. Frueauf JB, Ballicora MA, Preiss J (2001) Aspartate residue 142 is important for catalysis by ADP-glucose pyrophosphorylase from Escherichia coli. The Journal of biological chemistry 276: 46319–46325.
23. Laemmli UK (1970) Cleavage of structural proteins during the assembly of the head of bacteriophage T4. Nature 227: 680–685.
24. Gasteiger E, Gattiker A, Hoogland C, Ivanyi I, Appel RD, et al. (2003) ExPASy: The proteomics server for in-depth protein knowledge and analysis. Nucleic acids research 31: 3784–3788.
25. Fusari C, Demonte AM, Figueroa CM, Aleanzi M, Iglesias AA (2006) A colorimetric method for the assay of ADP-glucose pyrophosphorylase. Analytical biochemistry 352: 145–147.
26. Uttaro AD, Ugalde RA, Preiss J, Iglesias AA (1998) Cloning and expression of the glgC gene from Agrobacterium tumefaciens: purification and characterization of the ADPglucose synthetase. Archives of biochemistry and biophysics 357: 13–21.
27. Parsons TF, Preiss J (1978) Biosynthesis of bacterial glycogen. Isolation and characterization of the pyridoxal-P allosteric activator site and the ADP-glucose-protected pyridoxal-P binding site of Escherichia coli B ADP-glucose synthase. The Journal of biological chemistry 253: 7638–7645.
28. Parsons TF, Preiss J (1978) Biosynthesis of bacterial glycogen. Incorporation of pyridoxal phosphate into the allosteric activator site and an ADP-glucose-protected pyridoxal phosphate binding site of Escherichia coli B ADP-glucose synthase. The Journal of biological chemistry 253: 6197–6202.
29. Sheng J, Charng YY, Preiss J (1996) Site-directed mutagenesis of lysine382, the activator-binding site, of ADP-glucose pyrophosphorylase from Anabaena PCC 7120. Biochemistry 35: 3115–3121.
30. Preiss J, Romeo T (1989) Physiology, biochemistry and genetics of bacterial glycogen synthesis. Advances in microbial physiology 30: 183–238.
31. Gardiol A, Preiss J (1990) Escherichia coli E-39 ADPglucose synthetase has different activation kinetics from the wild-type allosteric enzyme. Archives of biochemistry and biophysics 280: 175–180.
32. Gomez-Casati DF, Igarashi RY, Berger CN, Brandt ME, Iglesias AA, et al. (2001) Identification of functionally important amino-terminal arginines of Agrobacterium tumefaciens ADP-glucose pyrophosphorylase by alanine scanning mutagenesis. Biochemistry 40: 10169–10178.
33. Bejar CM, Jin X, Ballicora MA, Preiss J (2006) Molecular architecture of the glucose 1-phosphate site in ADP-glucose pyrophosphorylases. The Journal of biological chemistry 281: 40473–40484.
34. Asencion Diez MD, Ebrecht AC, Martinez LI, Aleanzi MC, Guerrero SA, et al. (2013) A Chimeric UDP-Glucose Pyrophosphorylase Produced by Protein Engineering Exhibits Sensitivity to Allosteric Regulators. International journal of molecular sciences 14: 9703–9721.
35. Eidels L, Edelmann PL, Preiss J (1970) Biosynthesis of bacterial glycogen. 8. Activation and inhibition of the adenosine diphosphoglucose pyrophosphorylase of Rhodopseudomonas capsulata and of Agrobacterium tumefaciens. Archives of biochemistry and biophysics 140: 60–74.
36. Furlong CE, Preiss J (1969) Biosynthesis of bacterial glycogen. VII. Purification and properties of the adenosine diphosphoglucose pyrophosphorylase of Rhodospirillum rubrum. The Journal of biological chemistry 244: 2539–2548.

Carbon Transfer from the Host to *Tuber melanosporum* Mycorrhizas and Ascocarps Followed Using a ^{13}C Pulse-Labeling Technique

François Le Tacon[1,2], Bernd Zeller[3]*, Caroline Plain[4,5], Christian Hossann[4,5], Claude Bréchet[4,5], Christophe Robin[6,7]

1 INRA, UMR 1136, Interactions Arbres/Microorganismes (IAM), Centre INRA de Nancy, Champenoux, France, 2 Université de Lorraine, UMR 1136, Interactions Arbres/Microorganismes (IAM), Faculté des Sciences, Vandoeuvre les Nancy, France, 3 INRA, UR 1138, Biogéochimie des Ecosystèmes Forestiers (BEF), Centre INRA de Nancy, Champenoux, France, 4 INRA, UMR 1137, Ecologie et Ecophysiologie Forestières (EEF), Centre INRA de Nancy, Champenoux, France, 5 Université de Lorraine, UMR 1137, Ecologie et Ecophysiologie Forestières (EEF), Faculté des Sciences, Vandoeuvre les Nancy, France, 6 Université de Lorraine, UMR 1121 « Agronomie & Environnement » Nancy-Colmar, Vandoeuvre les Nancy, France, 7 INRA, UMR 1121 « Agronomie & Environnement » Nancy-Colmar, Centre INRA de Nancy, Vandœuvre les Nancy, France

Abstract

Truffles ascocarps need carbon to grow, but it is not known whether this carbon comes directly from the tree (heterotrophy) or from soil organic matter (saprotrophy). The objective of this work was to investigate the heterotrophic side of the ascocarp nutrition by assessing the allocation of carbon by the host to *Tuber melanosporum* mycorrhizas and ascocarps. In 2010, a single hazel tree selected for its high truffle (*Tuber melanosporum*) production and situated in the west part of the Vosges, France, was labeled with $^{13}CO_2$. The transfer of ^{13}C from the leaves to the fine roots and *T. melanosporum* mycorrhizas was very slow compared with the results found in the literature for herbaceous plants or other tree species. The fine roots primarily acted as a carbon conduit; they accumulated little ^{13}C and transferred it slowly to the mycorrhizas. The mycorrhizas first formed a carbon sink and accumulated ^{13}C prior to ascocarp development. Then, the mycorrhizas transferred ^{13}C to the ascocarps to provide constitutive carbon (1.7 mg of ^{13}C per day). The ascocarps accumulated host carbon until reaching complete maturity, 200 days after the first labeling and 150 days after the second labeling event. This role of the *Tuber* ascocarps as a carbon sink occurred several months after the end of carbon assimilation by the host and at low temperature. This finding suggests that carbon allocated to the ascocarps during winter was provided by reserve compounds stored in the wood and hydrolyzed during a period of frost. Almost all of the constitutive carbon allocated to the truffles (1% of the total carbon assimilated by the tree during the growing season) came from the host.

Editor: Peter Shaw, Roehampton University, United Kingdom

Funding: For this work, the authors utilized the online continuous flow CN analyzer (Carlo Erba NA1500) coupled with an isotope ratio mass spectrometer (Finnigan delta S) and DNA sequencing facilities at INRA-Nancy financed by INRA and the Regional Council of Lorraine. The pulse-labeling experiment was supported by the SYSTRUF programme (An integrated approach for sustainable management of ecosystems producing Black Truffle, *Tuber melanosporum*) financed by the French ANR (Agence Nationale de la Recherche; programme SYSTERRA, ANR-09-STRA-10-02). The funders had no role in study design, data collection and analysis, decision to publish, or preparation of the manuscript.

Competing Interests: The authors have declared that no competing interests exist.

* E-mail: zeller@nancy.inra.fr

Introduction

Despite their renown, the life cycle of the true truffles belonging to the genus *Tuber*, which are members of the Ascomycota, is not well understood. These species form ectomycorrhizas with different hosts [1,2]. For sexual reproduction, it is hypothesized that haploid mycorrhizas of one mating type form an antheridium producing male gametes and, through an ascogonial filament or cord, an ascogonial apparatus composed of ascogonial cells and a trichogyne. The female haploid trichogyne of one mating type is assumed to collect the male gametes of the opposite mating type, allowing the ascogonial apparatus to form, after plasmogamy, an ascogenous heterokaryotic tissue, which appears to be surrounded by homokaryotic maternal tissues. The growth of these tissues gives rise to the ascocarp.

In contrast to ectomycorrhizal basidiomycota sporocarps, such as those of *Boletus*, *Amanita* or *Laccaria*, which develop over a number of days directly from diploid mycorrhizas [3], *Tuber* ascocarps grow more slowly [4].

It takes at least six months between the production of the primordia and full ascocarp development. We thus hypothesize that the processes involved in ascocarp development and carbon acquisition are different from those of basidiocarps. It is not known whether the developing ascocarp is fed *via* a direct transfer of carbohydrates from the host tree through the mycorrhizas and the ascogonial structure or whether the ascocarp becomes independent of its host some weeks or months after its development. In the latter case, it is assumed that truffles might be able to use dead host tissues or soil organic matter as carbon (C) and nitrogen (N) sources, as indicated by some authors [5,6], through a saprotrophic process. This feeding behavior cannot be excluded because the truffle ascocarp can develop an external mycelium from its peridium. This mycelium could colonize dead cells from living roots, dead roots, other dead organic tissues or mineral structures

[6]. Similarly, in pure cultures, the mycelium of *T. melanosporum* might use cellulose, cellobiose, lignin, chitin and tannins as carbon sources [7]. However, sequencing of the *T. melanosporum* genome showed that this fungus exhibits a restricted repertoire of genes coding for Carbohydrate Active enZymes (CAZymes) that are able to degrade dead organic matter [8]. *T. melanosporum* presents many fewer GH-encoding genes than saprotrophs, and cellulases from families GH6 and GH7 are absent [8]. These findings suggest that the saprotrophic ability of *T. melanosporum* is weak. In addition, *T. melanosprum* has an invertase gene that allows plant-derived sucrose to be hydrolyzed, suggesting the ability to use simple sugars from the host.

Our previous results based on the natural abundance of ^{13}C and ^{15}N in the ascocarp indicate that *T. melanosporum* behaves like a true ectomycorrhizal fungus and that the ascocarp cannot be mainly supplied *via* saprotrophic pathways from surrounding soil organic matter or dead host tissues. Our previous findings also suggest that *T. melanosporum* ascocarps cannot be completely independent at any time during their development, even during late maturation [9]. Similarly, *Tuber* ascocarps never develop when separated from their host [10].

However, *in situ* ^{13}C and ^{15}N labeling experiments are the only way to definitively answer the questions regarding carbon and nitrogen allocation during *Tuber* ascocarp differentiation. The technical difficulties inherent in this methodology are numerous, which is likely why no convincing experiments have yet been conducted to address this topic.

Numerous *in situ* ^{13}CO$_2$ pulse-labeling experiments have been conducted on annual crops or grasslands. These studies all demonstrated a rapid carbon flux pathway from the host to the roots [11,12] and from the roots to the rhizosphere [12,13]. Some of these studies include arbuscular mycorrhizas (AM). For example, Johnson et al. [14] showed that between 5 and 8% of the carbon lost by plants was respired by the AM mycelium over the first 21 h after labeling.

Several studies on carbon allocation in trees have been performed using pulse labeling. Most of these analyses were conducted in microcosms or mesocosms under controlled conditions with young seedlings. ^{14}C pulse labeling followed by autoradiography or counting by scintillation was employed in these experiments [12,15–24]. To our knowledge, very few studies have been conducted *in situ* with adult trees [25–28]. None of these experiments have considered the fructification of the associated fungi.

The aim of our work was to assess the allocation of carbon by the host to *T. melanosporum* mycorrhizas and ascocarps. This assessment was achieved *via* an *in situ* ^{13}CO$_2$ pulse-labeling experiment performed on a 20-year-old hazel tree in a truffle orchard established in the northeast of France.

Materials and Methods

The Experimental Site

The experiment was performed in Rollainville, which is situated in the west part of the Vosges in France on a limestone plateau of the Jurassic period (latitude 48° 18′ 42″; longitude 5° 44′ 13″; elevation 360 m; annual rainfall 941 mm with a maximum in July; mean annual temperature 9.5°C). The soil is a brown calcisoil (WRB 2006) with a silty clay texture, a high alkaline pH (water pH 7.97), a moderate content of organic matter (9.4%) and a limestone content of 8.8%. This soil is poor in available phosphorus and moderate in available potassium and magnesium. It is free-draining, highly granular and aerated.

The truffle orchard in which the experiment was conducted was established in 1991 by one of us (Christophe Robin). It was previously a cultivated site. No protected species were sampled. Hazel trees inoculated with *T. melanosporum* (Vittad.) (black Perigord truffle) marketed by the Naudet nurseries (http://www. pepinieres-naudet.com/) were planted [29]. The first truffle harvest began in November 2005.

Labeling

In 2010, a single tree (A11, 4 m in height) was selected based on its high truffle producing. Two stainless steel scaffolds 6 m in height were built in parallel with one another on both sides (east and west) of the tree to install the labeling chamber. The two scaffolds were secured and attached to one another with stainless steel bars. The base of the chamber was sealed around the stems of the tree using adhesive tape with a width of 100 mm. The entire tree was enclosed in a 28-m^3 cylindrical 200-μm polyethane film chamber into which pure ^{13}CO$_2$ gas was injected. The hazel tree was pulse-labeled first on the 10th of July 2010 and a second time on the 1st of September 2010. In July, the tree was watered one day before labeling (30 mm of water under the crown). The air temperature and air humidity inside the chamber were recorded with a single probe (HMP50, Vaisala, Finland) and a datalogger (CR1000, Campbell UK) at 30 s intervals. The labeling chamber was closed at 6:36 UT for the first labeling and at 10:41 UT for the second. Prior to injection, the CO$_2$ concentration in the chamber was impoverished through leaf assimilation. Then, 15 l of ^{13}CO$_2$ (99 atom %, CORTECNET, France) were injected at a flow rate setting between 0.11 and 0.18 l min^{-1}. Injection was initiated when the CO$_2$ concentration reached 139 μmol.mol^{-1} for the first labeling (7:40 UT) and 150 μmol mol^{-1} for the second (10:57 UT). Total CO$_2$ was regulated at 380 vpm using a ^{13}CO$_2$/^{12}CO$_2$ IRGA (S710, SICK/MAIHAC, Germany), and the evolution of the ^{12}CO$_2$ and ^{13}CO$_2$ concentrations was recorded inside the chamber. The concentration of ^{13}CO$_2$ inside the chamber reached 300 μmol mol^{-1} in the two labelings. Then, it declined to 73 μmol mol^{-1} during the first labeling event and 63 μmol mol^{-1} during the second. Finally, the chamber was opened at 9:15 UT for the first labeling and 13:04 UT for the second. The two labeling periods lasted 01:45 h and 02:07 h, respectively.

Based on the obtained data, the tree assimilated a total of 16.7 g of ^{13}C during the two pulse-labeling periods. From October 2010 to March 2011, the crown of the tree was enclosed in a net to prevent any direct transfer of carbon to the soil through the falling of dead leaves, branches, nuts or catkins. These materials were collected regularly from the inside of the net. From December 2010 to February 2011, the soil under the tree was protected from frost using straw mulch (15 cm of straw enclosed in a plastic net).

Sampling

Four quadrats of 1 m^2 were positioned under the tree at the four cardinal directions (south, north, west and east) at a distance of one meter from the trunk. On eight dates (1, 4, 83, 101, 133, 168, 204 and 264 days from the first sampling performed on the 7th of July 2010), ascocarps, ectomycorrhizal root tips, fine roots, bulk soil, mycorrhizospheric soil and ascocarpic soil were collected from the 0 to 10 cm depth in each of the four squares. During the same period, ascocarps were collected under control trees. Leaves, branches, catkins and buds were also collected in different periods and at the four positions (south, north, west and east).

Tree fine roots and mycorrhizas. Tree fine roots (≤2 mm diameter) and mycorrhizas were carefully retrieved from the soil and washed in water under a dissecting microscope. *Tuber melanosporum* mycorrhizas were identified *via* morphotyping on the basis of color, mantle shape and surface texture and some also by molecular typing. Fine roots and mycorrhizas were then treated for ten minutes with 1 M hydrochloric acid and then washed with water to eliminate soil calcium carbonate.

Mycorrhizas were confirmed as being associated with *T. melanosporum* using molecular methods. Genomic DNA was extracted with the DNeasy Mini Kit (Qiagen SA, Courtaboeuf, France) following the manufacturer's instructions. *T. melanosporum* mycorrhizas were checked using species-specific ribosomal-DNA, internally transcribed-spacer (ITS) primers [30,31]. A microsatellite genotyping of *T. melanosporum* mycorrhizas was performed using primer pairs corresponding to ten SSR markers. This data were then used to analyze the fine scale spatial genetic structure of *T. melanosporum* at the Rollainville site [32].

Soil. Bulk soil was collected at least 10 cm from any mycorrhizas or ascocarps. Mycorrhizospheric soil was obtained by carefully shaking roots with mycorrhizas and using needles or forceps. Ascocarpic soil was obtained by removing the soil adhering to the ascocarps using needles and forceps.

The soil samples collected in the field were immediately placed in an icebox and transferred to the laboratory at 4°C. After separation of the ascocarpic and mycorhizospheric soil, all of the samples were maintained at –80°C. They were not treated with hydrochloric acid.

Soil water extracts. To perform ^{13}C and ^{12}C measurements in soil water extracts, we used a portion of the samples held at –80°C. All of the samples were cleaned by removing small stones and shells using forceps under a dissecting microscope but were not treated by hydrochloric acid or ground. Living and dead mycorrhizas were removed from the mycorrhizospheric soil using forceps under a dissecting microscope.

For each sample, approximately 100 mg of soil was introduced into an Eppendorf tube with 0.5 ml of distilled water. The samples were shaken at 4°C for 24 h and then centrifuged for 5 minutes at 10,000 g. A 400 µl aliquot of the obtained supernatant was removed and immediately stored at −80°C. The 400 µl sample was then reduced to 100 µl using a cryodessicator and dried in a metal capsule. Each metal capsule was weighed before use and after drying to obtain the dry weight of organic matter dissolved in 400 µl. The dry weight of the soil introduced into each Eppendorf was also determined.

Ascocarps. During the 2010–2011 period of truffle production, 24 ascocarps produced beneath the labeled tree were found by chance inside the four squares or located by a dog outside of the four squares when mature. The ascocarps were carefully retrieved from the soil using a small garden trowel, as in the four squares. Harvesting was performed at five different times: 83, 101, 133, 168 and 204 days after the first labeling. During the following period of production (2011–2012), 3 ascocarps produced beneath the labeled tree were harvested (558 days after the first labeling). Ascocarps were also harvested under non labeled trees at three different times during the 2010–2011 period of production and at one time (01 16 2012) during following period of production. The ascocarps were also confirmed as belonging to T. melanosporum using molecular methods. The fresh weight of all of the ascocarps was determined after cleaning. Ten ascocarps were oven dried to obtain the average dry weight percentage (58.9%).

All of the ascocarps were also described morphologically and microscopically and classified using the following criteria (Table 1):

The first stages (truffles of less than 1 g) could not be harvested. We attempted to quantify the constitutive carbon derived by the ascocarps from the host by assuming that, during the 2010–2011 period of production, the 24 ascocarps were present at the time of the first crop and that they all grew in a synchronized manner.

The ascocarps of the last crop were partly desynchronized from the previous ascocarps. Their growth was slowed by low temperatures. Consequently, we excluded the last harvest and considered only the 18 ascocarps harvested from 28 September to 22 December 2010. We also considered only the gleba in the calculations, as the weight of the peridium was negligible. We used the average C concentration in the 18 ascocarps cropped under the labeled tree (43.48% C). For each date, the accumulated weight of constitutive carbon in the ascocarps (ΣCW) was as follows:

$$\Sigma CW = CW * n$$

where CW is the average constitutive carbon on that date, and n is the total number of ascocarps harvested from the beginning of the study.

On each date, the weight of ^{13}C derived from the host (^{13}CW) was calculated as follows:

$$^{13}CW = \Sigma CW * (^{13}C_{labelled} - ^{13}C_{natural\ abundance})$$

where ΣCW is the accumulated weight of constitutive carbon in the ascocarps; $^{13}C_{labelled}$ is the measured ^{13}C abundance on each date; and $^{13}C_{natural\ abundance}$ is the natural abundance.

Leaves. On each date, ten leaves were collected around the crown in the middle part of the tree at the four cardinal points and pooled together to obtain one sample per cardinal point.

For each date, there were four replicates of each type of material (leaves, fine roots, mycorrhizas, soil, soil solutions), with the exception of ascocarps, the number of which depended on the harvest. The samples were first air dried, then dried at 60°C for 48 h and ground to a fine powder using a shaker with steel beads.

Isotopic analysis. The percentages of total C and the C isotopic compositions in the leaves, fine roots, mycorrhizas, ascocarps, bulk soil, mycorrhizospheric soil, ascocarpic soil and solutions of bulk, myco-rhizospheric and ascocarpic soil were determined at INRA Nancy using an online continuous flow CN analyzer (Carlo Erba NA 1500) coupled to an isotope ratio mass spectrometer (Finnigan delta S). Values were reported using standard notation ($\delta^{13}C$ ‰) relative to Vienna PeeDee Belemnite (VPDB), employing polyethylene foil (IAEA-CH-7) as a standard.

$\delta^{13}C$ values were calculated with the usual formula:

$$\delta^{13}C = (R_{sample}/R_{PDB} - 1) * 1000$$

where R is the molar ratio of $^{13}C/^{12}C$, and R_{PDB} is the molar ratio of PeeDee Belemnite. For ascocarps, the natural abundance $\delta^{13}C$ (‰) value was calculated by averaging the $\delta^{13}C$ values of ten ascocarps collected during the same period beneath an unlabeled hazel tree. For each organ, Excess (‰) $\delta^{13}C = \delta^{13}C_{labelled} - \delta^{13}C_{natural\ abundance}$. For soil samples, the amount of soluble ^{13}C was expressed in nanograms per 100 mg of dried soil.

Statistical Analyses

Analyses of variance for experimental data were conducted using the R software (R project for Statistical computing, http://www.R-project.org). Analyses of variance were performed using Type-II sum of squares (Anova function from package "car") when data were missing, causing unbalanced design. When necessary, data were transformed prior to the Anova using the Box-Cox method [33]. The criterion for statistical significance was set at $p < 0.05$.

Table 1. Description of the maturation stages of *T. melanosporum* ascocarps (modified from Giovanni Pacioni, personal communication).

Stage	Description	Size or Weight
5a (Ascospores beginning to form)	Fertile veins with asci and ascospores beginning to form.	1 cm
5b (Smooth ascospores)	No echinulated ascospores; Brown-black peridium; White gleba; No aroma.	1–2 g
6a (Ornamented ascospores)	White echinulated ascospores; Brown-to-black peridium. Numerous open cracks. New warts in formation; White-to-clear brown gleba; white veins clearly visible; Weak aroma.	2–5 g
6b (Ornamented brown ascospores)	Black-brown peridium with few closed cracks; Grey-black gleba between white veins; 80% echinulated ascospores, brown to dark brown; 20% of white echinulated ascopsores; Fairly developed aroma; Not completely mature.	5–15 g
6c (Brown-to-dark brown echinulated ascospores)	Black-brown peridium with few closed cracks; Black gleba between white veins; Very well developed aroma; Completely mature.	15–100 g and more

Results

The flux of pulse-derived ^{13}C from the leaves to the fine roots, mycorrhizas, ascocarps and soil was traced and quantified over a seven-month post-labeling period.

Leaves

The natural $\delta^{13}C$ abundance in the leaves presented an average value of -27.66 ‰ prior to the first labeling (Table 2). Leaf $\delta^{13}C$ reached a level of almost 300 ‰ just after the end of the first $^{13}CO_2$ injection, after which it decreased rapidly (to 35 ‰ after 5 days) but remained positive until the second labeling, when it peaked at 470 ‰. The $\delta^{13}C$ subsequently decreased until reaching a negative value at leaf fall. Dormant buds sampled during winter showed high ^{13}C abundance, as did branches formed in 2010 whose ^{13}C concentration was higher than in older branches. The $\delta^{13}C$ reached 76 ‰ in the following spring in the newly formed leaves, just after bud break.

Fine Roots and Mycorrhizas

Mycorrhizas were significantly more labeled than fine roots ($p < 0.01$) and there was a date effect ($p < 0.001$) but no interaction (Table 3). A nonsignificant increase of the $\delta^{13}C$ was visible in fine roots sampled 26 days after the first labeling. The $\delta^{13}C$ level in fine roots always remained below zero throughout the period following the first ^{13}C pulse. The $\delta^{13}C$ was higher after the second labeling, from the sampling at 149 days after the second pulse where it peaked at 9.87 ‰ (January 2011). There was a transfer of ^{13}C to the *T. melanosporum* mycorrhizas that became positive 80 days and 165 days after the first labeling (113 days after the second one). The mycorrhizal $\delta^{13}C$ level peaked at +22.75 ‰, 80 days after the first pulse and then decreased. It increased again after the second labeling, reaching a maximum of +55.35 ‰ prior to decreasing again.

Ascocarps

The first ascocarps beneath the labeled tree were harvested on the 28th of September 2010 and the last on the 27th of January, 2011 (Table 4). The first ascocarps were immature (stages 5b to 6a). They matured gradually, and in January 2011, all of the harvested ascocarps were fully ripened (stage 6c). The synchronization between the ascocarps was not complete, and there were some variations in the maturation stage on each date.

At the end of September, the average fresh weight of the ascocarps was less than two grams. The ascocarps continued to grow until the end of December to reach an average of 35 g. The ascocarps harvested at the end of January 2011 were smaller than those from December 2010, most likely due to low soil temperatures, differences in primordium production times and

Table 2. Kinetics of $\delta^{13}C$ (in ‰) in leaves of the hazel tree A11 after pulse labeling and $\delta^{13}C$ of buds and branches sampled during the winter following the pulses.

Time in days from the first labeling	−3	0	5	18	52	80	130	210	261
Time in days from the second labeling					0	28	78	158	209
Leaves (‰)	−27.66 (1.51) a	290.71 (173.00) b	35.26 (55.63) a	29.65 (57.88) a	469.07 (211.2) b	11.45 (12.75) a	−4.45 (12.45) a		76.70 (43.24) a
Dormant buds (‰)								45.53 (18.36)	
2010 branches (‰)								13.81 (10.88)	
2009 branches (‰)								−9.11 (4.88)	
2008 branches (‰)								11.96 (5.37)	
2005 branches (‰)								16.84 (2.66)	

One-way Anova has been performed using R; data have been raised to power minus 2 prior to Anova as suggested by the Box-Cox method in order to ensure the normality of residuals. Standard errors of means are given in brackets and mean values followed by a different letter are significantly different from the others at p<0.05 (Mean comparison Tukey's test).

Table 3. Kinetics of $\delta^{13}C$ (in ‰) in the fine roots and *T. melanosporum* mycorrhizas beneath hazel tree A11 in 2010–2011 after the pulse labelings of the leaves with $^{13}CO_2$.

Time in days from the first labeling	−3	1	26	80	130	165	201	261
Time in days from the second labeling				28	78	113	149	209
Fine roots (‰)	−27.62 (0.29) a	−26.62 (2.28) a	− 9.73 (33.52) ab	−13.15 (13.60) ab	−19.34 (3.51) ab	− 10.12 (12.06) ab	9.87 (41.20) ab	3.82 (17.52) b
Mycorrhizas (‰)	−27.60 (0.19) a	−24.09 (5.35) a	−24.60 (2.24) a	22.75 (47.97) b	−5.31 (19.33) b	26.3 (12.08) b	52.35 (35.06) b	18.85 (7.23) b

Two-way Anova has been performed using R; data have been raised to the power minus 2 prior to Anova as suggested by the Box-Cox method in order to ensure the normality of residuals. Anova showed a 'date' (p<0.01) and 'organ' (p<0.001) effects but no interaction. Standard errors of means are given (in brackets). Mean comparison has been made for simple effects (Tukey test): for d^{13}C in fine roots and d^{13}C in mycorrhizas respectively, means followed by a different letter are significantly different.

maturation desynchronization among the different ascocarps. The 24 collected ascocarps produced a total of 446 g fresh weight.

In the peridium and the gleba, the $\delta^{13}C$ was highest in the first harvest (87 and 125‰ respectively). It subsequently decreased and then increased again after the second labeling. The ^{13}C enrichment was significantly higher in the gleba than in the peridium (p = 0.034), and the $\delta^{13}C$ in the peridium and gleba were significantly higher than natural abundance, whatever the sampling date.

At the end of December, the gleba of the ascocarps, which had reached full growth, were 35% more enriched in ^{13}C than the mycorrhizas and almost three times more enriched than the fine roots. The $\delta^{13}C$ in the ascocarps harvested in 2012 under the labeled tree, 555 days after the first labeling, displayed a value identical to the natural abundance. In the controls, the $\delta^{13}C$ values in the ascocarps were not different between the gleba and the peridium and remained very stable between −25 and −26‰ throughout the period of maturation.

Constitutive ^{13}C Derived by the Ascocarps from the Host

From the 28th of October to the 22th of December, 2010, the ascocarps continuously absorbed ^{13}C from the host (Table 5). A total of 160.4 mg of ^{13}C was derived to provide constitutive carbon to the ascocarps at a flux of 1.7 mg per day until the 22nd of December, representing 0.96% of the ^{13}C assimilated by the tree during the two labeling periods. The ^{13}C allocated to the ascocarps to provide constitutive carbon was even higher, as there were only 18 fruiting bodies considered in this case, instead of 24. The increase in derived ^{13}C used as constitutive carbon in the ascocarps was similar to the increase in total constitutive carbon.

Table 4. (A) Maturity, numbers and fresh weight of ascocarps harvested beneath the labeled tree A11; (B) $\delta^{13}C$ (in ‰) in *T. melanosporum* ascocarps (peridium and gleba) beneath the labeled tree A11, and beneath non-labeled trees (natural abundance) at each sampling date from October 2010 to January 2011.

		2010				2011	2012
	Harvest date	**Sept-28**	**Oct-16**	**Nov-17**	**Dec-22**	**Jan-27**	**Jan-16**
A	Time in days from the first labeling	80	98	130	165	201	555
	Time in days from the second labeling	28	46	78	113	149	503
	Stage of maturity	5b to 6a	6a	6a to 6b	6b to 6c	6c	
	Average fresh weight (g)	1.95	10.6	15.3	35.5	19.4	
	Number of ascocarps harvested	4	5	3	6	6	
	Fresh weight harvested (g)	7.8	63	45.9	213	116	
	Accumulated fresh weight harvested (g)	7.8	70.8	116.7	330	446	
B	$\delta^{13}C$ in the peridium in ‰, labeled tree A11	+87.01 (34.23) b	+60.36 (12.67) b	+69.56 (34.30) b	+59.43 (6.35) b	+77.36 (26.08) b	−25.84 (0.37) b
	$\delta^{13}C$ in the gleba in ‰, labeled tree A11	+125.4 (29.93) c	+78.94 (12.33) b	+79.13 (22.37) b	+67.17 (15.67) b	+82.12 (27.87) b	−26.22 (0.46) a
	$\delta^{13}C$ in the peridium in ‰, Controls		−26.11 (0.32)		−25.65 (0.60)	−25.84 (0.87)	−26.16 (0.61)
	$\delta^{13}C$ in the gleba in ‰, Controls		−25.71 (0.18)		−24.93 (0.70)	−25.39 (0.34)	−25.79 (0.83)

n = 3 to 7 ascocarps harvested at each sampling date. Two-way Anova has been performed using R (table 4B); data have been log transformed prior to Anova as suggested by the Box-Cox method in order to ensure the normality of residuals. Anova showed a 'date' (p<0.05) and 'organ' (p<0.001) effects but no interaction. Standard errors of means are given (in brackets). Mean comparison has been made for simple effects (Tukey test): for $\delta^{13}C$ in the peridium and $\delta^{13}C$ in the gleba, means followed by a different letter are significantly different.

Table 5. Characteristics of the *T. melanosporum* ascocarps harvested in 2010–2011 beneath the labeled hazel tree A11 and estimations of the amounts of ascocarpic ^{13}C derived from the host tree.

	2010				2011
Harvest date	**Sept-28**	**Oct-16**	**Nov-17**	**Dec-22**	**Jan-27**
Number of ascocarps harvested	4	5	3	6	6
Average dry weight (g)	1.15	6.24	9.00	30.10	11.42
Ascocarp dry weight harvested (g; 24 ascocarps)	4.59	37.08	27.02	73.82	68.73
Accumulated ascocarp dry weight (g; 24 ascocarps)	4.59	41.67	68.69	183.97	252.70
Accumulated dry weight in g (18 ascocarps)	20.70	112.32	162.00	361.80	
Accumulated constitutive carbon in the ascocarps in g (SCW)(18 ascocarps)	9.00	48.84	70.44	157.38	
^{13}C derived from the host in mg (^{13}CW)(18 ascocarps)	14.81	55.59	82.66	160.37	

From Fine Roots to Ascocarps

Despite appearing nonstatistically significant, the transfer of ^{13}C was effective in the fine roots, which never accumulated ^{13}C (Figure 1). The transfer of ^{13}C from the fine roots to the mycorrhizas was delayed for several weeks. From the mycorrhizas to the ascocarps, the ^{13}C transfer was intensive 80 days after the first injection of $^{13}CO_2$. It then decreased, followed by increasing from days 113 to 149 after the second labeling. It was not possible to harvest truffles in early stages (less than 1 g). The early transfer of ^{13}C from the host towards the ascocarps probably occurred at least 60 days after the first labeling.

Soil and Soil Water Extracts

The $\delta^{13}C$ values did not differ significantly between the three soil compartments (bulk, mycorhizospheric and ascocarpic soil) and remained constant (average $-25.22‰$) from July 7 2010 (before the first labeling) to January 2011 (Table 6). There was a date effect ($p = 0.0011$), the last sampling date presenting higher $\delta^{13}C$ values than at 1, 98 and 130 days after the first labeling.

The C content in water soil extracts was higher in the mycorhizospheric soil (173.5 ng per 100 mg of dry soil) than in the bulk soil (107.9 ng per 100 mg of dried soil) ($p < 0.001$) or in the ascocarpic soil (131.7 ng per 100 mg of dried soil) ($p < 0.05$).

The $\delta^{13}C$ values in the soil water extracts did not differ significantly between the three compartments and remained constant throughout the period investigated, with an average value of $-24.45‰$ (Table 7).

Discussion and Conclusions

It is accepted that ectomycorrhizal fungi do not rely on dead organic matter as a carbon source. Using ^{14}C as a tracer in forest conditions (Oak Ridge Reservation, Tennessee), Treseder et al. [34] demonstrated that basidiomycota ectomycorrhizal fungi acquired most or all of their carbon from their hosts and that less than 2% of the carbon in the ectomycorrhizal biomass originated from the litter. Similarly, in several studies examining changes in photoperiods or photosynthesis rates, the defoliation or

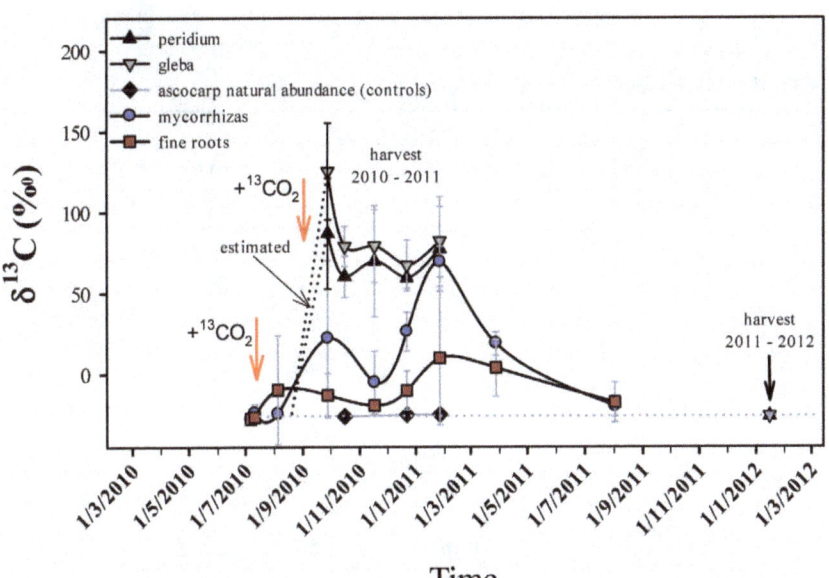

Figure 1. Temporal variation of $\delta^{13}C$ (‰) in the fine roots, mycorrhizas and peridium and gleba of ascocarps beneath hazel tree A11 in 2010–2011. Error bars represent the standard deviation of the means.

Table 6. $\delta^{13}C$ (‰) in the soil compartments (bulk soil, mycorhizospheric soil, soil adhering to the ascocarps) following the pulse labelings of the A11 hazel tree.

Harvest date	2010						2011	
	July-07	July-10	Sept-28	Oct-16	Nov-17	Dec-22	Jan-27	March-28
Time in days from the first labeling	−3	1	80	98	130	165	201	261
Time in days from the second labeling			28	46	78	113	149	209
$\delta^{13}C$ in the bulk soil (‰)	−24.46 (0.80)	−25.13 (0.80)	−25.71	−26.28 (1.26)	−25.98 (0.42)	−25.29 (0.89)	−25.15 (0.43)	−24.33 (0.39)
$\delta^{13}C$ in the myco-rhizospheric soil (‰)	−25.53 (1.29)	−26.51 (0.38)	−26.96 (0.38)		−26.17 (0.48)	−23.74		−23.55 (1.99)
$\delta^{13}C$ in the adherent ascocarpic soil (‰)				−25.84 (0.82)	−26.07 (0.73)	−24.39 (1.98)	−24.06 (3.14)	

Two-way Anova has been performed using R; data have been log transformed prior to Anova as suggested by the Box-Cox method in order to ensure the normality of residuals. Anova showed a 'date' effect (p = 0.0011) but no 'organ effect' and no interaction. Analysis of variance was performed using Type-II sum of squares (Anova function from package "car") because of unbalanced design resulting from missing data. When available, standard errors of means are given (in brackets).

girdling of the host suggested that basidiomycota sporocarps depend strongly on newly synthesized carbon from the host [35–39]. However, 75% defoliation of a *Pinus pinaster* stand affected the mycorrhizal community but did not decrease basidiomycota sporocarp biomass or abundance [40].

However, Hobbie et al. [41], using the ^{14}C signal from 1950s to 1960s thermonuclear testing, suggested that some ectomycorrhizal fungi like *Cortinarius*, *Leccinum* or *Tuber* could be able to use some insoluble soil organic matter.

In our experiment, the transfer of ^{13}C from the hazel tree leaves to the mycorrhizas via the fine roots was very slow. This result contrasts with what is observed in herbaceous plants, in which allocation to the roots is far more rapid [11,13]. Furthermore, the carbon allocation by the host to the ectomycorrhizas appeared to take place very slowly compared to what has been found in arbuscular mycorrhizal plants. For example, Leake et al. [18] observed peak transfer from cores colonized by AM mycelium 9–14 h after labeling. This transfer was also slower than the transfer observed by Högberg et al. [25] in a *Pinus sylvestris* forest, where the ^{13}C content peaked after 4–7 days in ectomycorrhizal pine roots.

In our experiment, the fine roots acted mainly as a conduit. They did not accumulate ^{13}C and transferred it slowly to the mycorrhizas. The mycorrhizas first formed a carbon sink and accumulated ^{13}C prior to ascocarp development. Then, the mycorrhizas transferred ^{13}C to the truffles, which accumulated carbon from the host until reaching complete maturity, 200 days after the first labeling and 150 days after the second labeling. This role of the *Tuber* ascocarps as a carbon sink was observed several months after the end of carbon assimilation by the host, which lost its leaves, and at low temperature. This finding suggests that the carbon allocated to the ascocarps during fall and winter was provided by reserve compounds stored in the trunk, branches, buds or thick roots. In the labeled tree, there was ^{13}C accumulation in the dormant buds and in the newly formed branches.

Tissues of deciduous trees store starch in autumn [42]. This starch is partly hydrolyzed during dormancy. In poplar trees during autumn and winter, starch hydrolysis results in a huge increase in sucrose and its galactosides [43]. This process represents a mechanism for protection against frost [44]. Starch resynthesis occurs at the end of dormancy, and new hydrolysis of starch into simple sugars is observed at bud break [45–47]. Carbon

Table 7. $\delta^{13}C$ (‰) in soil water extracts following the pulse labeling of the A11 hazel tree.

Harvest date	2010						2011	
	July-07	July-10	Sept-28	Oct-16	Nov-17	Dec-22	Jan-27	March-28
Time in days from the first labeling	−3	1	80	98	130	165	201	261
Time in days from the second labeling			28	46	78	113	149	209
$\delta^{13}C$ of the bulk soil solution (‰)	−25.20 (1.02)	−24.73 (1.73)	−24.97 (1.02)	−24.09 (1.94)	−24.89 0.79	−24.68 (1.97)	−25.75 (0.71)	−25.25 (0.40)
Soluble ^{13}C (ng for 100 mg soil)	90.5 (25.1)	83.6 (11.8)	105.2 (14.8)	159.5 (56.9)	86.1 (25.8)	69.2 (13.4)	83.7 (10.0)	185.9 (30.3)
$\delta^{13}C$ of the myco-rhizospheric soil solution (‰)	−25.24 (0.95)	−25.24 (0.28)	−23.49 (0.66)		−25.00 (0.61)	−25.55		−22.76 (4.15)
Soluble ^{13}C (ng for 100 mg soil)	177.5	134.2	253.0		168.6 (73.0)	103.6		221.1 (95.9)
$\delta^{13}C$ of the ascocarpic soil solution (‰)				−24.40 (0.67)	−24.08 (0.18)	−22.13 (3.88)	−22.68 (3.77)	
Soluble ^{13}C (ng for 100 mg soil)				141.9 (7.7)	112.6 (23.9)	146.6 (50.3)	125.9 (27.9)	

Two-way Anova has been performed using R; data have been log transformed prior to Anova as suggested by the Box-Cox method in order to ensure the normality of residuals. Anova showed no main effects and no interaction. Analysis of variance was performed using Type-II sum of squares (Anova function from package "car") because of unbalanced design resulting from missing data. Anova showed a slight date effect (p = 0.046) and no interaction. When available, standard errors of means are given (in brackets).

remobilization in deciduous trees could provide up to approximately 40% of the C used for new tissue formation and can contribute to early wood formation [48–50]. Under the conditions of the present study, we can assume that the carbon allocated by the host to *T. melanosporum* ascocarps at the end of their development is in the form of simple sugars produced *via* starch hydrolysis during tree dormancy, rather than from amylase activity prior to or during bud breaking. Nevertheless, under a Mediterranean climate, the carbon used for growth of *T. melanosporum* ascocarps could enter into competition with the carbon necessary for bud bursting or early wood formation. However, the quantity of carbon necessary for ascocarp development (approximately 1% of the assimilated C) is not comparable to the quantity of C necessary for bud breaking. In 2012, when the ascocarp $\delta^{13}C$ content was found to be equal to the natural abundance, it became clear that the carbon allocated to the fruiting bodies was only coming from the carbon assimilated by the tree during the growing season.

Using *Pinus densiflora* seedlings in mycorrhizal association with *Laccaria amethystina* and labeled with $^{14}CO_2$, Teramoto et al. [3] showed that there was transfer of host carbon to the sporocarps over one or a number of days in rhizoboxes based on autoradiography and radioactivity counting. This finding indicates that the fungus primarily used recently assimilated carbon. This result obtained in an ectomycorhizal member of Basidiomycotina producing fruitbodies over a number of days contrasts with our results obtained with an ectomycorrhizal member of the Ascomycotina, in which the development of ascocarps requires carbon stored in the trunk or roots of the host, and the process takes several weeks/months.

These results demonstrate, for the first time under field conditions, that *Tuber* mycorrhizas provide a slow, but dominant pathway for carbon flux from trees to ascocarps. From September 28 2010 until December 22 2010, 1.7 mg ^{13}C was transferred per day from the mycorrhizas to the ascocarps to provide constitutive carbon. The total amount of constitutive ^{13}C transferred was approximately 1% of the ^{13}C assimilated by the tree during the two labeling periods. These data do not include the carbon respired by the ascocarps.

This $^{13}CO_2$ pulse-labeling experiment corroborates our previous results based on natural ^{13}C and ^{15}N abundance showing that carbon allocation for ascocarp development could not be supplied mainly *via* saprotrophic pathways [9]. According to our pulse-labeling experiment, almost all of the carbon allocated to the truffle ascocarps came from the host. These findings also corroborate the results of sequencing of the *T. melanosporum* genome, which showed that this fungus has a limited repertoire of genes coding for CAZymes [8]. Nevertheless, we cannot exclude the possibility of weak carbon allocation to the ascocarps from soil organic matter. Soil labeling experiments (^{15}N and ^{13}C) are necessary to determine whether truffle ascocarps can also use dead host tissues or soil organic matter as carbon and nitrogen sources.

Several questions remain concerning the mode of carbon transfer between the host tree and developing ascocarps. Based on our results, it is clear that this transfer cannot occur through the soil. The soil and the soil solutions never appeared to be enriched in ^{13}C, regardless of the compartment considered. Epron et al. [26] also observed an absence of bulk soil enrichment after $^{13}CO_2$ pulse labeling of three species (*Fagus sylvatica*, *Quercus petraea* and *Pinus pinaster*).

This transfer also cannot take place *via* the external mycelium of the mycorrhizas. This mycelium extends to a maximum of a few millimeters from the mantle, and the ascocarps generally develop at least several cm from the mycorrhizas. The most likely hypothesis to explain this type of carbon transfer is a transport through the ascogonial structure, which could provide a direct connection between mycorrhizas and ascocarps.

In conclusion, it appears evident that *Tuber* ascocarps are dependent on their hosts throughout their development. These results contradict the statements of well-recognized truffle handbooks and could be of some importance for improvement of truffle cultivation methods, for example, through using caution regarding tree pruning at truffle primordium production and during truffle growth.

Acknowledgments

We want to thank Biela, our dog, which hunted the labeled truffles with success. Jean Villerd, (UMR 1121, Université de Lorraine) for his help with the statistical analysis. Erik Hobbie, for his constructive comments, which helped us to improve the manuscript.

Author Contributions

Conceived and designed the experiments: FLT BZ. Performed the experiments: BZ CP CR FLT. Analyzed the data: CP BZ FLT CR. Contributed reagents/materials/analysis tools: CH CB CP. Wrote the paper: FLT. Edited the manuscript: FLT CR CP BZ.

References

1. Paolocci F, Rubini A, Riccioni C, Arcioni S (2006) Re-evaluation of the life cycle of Tuber magnatum. Appl Environ Microbiol 72: 2390–2393.
2. Riccioni C, Belfiori B, Rubini A, Passeri V, Arcioni S, et al. (2008) Tuber melanosporum outcrosses: analysis of the genetic diversity within and among its natural populations under this new scenario. New Phytol 180: 466–478.
3. Teramoto M, Wu B, Hogetsu T (2012) Transfer of ^{14}C-photosynthate to the sporocarp of an ectomycorrhizal fungus Laccaria amethystina. Mycorrhiza 22: 219–225.
4. Olivier J-M, Savignac J-C, Sourzat P (2012) Truffe et trufficulture, Fanlac editor, Périgueux, France.
5. Barry D, Staunton S, Callot G (1994) Mode of the absorption of water and nutrients by ascocarps of Tuber melanosporum and Tuber aestivum. A radioactive tracer technique. Can J Bot 72: 317–322.
6. Callot G, Byé P, Raymond M, Fernandez D, Pargney JC, et al. (1999) La truffe, la terre, la vie: Quae, INRA. 210 p.
7. Mamoun M, Olivier JM (1991) Influence du substrat carboné et de la forme d'azote minéral sur la croissance de T. melanosporum Vittad. en culture pure. Application à la production de biomasse mycélienne. Agronomie 11: 521–527.
8. Martin F, Kohler A, Murat C, Balestrini R, Coutinho PM, et al. (2010) Perigord black truffle genome uncovers evolutionary origins and mechanisms of symbiosis. Nature 464: 1033–1038.
9. Zeller B, Bréchet C, Maurice JP, Le Tacon F (2008) Saprotrophic versus symbiotic strategy during truffle ascocarp development under holm oak. A response based on ^{13}C and ^{15}N natural abundance. Ann For Sci 65: 607–614.
10. Rouquerolle T, Payre H (1975) Conséquences de quelques particularités biologiques des Tuber sur les caractères des cultures de mycélium et sur la formation des truffes. Rev Mycol 29: 213–224.
11. Robin C, Vaillant V, Vansuyt G, Zinsou C (1990) Assimilate partitioning in Pachyrhizus erosus L. during long-day vegetative development. Plant Physiol Biochem 28: 343–349.
12. Leake JR, Ostle NJ, Rangel-Castro IJ, Johnson D (2006) Carbon fluxes from plants through soil organisms determined by field $^{13}CO_2$ pulse-labelling in an upland grassland. Applied Soil Ecology 33: 152–175.
13. NGuyen C, Todorovic C, Robin C, Christophe A, Guckert A (1999) Continuous monitoring of rhizosphere respiration after labelling of plant shoots with $^{14}CO_2$. Plant Soil 212: 189–199.
14. Johnson D, Leake JR, Ostle N, Ineson P, Read DJ (2002) In situ $^{13}CO_2$ pulse-labelling of upland grassland demonstrates a rapid pathway of carbon flux from arbuscular mycorrhizal mycelia to the soil. New Phytol 153: 327–334.
15. Ek H (1997) The influence of nitrogen fertilization on the carbon economy of Paxillus involutus in ectomycorrhizal association with Betula pendula. New Phytol 135: 133–142.
16. Finlay RD, Read DJ (1986) The structure and function of the vegetative mycelium of ectomycorrhizal plants. 1. The translocation of ^{14}C-labelled carbon

between plants interconnected by a common mycelium. New Phytol 103: 143–156.

17. Horwath WR, Pregitzer KS, Paul EA (1994) [14]C Allocation in tree-soil systems. Tree Physiol 14: 1163–1176.

18. Leake JR, Donnelly DP, Saunders EM, Boddy L, Read DJ (2001) Rates and quantities of carbon flux to ectomycorrhizal mycelium following ^{14}C pulse labeling of Pinus sylvestris seedlings: effects of litter patches and interaction with a wood-decomposer fungus. Tree Physiol 21: 71–82.

19. Leanne JP, Simard SW (2008) Minimum pulses of stable and radioactive carbon isotopes to detect belowground carbon transfer between plants. Plant Soil 308: 23–35.

20. Miller SL, Durall DM, Rygiewicz PT (1989) Temporal allocation of ^{14}C to extramatrical hyphae of ectomycorrhizal ponderosa pine seedlings. Tree Physiol 5: 239–249.

21. Norton JM, Smith JL, Firestone MK (1990) Carbon flow in the rhizosphere of Ponderosa pine seedlings. Soil Biol Biochem 22: 449–455.

22. Ostle N, Whiteley AS, Bailey MJ, Sleep D, Ineson P, et al. (2003) Active microbial RNA turnover in a grassland soil estimated using a [13]CO$_2$ spike. Soil Biol Biochem 35: 877–885.

23. Rygiewicz PT, Andersen CP (1994) Mycorrhizae alter quality and quantity of carbon allocated below ground. Nature 369: 58–60.

24. Simard SW, Durall DM, Jones MD (1997) Carbon allocation and carbon transfer between Betula papyrifera and Pseudotsuga menziesii seedlings using a ^{13}C pulse-labeling method. Plant Soil 191: 41–55.

25. Högberg P, Högberg MN, Gottlicher SG, Betson NR, Keel SG, et al. (2008) High temporal resolution tracing of photosynthate carbon from the tree canopy to forest soil microorganisms. New Phytol 177: 220–228.

26. Epron D, Ngao J, Dannoura M, Bakker MR, Zeller B, et al. (2011) Seasonal variations of belowground carbon transfer assessed by in situ [13]CO$_2$ pulse labelling of trees. Biogeosciences 8: 1153–1168.

27. Plain C, Gerant D, Maillard P, Dannoura M, Dong Y, et al. (2009) Tracing of recently assimilated carbon in respiration at high temporal resolution in the field with a tuneable diode laser absorption spectrometer after in situ [13]CO$_2$ pulse labelling of 20-year-old beech trees. Tree Physiol 29: 1433–1445.

28. Subke JA, Vallack HW, Magnusson T, Keel SG, Metcalfe DB, et al. (2009) Short-term dynamics of abiotic and biotic soil [13]CO$_2$ effluxes after in situ [13]CO$_2$ pulse labelling of a boreal pine forest. New Phytol 183: 349–357.

29. Chevalier G, Poitou N (1990) Study of important factors affecting the mycorrhizal development of the truffle fungus in the field using plants inoculated in nurseries. Agric Ecosyst Environ 28: 75–77.

30. Paolocci F, Rubini A, Granetti B, Arcioni S (1999) Rapid molecular approach for a reliable identification of Tuber spp. ectomycorrhizae. FEMS Microbiol Ecol 28: 23–30.

31. Rubini A, Paolocci F, Granetti B, Arcioni S (1998) Single step molecular characterization of morphologically similar black truffle species. FEMS Microbiol Lett 164: 7–12.

32. Murat C, Rubini A, Riccioni C, De la Varga H, Akroume E, et al. (2013) Fine scale spatial genetic structure of the black truffle (Tuber melanosporum) investigated with neutral microsatellites and functional mating type genes. New Phytol doi: 10.1111/nph.12264.

33. Box GEP, Cox DR (1964) An analysis of transformation, Journal of the Royal Statistical Society. Series B (Methodological).26–2: 211–252.

34. Treseder KK, Torn MS, Masiello CA (2006) An ecosystem-scale radiocarbon tracer to test use of litter carbon by ectomycorrhizal fungi. Soil Biol Biochem 38: 1077–1082.

35. Godbout C, Fortin JA (1992) Effects of nitrogen fertilization and photoperiod on basidiome formation of Laccaria bicolor associated with container-grown jack pine seedlings. Can J Bot 70: 181–185.

36. Högberg P, Nordgren A, Buchmann N, Taylor AFS, Ekblad A, et al. (2001) Large-scale forest girdling shows that current photosynthesis drives soil respiration. Nature 41: 789–792.

37. Kuikka K, Härmä E, Markkola A, Rautio P, Roitto M, et al. (2003) Severe defoliation of Scots pine reduces reproductive investment by ectomycorrhizal symbionts. Ecology 84: 2051–2061.

38. Lamhamedi MS, Godbout C, Fortin JA (1994) Dependence of Laccaria bicolor basidiome development on current photosyn- thesis of Pinus strobus seedlings. Can J For Res 24: 1797–1804.

39. Last FT, Pelham J, Mason PA, Ingleby K (1979) Influence of leaves on sporophore production by fungi forming sheathing mycorrhizas with Betula spp. Nature 280: 168–169.

40. Pestaña M, Santolamazza-Carbone S (2011) Defoliation negatively affects plant growth and the ectomycorrhizal community of Pinus pinaster in Spain. Oecologia 165: 723–733.

41. Hobbie EA, Quimette AP, Schuur EAG, Kierstead D, Trappe JM, et al. (2012) Radiocarbon evidence for the mining of organic nitrogen from soil by mycorrhizal fungi. Biogeochemistry doi: 10.1007/s10533-012-9779-z.

42. Essiamah S, Eschrich W (1985) Changes in starch content in the storage tissues of deciduous trees during winter and spring. IAWA Bulletin 6: 97–106.

43. Sauter JJ, van Cleve B (1994) Storage, mobilization and interrelations of starch, sugars, protein and fat in the ray storage tissue of poplar trees. Trees 8: 297–304.

44. Palonen P, Buszard D, Donnelly D (2000) Changes in carbohydrates and freezing tolerance during cold acclimation of red raspberry cultivars grown in vitro and in vivo. Physiol Plant 110: 393–401.

45. Bollmark L, Sennerby-Forsse L, Ericsson T (1999) Seasonal dynamics and effects of nitrogen supply rate on nitrogen and carbohydrate reserves in cutting-derived Salix viminalis plants. Can J For Res 29: 85–94.

46. Johansson T (1993) Seasonal changes in contents of root starch and soluble carbohydrates in 4–6 year old Betula pubescens and Populus tremula. Scand J For Res 8: 94–106.

47. Landhäusser SM, Lieffers VJ (2003) Seasonal changes in carbohydrate reserves in mature northern Populus tremuloides clones. Trees - Structure and Function 17: 471–476.

48. Kagawa A, Sugimoto A, Maximov TC (2006a) [13]CO$_2$ pulse-labelling of photoassimilates reveals carbon allocation within and between tree rings. Plant Cell Environ 29: 1571–1584.

49. Kagawa A, Sugimoto A, Maximov TC (2006b) Seasonal course of translocation, storage and remobilization of ^{13}C pulse-labeled photoassimilate in naturally growing Larix gmelinii saplings. New Phytol 171: 793–803.

50. Maurel K, Leite GB, Bonhomme M, Guilliot A, Rageau R, et al. (2004) Trophic control of bud break in peach (Prunus persica) trees: a possible role of hexoses. Tree Physiol 24: 579–588.

pSiM24 Is a Novel Versatile Gene Expression Vector for Transient Assays As Well As Stable Expression of Foreign Genes in Plants

Dipak Kumar Sahoo[1]*, Nrisingha Dey[2], Indu Bhushan Maiti[1]*

1 KTRDC, College of Agriculture, Food and Environment, University of Kentucky, Lexington, Kentucky, United States of America, 2 Department of Gene Function and Regulation, Institute of Life Sciences, Bhubaneswar, Odisha, India

Abstract

We have constructed a small and highly efficient binary Ti vector pSiM24 for plant transformation with maximum efficacy. In the pSiM24 vector, the size of the backbone of the early binary vector pKYLXM24 (GenBank Accession No. HM036220; a derivative of pKYLX71) was reduced from 12.8 kb to 7.1 kb. The binary vector pSiM24 is composed of the following genetic elements: left and right T-DNA borders, a modified full-length transcript promoter (M24) of *Mirabilis mosaic virus* with duplicated enhancer domains, three multiple cloning sites, a 3′rbcSE9 terminator, replication functions for *Escherichia coli* (ColE1) and *Agrobacterium tumefaciens* (pRK2-OriV) and the replicase trfA gene, selectable marker genes for kanamycin resistance (nptII) and ampicillin resistance (bla). The pSiM24 plasmid offers a wide selection of cloning sites, high copy numbers in *E. coli* and a high cloning capacity for easily manipulating different genetic elements. It has been fully tested in transferring transgenes such as green fluorescent protein (GFP) and β-glucuronidase (GUS) both transiently (agro-infiltration, protoplast electroporation and biolistic) and stably in plant systems (*Arabidopsis* and tobacco) using both agrobacterium-mediated transformation and biolistic procedures. Not only reporter genes, several other introduced genes were also effectively expressed using pSiM24 expression vector. Hence, the pSiM24 vector would be useful for various plant biotechnological applications. In addition, the pSiM24 plasmid can act as a platform for other applications, such as gene expression studies and different promoter expressional analyses.

Editor: Jin-Song Zhang, Institute of Genetics and Developmental Biology, Chinese Academy of Sciences, China

Funding: This work was fully supported by the KY state KTRDC grant (no. 1235411320) to IBM. The funders had no role in study design, data collection and analysis, decision to publish, or preparation of the manuscript.

Competing Interests: The authors have declared that no competing interests exist.

* E-mail: dipak_sahoo11@rediffmail.com (DKS); imaiti@uky.edu (IBM)

Introduction

The transfer of foreign genes into higher plants mediated either by *Agrobacterium tumefaciens* or by employing a biolistic process is the core technique used in genetic engineering-based plant modification. Many useful and versatile vectors have been constructed since the birth of the concept and the first generation of binary vectors for plant transformation [1–6]. The general trend in the binary vector development has been to increase the plasmid stability during a long co-cultivation period of *A. tumefaciens* with the target host plant tissues and also to understand the molecular mechanism of broad host-range replication, and to use it to reduce the size of plasmid for ease in cloning and for higher plasmid yield in *Escherichia coli* [7,8]. A number of large (>10 kb), first-generation binary vectors have been constructed for plant transformation, including Ti plasmid [2], pBin19 [1], pKYLX7 [4] and other expression vectors [3]. One of the binary vectors, pBin19 [1], has been modified to pBI121 and pIG121Hm [9,10] to use the β-glucuronidase (*GUS*) reporter gene in plant transformation. Binary vectors include pKYLX expression vectors containing 35S and rbcS promoters that are suitable for constitutive or light-regulated expression of foreign genes [4]. These vectors and their derivatives were soon widely distributed among plant scientists. In addition, another widely used series of

vectors includes pPZP vectors [11] and their modified form, pCAMBIA vectors (www.cambia.org). Xiang et al., (1999) constructed a pCB mini-binary vector series [12] from the relatively large, first-generation binary plasmid pBin [1]. Over time, vector technology evolved, and new generations of plant transformation vectors with improved cloning and delivery strategy were introduced, for example, pGreen vectors [13]; pGD or pSITE vectors, which are suitable for the stable integration or transient expression of various autofluorescent protein fusions in plant cells [14,15]; the pCLEAN binary vector system [16]; the pHUGE binary vector system [17]; and binary bacterial artificial chromosome BIBAC vectors [18]. The TMV RNA-based vector pJLTRBO [19] and its derivative pPZPTRBO [20] were reported to produce recombinant proteins in plants without using the RNA-silencing inhibitor P19. Similar expression levels were provided by the pEAQ-HT vector which has an integrated P19 expression cassette [21]. A bean yellow dwarf virus single-stranded DNA-based vector, pBY030-2R was reported to produce high amount of recombinant proteins [22] while the pMAA-Red vector was known for easy production of transgenic *Arabidopsis* overexpression lines with strong expression levels of the gene of interest [23].

The binary vectors widely used for plant transformations vary in size, origin of replication, bacterial selectable markers, T-DNA borders and overall structure. Recent modifications of binary vectors provide a number of user-friendly features, such as a wide selection of cloning sites, high copy numbers in E. coli, improved compatibility with strains of choice, a wide pool of selectable markers for plants and a high frequency of plant transformation. Although recent improvements are very useful, the classic vector configuration still appears to be good enough in many occasions. Plasmid manipulations are also easier if the vector replicates in E. coli to high copy numbers. Moreover, the efficiency of in vitro recombination procedures is inversely proportional to the size of the vector DNA [24]. With an increased requirement for the transfer of large pieces of DNA into plants, the size of binary vectors should be kept to a minimum. The availability of low-molecular-weight, versatile plant expression vectors is currently insufficient in plant molecular biology. For these reasons, we designed a smaller binary vector, pSiM24, which offers a wide selection of cloning sites, high copy numbers in E. coli and is fully functional in the transient (using both the gene-gun or Agro-infiltration methods) as well as stable transformation of plants.

Materials and Methods

Chemicals, enzymes, bacterial strains and plasmids

Antibiotics (ampicillin, kanamycin, rifampicin, tetracycline, hygromycin) and chemicals were purchased from Sigma-Aldrich (St. Louis, MO, USA) or Thermo Fisher Scientific (Waltham, MA, USA). All restriction endo-nucleases and DNA-modifying enzymes were obtained from New England Biolab (Beverly, MA, USA) or Invitrogen-Life Technologies (Grand Island, NY USA). The TB1 strain of E. coli [25] and the C58C1 [GV3850] strain of A. tumefaciens [26] were used. The plasmids pBluescriptIIKS(+) (Genbank Accession no. X52327) from Stratagene (la Jolla, CA, USA) and pKYLX7 [4] or its derivative pKM24KH (GenBank accession no. HM036220.1) were used. Cultures of E. coli transformed with pUC-based vectors were grown in the presence of ampicillin (100 µg/ml). Transformed agrobacterium was grown in the presence of kanamycin (25 µg/ml) and rifampicin (100 µg/ml). The Vip3A and KMP-11 antibodies were provided by Dr Raj K. Bhatnagar, ICGEB, New Delhi, India and Dr Shyamal Roy, IICB, Kolkata, India. The IL-10 antibody was obtained from Imgenex, Bhubaneswar, India.

In vitro cloning procedures and DNA sequencing

All in vitro recombination techniques were employed using previously described standard methods [27,28]. For DNA sequencing, a dye terminator labeling procedure was followed using a Genome Lab DTCS-Quick Start kit (Beckman Coulter, USA), and an automated sequencing machine (Beckman Coulter CEQ 8000 Genetic Analysis System, USA) was used in accordance with the manufacturer's instructions.

Construction of plasmid vector pBTdna, pBTdna-rbcT and pBTdna-rbcT-KanR

We designed and generated a 522-bp synthetic DNA fragment containing left and right T-DNA borders and three multiple cloning sites (MCS) of general the structure 5'-BssHI-KpnI-left T-DNA(147-bp)-MCS1(BstXI-StuI-FspAI-PasI-SanDI-BstZ171-SmaI)-MCS2 (EcoRI-HindIII-BamHI-XhoI-HpaI-MluI-SalI-SstI-PstI-XbaI)-MCS3 (ClaI-SpeI-BglII-BstEII-EcoNI-FseI-SwaI-NruI-PacI-right T-DNA border (162-bp)-EcoRV-BssHII-3'. This fragment was synthe-sized by GenArt-Life Technologies (Carlsbad, CA, USA). The 5'-BssHII-BssHII-3'fragment was cloned into the corresponding sites of pBluescriptIIKS(+), and the resulting plasmid was named pBTdna.

A 657-bp poly(A) signal 3'-rbcS-E9 of the general structure 5'-ClaI-3'-rbcSE9-XbaI-3' was isolated from the binary vector pKM24KH (GenBank Accession No. HM036220) and was inserted into the corresponding site of pBTdna to generate the plasmid pBTdna-rbcT.

A 1343-bp synthetic neomycin phosphotransferase gene/kana-mycin resistance gene (nptII/KanR) of the general structure 5'-BglII-Nos-promoter-KanR cDNA-Nos-terminator-SpeI-3' was obtained from GenArt-Life Technologies (Carlsbad, CA, USA). The open reading frame of the KanR gene was optimized for plant codon bias. The KanR gene, with the structure 5'-BglII-SpeI-3', was cloned into the corresponding sites of pBTdna-rbcT to create the plasmid pBTdna-rbcT-KanR (also called pBTRK). The plasmid contains a 2498-bp micro-Tdna fragment of the general structure 5'-BssHII-KpnI-left T-DNA border-MCS1-MCS-2-XbaI-3'rbcS Terminator-ClaI-SpeI-KanR gene (comple-ment)-BglII-MCS3-right T-DNA border- EcoRV-BssHII-3'. The sequence integrity of the fragment was confirmed before further use. The sequence information of these genetic elements is provided in the NCBI database (GenBank accession no. KF032933).

Construction of non-T-DNA plasmids: pBtrfA, pB-oriV-trfA, pBAmpR-ColEI-oriV-trfA

We designed a non-T-DNA plasmid of the general structure 5'-BssHII-KpnI-AmpR gene-ColEI (origin of replication of pMB1)-ApaI-oriV of pRK2-SalI-trfA gene-EcoRV-BssHII-3'. A synthetic DNA fragment of the physical map 5'-BssHII-KpnI-ApaI-SalI-trfA gene-EcoRV-BssHII was obtained from a commercial supplier (GenArt-Life Technologies, CA, USA). The open reading frame of the trfA gene was optimized with the bias codon of A. tumefaciens. This 5'-BssHII-BssHII-3' fragment was cloned into the corresponding sites of pBluescriptIIKS(+) to create the plasmid pBtrfA.

A 642-bp fragment of the replicon OriV of pRK2 was PCR-amplified using appropriately designed forward and reverse primers to insert an ApaI site at the 5'-end and a SalI site at the 3'-end. The gel-purified PCR fragment 5'-ApaI-OriV-SalI-3' was inserted into the corresponding site of pBtrfA to form the plasmid pB-oriV-trfA. A fragment of 1803 bp containing the AmpR gene and the ColEI replicon in pMA (GeneArt vector, Registry part no. K157000), a pUC derivative, was PCR-amplified using appropriately designed forward and reverse primers to insert the KpnI site at the 5'-end and the ApaI site at the 3'-end. The PCR fragment was digested with KpnI and ApaI, and the gel-purified fragment 5'-KpnI-ApaI-3' was cloned into the corresponding site of pB-oriV-trfA to generate the plasmid pBAmpR-ColEI-oriV-trfA.

Construction of pSi and pSiM24

The T-DNA portion was isolated from pBTdna-rbcT-KanR. First, the pBTdna-rbcT-KanR plasmid was digested with PvuII; the larger band (4235-bp fragment) was isolated and further digested with BssHII to generate a 2492-bp fragment of the general structure (5'-BssHII-KpnI-left T-DNA border-MCS-1-MCS-2-XbaI-3'rbcS Terminator-ClaI-MCS-3-right T-DNA bor-der-EcoRV-BssHII-3'). The non-T-DNA portion of a 3830-bp fragment of the general structure (5'-BssHII-KpnI-AmpRI-ColEI-ApaI-OriV-SalI-trfA-EcoRV-BssHII-3') was isolated from pBAmpR-ColEI-oriV-trfA. Two fragments (T-DNA and nonT-DNA portions) were ligated and circularized to produce the binary vector pSi. The modified full-length transcript promoter (M24) of the Mirabilis mosaic virus [27,29] was inserted as 5'-EcoRI-HindIII-

3′ into the corresponding sites of pSi, and the resulting plasmid was named pSiM24. The fully annotated sequence of pSiM24 is available in the NCBI database (GenBank accession no. KF032933).

Construction of plant expression vectors with green fluorescent protein (GFP) and β-glucuronidase (GUS) reporter genes

The M24 promoter fragment 5′-EcoR1-M24-HindIII-3′ and the reporter gene 5′-XhoI-GUS-SstI-3′ or 5′-XhoI-GFP-SstI-3′ were inserted into the corresponding sites of pBTRK and pSi to generate expression constructs pBTRK-M24-GUS/GFP and pSiM24-GUS/GFP, respectively.

Tobacco plant transformation

The plant expression constructs pSiM24-GUS and pSiM24 were introduced into the *A. tumefaciens* strain GV3850 by the freeze-thaw method [30]. Tobacco plants (*Nicotiana tabacum* cv. SamsunNN) were transformed with *Agrobacterium* harboring pSiM24-GUS and pSiM24 constructs as described previously [31] or by the gene-gun method using pSiM24-GUS and pSiM24 constructs [32]. Tobacco shoots and then roots were regenerated from kanamycin-resistant calli derived from independent leaf discs. Ten independent kanamycin-resistant plant lines (R_0 generation, 1st progeny) were generated for the constructs pSiM24-GUS and pSiM24 and were maintained under greenhouse conditions ($30 \pm 5°C$ with both natural and supplementary lighting of minimum photon flux density, 300 µmole/m^2/s, 17 h day/7 h night cycle). Seeds were collected from self-pollinated primary transformants. Transgenic tobacco seeds (R1 progeny, 2nd generation) were germinated in the presence of Kanamycin (250 mg/L). Positive transformants with a KanR:KanS ratio of 3:1 progeny segregation were selected for further analysis. Transgenic lines (R_1 and R_2 progeny, second and third generation) were screened for gene integration, transcription and translation by polymerase chain reaction (PCR), reverse transcriptase-PCR (RT-PCR), real-time quantitative RT-PCR (qRT-PCR), enzymatic assays and GUS histochemical analysis.

Generation of transgenic *Arabidopsis* plants

The pSiM24 and pSiM24-GUS plasmids introduced into *A. tumefaciens* GV3850 were used to transfer each of these constructs into *Arabidopsis* (*Arabidopsis thaliana* ecotype Columbia-0) by the floral dip method [33]. The transgenic *Arabidopsis* plants were selected and maintained as described previously [34].

Transient Agro-infiltration assay of pSiM24-GUS in tobacco leaves

Suspensions of the *A. tumefaciens* strain GV3850 bearing pSiM24 and pSiM24-GUS constructs were infiltrated into leaves of *Nicotiana benthamiana* as described previously [35]. After two days of agro-infiltration, the transient GUS expression was evaluated by the histochemical GUS staining method [9].

Transient expression analysis in tobacco protoplasts

The isolation of tobacco protoplasts from the suspension cell cultures of *N. tabacum* L. cv Xanthi-Brad and electroporation of tobacco protoplasts with supercoiled plasmid pBTRKM24-GUS/GFP and pBTRKM24 constructs were performed as described previously [36]. After 20 h, protoplasts were harvested for fluorometric GUS enzymatic assay [9]. GUS expression levels were within ±10% for a given construct in this study. All constructs were tested in at least five independent experiments.

Biolistic-onion peel transient assay

Onion tissues were prepared and bombarded with pBTRKM24, pBTRKM24-GUS, pSiM24 and pSiM24-GUS plasmids following a standard protocol [37]. After two days, transient GUS expression was detected by a histochemical method [9] and visualized under an Olympus SZX12 bright-field microscope.

Real-time quantitative reverse transcription polymerase chain reaction (qRT-PCR)

The expression levels of GUS mRNA in transgenic tobacco and *Arabidopsis* plants developed for the plasmids pKCaMV35SGUS and pSiM24GUS were evaluated by real-time quantitative RT-PCR [38] using GUS-specific forward (5′-d-TTACGTCCTGTA-GAAACCCCA-3′) and reverse (5′-d-ACTGCCTGGCACAG-CAAT TGC-3′) primers. The qPCR assays were performed using the iTaq SYBR Green Supermix with ROX (Bio-Rad, USA) according to the manufacturer's instructions. Tobacco tubulin (by using forward 5′-d-ATGAGAGAGTGCATATCGAT-3′ and reverse 5′-d-TTCACTGAAGAAGGTGTTGAA-3′ primers) was used as an internal control to normalize the expression of GUS. The comparative threshold cycle (Ct) method (Applied Biosystems bulletin, part No. 4376784 Rev. C, 04/2007) was used to evaluate the relative expression levels of the transcripts. The threshold cycle was automatically determined for each reaction by the system set with default parameters (Step One Real-Time PCR System, Applied Biosystems). The specificity of the PCR was determined by melting curve analysis of the amplified products using the standard method installed in the system (Step One Real-Time PCR System, Applied Biosystems).

β-Glucuronidase (GUS) assay and histochemical GUS staining

Fluorometric GUS enzymatic assays for measuring GUS activities in tobacco protoplast extracts, *Arabidopsis* and tobacco plant extracts were performed as described previously [9,39]. The total protein content in protoplast and plant extracts was estimated by the Bradford method using BSA as a standard [40]. Histochemical GUS staining was carried out in plants following the published protocol [9,34], and photographs were taken under a bright-field microscope (Olympus SZX12).

GFP detection

GFP fluorescence images of electroporated tobacco protoplasts, onion epidermal cells and transgenic *Arabidopsis* leaves expressing GFP were analyzed using a confocal laser scanning microscope (TCS SP5; Leica Microsystems CMS GmbH, D-68165 Mannheim, Germany) using LAS AF (Leica Application Suite Advanced Fluorescence) 1.8.1 build 1390 software under a PL FLUOTAR objective (10.0X/N.A.0.3 DRY) using a confocal pinhole set of 1 airy unit and a 1× zoom factor for improved 8-bit resolution, as described previously [28,38]. To excite the expressed GFP in transgenic plants, a 488-nm argon laser (30%) with an AOTF (allowing for 40% transmission) was used, and fluorescence emission spectra were collected between 501 and 580 nm with the photomultiplier tube (PMT) detector gain set to 1050 V [28].

Transient expression of GUS using pSiM24 vector through vacuum infiltration method

Suspensions of the *A. tumefaciens* strain GV3850 bearing pSiM24 and pSiM24-GUS constructs were prepared as previously described [35], and the infiltration procedure was conducted following a previously reported protocol [41]. Leaves of *N.*

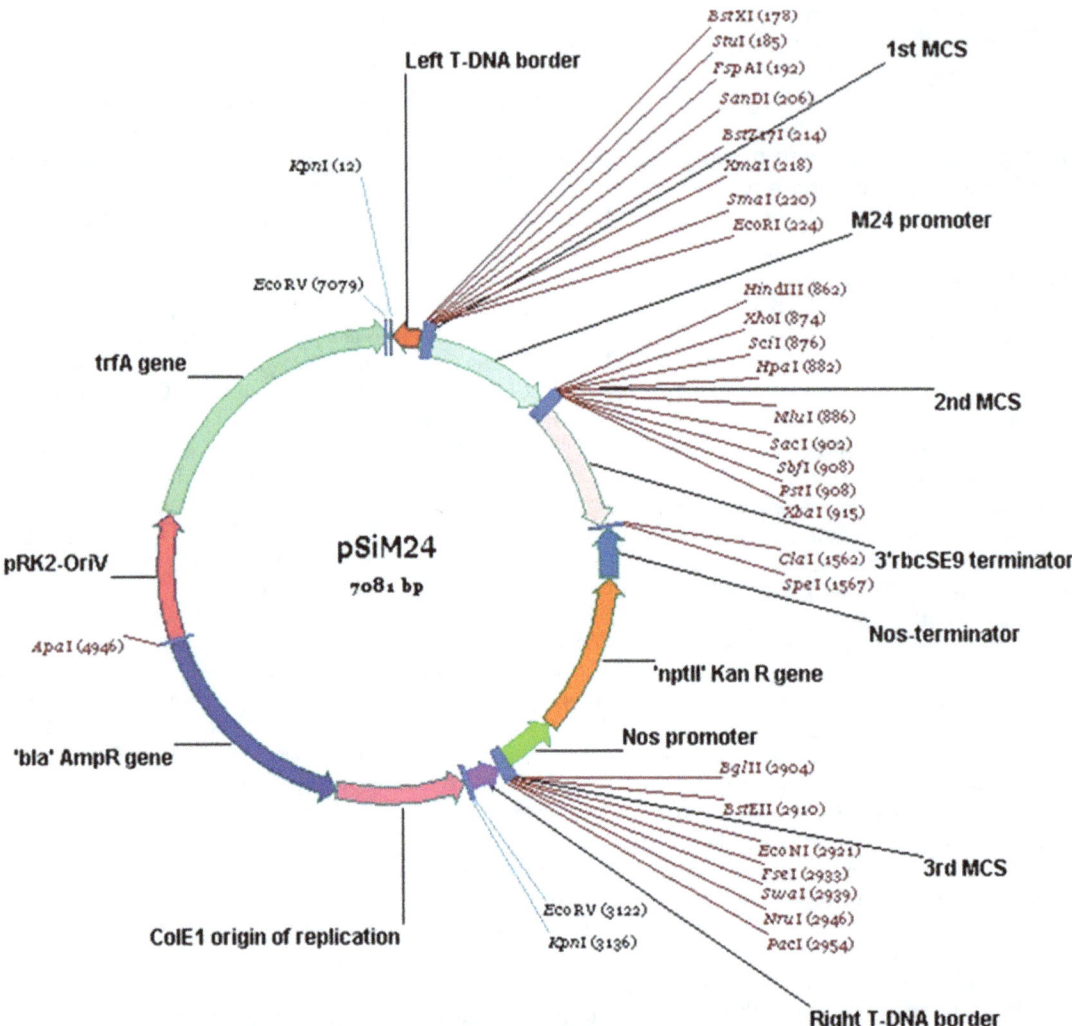

Figure 1. Schematic presentation of binary vector pSiM24. The backbone structure of binary vector pSiM24 (7081-bp) containing the modified full-length transcript promoter (M24) of the *Mirabilis mosaic virus*, which directs the coding sequences of the gene of interest; left T-DNA and right T-DNA borders (Left T-DNA, Right T-DNA); selection marker genes (KanR, neomycin phosphotransferase II, nptII) directed by nopaline synthase promoter (Nos promoter); terminator sequences of ribulose bisphosphate carboxylase small subunits (3'rbcSE9); nopaline synthase terminator (Nos terminator); multiple cloning sites (first MCS, second MCS and third MCS) with various restriction sites; replicon unit pRK2 oriV; trfA gene for agrobacterium; ColE1 origin of replication for *E. coli*; and 'bla' AmpR gene for resistance to ampicillin.

benthamiana were weighed and submerged in a suspension of *A. tumefaciens* strain GV3850 bearing pSiM24 or pSiM24-GUS plasmids. A vacuum level of 760 mm Hg was applied and released several times until the leaves became translucent [41]. Leaves were transferred into MS-media-containing plates and incubated at room temperature for two days. GUS expression in the infiltrated leaves was evaluated by the GUS histochemical staining method and GUS assay [9,39].

Table 1. Transformation frequencies of *Escherichia coli* strain TB1 for pSiM24 binary vector.

Plasmid	Total size (kb)	Amount DNA used (pg)	Average colony (n>4)	Ratio to pCAMBIA control	Ratio to pKM24KH control
pSiM24	7.08	100	634 ± 54^b	4.73	5.66
pSiM24-GUS	8.89	125	619 ± 46^b	4.62	5.53
pSiM24-GFP	7.8	110	608 ± 38^b	4.54	5.43
pCAMBIA2300	8.74	123.4	134 ± 12^a	1	1.2
pKM24KH	12.94	183	112 ± 13^a	0.84	1

The bacteria were transformed with equal molar amounts of each of the plasmid DNA. Statistical analysis of the data was performed adopting one way ANOVA analysis (using GraphPad Prism version 5.01) and presented as the means ± S.D. A *P* value of less than 0.05 was considered significant indicated by different superscript letters.

Table 2. Transformation frequencies of *Agrobacterium tumefaciens* strain GV3850 for pSiM24 binary vector.

Plasmid	Total size (kb)	Amount DNA used (µg)	Average colony (n>4)	Ratio to pCAMBIA control	Ratio to pKM24KH control
pSiM24	7.08	1	546±42[b]	1.4	1.81
pSiM24-GUS	8.89	1.25	534±48[b]	1.37	1.77
pSiM24-GFP	7.8	1.1	542±36[b]	1.39	1.79
pCAMBIA2300	8.74	1.23	391±33[a]	1	1.29
pKM24KH	12.94	1.83	302±20[c]	0.77	1

The bacteria were transformed with equal molar amounts of each of the plasmid DNA. Statistical analysis of the data was performed adopting one way ANOVA analysis (using GraphPad Prism version 5.01) and presented as the means ± S.D. A P value of less than 0.05 was considered significant indicated by different superscript letters.

Analysis GFP-*AtCESA3*[ixr1-2], Vip3A(a), KMP-11, IL-10 and nat-T-phyllo-GFP after transient expression in tobacco using pSiM24 vector

The GFP fused *Arabidopsis* mutated CESA3 (GFP-*AtCESA3*[ixr1-2]) fragment with Xho I and Sst I sites was obtained from pKM24KH-MD1 (GenBank accession no. JX996118) [42–43] by restriction digestions. Likewise, the native T-phylloplanin fused GFP with the apoplast targeting sequence (nat-T-phyllo-GFP) fragment with Xho I and Sst I sites was obtained from pKM24-ibm8 (GenBank accession no. KF951257) [44] by restriction digestions. Both these fragments were cloned in pSiM24 following standard protocols [28] and the resulted plasmids were named as pSiM24-GFP-*AtCESA3*[ixr1-2] and pSiM24-nat-T-phyllo-GFP. Suspension of *A. tumefaciens* strain pGV3850 harboring pSiM24, pSiM24-GFP-*AtCESA3*[ixr1-2] and pSiM24-nat-T-phyllo-GFP constructs were infiltrated into leaves of tobacco plants (*N. tabacum* cv. SamsunNN) as described earlier [35]. After two days of Agro-infiltration the transient *AtCESA3*[ixr1-2] expression was evaluated by RT-PCR by using gene specific primers and also by Western blotting using *AtCESA3*[ixr1-2] polyclonal antibody as described earlier [42]. The transient nat-T-phyllo-GFP expression was evaluated by RT-PCR by using gene specific primers and also by confocal microscopy as previously described [44].

Vegetative insecticidal gene, *vip3A(a)* [45,46], kinetoplastid membrane protein-11 (KMP-11) [47] and interleukin-10 (IL-10) [48] were cloned at XhoI and SstI sites in pSiM24 vector to generate pSiM24-vip3A(a), pSiM24-KMP-11 and pSiM24-IL-10 plasmids for transient expression assay in tobacco protoplasts. The isolation of tobacco protoplasts from the suspension cell cultures of *N. tabacum* L. cv Xanthi-Brad and electroporation of tobacco

protoplasts with pSiM24-vip3A(a), pSiM24-KMP-11 and pSiM24-IL-10 constructs were performed as described previously [28]. Electroporated protoplasts were incubated for 48 hours and harvested in protein extraction buffer (1X PBS, 0.1% Tween 20 and 1 mM PMSF). Protein samples from pSiM24, pSiM24-vip3A(a), pSiM24-KMP-11 and pSiM24-IL-10 transfected protoplasts were lyophilized and dissolved in protein extraction buffer. The transient *Vip3A(a)* expression was evaluated by Western blotting using Vip3A-specific polyclonal antibody as described earlier [46]. Concentration of transiently expressed KMP-11 and IL-10 was estimated by using anti-KMP-11 and anti-IL-10 antibody following indirect enzyme-linked immunosorbance assay (ELISA) protocol [49].

Results

Features of assembled binary expression vector pSiM24

The binary expression vector pSiM24 was designed to reduce the size of the vector backbone by eliminating non-essential elements of our previous vector pKM24KH (size 12,945-bp, GenBank accession no. HM036220), a derivative of pKYLX7 [4]. The pKM24KH vector is a low-copy-number plasmid. We replaced the *E. coli* replication unit with a high-copy-number replicon ColEI in pSiM24, making the identification and characterization of gene inserts easier. We also modified the agrobacterium replicon unit (Oriv-trfA of pRK2) by optimizing the trfA open reading frame for better expression. The overall DNA yields and transformation frequency of the new vector pSiM24 were several times greater than those of the previous vector pKM24KH in both *E. coli* and *A. tumefaciens*. The binary vector pSiM24 (Figure 1; GenBank Accession no. KF032933) has

Table 3. Binary Ti vectors pSiM24 produced higher plasmid DNA yields in *Escherichia coli* strain TB1 over pCAMBIA.

Plasmid	Average DNA yield (µg) (n>3)	Ratio to pCAMBIA control	Ratio to pKM24KH control
pSiM24	113.5±21.3[b]	3.47	8.11
pSiM24-GUS	96.5±13.2[b]	2.95	6.89
pSiM24-GFP	105.7±16.3[b]	3.21	7.55
pCAMBIA2300	32.74±4.0[a]	1	2.34
pKM24KH	14±1.6[c]	0.43	1

Three single colonies of each plasmid constructs were grown for 16 hrs in 25 ml LB media with 100 mg/L ampicillin (for pSiM24) or 50 mg/L kanamycin (for pCAMBIA) or 15 mg/L tetracycline (for pKM24KH). Plasmid DNA was purified using QIA Midiprep columns. DNA yields represent the average of three independent samples. Statistical analysis of the data was performed adopting one way ANOVA analysis (using GraphPad Prism version 5.01) and presented as the means ± S.D. A P value of less than 0.05 was considered significant indicated by different superscript letters.

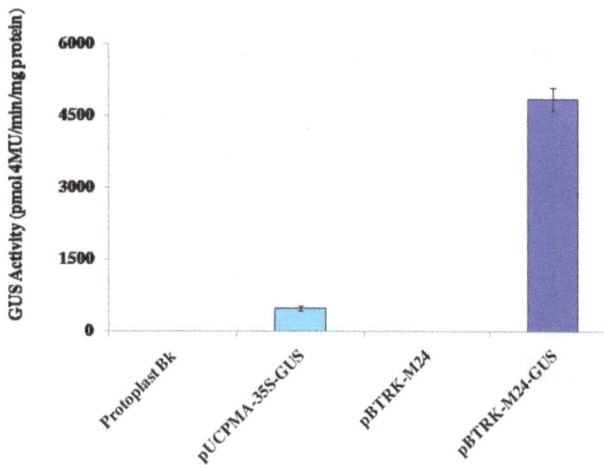

Figure 2. The transient GUS expression analysis of T-DNA assembled fragment of pSiM24 in tobacco protoplast system. Transient GUS expression analysis of pBTRK-M24 (T-DNA assembled fragment of pSiM24), pBTRK-M24-GUS (pBTRK-M24 with GUS reporter gene) constructs in tobacco protoplast. The pUCPMA-35S-GUS construct carries the constitutive CaMV35S promoter. The average GUS activity ± SD is presented in the histogram of three independent experiments replicated three times for each construct. The values significantly differ between tobacco protoplasts with pBTRK-M24-GUS from others at P<0.01 based on Student's t-test.

the following genetic elements: left T-DNA border (coordinates 65 to 90, complement); M24 promoter (223 to 860); three multiple cloning sites; 3'rbcS terminator (915 to 1565); KanR(nptII) gene (1566 to 2911, complement); KanR-terminator (1566 to 1820, complement); Kan R-cDNA (1821 to 2618, complement); KanR-promoter (2619 to 2903, complement); right T-DNA (2957 to 3118), right T-DNA border (3034 to 3059, complement); non-T-DNA portion, ColE1 origin of replication (3137 to 3804, complement); AmpR(bla) gene region (3805 to 4940, complement); terminator (3805 to 3951, complement); AmpR-cDNA (3952 to 4812, complement); AmpR-promoter (4813 to 4940, complement); pRK2-Ori V (coordinates 4947 to 5567); pRK2-trfA gene (5577 to 7081); promoter (5577 to 5880); cDNA (5881 to 7029) and terminator (7030 to 7081).

DNA yield and transformation frequencies of *E. coli* and *A. tumefaciens* with pSiM24 binary vectors

DNA yield and transformation frequencies in *E. coli* and *A. tumefaciens* were evaluated and presented (Tables 1–3). Transformations were performed with equal molar amounts of each of the plasmids to normalize increasing plasmid size as previously described [50]. The transformation frequencies of pSiM24 vectors are four- to six-fold higher than pCAMBIA and pKM24KH vectors, in *E. coli* (Table 1). The transformation frequency of the pSiM24 vector is 1.4- to 1.8-fold higher than conventional pCAMBIA and pKM24KH vectors, in *A. tumefaciens*, although the effect is not as marked as in *E. coli* (Table 2). The DNA yields of pSiM24 vectors were approximately three-fold greater than those of the pCAMBIA vector and seven to eight-fold higher than those of pKM24KH in *E. coli* (Table 3).

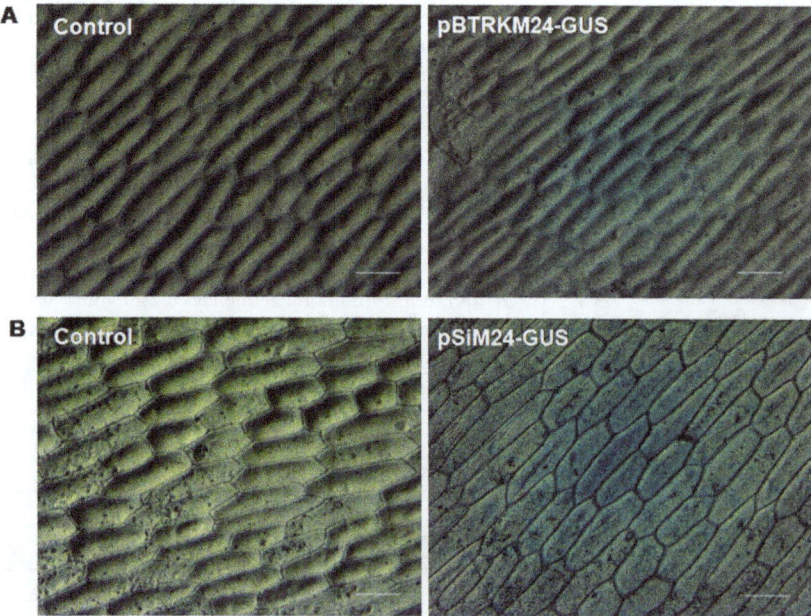

Figure 3. The transient GUS expression analysis of T-DNA assembled fragment of pSiM24 containing GUS and pSiM24-GUS in onion epidermal cells. (A) Light microscopy images of X-gluc-treated onion epidermal cells bombarded with pBTRKM24-GUS (T-DNA assembled fragment of pSiM24 with GUS reporter gene) construct DNA-loaded gold particles are presented. Control represents untransformed onion epidermal cells. These micrographs are representative of data collected after examination of onion epidermal cells from a minimum of four independent experiments. Scale bar, 100 μm. (B) Light microscopy images of X-gluc-treated onion epidermal cells bombarded with pSiM24-GUS (pSiM24 with GUS reporter gene) construct DNA-loaded gold particles are presented. Control represents onion epidermal cells with pSiM24 without GUS reporter gene. These micrographs are representative of data collected after examination of onion epidermal cells from a minimum of four independent plant experiments. Scale bar represents 100 μm.

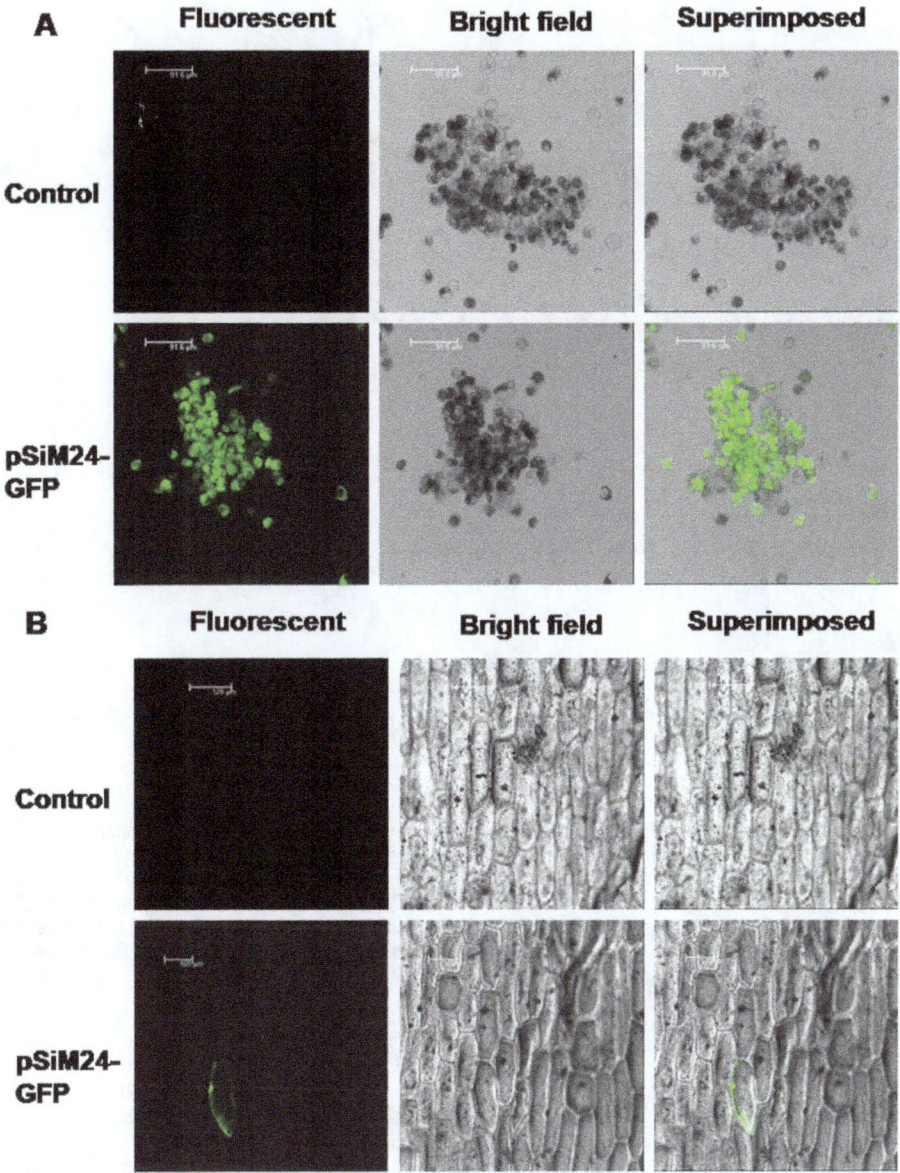

Figure 4. Transient expression of pSiM24-GFP in tobacco protoplasts and onion epidermal cells. (A) Protoplasts were transfected with plasmids pSiM24 (Control) and pSiM24-GFP (having GFP reporter gene). Transformation efficiencies were determined by analyzing the protoplasts with fluorescence after incubation for overnight. Fluorescent, bright-field and superimposed (bright-field and green fluorescent) confocal laser scanning micrographs of tobacco protoplasts are presented. Scale bar, 92 μm. (B) Fluorescent, bright-field and superimposed (bright-field and green fluorescent) confocal laser scanning micrographs of onion epidermal cells bombarded with pSiM24-GFP construct DNA-loaded gold particles are presented. Control represents onion epidermal cells with pSiM24 visualized under CLSM. Scale bar, 100 μm.

Transient expression of the pBTRKM24-GUS and pSiM24-GUS/GFP constructs

The pBluescript-based constructs pBTRKM24-GUS with an M24 promoter and pUCPMA35S-GUS [27] with a 35S promoter were compared by tobacco protoplast transient assay. The M24 promoter showed approximately 10 times higher GUS activity than the CaMV 35S promoter (Figure 2). The pBTRKM24-GUS construct was also evaluated by the biolistic bombardment of epidermal cells of onion peels, showing strong GUS expression, as detected histochemically (Figure 3).

The pSiM24-GFP (with a different reporter gene, i.e., GFP) was studied in a tobacco protoplast system, where GFP fluorescence was visualized by confocal microscopy (Figure 4). The pSiM24-GUS construct was tested in an *Agrobacterium* infiltration assay in *N. benthamiana* leaves. The *A. tumefaciens* (strain C58C1-GV3850) carrying pSiM24 (empty vector), pK-CaMV35S-GUS and pSiM24-GUS constructs was used for agro-infiltration. Transient GUS expression detected histochemically, showed stronger GUS expression in agro-infiltrated patches for pSiM24-GUS construct than for pK-CaMV35S-GUS (Figure 5). The pSiM24-GUS/GFP plasmids were also bombarded in onion cells, and strong GUS or GFP expression was observed in transformed onion epidermal cells (Figure 3-4).

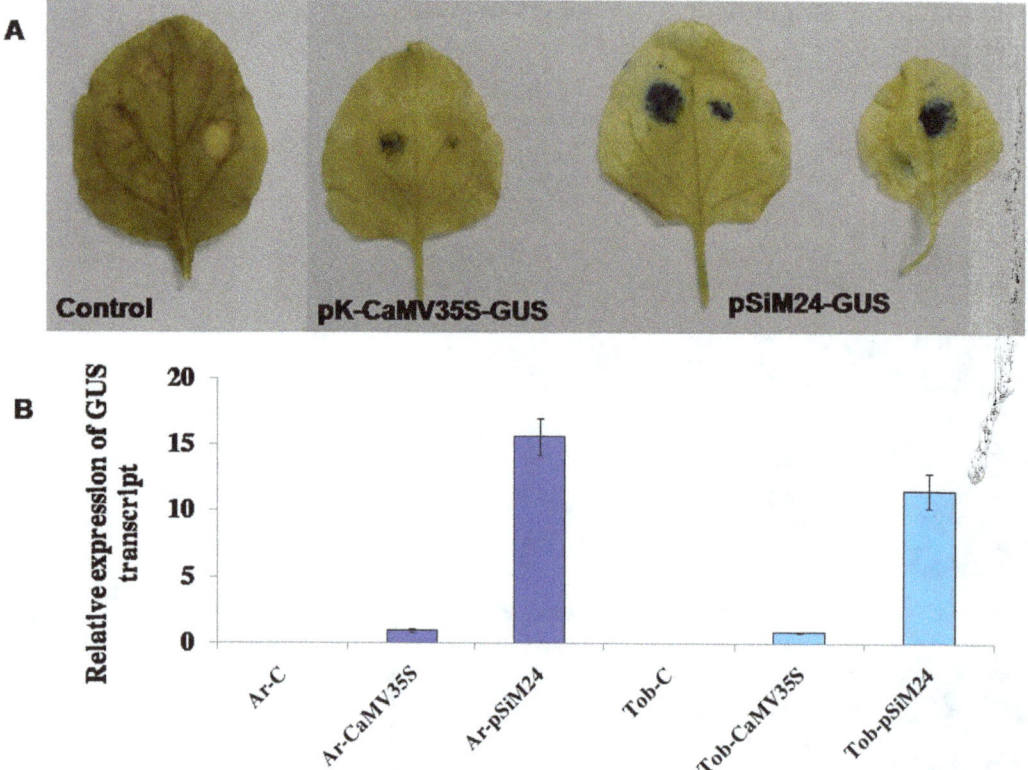

Figure 5. GUS expression analysis of pSiM24-GUS and pKCaMV35S-GUS in transient and transgenic systems. (A) Representative transient GUS expression levels from agrobacterium infiltration assays in *N. benthamiana* leaves are shown for pKCaMV35S-GUS and pSiM24-GUS constructs. GUS was detected histochemically. Control represents pSiM24 without GUS reporter gene. (B) Relative expression of GUS specific transcripts was measured in whole transgenic *Arabidopsis* (second generation, two weeks old; Ar-CaMV35S with pKCaMV35S and Ar-pSiM24 with pSiM24) and tobacco (second generation, three weeks old; Tob-CaMV35S with pKCaMV35S and Tob-pSiM24 with pSiM24) plants. The data represent relative expression of GUS transcript ± S.D. of four independent lines (n = 4) for each construct in which five plants per line were analyzed. The values significantly differ between control and transgenic plants at P<0.01 based on Student's *t*-test.

Transformation ability of pSiM24 vector in tobacco and *Arabidopsis*

Tobacco leaf discs were co-cultivated with *A. tumefaciens* for three days, and transformed leaf discs were selected in the presence of 250 mg/L kanamycin and 500 mg/L of cefotaxime for four weeks. The increase in fresh weight in transformed leaf discs was evaluated as described previously [51]. The increases in fresh weight of four-week-old leaf discs were compared and are presented in Table 4. Leaf discs treated with binary vectors showed a seven- to eight-fold increase in fresh weight over the vector-less control, remaining green with multiple regenerating shoots. In the negative vector-less control, leaf discs did not induce callus and turned yellow within two weeks of antibiotic selection. Thus, the percentage of leaf discs showing an increase in fresh

Table 4. Binary Ti vectors pSiM24 and pSiM24-GUS/GFP conferred a kanamycin-resistant fresh weight (FW) increase in tobacco leaf discs after transformation with *A. tumefaciens*.

Binary Ti Vectors	Mean FW/plate, g	Increase in FW in g per plate over vector-less control	% Leaf discs with increased FW
Vector-less control	0.83±0.09[a]	0[c]	0
pSiM24	8.12±2.1[b]	7.29±2.13[d]	88
pSiM24-GUS	6.85±1.8[b]	6.02±1.79[d]	98
pSiM24-GFP	7.32±2.4[b]	6.49±2.42[d]	79
pCAMBIA2300	7.85±1.3[b]	7.02±1.33[d]	83
pKM24KH	7.43±1.7[b]	6.6±1.68[d]	81

After the co-cultivation with *A. tumefaciens* strain GV3850 at 25°C for 3 days, leaf discs were selected on shooting medium containing 250 mg/L of kanamycin and 500 mg/L of cefotaxime for four weeks. Each treatment involved 5 plates with 10 leaf discs per plate. The same experiment was repeated two more times. Statistical analysis of the data was performed adopting one way ANOVA analysis (using GraphPad Prism version 5.01) and presented as the means ± S.D. A *P* value of less than 0.05 was considered significant indicated by different superscript letters.

Table 5. Effect of binary Ti vectors pSiM24 and pSiM24-GUS/ GFP on transformation frequencies in *A. thaliana*.

Binary Ti Vectors	Transformation frequency (%)
Vector-less control	0[a]
pSiM24	2.68±0.29[b]
pSiM24-GUS	2.45±0.26[b]
pSiM24-GFP	2.38±0.21[b]
pCAMBIA2300	1.32±0.12[c]
pKM24KH	1.18±0.2[c]

A. tumefaciens harboring binary vectors were cultured and floral dipping of *A. thaliana* plants was subsequently performed as described in materials and methods. T_1 seedlings were selected on solid MS medium containing kanamycin. The data was obtained from at least four independent lines developed for each construct. Statistical analysis of the data was performed adopting one way ANOVA analysis (using GraphPad Prism version 5.01) and presented as the means ± S.D. A *P* value of less than 0.05 was considered significant indicated by different superscript letters.

weight over the negative vector-less control indicates the proportion of putatively transformed leaf discs, which ranged from 79 to 98%. There appeared to be no detectable difference between pSiM24 binary vectors and the positive control pCAMBIA and pKM24KH vectors. The effect of pSiM24 binary vector on transformation frequency was also studied in *A. thaliana*. The pSiM24 and pSiM24-GUS/GFP vectors exhibited approximately two-fold more transformation frequency in *A. thaliana* than pCAMBIA2300 and pKM24KH vectors (Table 5).

Expression analysis of pSiM24-GUS/GFP in transgenic plants

Agrobacterium carrying the pSiM24-GUS reporter gene was used to transform *Arabidopsis* and tobacco plants. GUS histochemical analysis confirmed that the pSiM24 vector successfully expressed GUS genes in transgenic *Arabidopsis* and tobacco (both by agrobacterium-mediated transformation as well as by gene-gun methods) plants (Figure 6–7). The *Arabidopsis* pSiM24-GUS and pSiM24-GFP transgenic plants successfully expressed GUS and GFP proteins, as detected by GUS histochemical staining and confocal microscopy of GFP (Figure 7–8). Furthermore, GUS analysis of second-generation plants confirmed the successful

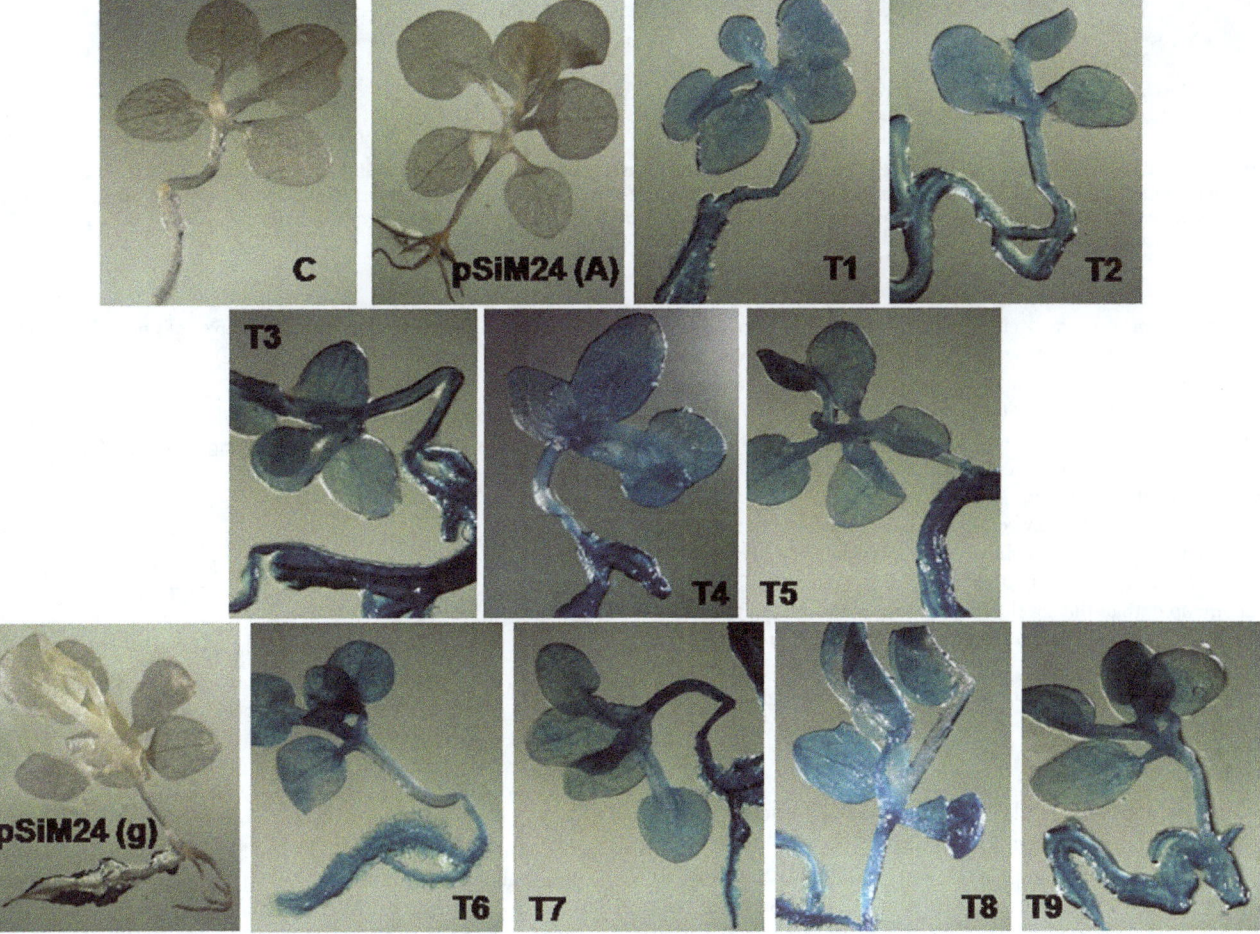

Figure 6. GUS expression in transgenic tobacco plants generated for constructs pSiM24 and pSiM24-GUS. Representative transgenic tobacco plants (second generation, three weeks old) generated by agrobacterium-mediated transformation (T1 to T5 lines) and transformation using gene-gun (T6 to T9 lines) were stained to determine GUS histochemical activity. C: Untransformed tobacco plant; pSiM24 (A): pSiM24 transgenic tobacco plants generated by agrobacterium-mediated transformation; pSiM24 (g): pSiM24 transgenic tobacco plants generated by biolistic bombardment method.

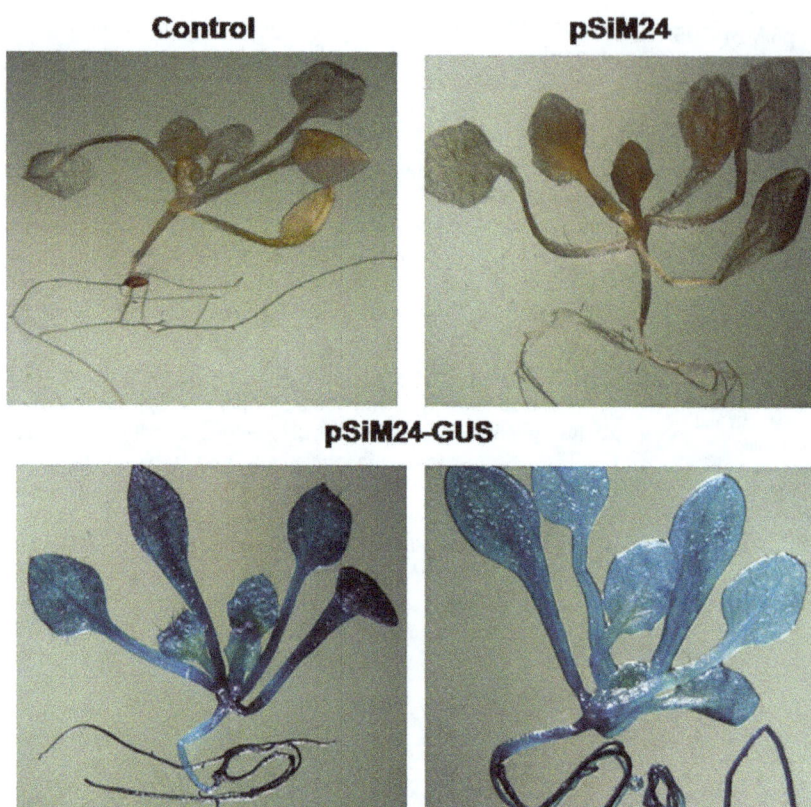

Figure 7. GUS expression in transgenic _Arabidopsis_ plants generated for constructs pSiM24 and pSiM24-GUS. Representative transgenic _Arabidopsis_ plants (second generation, two weeks old) generated by agrobacterium-mediated transformation were stained to determine GUS histochemical activity. Control: Untransformed _Arabidopsis_ plant; pSiM24: pSiM24 transgenic _Arabidopsis_ plants; pSiM24-GUS: pSiM24-GUS transgenic _Arabidopsis_ plants.

inheritance of the transgene from one generation to another generation in both transgenic tobacco and _Arabidopsis_ plants (Figure 9–10). The GUS activity was estimated biochemically in R2-generation transgenic _Arabidopsis_ and tobacco plants; it was observed that approximately two times higher GUS activity accumulated in leaf and stem tissues of _Arabidopsis_ plants than in tobacco plants (Figure 9). The expression of GUS activities in different tissues of transgenic _Arabidopsis_ and tobacco plants containing pSiM24-GUS showed the following pattern: Root > Leaf > Stem (Figure 9). In addition, the histological GUS staining documented that the level of GUS expression was high in the reproductive tissues of transgenic pSiM24-GUS tobacco and _Arabidopsis_ plants (Figure 10). Both in tobacco and _Arabidopsis_ transgenic plants, GUS transcript levels were higher for the pSiM24 binary vector than for the pKYLX-based expression vector pKCaMV35 (Figure 5).

Transient expression of GUS using pSiM24 vector through vacuum infiltration method

A. tumefaciens carrying pSiM24 and pSiM24-GUS constructs infiltrated _N. benthamiana_ leaves were assayed and histochemically stained for GUS enzyme. The completely infiltrated leaves showed approximately 1800 GUS units, whereas the partially infiltrated leaves exhibited approximately 850 GUS units (Figure 11A). One unit of GUS activity was defined as the amount of enzyme that liberated 1 p mol 4-methylumbelliferone mg^{-1} protein min^{-1}[52]. The agro-infiltrated leaves showed strong GUS expression, as

detected by histochemical staining, in leaves of both _N. benthamiana_ and _Zea mays_ (Figure 11B and Figure 12).

Transient expression of GFP-_AtCESA3_$^{ixr1-2}$, Vip3A(a), KMP-11, IL-10 and nat-T-phyllo-GFP genes using pSiM24 vector

Western blot analysis of pSiM24-GFP-_AtCESA3_$^{ixr1-2}$ agroinfiltrated leaf samples showed the expected bands of size 145 kD for _GFP-AtCESA3_$^{ixr1-2}$ as detected with AtCESA3-specific polyclonal antibodies (Figure 13A). In addition, RT-PCR analysis of agroinfiltrated leaf samples exhibited expected 1318 bp band for a portion of _GFP-AtCESA3_$^{ixr1-2}$ (Figure 13A). The transient expression of Vip3A(a) using pSiM24 expression vector in tobacco protoplast was detected by Vip3A-specific polyclonal antibodies that showed the expected bands of size 88 kD (Figure 13B). Using pSiM24 expression vector KMP-11 and IL-10 were also expressed transiently in tobacco protoplasts and showed expression up to 0.03 mg of KMP-11 and 0.08 mg of IL-10 per mg of protoplast protein samples by indirect ELISA (Figure 13C). The RT-PCR analysis and localization analysis of apoplast targeted nat-T-phyllo-GFP by confocal laser scanning microscopy showed the successful expression of _nat-T-phyllo-GFP_ using pSiM24 expression vector (Figure 14).

Discussion

A binary vector, used as a standard tool in the transformation of higher plants mediated by _A. tumefaciens_, consists of T-DNA and

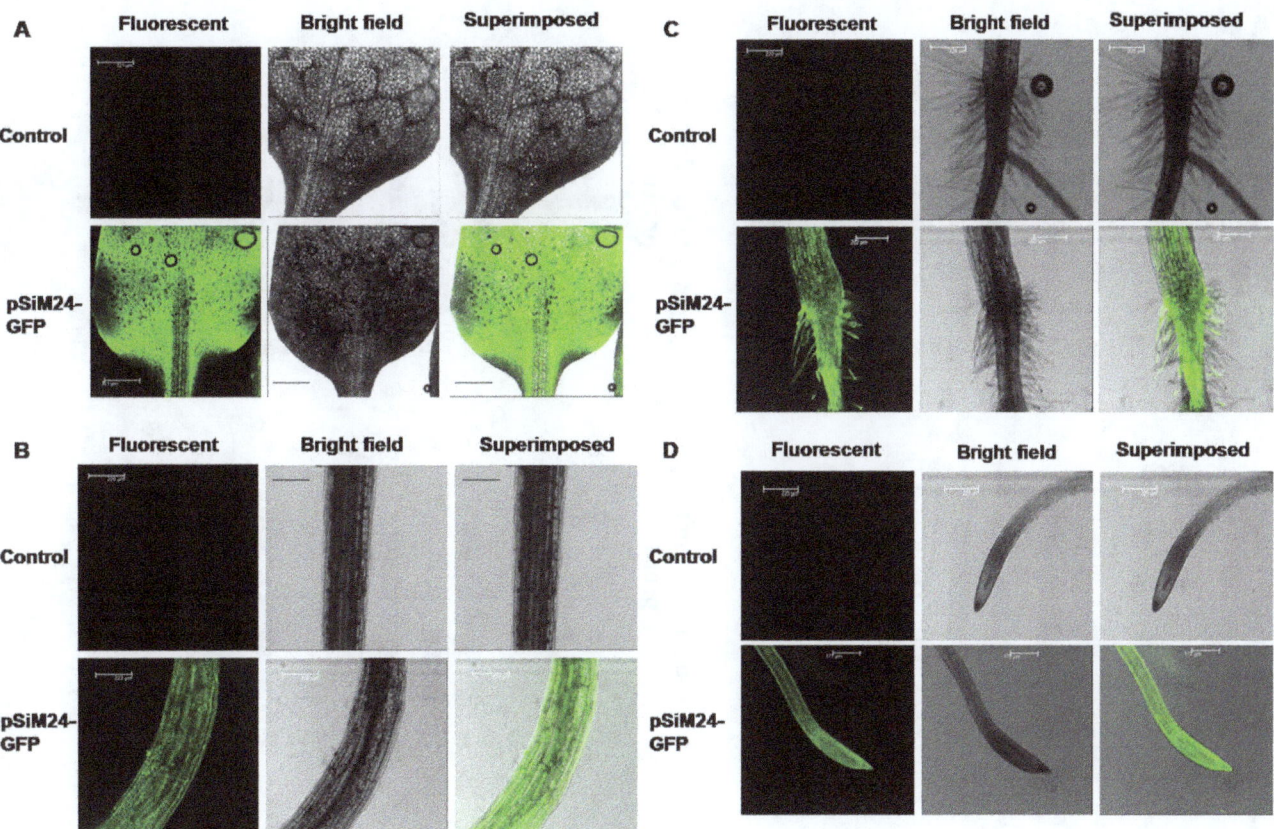

Figure 8. GFP expression in leaf, stem and roots of transgenic *Arabidopsis* plants generated for construct pSiM24-GFP. (A) Representative transgenic *Arabidopsis* plant leaves (second generation, two weeks old) generated by agrobacterium-mediated transformation were imaged to determine GFP activity. Fluorescent, bright-field and superimposed (bright-field and green fluorescent) confocal laser scanning micrographs of transgenic *Arabidopsis* leaves. Scale bar represents 320 μm. (B) Representative transgenic *Arabidopsis* plant stems (second generation, two weeks old) generated by agrobacterium-mediated transformation were imaged to determine GFP activity. Fluorescent, bright-field and superimposed (bright-field and green fluorescent) confocal laser scanning micrographs of transgenic *Arabidopsis* stems. Scale bar represents 220 μm. (C) Representative transgenic *Arabidopsis* plant stem-root junctions (second generation, two weeks old) generated by agrobacterium-mediated transformation were imaged to determine GFP activity. Fluorescent, bright-field and superimposed (bright-field and green fluorescent) confocal laser scanning micrographs of transgenic *Arabidopsis* stem-root junctions. Scale bar represents 220 μm. (D) Representative transgenic *Arabidopsis* plant roots (second generation, two weeks old) generated by agrobacterium-mediated transformation were imaged to determine GFP activity. Fluorescent, bright-field and superimposed (bright-field and green fluorescent) confocal laser scanning micrographs of transgenic *Arabidopsis* roots. Scale bar represents 225 μm.

the vector backbone. T-DNA is the segment delimited by the border sequences, the right border (RB) and the left border (LB), and may contain multiple cloning sites, a selectable marker gene for plants, a reporter gene and other genes of interest [7,53]. The vector backbone carries plasmid replication functions for *E. coli* and *A. tumefaciens*, selectable marker genes for the bacteria, optionally a function for plasmid mobilization between bacteria and other accessory components [7,8].

The binary vector pSiM24 has an overall size of 7.08 kb and carries a plant-gene expression cassette containing a highly active, constitutive promoter (M24) (GenBank Accession no. KF032933). The size of the pSiM24 vector is approximately 2000 bp shorter than the commercially available pCAMBIA vectors (www.cambia. org) and approximately 6000 bp shorter than pKYLX-based vectors [4]. In the pSiM24 binary vector, only the necessary elements were included to attain a minimum size. The right border (RB) and the left border (LB) of pSiM24 are imperfect, direct repeats of 25 bases. The RB and LB are considered to be the only essential cis-elements for T-DNA transfer [54]. The promoter carried by the expression cassettes described here has been studied

in transgenic plants (present study) and is also functional in plants such as tobacco [41–44,55], *Arabidopsis* and corn (Sahoo and Maiti, Unpublished Data). It has been documented that the *Mirabilis mosaic virus* full-length transcript promoter is constitutive in nature, exhibiting 14 to 25 times stronger activity than CaMV35S in the tobacco protoplast transient system and transgenic tobacco plants, respectively [27,34,42]. The modified full-length transcript promoter (M24) of the *Mirabilis mosaic virus* with duplicated enhancer domains [27,29,41–44,55] was used in the pSiM24 vector to evaluate gene expression in plants. In the pSiM24-GUS vector, the coding sequence of GUS was placed between the heterologous M24 promoter and the terminator sequence from the rbcSE9 gene (Figure 1) [43–44]. We showed that microT-DNAs in pSiM24 containing a kanamycin resistance gene and reporter gene (GUS or GFP) were integrated stably in the nuclear chromosomal DNA of transgenic plants for successive generation.

Selectable markers need to be expressed in calli, in cells from those plants that are being regenerated or in germinating embryos to facilitate plant transformation. Therefore, promoters for constitutive expression are preferred. In pSiM24, the Nos

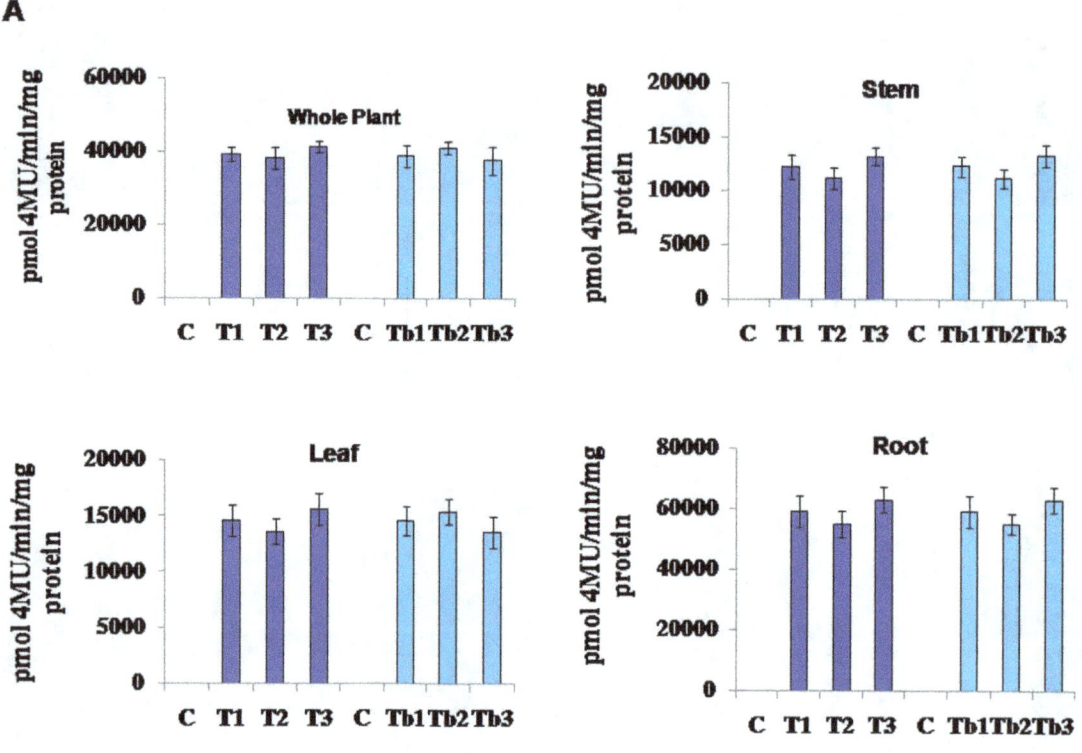

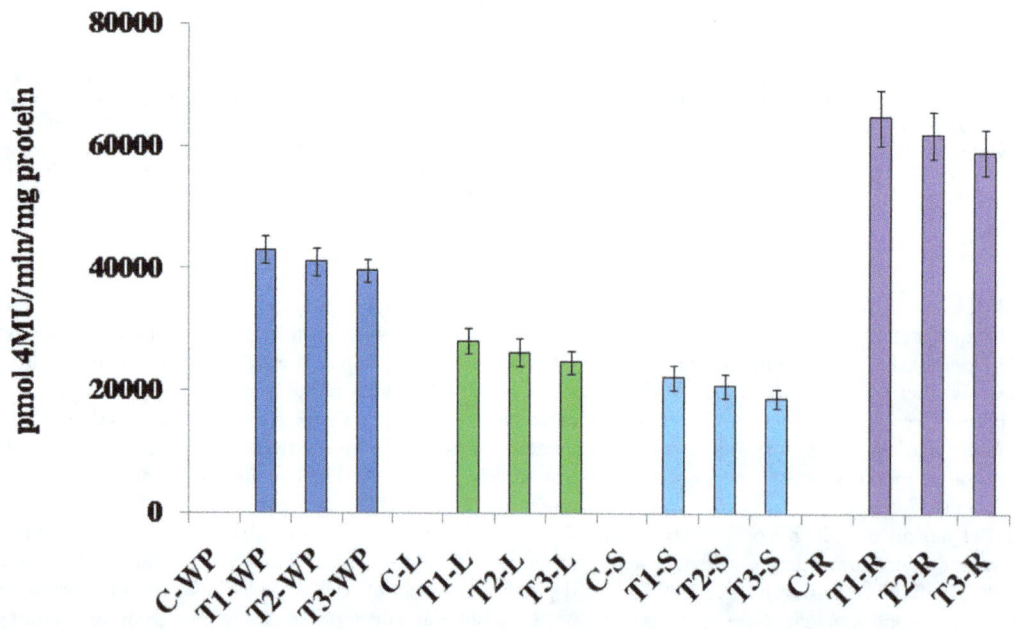

Figure 9. GUS expression in transgenic tobacco and *Arabidopsis* plants generated for constructs pSiM24 and pSiM24-GUS. (A) GUS enzymatic activity of pSiM24-GUS transgenic tobacco (second generation, 3 weeks old) lines was measured in whole plant, leaf, stem and root tissues. Soluble protein extracts isolated from different tissues of plants were used for GUS assay along with the wild-type plants (C). The data represent means ± S.D. of four second generation individuals from one line for each tissue (n = 4). The values significantly differ between control and transgenic plants at P<0.01 based on Student's *t*-test. T1, T2 and T3: Representative transgenic lines generated by agrobacterium-mediated plant transformation procedure; Tb1, Tb2 and Tb3: Representative transgenic lines generated by biolistic plant transformation procedure. (B) GUS enzymatic activity of pSiM24-GUS transgenic *Arabidopsis* (second generation, two weeks old) lines was measured in whole plant (WP), leaf (L), stem (S) and root (R) tissues. Soluble protein extracts isolated from different tissues of plants were used for GUS assay along with the wild-type plants (C).

The data represent means ± S.D. of four second generation individuals from one line for each tissue (n = 4). The values significantly differ between control and transgenic plants at P<0.01 based on Student's *t*-test. T1, T2 and T3: Representative transgenic lines generated by floral-dip plant transformation procedure.

promoter derived from nopaline synthase (Nos) of *A. tumefaciens* [56] was used to express the selectable marker gene (Figure 1). The choice of promoters responsible for selectable marker gene expression also plays an important role in the efficiency of transformation [57–59]. The use of weak promoters may not always be a bad idea because the levels of expression of marker genes and genes of interest are often linked, and the selection of transformants with weak selectable markers may cause strong expression of the gene of interest to be obtained [8]. It is generally recommended that different promoters be used for the selectable marker and expressing the gene of interest [57–59], as in the pSiM24 vector (Figure 1), which carries the M24 promoter for the expression of the gene of interest (here GUS in pSiM24-GUS) and Nos for the selectable marker (here, KanR). Homology-based gene silencing has been reported to occur extensively in transgenic plants [60]. Gene silencing due to promoter homology can be avoided by either using diverse promoters isolated from different plant and viral genomes or by designing synthetic promoters [27,28,34,38,61–68].

Depending on the plant species to be transformed, the choice of selectable markers greatly affects the efficiency of transformation, and permissive concentrations of selective agents vary considerably among plant species. Genes resistant to antibiotics or herbicides, such as kanamycin, hygromycin, phosphinothricin and glyphosate, are very popular. Kanamycin resistance has been most frequently exploited in the transformation of many dicotyledonous plants such as tobacco, tomato, potato and *Arabidopsis* [34,42,69]. The pSiM24 binary vector contains a synthetic 'nptII' KanR gene (nos

promoter-KanR cDNA-Nos terminator), the open reading frame of which is optimized for plant codon bias; hence, the nptII gene serves both as a selectable marker for the regeneration of plantlets on kanamycin-containing medium (for tobacco 250–300 µg/ml) and as a screenable marker for agrobacterium (25 µg/ml). In the present study, pSiM24-containing nptII gene was used to select the transformed *Arabidopsis* and tobacco plants in 30 µg/ml and 250 µg/ml kanamycin, respectively, (Figure 6–7). Choice of antibiotics is an important factor in plant transformation. For example, kanamycin may not suitable for rice and maize cells, whereas hygromycin resistance (hpt) is very good for rice transformation [10], and the phosphinothricin resistance gene (bar) is efficient for maize and other cereals [70,71]. We also developed a binary vector pKDH, which has a structure similar to that of pSiM24, but the selection marker KanR gene was replaced with a hygromycin resistance (HygR, Hygromycin B transferase, HPH) gene for the selection of transgenic monocot plants, and the sequence information of the binary vector pKDH was provided (Genebank Accession no. KF041008).

The components of the pSiM24 expression system vector are arranged in a modular configuration in which the promoter, terminator and MCS cassettes are flanked by unique restriction endonuclease cleavage sites. The pSiM24 vector provides nine unique cloning sites in the first multiple cloning site (MCS) between the left T-DNA border and the M24 promoter (BstXI, StuI, EspAI, PasI, KflI, Bstz17I, SmaI, XmaI and EcoRI), twelve unique cloning sites in the second MCS between the M24 promoter and the pea rbcSE9 terminator (HindIII, AbsI, PspXI,

Figure 10. GUS expression in flowers of transgenic tobacco and *Arabidopsis* plants generated for constructs pSiM24 and pSiM24-GUS. Histological GUS staining in different floral tissues of untransformed (control), pSiM24 and pSiM24-GUS plants. Histological GUS staining shows strong GUS expression in all floral tissues of transgenic pSiM24-GUS *Arabidopsis* (upper panel) and tobacco (lower panel) plants.

A

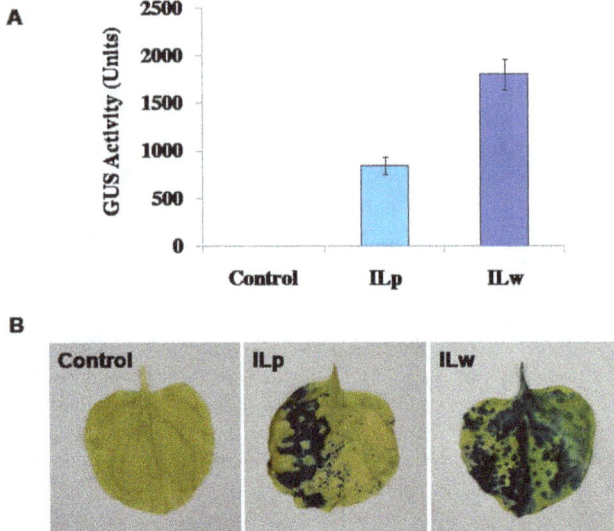

B

Figure 11. Transient expression of GUS in pSiM24-GUS Agro-infiltrated *N. benthamiana* leaves using vacuum infiltration method. (A) GUS enzymatic activity of pSiM24 and pSiM24-GUS *A. tumefaciens*-infiltrated *N. benthamiana* leaves was measured, and one unit of GUS activity was defined as the amount of enzyme that liberated 1 pmol 4-methylumbelliferone mg^{-1} protein min^{-1}. The data represent means ± S.D. of four biological replicates for each construct (n = 4). The values significantly differ between control (pSiM24) and pSiM24-GUS agro-infiltrated leaf samples at $P<0.01$ based on Student's *t*-test. ILw: Whole Infiltrated Leaf; ILp: Partial Infiltrated Leaf. (B) The pSiM24 and pSiM24-GUS *A. tumefaciens*-infiltrated *N. benthamiana* leaves were histochemically stained for GUS enzyme. The pSiM24-GUS agro-infiltrated leaves showed stronger GUS expression, as detected by histochemical staining. Control: *N. benthamiana* leaf infiltrated with *A. tumefaciens* harboring pSiM24 construct; ILw: Whole Infiltrated Leaf; ILp: Partial Infiltrated Leaf. Both ILw and ILp are *N. benthamiana* leaves infiltrated with *A. tumefaciens* carrying pSiM24-GUS construct.

SciI, XhoI, HpaI, MluI, Eco53kI, SacI, SbfI, PstI, and XbaI) and seven unique cloning sites in the third MCS between the Nos promoter and the right T-DNA border (BglII, BstEII, EcoNI, FseI, SwaI, NruI, and PacI). This configuration facilitates the modification or replacement of individual components in the pSiM24 vector. The MCS in pSiM24 contains more additional cleavage sites than that of pUC19. It should be noted that the orientation of the MCS in the pSiM24 plasmid, relative to the rbcS and M24 promoters, is opposite that in pUC19, relative to the lac promoter. The presence of a number of cloning sites unique to the three MCS allow for gene-stacking applications to introduce multiple gene with additional sequences, such as translational initiation, signal and transit peptide sequences and translational termination, into these plasmids. The pSiM24 vector provides a number of options for cloning, transformation and expression strategies. The M24 promoter in the pSiM24 plasmid can be easily replaced with other promoters as EcoRI-HindIII cassettes, thus making different strategies for the regulated expression of foreign genes possible.

Reporter genes, whose expression can be easily monitored, are useful in many ways in plant transformation. Strength and temporal, spatial and other types of regulation of promoters and elements may be conveniently assayed by connecting these elements to the reporter genes. Genes for β-glucuronidase (GUS) [9], luciferase [72] and GFP [73] are popular examples. In the present study, two different reporter genes, GUS and GFP, were introduced into the pSiM24 vector to monitor and analyze their expression under the M24 promoter in both stable and transient

Figure 12. Transient expression of GUS in pSiM24-GUS Agro-infiltrated *Zea mays* leaves using vacuum infiltration method. The representative pSiM24-GUS agro-infiltrated leaf showed strong GUS expression, as detected by histochemical staining. Control: *Z. mays* leaf infiltrated with *A. tumefaciens* carrying pSiM24; pSiM24-GUS: *Z. mays* leaf infiltrated with *A. tumefaciens* harboring pSiM24-GUS construct.

systems. Reporter genes that are connected to constitutive promoters may be used to monitor the process of transformation. The expression of the reporter genes soon after the inoculation of plant cells with *A. tumefaciens*, which is referred to as "transient expression", is a good indication of the transfer of the T-DNA from the bacteria to the nuclei of plant cells. The expression of the reporter genes later in a cluster of cells growing on selection media provides evidence of the integration of the T-DNA in plant chromosomes. A binary vector that carries a constitutive selectable marker and a constitutive reporter is very useful as a control vector both in transformation experiments and in assays of gene expression. Hence, in pSiM24, both "nptII" and GUS/GFP were constitutively expressed by using two different constitutive promoters, i.e., Nos for nptII and M24 for GUS/GFP, for expression in transgenic plants (Figure 6–10).

The rbcSE9 polyadenylation signal used in the pSiM24 vector has previously been used to direct efficient mRNA3′ end formation from chimeric genes in transformed tobacco [74–76]. These 3′ ends are identical to those observed in pea, indicating that this signal is suitable for the predictable expression of foreign genes in plants. The 3′ regions of the cauliflower mosaic virus 35S transcript and the nopaline synthase gene in the wild-type T-DNA of *A. tumefaciens* are frequently used as a 3′ signal to direct selectable marker genes expression.

In pSiM24, the "bla" AmpR gene, which confers resistance to ampicillin, was used as the marker for bacterial selection for *E. coli*. The selectable marker for plants, Nos-nptII, in pSiM24 also provides fair levels of resistance to both *E. coli* and *A. tumefaciens*. Binary vectors need to be replicated both in *E. coli* and *A. tumefaciens*. Hence, the pSiM24 vector carries all of the functions necessary for replication and transfer in *Escherichia coli* and *A. tumefaciens*, which includes a ColE1-replicon and an RK2-replicon derived from pRK2013 [77]. The pSiM24 binary vector carries the origin of vegetative replication (OriV) and the transacting

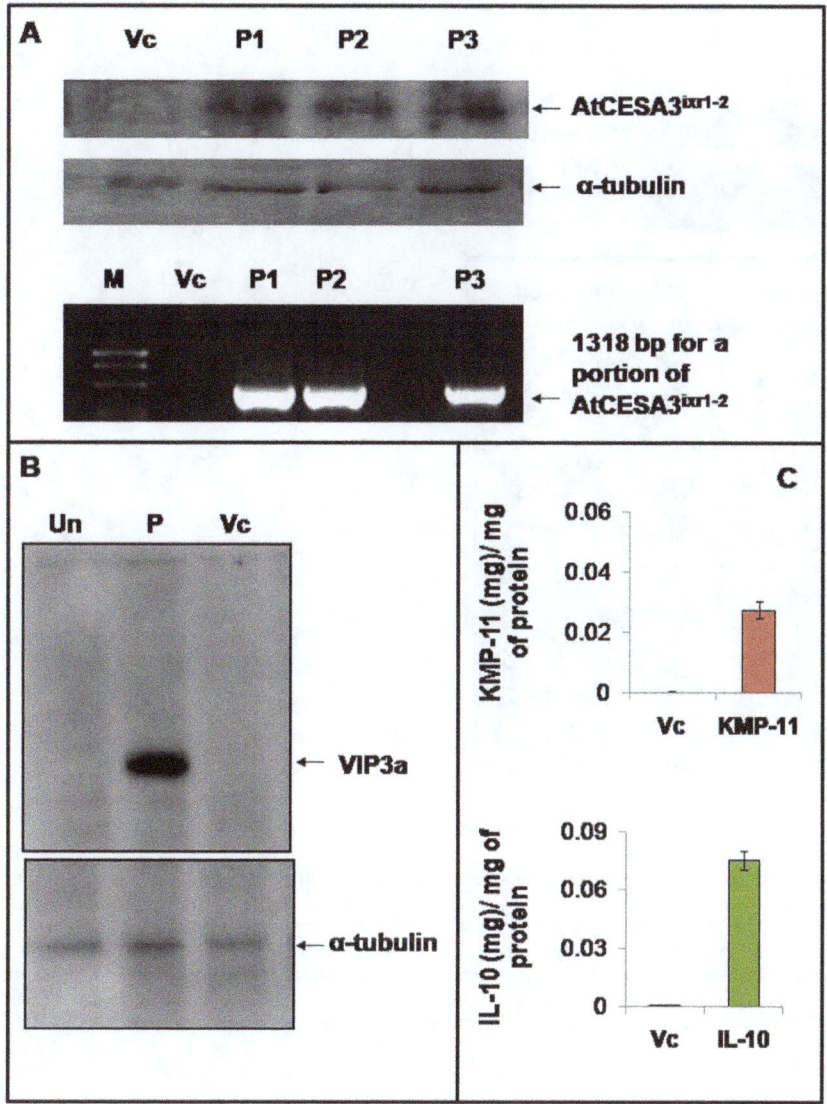

Figure 13. Transient expression of *GFP-AtCESA3^{ixr1-2}*, Vip3A(a), KMP-11 and IL-10 genes using pSiM24 expression vector. (A) Western blot analysis of transiently expressed *GFP* fused *Arabidopsis CESA3^{ixr1-2}* (*GFP-AtCESA3^{ixr1-2}*) detected with AtCESA3-specific polyclonal antibodies showed the expected bands of size 145 kD (Upper panel). Signals were quantitated, normalized to the α-tubulin loading control (Middle panel). RT-PCR products for a portion of *GFP-AtCESA3^{ixr1-2}* with the expected 1318 bp band are shown (Lower panel). Transiently *GFP-AtCESA3^{ixr1-2}* expressed agro-infiltrated *N. tabacum* L. variety Samsun NN leaf samples using pSiM24-*GFP-AtCESA3^{ixr1-2}* (P1, P2 and P3) and pSiM24 (Vc; empty vector control) are presented. (B) Western blot of vegetative insecticidal protein, Vip3A(a) expression in tobacco protoplast. Detection by Vip3A-specific polyclonal antibodies showed the expected bands of size 88 kD (Upper panel). Un: Untransfected control; P: pSiM24-Vip3A(a) transfected protoplast; Vc: pSiM24 vector control transfected protoplast. Signals were quantitated, normalized to the α-tubulin loading control (Lower panel). (C) Estimation of KMP-11 (Kinetoplastid membrane protein-11) concentration (expressed as mg KMP-11 per mg protein) in tobacco protoplasts by ELISA. KMP-11: pSiM24-KMP-11 transfected protoplast; Vc: pSiM24 vector control transfected protoplast. (D) Estimation of Interleukin-10 (IL-10) concentration (expressed as mg IL-10 per mg protein) in tobacco protoplasts by ELISA. IL-10: pSiM24-IL-10 transfected protoplast; Vc: pSiM24 vector control transfected protoplast.

replication functions (Trf) of plasmid incompatibility group P (IncP) plasmids [78]. The types of replication functions exploited determine the copy numbers and stability of the plasmids in bacterial cells. *E. coli* exhibited a transformation frequency up to five- to six-fold higher with pSiM24 than with conventional pCAMBIA and pKM24KH vectors, (Table 1) and the plasmid DNA yields of pSiM24 binary Ti vectors were three-fold and seven- to eight-fold higher in *E. coli* than those of conventional pCAMBIA 2300 and pKM24KH, respectively (Table 3). The pSiM24 binary vector contains the ColE1 replicon without a bom (basis of mobility) sequence, which again reduces its size. The bom

function is necessary for plasmid mobilization from *E. coli* to *A. tumefaciens* [79]. This function is not necessary when vectors are introduced into *A. tumefaciens* by electroporation or freeze-thaw methods.

Not only reporter genes, other introduced genes of size up to 4 kb were also effectively expressed using pSiM24 expression vector. In the present study, nat-T-phyllo-GFP [44] was expressed transiently using pSiM24 expression vector in tobacco leaves. T-phylloplanins have antimicrobial properties and are known to inhibit blue mold disease caused by *Peronospora tabacina* [41,44,80,81]. Both the native and mature tobacco phylloplanin

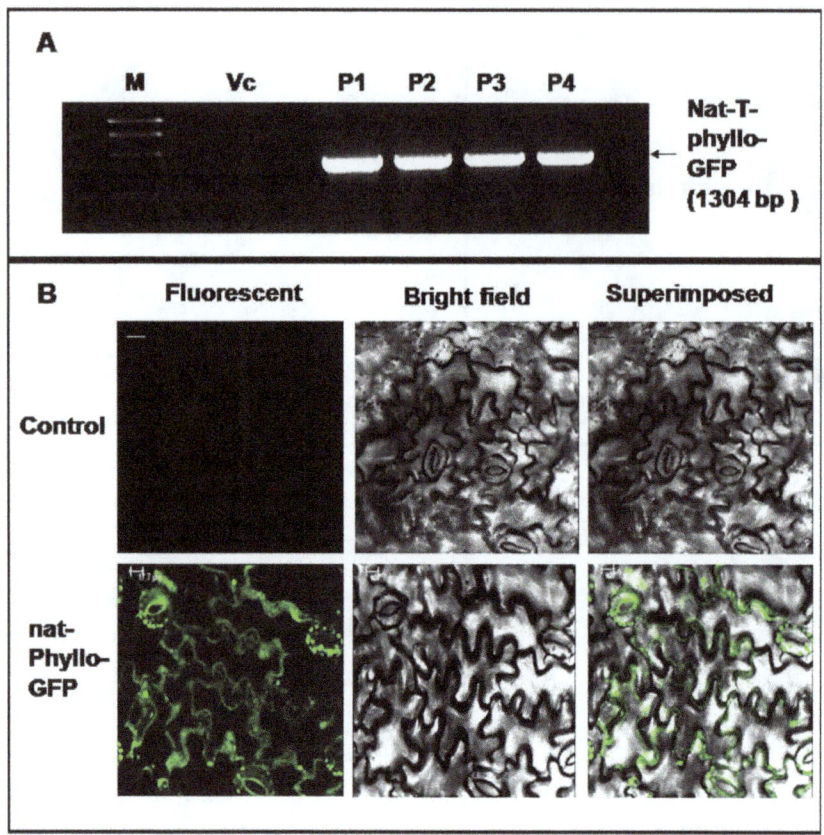

Figure 14. Transient expression of *nat-T-phyllo-GFP* using pSiM24 expression vector. (A) RT-PCR products for native T-phylloplanin fused GFP (*nat-T-phyllo-GFP*) with the expected 1304 bp band are shown. Transiently *nat-T-phyllo-GFP* expressed agro-infiltrated *N. tabacum* L. variety Samsun NN leaf samples using pSiM24-*nat-T-phyllo-GFP* (P1, P2, P3 and P4) and pSiM24 (Vc; empty vector control) are presented. (B) Localization analysis of apoplast targeted nat-T-phyllo-GFP by confocal laser scanning microscopy. Agro-infiltrated *N. tabacum* L. variety Samsun NN leaf cells for pSiM24-nat-T-phyllo-GFP construct expressing GFP (green fluorescence) was visualized by confocal laser scanning microscopy. No GFP fluorescence was detected in agro-infiltrated tobacco leaf cells for pSiM24 construct (Control). Scale bar represents 20 μm on all images. Fluorescent, bright-field and superimposed (bright-field and green fluorescent) confocal laser scanning micrographs of leaf sections are presented.

gene fused with GFP targeted to the apoplasm increases resistance to blue mold disease in tobacco [41,44]. Here, the expression of nat-T-phyllo-GFP using pSiM24 vector was confirmed by the GFP fluorescence in apoplast region of agroinfiltrated plant leaves (Figure 14). Another, chimeric gene (GFP-*AtCESA3*$^{ixr1-2}$) of size 4086-bp fragment was also successfully expressed transiently using pSiM24 following agro-infiltration procedure. The overexpression of GFP fused to the *Arabidopsis CESA3*$^{ixr1-2}$ (GFP-*AtCESA3*$^{ixr1-2}$) gene in transgenic tobacco was known for increasing cellulose digestibility and biomass saccharification [42,43]. Further genes like Vip3A(a), KMP-11 and IL-10 were also successfully expressed transiently in tobacco protoplasts using pSiM24 expression vector (Figures 13). The gene product of novel vegetative insecticidal gene, vip3A(a) shows activity against lepidopteran insects [45,46]. KMP-11, a flagellar protein is known to play an essential role in regulating cytokinesis in both amastigote and promastigote forms of leishmania [47] and is a potential stimulator of T-lymphocyte proliferation [82]. Interleukin-10 (IL-10), an anti-inflammatory cytokine secreted under different conditions of immune activation by a variety of cell types, including T cells, B cells, and monocytes/ macrophages [83,84] has been shown to suppress a broad range of inflammatory responses and as an important factor in maintaining homeostasis of overall immune responses [85,86] and thus has been used for developing novel therapies for several human diseases such as allergic responses and autoimmune diseases [87].

The effect of pSiM24 binary vector on transformation frequency studied in *A. thaliana* verified the pSiM24 and pSiM24-GUS/GFP vectors exhibited more transformation frequency in *A. thaliana* than pCAMBIA2300 and pKM24KH vectors (Table 5) however strong GUS transgene expression in pSiM24-GUS transgenic tobacco and *Arabidopsis* than pK-CaMV35S-GUS transgenic plants depends upon the strong M24 promoter (Figure 5) [41,42].

The pSiM24 vector was observed to be active in transferring the transgene both transiently (Figure 3–5, Figure 11–14) and stably (Figure 6–10) in plant systems, making it useful for various plant biotechnological applications. This plasmid has multiple cloning sites and can act as a platform for various applications, such as gene expression studies and different promoter expressional analyses. In addition, pSiM24 offers a wide selection of cloning sites and high copy numbers in *E. coli* for the facile manipulation of different genetic elements. Thus, the pSiM24 binary vector system described in this study has a high degree of flexibility and may serve as a useful tool for the transformation of plants, making it apt for future use in field release experiments.

Acknowledgments

We are very much indebted to the Kentucky Tobacco Research and Development Center (KTRDC) for the provided facilities and support.

The authors would like to thank Ms. Bonnie Kinney for her excellent care of the transgenic tobacco plants and Prof. Arthur G. Hunt, University of Kentucky, Lexington, USA, for critically reading and making necessary corrections to the manuscript. The information reported in this paper (No. 14-17-039) is part of a project of the Kentucky Agricultural Experiment Station and is published with the approval of the Director.

Author Contributions

Conceived and designed the experiments: IBM DKS ND. Performed the experiments: DKS. Analyzed the data: IBM DKS ND. Contributed reagents/materials/analysis tools: IBM DKS ND. Wrote the paper: IBM DKS.

References

1. Bevan M (1984) Binary Agrobacterium vectors for plant transformation. Nucleic Acids Res 12: 8711–8721.
2. Hoekema A, Hirsch PR, Hooykaas PJJ, Schilperoort RA (1983) A Binary Plant Vector Strategy Based on Separation of Vir-Region and T-Region of the Agrobacterium-Tumefaciens Ti-Plasmid. Nature 303: 179–180.
3. Klee HJ, Yanofsky MF, Nester EW (1985) Vectors for Transformation of Higher-Plants. Bio-Technology 3: 637–642.
4. Schardl CL, Byrd AD, Benzion G, Altschuler MA, Hildebrand DF, et al. (1987) Design and Construction of a Versatile System for the Expression of Foreign Genes in Plants. Gene 61: 1–11.
5. Xiang C, Han P, Lutziger I, Wang K, Oliver DJ (1999) A mini binary vector series for plant transformation. Plant Mol Biol 40: 711–717.
6. Zupan JR, Zambryski P (1995) Transfer of T-DNA from Agrobacterium to the plant cell. Plant Physiol 107: 1041–1047.
7. Hellens R, Mullineaux P, Klee H (2000) Technical Focus:a guide to Agrobacterium binary Ti vectors. Trends Plant Sci 5: 446–451.
8. Komari T, Takakura Y, Ueki J, Kato N, Ishida Y, et al. (2006) Binary vectors and super-binary vectors. Methods Mol Biol 343: 15–41.
9. Jefferson RA, Klass M, Wolf N, Hirsh D (1987) Expression of chimeric genes in Caenorhabditis elegans. J Mol Biol 193: 41–46.
10. Hiei Y, Ohta S, Komari T, Kumashiro T (1994) Efficient transformation of rice (Oryza sativa L.) mediated by Agrobacterium and sequence analysis of the boundaries of the T-DNA. Plant J 6: 271–282.
11. Hajdukiewicz P, Svab Z, Maliga P (1994) The small, versatile pPZP family of Agrobacterium binary vectors for plant transformation. Plant Mol Biol 25: 989–994.
12. Xiang CB, Han P, Lutziger I, Wang K, Oliver DJ (1999) A mini binary vector series for plant transformation. Plant Mol Biol 40: 711–717.
13. Hellens RP, Edwards EA, Leyland NR, Bean S, Mullineaux PM (2000) pGreen: a versatile and flexible binary Ti vector for Agrobacterium-mediated plant transformation. Plant Mol Biol 42: 819–832.
14. Chakrabarty R, Banerjee R, Chung SM, Farman M, Citovsky V, et al. (2007) pSITE vectors for stable integration or transient expression of autofluorescent protein fusions in plants: Probing Nicotiana benthamiana-virus interactions. Molecular Plant-Microbe Interactions 20: 740–750.
15. Goodin MM, Dietzgen RG, Schichnes D, Ruzin S, Jackson AO (2002) pGD vectors: versatile tools for the expression of green and red fluorescent protein fusions in agroinfiltrated plant leaves. Plant Journal 31: 375–383.
16. Thole V, Worland B, Snape JW, Vain P (2007) The pCLEAN dual binary vector system for Agrobacterium-mediated plant transformation. Plant Physiol 145: 1211–1219.
17. Untergasser A, Bijl GJ, Liu W, Bisseling T, Schaart JG, et al. (2012) One-step Agrobacterium mediated transformation of eight genes essential for rhizobium symbiotic signaling using the novel binary vector system pHUGE. PLoS One 7: e47885.
18. Takken FLW, van Wijk R, Michielse CB, Houterman PM, Ram AFJ, et al. (2004) A one-step method to convert vectors into binary vectors suited for Agrobacterium-mediated transformation. Current Genetics 45: 242–248.
19. Lindbo JA (2007) TRBO: A high-efficiency tobacco mosaic virus RNA-Based overexpression vector. Plant Physiology 145: 1232–1240.
20. Shah KH, Almaghrabi B, Bohlmann H (2013) Comparison of Expression Vectors for Transient Expression of Recombinant Proteins in Plants. Plant Molecular Biology Reporter 31: 1529–1538.
21. Sainsbury F, Thuenemann EC, Lomonossoff GP (2009) pEAQ: versatile expression vectors for easy and quick transient expression of heterologous proteins in plants. Plant Biotechnology Journal 7: 682–693.
22. Huang Z, Chen Q, Hjelm B, Arntzen C, Mason H (2009) A DNA Replicon System for Rapid High-Level Production of Virus-Like Particles in Plants. Biotechnology and Bioengineering 103: 706–714.
23. Ali MA, Shah KH, Bohlmann H (2012) pMAA-Red: a new pPZP-derived vector for fast visual screening of transgenic Arabidopsis plants at the seed stage. Bmc Biotechnology 12.
24. Wang CT, Yin XL, Kong XX, Li WS, Ma L, et al. (2013) A Series of TA-Based and Zero-Background Vectors for Plant Functional Genomics. Plos One 8.
25. Vieira J, Messing J (1982) The Puc Plasmids, an M13mp7-Derived System for Insertion Mutagenesis and Sequencing with Synthetic Universal Primers. Gene 19: 259–268.
26. Zambryski P, Joos H, Genetello C, Leemans J, Vanmontagu M, et al. (1983) Ti-Plasmid Vector for the Introduction of DNA into Plant-Cells without Alteration of Their Normal Regeneration Capacity. Embo Journal 2: 2143–2150.
27. Dey N, Maiti IB (1999) Structure and promoter/leader deletion analysis of mirabilis mosaic virus (MMV) full-length transcript in transgenic plants. Plant Mol Biol 40: 771–782.
28. Sahoo DK, Ranjan R, Kumar D, Kumar A, Sahoo BS, et al. (2009) An alternative method of promoter assessment by confocal laser scanning microscopy. J Virol Methods 161: 114–121.
29. Dey N, Maiti IB (1999) Further characterization and expression analysis of mirabilis mosaic caulimovirus (MMV) full-length transcript promoter with single and double enhancer domains in transgenic plants. Transgenics 3: 61–+.
30. Hofgen R, Willmitzer L (1988) Storage of Competent Cells for Agrobacterium Transformation. Nucleic Acids Res 16: 9877–9877.
31. Maiti IB, Murphy JF, Shaw JG, Hunt AG (1993) Plants that express a potyvirus proteinase gene are resistant to virus infection. Proc Natl Acad Sci U S A 90: 6110–6114.
32. Svab Z, Hajdukiewicz P, Maliga P (1990) Stable transformation of plastids in higher plants. Proc Natl Acad Sci U S A 87: 8526–8530.
33. Zhang XR, Henriques R, Lin SS, Niu QW, Chua NH (2006) Agrobacterium-mediated transformation of Arabidopsis thaliana using the floral dip method. Nature Protocols 1: 641–646.
34. Kumar D, Patro S, Ranjan R, Sahoo DK, Maiti IB, et al. (2011) Development of useful recombinant promoter and its expression analysis in different plant cells using confocal laser scanning microscopy. PLoS One 6: e24627.
35. Voinnet O, Rivas S, Mestre P, Baulcombe D (2003) An enhanced transient expression system in plants based on suppression of gene silencing by the p19 protein of tomato bushy stunt virus. Plant Journal 33: 949–956.
36. Maiti IB, Gowda S, Kiernan J, Ghosh SK, Shepherd RJ (1997) Promoter/leader deletion analysis and plant expression vectors with the figwort mosaic virus (FMV) full length transcript (FLt) promoter containing single or double enhancer domains. Transgenic Res 6: 143–156.
37. Lu Y, Chen X, Wu Y, Wang Y, He Y (2013) Directly transforming PCR-amplified DNA fragments into plant cells is a versatile system that facilitates the transient expression assay. PLoS One 8: e57171.
38. Banerjee J, Sahoo DK, Dey N, Houtz RL, Maiti IB (2013) An Intergenic Region Shared by At4g35985 and At4g35987 in Arabidopsis thaliana Is a Tissue Specific and Stress Inducible Bidirectional Promoter Analyzed in Transgenic Arabidopsis and Tobacco Plants. PLoS One 8: e79622.
39. Jefferson RA, Kavanagh TA, Bevan MW (1987) Gus Fusions - Beta-Glucuronidase as a Sensitive and Versatile Gene Fusion Marker in Higher-Plants. Embo Journal 6: 3901–3907.
40. Bradford MM (1976) A rapid and sensitive method for the quantitation of microgram quantities of protein utilizing the principle of protein-dye binding. Anal Biochem 72: 248–254.
41. Kroumova AB, Sahoo DK, Raha S, Goodin M, Maiti IB, et al. (2013) Expression of an apoplast-directed, T-phylloplanin-GFP fusion gene confers resistance against Peronospora tabacina disease in a susceptible tobacco. Plant Cell Reports 32: 1771–1782.
42. Sahoo DK, Stork J, DeBolt S, Maiti IB (2013) Manipulating cellulose biosynthesis by expression of mutant Arabidopsis proM24::CESA3(ixr1-2) gene in transgenic tobacco. Plant Biotechnol J 11: 362–372.
43. Sahoo DK, Maiti IB (2014) Biomass derived from transgenic tobacco expressing the *Arabidopsis CESA3^{ixr1-2}* gene exhibits improved saccharification. Acta Biologica Hungarica 65(2): 189–204 (In Press).
44. Sahoo DK, Raha S, Hall JT, Maiti IB (2014) Over-expression of the synthetic chimeric native-T-phylloplanin-GFP genes optimized for monocot and dicot plants renders enhanced resistance to blue mold disease in tobacco (*N. tabacum* L.). The Scientific World Journal 2014: Article ID 601314, 12 pages. DOI: http://dx.doi.org/10.1155/2014/601314.
45. Estruch JJ, Warren GW, Mullins MA, Nye GJ, Craig JA, et al. (1996) Vip3A, a novel Bacillus thuringiensis vegetative insecticidal protein with a wide spectrum of activities against lepidopteran insects. Proc Natl Acad Sci U S A 93: 5389–5394.
46. Selvapandiyan A, Arora N, Rajagopal R, Jalali SK, Venkatesan T, et al. (2001) Toxicity analysis of N- and C-terminus-deleted vegetative insecticidal protein from Bacillus thuringiensis. Appl Environ Microbiol 67: 5855–5858.
47. Li Z, Wang CC (2008) KMP-11, a basal body and flagellar protein, is required for cell division in Trypanosoma brucei. Eukaryot Cell 7: 1941–1950.
48. Minter RM, Ferry MA, Rectenwald JE, Bahjat FR, Oberholzer A, et al. (2001) Extended lung expression and increased tissue localization of viral IL-10 with adenoviral gene therapy. Proc Natl Acad Sci U S A 98: 277–282.
49. Song F, Sun X, Wang X, Nai Y, Liu Z (2014) Early diagnosis of tuberculous meningitis by an indirect ELISA protocol based on the detection of the antigen ESAT-6 in cerebrospinal fluid. Ir J Med Sci 183: 85–88.
50. Chan V, Dreolini LF, Flintoff KA, Lloyd SJ, Mattenley AA (2002) The effect of increasing plasmid size on transformation efficiency in Escherichia coli. Journal of Experimental Microbiology and Immunology 2: 207–223.
51. Lee S, Su G, Lasserre E, Aghazadeh MA, Murai N (2012) Small high-yielding binary Ti vectors pLSU with co-directional replicons for Agrobacterium

tumefaciens-mediated transformation of higher plants. Plant Science 187: 49–58.

52. Kusaba M, Takahashi Y, Nagata T (1996) A multiple-stimuli-responsive as-1-related element of parA gene confers responsiveness to cadmium but not to copper. Plant Physiology 111: 1161–1167.

53. Murai N (2013) Review: Plant Binary Vectors of Ti Plasmid in *Agrobacterium tumefaciens* with a Broad Host-Range Replicon of pRK2, pRi, pSa or pVS1. American Journal of Plant Sciences 4: 932–939.

54. Yadav NS, Vanderleyden J, Bennett DR, Barnes WM, Chilton MD (1982) Short Direct Repeats Flank the T-DNA on a Nopaline Ti Plasmid. Proceedings of the National Academy of Sciences of the United States of America-Biological Sciences 79: 6322–6326.

55. Chatterjee A, Das NC, Raha S, Babbit R, Huang QW, et al. (2010) Production of xylanase in transgenic tobacco for industrial use in bioenergy and biofuel applications. In Vitro Cellular & Developmental Biology-Plant 46: 198–209.

56. Breyne P, Gheysen G, Jacobs A, Vanmontagu M, Depicker A (1992) Effect of T-DNA Configuration on Transgene Expression. Molecular & General Genetics 235: 389–396.

57. Hiei Y, Komari T (2006) Improved protocols for transformation of indica rice mediated by Agrobacterium tumefaciens. Plant Cell Tissue and Organ Culture 85: 271–283.

58. Komari T, Hiei Y, Saito Y, Murai N, Kumashiro T (1996) Vectors carrying two separate T-DNAs for co-transformation of higher plants mediated by Agrobacterium tumefaciens and segregation of transformants free from selection markers. Plant Journal 10: 165–174.

59. Komori T, Imayama T, Kato N, Ishida Y, Ueki J, et al. (2007) Current status of binary vectors and superbinary vectors. Plant Physiol 145: 1155–1160.

60. Vaucheret H, Fagard M (2001) Transcriptional gene silencing in plants: targets, inducers and regulators. Trends in Genetics 17: 29–35.

61. Acharya S, Ranjan R, Pattanaik S, Maiti IB, Dey N (2013) Efficient chimeric plant promoters derived from plant infecting viral promoter sequences. Planta.

62. Bhattacharyya S, Dey N, Maiti IB (2002) Analysis of cis-sequence of subgenomic transcript promoter from the Figwort mosaic virus and comparison of promoter activity with the cauliflower mosaic virus promoters in monocot and dicot cells. Virus Research 90: 47–62.

63. Bhullar S, Chakravarthy S, Advani S, Datta S, Pental D, et al. (2003) Strategies for development of functionally equivalent promoters with minimum sequence homology for transgene expression in plants: cis-elements in a novel DNA context versus domain swapping. Plant Physiology 132: 988–998.

64. Kumar D, Patro S, Ghosh J, Das A, Maiti IB, et al. (2012) Development of a salicylic acid inducible minimal sub-genomic transcript promoter from Figwort mosaic virus with enhanced root- and leaf-activity using TGACG motif rearrangement. Gene 503: 36–47.

65. Patro S, Maiti IB, Dey N (2013) Development of an efficient bi-directional promoter with tripartite enhancer employing three viral promoters. Journal of Biotechnology 163: 311–317.

66. Pattanaik S, Dey N, Bhattacharyya S, Maiti IB (2004) Isolation of full-length transcript promoter from the Strawberry vein banding virus (SVBV) and expression analysis by protoplasts transient assays and in transgenic plants. Plant Science 167: 427–438.

67. Ranjan R, Patro S, Kumari S, Kumar D, Dey N, et al. (2011) Efficient chimeric promoters derived from full-length and sub-genomic transcript promoters of Figwort mosaic virus (FMV). Journal of Biotechnology 152: 58–62.

68. Ranjan R, Patro S, Pradhan B, Kumar A, Maiti IB, et al. (2012) Development and Functional Analysis of Novel Genetic Promoters Using DNA Shuffling, Hybridization and a Combination Thereof. PLoS One 7.

69. An G, Watson BD, Chiang CC (1986) Transformation of Tobacco, Tomato, Potato, and Arabidopsis-Thaliana Using a Binary Ti Vector System. Plant Physiol 81: 301–305.

70. Ishida Y, Saito H, Ohta S, Hiei Y, Komari T, et al. (1996) High efficiency transformation of maize (Zea mays L.) mediated by Agrobacterium tumefaciens. Nat Biotechnol 14: 745–750.

71. Vasil IK (1996) Milestones in crop biotechnology—transgenic cassava and Agrobacterium-mediated transformation of maize. Nat Biotechnol 14: 702–703.

72. Ow DW, De Wet JR, Helinski DR, Howell SH, Wood KV, et al. (1986) Transient and stable expression of the firefly luciferase gene in plant cells and transgenic plants. Science 234: 856–859.

73. Pang SZ, DeBoer DL, Wan Y, Ye G, Layton JG, et al. (1996) An improved green fluorescent protein gene as a vital marker in plants. Plant Physiol 112: 893–900.

74. Fluhr R, Chua NH (1986) Developmental Regulation of 2 Genes Encoding Ribulose-Bisphosphate Carboxylase Small Subunit in Pea and Transgenic Petunia Plants - Phytochrome Response and Blue-Light Induction. Proc Natl Acad Sci U S A 83: 2358–2362.

75. Fluhr R, Kuhlemeier C, Nagy F, Chua NH (1986) Organ-Specific and Light-Induced Expression of Plant Genes. Science 232: 1106–1112.

76. Fluhr R, Moses P, Morelli G, Coruzzi G, Chua NH (1986) Expression Dynamics of the Pea Rbcs Multigene Family and Organ Distribution of the Transcripts. Embo Journal 5: 2063–2071.

77. Koncz C, Schell J (1986) The Promoter of TI-DNA Gene 5 Controls the Tissue-Specific Expression of Chimeric Genes Carried by a Novel Type of Agrobacterium Binary Vector. Molecular & General Genetics 204: 383–396.

78. Pansegrau W, Lanka E, Barth PT, Figurski DH, Guiney DG, et al. (1994) Complete Nucleotide-Sequence of Birmingham Incp-Alpha Plasmids - Compilation and Comparative-Analysis. J Mol Biol 239: 623–663.

79. Lemos ML, Crosa JH (1992) Highly Preferred Site of Insertion of Tn7 into the Chromosome of Vibrio-Anguillarum. Plasmid 27: 161–163.

80. Kroumova AB, Shepherd RW, Wagner GJ (2007) Impacts of T-Phylloplanin gene knockdown and of Helianthus and Datura phylloplanins on Peronospora tabacina spore germination and disease potential. Plant Physiol 144: 1843–1851.

81. Shepherd RW, Bass WT, Houtz RL, Wagner GJ (2005) Phylloplanins of tobacco are defensive proteins deployed on aerial surfaces by short glandular trichomes. Plant Cell 17: 1851–1861.

82. Tolson DL, Jardim A, Schnur LF, Stebeck C, Tuckey C, et al. (1994) The kinetoplastid membrane protein 11 of Leishmania donovani and African trypanosomes is a potent stimulator of T-lymphocyte proliferation. Infect Immun 62: 4893–4899.

83. Filippi CM, von Herrath MG (2008) IL-10 and the resolution of infections. J Pathol 214: 224–230.

84. Sabat R, Grutz G, Warszawska K, Kirsch S, Witte E, et al. (2010) Biology of interleukin-10. Cytokine Growth Factor Rev 21: 331–344.

85. Stober CB, Lange UG, Roberts MT, Alcami A, Blackwell JM (2005) IL-10 from regulatory T cells determines vaccine efficacy in murine Leishmania major infection. J Immunol 175: 2517–2524.

86. Villalta SA, Rinaldi C, Deng B, Liu G, Fedor B, et al. (2011) Interleukin-10 reduces the pathology of mdx muscular dystrophy by deactivating M1 macrophages and modulating macrophage phenotype. Hum Mol Genet 20: 790–805.

87. Gelderblom H, Schmidt J, Londono D, Bai Y, Quandt J, et al. (2007) Role of interleukin 10 during persistent infection with the relapsing fever Spirochete Borrelia turicatae. Am J Pathol 170: 251–262.

Determining the GmRIN4 Requirements of the Soybean Disease Resistance Proteins Rpg1b and Rpg1r Using a *Nicotiana glutinosa*-Based Agroinfiltration System

Ryan Kessens, Tom Ashfield, Sang Hee Kim¤, Roger W. Innes*

Department of Biology, Indiana University, Bloomington, Indiana, United States of America

Abstract

Rpg1b and *Rpg1r* are soybean disease resistance (*R*) genes responsible for conferring resistance to *Pseudomonas syringae* strains expressing the effectors AvrB and AvrRpm1, respectively. The study of these cloned genes would be greatly facilitated by the availability of a suitable transient expression system. The commonly used *Niciotiana benthamiana*-based system is not suitable for studying *Rpg1b* and *Rpg1r* function, however, because expression of AvrB or AvrRpm1 alone induces a hypersensitive response (HR), indicating that *N. benthamiana* contains endogenous *R* genes that recognize these effectors. To identify a suitable alternative host for transient expression assays, we screened 13 species of *Nicotiana* along with 11 accessions of *N. tabacum* for lack of response to transient expression of *AvrB* and *AvrRpm1*. We found that *N. glutinosa* did not respond to either effector and was readily transformable as determined by transient expression of β-glucuronidase. Using this system, we determined that *Rpg1b*-mediated HR in *N. glutinosa* required co-expression of *avrB* and a soybean ortholog of the Arabidopsis *RIN4* gene. All four soybean *RIN4* orthologs tested worked in the assay. In contrast, Rpg1r did not require co-expression of a soybean *RIN4* ortholog to recognize AvrRpm1, but recognition was suppressed by co-expression with AvrRpt2. These observations suggest that an endogenous *RIN4* gene in *N. glutinosa* can substitute for the soybean *RIN4* ortholog in the recognition of AvrRpm1 by Rpg1r.

Editor: Hua Lu, UMBC, United States of America

Funding: RK was the recipient of a summer research grant from the Hutton Honors College of Indiana University (http://www.indiana.edu/~iubhonor/). This work was supported by the National Institute of General Medical Sciences of the National Institutes of Health grant no. R01 GM046451 to RWI (http://www.nigms.nih.gov/Pages/default.aspx). The funders had no role in study design, data collection and analysis, decision to publish, or preparation of the manuscript.

Competing Interests: The authors have declared that no competing interests exist.

* Email: rinnes@indiana.edu

¤ Current address: Division of Plant Sciences, University of Missouri, Columbia, Missouri, United States of America

Introduction

Effector-triggered immunity (ETI) depends on the expression of disease resistance (*R*) genes, the majority of which encode nucleotide binding-leucine rich repeat (NB-LRR) proteins. NB-LRR proteins have been shown to detect effectors by both direct and indirect mechanisms [1]. Indirect recognition requires an intermediate protein that serves as a substrate for the effector. These intermediate proteins are targeted by effectors and can be modified in ways such as proteolytic cleavage or phosphorylation. This allows R proteins to indirectly detect the presence of effectors by monitoring the status of the effector targets [2].

AvrB and AvrRpm1 are two effectors found in certain strains of *Pseudomonas syringae* [3,4], the causative agent of soybean bacterial blight and related diseases in other plant species. In *Arabidopsis thaliana*, the *R* gene *RPM1* confers resistance to bacteria expressing both of these effectors, but only if a functional copy of *RIN4* is also present in the genome [5,6]. This is considered a classic example of indirect recognition. The R protein (RPM1) guards the effector target (RIN4), which is modified in the presence of both AvrB and AvrRpm1 [6]. Both of these effectors are thought to modify RIN4 through phosphory-lation, but whether this is direct or indirect remains unclear [7] [8]. Phosphorylation of RIN4 is detected by RPM1, which then initiates a signaling cascade that leads to a form of programmed cell death known as the hypersensitive response (HR) that inhibits the spread of the invading pathogen [9].

While recognition of AvrB and AvrRpm1 in soybean likely has similarities to that in Arabidopsis, there are at least two key differences. Unlike Arabidopsis, which requires just one *R* gene to confer resistance to both AvrB and AvrRpm1, soybean requires two *R* genes. These *R* genes are *Rpg1b* and *Rpg1r*, which confer resistance to AvrB and AvrRpm1, respectively [10]. Another difference is that the Arabidopsis genome only encodes one *RIN4* gene while soybean contains four *RIN4*-homologues (*GmRIN4a*, *GmRIN4b*, *GmRIN4c*, and *GmRIN4d*) [11]. At least two of these family members, *GmRIN4a* and *GmRIN4b*, appear to be necessary for *Rpg1b* to confer resistance to AvrB [12].

Several *R* gene recognition systems have been reconstituted in *Nicotiana benthamiana* by infiltrating leaves with a mixture of *Agrobacterium tumefaciens* strains that transfer genes coding for an effector, R protein, and any intermediate protein(s) that might be necessary for an R protein to recognize the effector [13-15]. If an

R gene confers resistance to a particular effector, it will initiate a signaling cascade leading to HR, which can be observed on plant leaves as brown discoloration and/or leaf collapse. While *N. benthamiana* is a useful transient system for investigating many *R* genes in this manner, it has limitations for the study of AvrB and AvrRpm1-specific *R* genes. Specifically, *N. benthamiana* contains endogenous *R* gene(s) able to detect these pathogen effectors. This makes it difficult to use *N. benthamiana* to study *R* genes from other species that recognize AvrB and/or AvrRpm1 because expression of these effectors alone triggers HR. While this can be partially mitigated by careful titration of the density at which the Agrobacterium strains are infiltrated [8], it would be useful to identify an alternative plant species for these experiments that does not respond to AvrB and AvrRpm1 [8,16].

The goal of this study was to identify an alternative species to *N. benthamiana* for transient expression studies involving soybean Rpg1b and Rpg1r and to determine the *GmRIN4* requirements of each of these R proteins. An ideal species would have many of the characteristics that make *N. benthamiana* a good system such as ease of injection and high-transformation efficiency, but would not display signs of HR when *AvrB* and *AvrRpm1* were expressed alone. To achieve this goal, we screened *Nicotiana* germplasm, including 13 distinct species and 11 accessions of *N. tabacum*, for their response to transiently expressed *AvrB* and *AvrRpm1*. The transformation efficiency of each species was also assessed using a *GUS* reporter gene to ensure that lack of HR was not simply due to low levels of gene expression. The well-studied *RPS5*-mediated HR pathway was reconstituted in the most promising genotype (*N. glutinosa*) to test its efficacy for reconstructing an *R* gene pathway. This was accomplished by co-expressing the *P. syringae* effector *AvrPphB*, its target *PBS1*, and the *R* gene *RPS5* [17]. The final step was to co-express each of the soybean *R* genes with their corresponding effectors and one or more of the *GmRIN4*s to determine which, if any, *GmRIN4*s were required by either R protein to detect its corresponding effector.

Results

Screening *Nicotiana* germplasm for accessions in which AvrB and AvrRpm1 do not trigger an HR

To identify *Nicotiana* species/accessions that do not respond to transient expression of the effectors *AvrB* and *AvrRpm1* with an HR, we used a two-step approach. Thirteen distinct species of *Nicotiana* and 11 accessions of *N. tabacum* were visually assessed for signs of effector recognition upon transient expression of *AvrB* or *AvrRpm1*. Effector recognition was determined by looking for morphological changes on plant leaves expressing either effector. These changes in morphology included brown discoloration on the adaxial and abaxial surface of leaves, a "shiny" phenotype on the abaxial surface, or full leaf collapse (Fig. S1). As a negative control, each plant was infiltrated with *Agrobacterium* carrying a plasmid with a *GUS* reporter gene. One leaf on each plant was transformed with *GUS* and compared to another leaf on the same plant expressing *AvrB* on one half and *AvrRpm1* on the other. Lack of effector recognition was assumed when the response to infiltration with the AvrB and AvrRpm1-containing strains was no stronger than the response to transformation with *GUS*. Six species of *Nicotiana* did not respond to either effector: *N. alata, N. glutinosa, N. knightiana, N. nudicaulis, N. rotundifolia,* and *N. tomentosiformis* (Table 1). All 11 accessions of *N. tabacum* screened exhibited signs of HR when expressing at least one of the effectors (Table 1).

Before proceeding, we eliminated *N. alata* and *N. rotundifolia* as potential transient systems because of problems inherent to both

species. *N. alata* displayed poor seed germination while *N. rotundifolia* was difficult to infiltrate with Agrobacterium. The transformation efficiency of the remaining species was determined to ensure that a lack of HR was not due to poor transgene expression. This was accomplished by quantifying β-glucuronidase activity in leaf tissue following infiltration with the Agrobacterium strain containing the *GUS* reporter gene. For comparison, *N. benthamiana* leaves were also transiently transformed with the *GUS*-containing strain, leaves harvested, and enzyme activity assayed in conjunction with each of the non-responding species. The results indicated that transgene expression levels in all of the species, including *N. benthamiana*, were quite variable between different individuals of the same genotype. This variation occurred even though all of the plants used in a given experiment were grown under the same conditions and great effort was taken to ensure conditions were consistent between experiments. Despite this variation, we found that *N. benthamiana* and *N. glutinosa* consistently yielded the highest *GUS* expression levels (Fig. S2). We also assayed the transformation efficiency of *N. knightiana* and *N. nudicaulis*, which consistently resulted in a GUS activity 5 to 10 fold lower than that observed in *N. benthamiana*.

Reconstituting the *RPS5, Rpg1b,* and *Rpg1r*-mediated disease resistance pathways

To assess whether *N. glutinosa* can be used as a transient system to reconstitute NB-LRR signaling pathways, we first tested the *RPS5* pathway from Arabidopsis, as this pathway has previously been successfully reconstituted in *N. benthamiana* [13]. RPS5 is an NB-LRR disease resistance protein from Arabidopsis that confers resistance to *P. syringae* strains expressing the effector gene *AvrPphB* [17]. Recognition of AvrPphB by RPS5 requires another Arabidopsis protein, PBS1, which is proteolytically cleaved by AvrPphB [18]. The N and C-terminal cleavage products of PBS1 bind to and activate RPS5 resulting in an HR [19]. We reconstituted the *RPS5*-mediated defense pathway by co-expressing *AvrPphB, PBS1,* and *RPS5* in *N. glutinosa*. As expected, strong leaf collapse was observed at the site of Agrobacterium infiltration when all three of the genes were expressed, but not when leaves lacked expression of any one component of the pathway (Fig. 1).

After successfully reconstituting the *RPS5* pathway, we investigated which components were necessary to reconstitute the *Rpg1b* pathway. This was accomplished by co-transforming *N. glutinosa* leaves with combinations of *AvrB, Rpg1b,* and the *GmRIN4*s. Leaves expressing a combination of *AvrB, Rpg1b,* and at least one of the *GmRIN4*s consistently gave a stronger response than leaves expressing *AvrB* and *Rpg1b* or *AvrB* and a *GmRIN4*. Figure 2a shows representative leaves expressing each combination, while Figure 2b shows an assessment of the strength of HR for each combination.

The finding that Rpg1b required co-expression of a GmRIN4 to detect AvrB raised the question of whether Rpg1r would similarly require a GmRIN4 to detect AvrRpm1. We have recently cloned Rpg1r (Genbank accession number KF958751; [20]). Transient overexpression of Rpg1r by itself in *N. glutinosa* induced a visible collapse, indicating that Rpg1r possesses autoactivity when overexpressed [20]. We found, however, that this autoactivity could be nearly eliminated by fusing super yellow fluorescent protein onto the C-terminus of Rpg1r [20]. Using this Rpg1r-sYFP construct, we were able to assess whether Rpg1r required co-expression of a GmRIN4 to induce HR in response to AvrRpm1. Unlike Rpg1b, co-expression of Rpg1r-sYFP and AvrRpm1 in the absence of GmRIN4 was sufficient to induce leaf collapse in *N. glutinosa* [20]. Importantly, co-expression of Rpg1r

Table 1. *Nicotiana* accessions used in this study and their responses to AvrB and AvrRpm1.

		Response to AvrB		Response to AvrRpm1	
PI number	Type	Response Characteristics	Effector-dependent response	Response Characteristics	Effector-dependent response
N. tabacum **accessions**					
552452	Maryland	s, b	+	s	+
404956	Oriental	s, b	+	s, b	+
378072	Oriental	s, b, lc	+	s, b	+
405603	Oriental	b	+	s	+
292205	Cigar filler	b	+	s	+
405604	Cigar filler	s, b	+	s, b	+
552348	Cigar binder	s, b	+	s	+
552619	Cigar wrapper	s, b, lc	+	s, b	+
552453	Flume cured	b, lc	+	s, b	+
543792	Burley	b, lc	+	b	+
551280	Burley	b, lc	+	b, lc	+
Nicotiana **species**					
42337	*N. langsdorffii*	b	+	nr	-
555531	*N. longiflora*	s, b	+	s, b	+
241768	*N. glutinosa*	s	-	s	-
555553	*N. rotundifolia*	nr	-	nr	-
555527	*N. knightiana*	nr	-	nr	-
555570	*N. sylvestris*	nr	-	s, b	+
555554	*N. rustica*	s, b	+	s, b	+
555552	*N. repanda*	s, b	+	nr	-
42334	*N. alata*	nr	-	nr	-
503323	*N. debneyi*	b	+	b	+
555540	*N. nudicaulis*	nr	-	nr	-
555572	*N. tomentosiformis*	nr	-	nr	-

A (+) sign indicates that leaves expressing a given effector gave a stronger response than leaves expressing *GUS*, while a (-) sign indicates there was no difference between leaves expressing an effector and those expressing *GUS*. The observed morphologies in response to *Agrobacterium*-mediated transformation were leaf browning (b), shininess on the abaxial surface (s), leaf collapse (lc), and no response (nr). See Figure S1 for photographs of phenotypes. Each plant species/accession was tested at least 3 times with similar results.

with AvrB in *N. glutinosa* did not induce HR regardless of whether a GmRIN4 was co-expressed, which established that the specificities of Rpg1b and Rpg1r are retained in this system [20].

Notably, co-expression of GmRIN4 with untagged Rpg1r did not suppress its autoactivity [20], indicating that Rpg1r autoactivity is

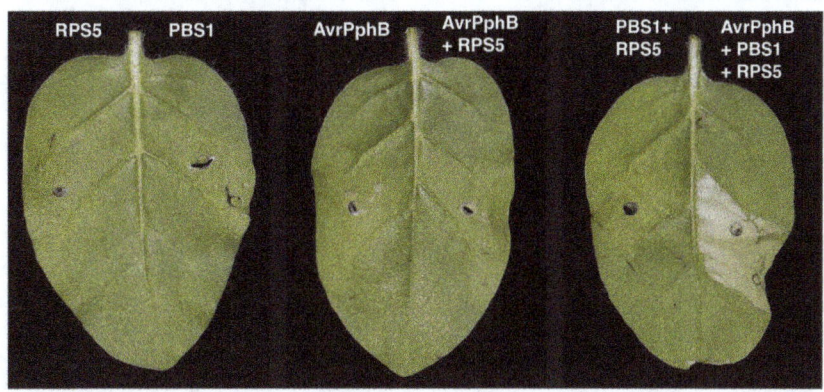

Figure 1. Reconstituting the *RPS5*-mediated defense pathway in *N. glutinosa*. The left and right side of each leaf were transiently transformed with the gene(s) listed above each image. Leaves were detached and photographed 24 hours after dexamethasone induction.

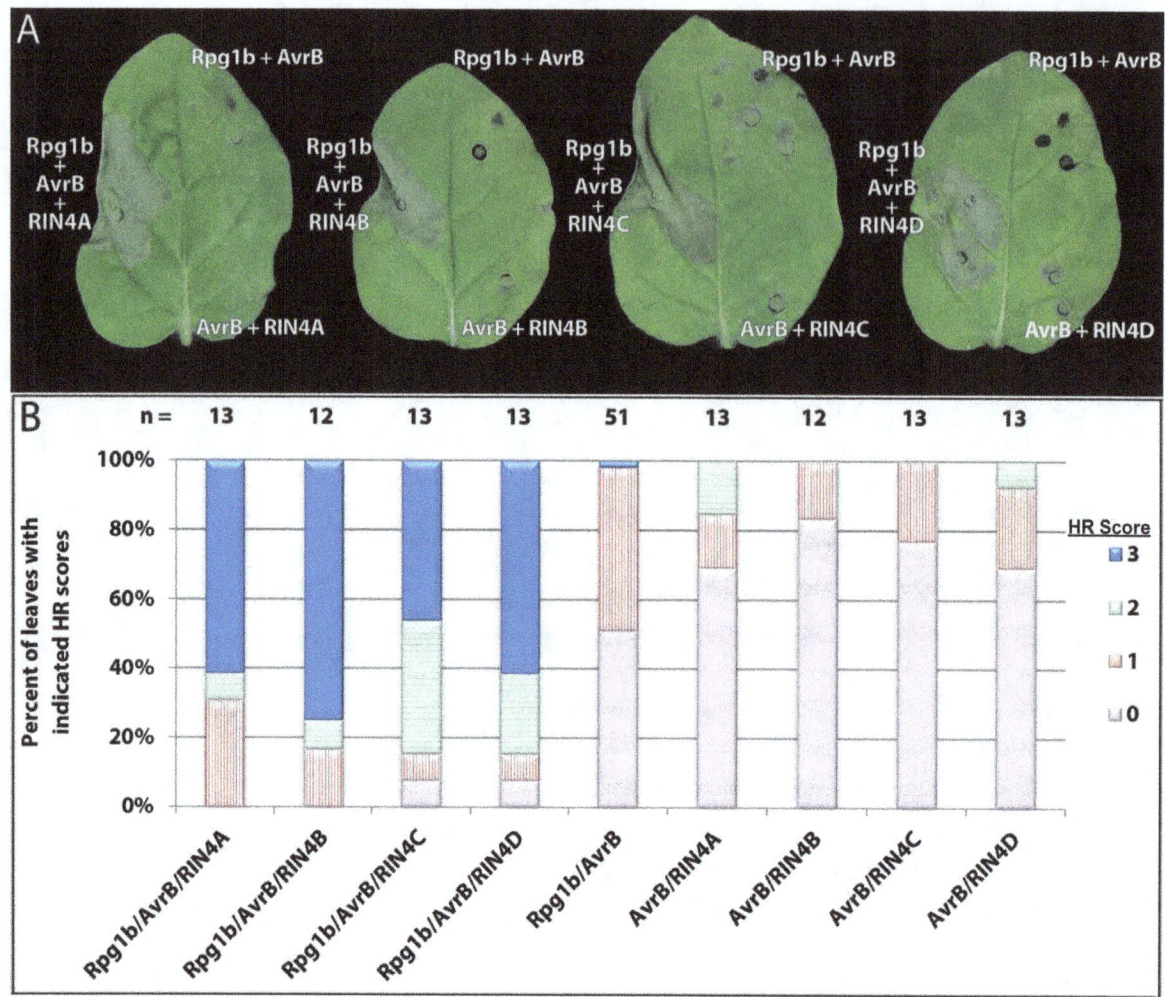

Figure 2. Reconstituting the *Rpg1b*-mediated defense pathway in *N. glutinosa*. (A) Activation of Rpg1b by AvrB requires co-expression of a *GmRIN4* gene. The images shown are of typical responses displayed by *N. glutinosa* leaves expressing the combination of genes labeled on each image. **(B)** Quantification of *Rpg1b*-mediated HR when co-expressed with various combinations of *AvrB* and *GmRIN4* genes. The strength of Rpg1b-mediated HR was determined by the extent of leaf collapse in the infiltrated area. Based on the extent of leaf collapse in the infiltrated area, plant leaves were categorized into 4 classes: 0 (no collapse); 1 (less than one third collapsed); 2 (one third to two thirds collapsed); 3 (greater than two thirds collapsed). Images were taken and plant leaves were scored approximately 2 days after transgene induction. The number of leaves infiltrated and scored for each combination (n) is listed above each bar. This experiment was repeated 3 times with similar results.

not a consequence of activation of Rpg1r by loss of RIN4, as has been reported for the Arabidopsis RPS2 protein [21].

Since Rpg1r did not require co-expression of a GmRIN4 to recognize AvrRpm1 in *N. glutinosa*, we hypothesized that it may be using an endogenous RIN4 protein for this purpose. To test this hypothesis, we employed the *P. syringae* effector AvrRpt2, which is a cysteine protease that has been shown to cleave Arabidopsis RIN4, leading to its degradation [22]. A BLAST search of the *N. benthamiana* genome using the Arabidopsis RIN4 amino acid sequence as the query revealed two predicted full length proteins with high sequence similarity and conserved AvrRpt2 cleavage sites [23] (Fig. S3). By expressing *AvrRpt2* under the constitutively active CaMV 35S promoter, we hoped to eliminate any endogenous RIN4 homologues in *N. glutinosa* before inducing the expression of *Rpg1r* and *AvrRpm1*, both of which were under the control of a DEX-inducible promoter. Co-expression of *AvrRpt2* with *Rpg1r* and *AvrRpm1* led to a reduction in the severity of HR, as indicated by reduced leaf collapse, compared to co-expression of the proteolytically inactive *AvrRpt2 (C122A)*

mutant with *Rpg1r* and *AvrRpm1*. The images in Fig. 3a are representative of the typical responses displayed by leaves expressing each combination while Figure 3b is an assessment of the strength of HR for each combination. These data suggest that Rpg1r is employing an endogenous copy of RIN4 for recognition of AvrRpm1.

Discussion

P. syringae strains have a wide host range and include pathovars such as *P. syringae* pv. *tabaci* that infect *Nicotiana* species. It is thus not surprising that *N. benthamiana* has evolved the ability to recognize specific *P. syringae* effector proteins. Individual *P. syringae* strains express numerous effectors, with great variation in specific effector repertoire between strains [24,25]. This large effector complement is likely the result of a co-evolutionary arms race between *P. syringae* and its host plants. The goal of this study was to identify a *Nicotiana* species that lacked endogenous *R* genes with the ability to recognize *AvrB* and

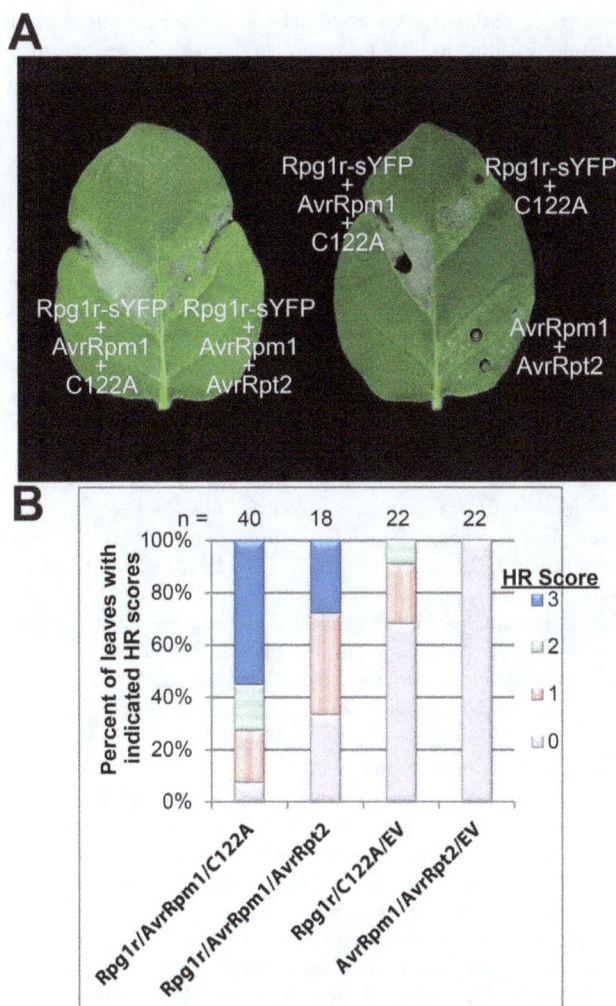

Figure 3. Reconstituting the *Rpg1r*-mediated defense pathway in *N. glutinosa*. (A) *Rpg1r-sYFP*-mediated HR does not require co-expression of a *GmRIN4* gene, but is suppressed by *avrRpt2*. The images shown are of typical responses displayed by *N. glutinosa* leaves expressing the combination of genes labeled on each image. **(B)** Quantification of *Rpg1r-sYFP* mediated HR. Responses were categorized as described in Figure 2. C122A indicates the protease inactive form of AvrRpt2. Images were taken and plant leaves were scored approximately 2 days after transgene induction. The number of leaves infiltrated and scored for each combination (n) is listed above each bar. This experiment was repeated 3 times with similar results.

AvrRpm1 expression and use this species to determine the *RIN4* requirements of *Rpg1b* and *Rpg1r*.

While *N. benthamiana* is widely used by plant biologists for transient gene expression, its ability to recognize the *P. syringae* effectors AvrB and AvrRpm1 makes it unsuitable for structure/function studies on the corresponding R proteins responsible for detecting these effectors. The benefits of *N. benthamiana* as a transient expression system, along with the great species diversity of *Nicotiana*, make this genus a good candidate for finding other species that could serve as suitable transient expression systems. While the six species highlighted in Table 1 did not respond to either effector in the initial screen, quantifying GUS transgene expression showed that a lack of a response in many of the species could be attributed to poor transformation efficiency. Even *N. glutinosa*, the species that gave the highest and most consistent

transformation efficiency, gave variable results within and between experimental replicates. Important factors for obtaining efficient transformation included using young plants (~4 weeks old), avoiding the youngest and oldest leaves (typically the 3rd and 4th true leaves were injected), and using a transformation protocol that included acetosyringone in the infiltration medium.

Plants used in the effector screen were raised under long-day conditions (16 hr light/8 hr dark) to promote faster growth. Subsequently, when performing HR assays, we found that plants raised under short-day conditions (9 hr light, 15 hr dark) produced broader and thinner leaves that gave a more distinct and reproducible HRs. However, when *N. glutinosa* were raised solely under short-day conditions, the seedlings would sometimes develop poorly with excessively long hypocotyls. Therefore, the *N. glutinosa* used in HR assays were germinated and grown under long-day conditions for 12-14 days before being transferred to short-day conditions until being injected.

Phylogenetic analysis of the NBS region from Rpg1b has previously shown that RPM1 is not orthologous to Rpg1b, indicating that their common ability to recognize AvrB is due to convergent evolution [26]. The findings by Selote and Kachroo (2010), along with the findings from this study, reveal that not only have these two *R* genes independently evolved the ability to confer resistance to *AvrB*-expressing *P. syringae* strains, but they have also independently evolved the need for a functional *RIN4*-like protein to confer this resistance.

Through the use of virus-induced gene silencing (VIGS) of soybean *RIN4* genes, Selote and colleagues have previously determined that both *GmRIN4A* and *GmRIN4B* are required for *Rpg1b*-mediated resistance to *P. syringae* strains expressing *AvrB* [12], while *GmRIN4C* and *GmRIN4D* are not [16]. Contrary to their findings, our findings suggest that each of the individual GmRIN4 proteins can be used by Rpg1b to recognize AvrB, as co-expression of any *GmRIN4* with *Rpg1b* and *AvrB* in *N. glutinosa* resulted in an HR. However, a major difference between Selote's work and ours is that they used VIGS to silence native genes while we used transient expression to over-express foreign genes. If there is a GmRIN4 expression level threshold in soybean required for Rpg1b function, then transient overexpression would likely exceed this threshold. Therefore, it is possible that *GmRIN4C* and *GmRIN4D* could be used for *Rpg1b*-mediated resistance in soybean if they were expressed at a high enough level.

As is the case with *Rpg1b*, phylogenetic evidence and amino acid sequence comparisons indicate that *Rpg1r* and *RPM1* are not orthologous and have very little amino acid sequence similarity outside the conserved NB-ARC domain [20]. Our initial observation that co-expression of *Rpg1r* with *AvrRpm1* was sufficient to trigger HR led us to hypothesize that *Rpg1r* did not have a *GmRIN4* requirement. However, co-expression of *AvrRpt2* with *Rpg1r* and *AvrRpm1* was able to reduce the leaf collapse associated with HR. While this suggests Rpg1r requires a RIN4 homologue to detect AvrRpm1, it is not definitive. It is possible that AvrRpt2 is targeting another component of the pathway required for effector recognition or is targeting a step downstream of effector recognition. If Rpg1r does indeed use one or more *GmRIN4s* to detect AvrRpm1, this would indicate that Rpg1r, Rpg1b, RPM1, and RPS2 have all evolved the ability to detect pathogen effectors by monitoring the status of a RIN4 homolog, suggesting that RIN4 represents a common effector target across plant species, and thus a hub guarded by multiple NB-LRR proteins.

By reconstituting the *RPS5*, *Rpg1b*, and *Rpg1r* pathways in *N. glutinosa*, we have demonstrated that this system can be used to study the molecular requirements of a variety of R proteins. With the recently published draft sequence of the *N. benthamiana* genome [27], the ability to find homologous genes involved in these pathways is as simple as performing a BLAST search. The development of the *N. glutinosa* transient system will be especially useful for performing structure/function studies on Rpg1b and Rpg1r and assessing how their ability to distinguish between AvrB and AvrRpm1 is determined.

Materials and Methods

Plant material

All *Nicotiana* seeds were obtained from the USDA National Plant Germplasm System Nicotiana Collection at North Carolina State University in Raleigh, NC and grown in Metro-Mix 360 potting soil. Plants used for the effector screen and MUG assay (described below) were grown in a growth chamber under long-day conditions (16 hr light/8 hr dark) at 24°C. These plants were grown for 3-4 weeks before transient transformation. *N. glutinosa* plants used in subsequent HR assays were germinated under long-day conditions for 12-14 days then transferred to short-day conditions (9 hr light/16 hr dark) for ~2 more weeks before transformation, as these growth conditions produced leaves that were easier to infiltrate, giving HR phenotypes that were more distinct. The plants were grown at 22-24°C under both long and short-day conditions.

Agrobacterium-mediated transformation

Agrobacterium tumefaciens strain GV3101 (pMP90) was used in all experiments. All Agrobacterium strains, except for those carrying plasmids for the transfer of *AvrRpt2* and *AvrRpt2 (C122A)*, were grown overnight at 30°C in LB media with 50 μg/mL of kanamycin and 50 μg/mL of gentamycin. The strains harboring *AvrRpt2* and *AvrRpt2 (C122A)* were selected with 5 μg/mL of tetracycline. For the effector screen, a subculture was prepared the next day by inoculating fresh LB media, plus appropriate antibiotics, with overnight culture in a 1:10 (overnight culture:fresh media) ratio. The subculture was incubated for approximately 5 hours at 30°C with shaking after which it was centrifuged for 8 minutes at 5000 rpm. The bacterial pellet was resuspended in sterile deionized water for infiltration. For subsequent experiments a modified procedure, optimized for the efficient transformation of *N. glutinosa*, was used. For this procedure, the overnight culture was grown until saturated (~16 hrs) before the bacteria were pelleted and washed with 5 ml of 10 mM $MgCl_2$. The pellet was then resuspended in 3 ml of a solution containing 10 mM $MgCl_2$ and 100 μM acetosyringone (Sigma). The suspension was then incubated at room temperature for at least 2 hrs before being diluted to the appropriate density for injection using 10 mM $MgCl_2$ and 100 μM acetosyringone. Using this modified procedure for transforming *N. glutinosa* reduced the non-specific response to the Agrobacterium and gave more consistent results. Avoiding injecting Agrobacterium strains at an OD_{600} >0.3 also reduced the non-specific response to Agrobacterium sometimes observed in *N. glutinosa*. Important factors for obtaining efficient transformation included using young plants (~4 weeks old), avoiding the youngest and oldest leaves (typically the 3rd and 4th true leaves were injected), and using a transformation protocol that included acetosyringone in the infiltration medium.

For the effector screen and MUG assays, each Agrobacterium strain was infiltrated at an OD_{600} of 0.3. For the HR assays in *N. glutinosa*, combinations of up to 3 strains were co-infiltrated with each strain represented at an OD_{600} of 0.1. In these mixed inoculations the total Agrobacterium concentration remained at an OD_{600} of 0.3. An Agrobacterium strain with an empty vector plasmid was used as filler for combinations with fewer than three strains. A 1.0 mL needleless syringe was used to infiltrate the appropriate Agrobacterium strain(s). When necessary, a needle or razor blade was used to make a hole/nick at the intended injection site to facilitate subsequent injection with the needleless syringe.

Plasmids

The *P. syringae* effector genes *AvrPphB*, *AvrB* and *AvrRpm1* were cloned in the pTA7002 plasmid, which places the transgene under control of a dexamethasone (DEX) inducible promoter [28]. The empty vector, *RPS5*, *PBS1*, *Rpg1b*, *Rpg1r* and *GmRIN4* constructs also employed the pTA7002 vector. The *RPS5* and *PBS1* constructs contained C-terminal 5x-Myc and 3x-HA tags, respectively. The *GmRIN4* constructs contained an N-terminal 5x-Myc tag, while the *Rpg1r* construct contained a C-terminal sYFP tag. The *GUS* reporter gene was in the pCNL65 plasmid, which places the transgene under control of the cauliflower mosaic virus 35S promoter [29]. The effector genes *AvrRpt2* and *AvrRpt2 (C122A)* were also under CaMV 35S control in the pMD1 vector and each had a C-terminal 3x-HA tag [22]. Expression of DEX-inducible constructs was achieved by spraying transiently transformed plants with a solution of 50 μM dexamethasone (Sigma-Aldrich) and 0.02% Silwet-L77 (Momentive, Albany, NY) approximately 40 hours post-infiltration.

MUG fluorometric assay

The MUG fluorometric assay for β-glucuronidase (*GUS*) activity was adapted from [30]. Unless indicated otherwise, all reagents were obtained from Sigma-Aldrich. The third youngest leaf of each plant was transiently transformed with the *GUS* reporter gene. Six leaf discs (0.6 cm in diameter) were collected from each plant approximately 40 hr post-infiltration and ground in a 1.5 mL microfuge tube with 450 μL of extraction buffer (10 mM EDTA, 0.1% Triton X-100, 0.1% sodium lauryl sarcosine, and 10 mM β-mercaptoethanol, 50 mM phosphate buffer at pH 7). Five microliters of tissue extract were added to 500 μL of 1 mM MUG reaction buffer (4-methylumelliferone-β-D- glucuronide dissolved in extraction buffer). The reaction was incubated at 37°C and 40 μL aliquots were removed and added to 160 μL of stop buffer (0.2 M Na_2CO_3) in a black microtiter plate at zero time and subsequent time points. Forty microliters of each 4-methylumelliferone (MU) standard (20-100 μM) were also added to 160 μL of stop buffer in the microtiter plate. A Bradford assay kit (Biorad) was used to normalize each sample by calculating the protein concentration according to the manufacturer's instructions.

Fluorescence and absorbance measurements were made using a Thermo Scientific Appliskan microplate reader. A 340 nm excitation filter and 500 nm emission filter were used to measure the fluorescence from the MUG assay samples. A 595 nm filter was used to measure the absorbance of the samples from the Bradford assay.

Supporting Information

Figure S1 Examples of leaf morphologies observed in *Nicotiana* species expressing *AvrB* or *AvrRpm1*. The left image is a *N. tabacum* leaf exhibiting tissue browning from *AvrRpm1* (-) and *AvrB* (+) expression. In the center, the abaxial surface of a *N. glutinosa* leaf is exhibiting a "shiny" phenotype from both *AvrRpm1* and *AvrB* expression (a similar response was also observed in response to the *GUS* containing strain). The right image is an example of full leaf collapse in a *N. benthamiana* leaf expressing *AvrB*.

Figure S2   Box and whisker plot showing quantification of transformation efficiency as determined by a MUG fluorometric assay. The boxplot was generated from data compiled from 4 independent experiments with a total sample size of n = 28 for *N. benthamiana* and n = 29 for *N. glutinosa*. The whiskers represent minimum and maximum values of the data. The (•) symbol above the *N. benthamiana* boxplot indicates an outlying data point. Statistical significance was assessed using a two-tailed Student's *t*-test: * indicates $P = 0.001$.

Figure S3   Amino acid sequence alignment of *Arabidopsis* RIN4, the soybean RIN4s, and two putative RIN4

homologs from *N. benthamiana.* Each AvrRpt2 RIN4 cleavage site (RCS) of *Arabidopsis* RIN4 is indicated [21].

Acknowledgments

We thank the Nicotiana Collection at North Carolina State for providing seed of *Nicotiana* accessions.

Author Contributions

Conceived and designed the experiments: RK TA SHK RWI. Performed the experiments: RK TA SHK. Analyzed the data: RK TA SHK RWI. Contributed reagents/materials/analysis tools: RK TA SHK RWI. Wrote the paper: RK TA SHK RWI.

References

1. DeYoung BJ, Innes RW (2006) Plant NBS-LRR proteins in pathogen sensing and host defense. Nat Immunol 7: 1243-1249.
2. Dangl JL, Jones JD (2001) Plant pathogens and integrated defence responses to infection. Nature 411: 826-833.
3. Huynh TV, Dahlbeck D, Staskawicz BJ (1989) Bacterial blight of soybean: regulation of a pathogen gene determining host cultivar specificity. Science 245: 1374-1377.
4. Debener T, Lehnackers H, Arnold M, Dangl JL (1991) Identification and molecular mapping of a single *Arabidopsis thaliana* locus determining resistance to a phytopathogenic *Pseudomonas syringae* isolate. Plant Journal 1: 289-302.
5. Grant MR, Godiard L, Straube E, Ashfield T, Lewald J, et al. (1995) Structure of the Arabidopsis *RPM1* gene enabling dual specificity disease resistance. Science 269: 843-846.
6. Mackey D, Holt BF, 3rd, Wiig A, Dangl JL (2002) RIN4 interacts with *Pseudomonas syringae* type III effector molecules and is required for RPM1-mediated resistance in Arabidopsis. Cell 108: 743-754.
7. Liu J, Elmore JM, Lin ZJ, Coaker G (2011) A receptor-like cytoplasmic kinase phosphorylates the host target RIN4, leading to the activation of a plant innate immune receptor. Cell Host Microbe 9: 137-146.
8. Chung EH, da Cunha L, Wu AJ, Gao Z, Cherkis K, et al. (2011) Specific threonine phosphorylation of a host target by two unrelated type III effectors activates a host innate immune receptor in plants. Cell Host Microbe 9: 125-136.
9. Wright CA, Beattie GA (2004) *Pseudomonas syringae* pv. *tomato* cells encounter inhibitory levels of water stress during the hypersensitive response of *Arabidopsis thaliana*. Proc Natl Acad Sci U S A 101: 3269-3274.
10. Ashfield T, Keen NT, Buzzell RI, Innes RW (1995) Soybean resistance genes specific for different *Pseudomonas syringae* avirulence genes are allelic, or closely linked, at the *RPG1* locus. Genetics 141: 1597-1604.
11. Chen NW, Sevignac M, Thareau V, Magdelenat G, David P, et al. (2010) Specific resistances against *Pseudomonas syringae* effectors AvrB and AvrRpm1 have evolved differently in common bean (*Phaseolus vulgaris*), soybean (*Glycine max*), and *Arabidopsis thaliana*. New Phytol 187: 941-956.
12. Selote D, Kachroo A (2010) RPG1-B-derived resistance to AvrB-expressing *Pseudomonas syringae* requires RIN4-like proteins in soybean. Plant Physiol 153: 1199-1211.
13. Ade J, DeYoung BJ, Golstein C, Innes RW (2007) Indirect activation of a plant nucleotide binding site-leucine-rich repeat protein by a bacterial protease. Proc Natl Acad Sci U S A 104: 2531-2536.
14. de Vries JS, Andriotis VM, Wu AJ, Rathjen JP (2006) Tomato Pto encodes a functional N-myristoylation motif that is required for signal transduction in *Nicotiana benthamiana*. Plant J 45: 31-45.
15. Gao Z, Chung EH, Eitas TK, Dangl JL (2011) Plant intracellular innate immune receptor Resistance to *Pseudomonas syringae* pv. *maculicola* 1 (RPM1) is activated at, and functions on, the plasma membrane. Proc Natl Acad Sci U S A 108: 7619-7624.
16. Selote D, Robin GP, Kachroo A (2013) GmRIN4 protein family members function nonredundantly in soybean race-specific resistance against *Pseudomonas syringae*. New Phytol 197: 1225-1235.
17. Simonich MT, Innes RW (1995) A disease resistance gene in Arabidopsis with specificity for the *avrPph3* gene of *Pseudomonas syringae* pv. *phaseolicola*. Mol Plant Microbe Interact 8: 637-640.
18. Shao F, Golstein C, Ade J, Stoutemyer M, Dixon JE, et al. (2003) Cleavage of Arabidopsis PBS1 by a bacterial type III effector. Science 301: 1230-1233.
19. DeYoung BJ, Qi D, Kim SH, Burke TP, Innes RW (2012) Activation of a plant nucleotide binding rich repeat disease resistance protein by a modified self protein. Cell Microbiol 14: 1071-1084.
20. Ashfield T, Redditt T, Russell A, Kessens R, Rodibaugh N, et al. (2014) Evolutionary relationship of disease resistance genes in soybean and Arabidopsis specific for the *Pseudomonas syringae* effectors AvrB and AvrRpm1. Plant Physiol.
21. Axtell MJ, Chisholm ST, Dahlbeck D, Staskawicz BJ (2003) Genetic and molecular evidence that the *Pseudomonas syringae* type III effector protein AvrRpt2 is a cysteine protease. Mol Microbiol 49: 1537-1546.
22. Axtell MJ, Staskawicz BJ (2003) Initiation of RPS2-specified disease resistance in Arabidopsis is coupled to the AvrRpt2-directed elimination of RIN4. Cell 112: 369-377.
23. Chisholm ST, Dahlbeck D, Krishnamurthy N, Day B, Sjolander K, et al. (2005) Molecular characterization of proteolytic cleavage sites of the *Pseudomonas syringae* effector AvrRpt2. Proc Natl Acad Sci U S A 102: 2087-2092.
24. Lindeberg M, Cunnac S, Collmer A (2009) The evolution of *Pseudomonas syringae* host specificity and type III effector repertoires. Mol Plant Pathol 10: 767-775.
25. Baltrus DA, Nishimura MT, Romanchuk A, Chang JH, Mukhtar MS, et al. (2011) Dynamic evolution of pathogenicity revealed by sequencing and comparative genomics of 19 *Pseudomonas syringae* isolates. PLoS Pathog 7: e1002132.
26. Ashfield T, Ong LE, Nobuta K, Schneider CM, Innes RW (2004) Convergent evolution of disease resistance gene specificity in two flowering plant families. Plant Cell 16: 309-318.
27. Bombarely A, Rosli HG, Vrebalov J, Moffett P, Mueller LA, et al. (2012) A draft genome sequence of *Nicotiana benthamiana* to enhance molecular plant-microbe biology research. Mol Plant Microbe Interact 25: 1523-1530.
28. Aoyama T, Chua NH (1997) A glucocorticoid-mediated transcriptional induction system in transgenic plants. Plant J 11: 605-612.
29. Liu CN, Li XQ, Gelvin SB (1992) Multiple copies of *virG* enhance the transient transformation of celery, carrot and rice tissues by *Agrobacterium tumefaciens*. Plant Mol Biol 20: 1071-1087.
30. Jefferson RA, Bevan M, Kavanagh T (1987) The use of the *Escherichia coli* beta-glucuronidase as a gene fusion marker for studies of gene expression in higher plants. Biochem Soc Trans 15: 17-18.

Interaction of *Medicago truncatula* Lysin Motif Receptor-Like Kinases, NFP and LYK3, Produced in *Nicotiana benthamiana* Induces Defence-Like Responses

Anna Pietraszewska-Bogiel[1], Benoit Lefebvre[2,3], Maria A. Koini[1], Dörte Klaus-Heisen[2,3], Frank L. W. Takken[4], René Geurts[5], Julie V. Cullimore[2,3], Theodorus W.J. Gadella[1]*

1 Section of Molecular Cytology, Swammerdam Institute for Life Sciences, University of Amsterdam, Amsterdam, The Netherlands, 2 INRA, Laboratoire des Interactions Plantes-Microorganismes (LIPM), UMR441, F-31326 Castanet-Tolosan, France, 3 CNRS, Laboratoire des Interactions Plantes-Microorganismes (LIPM), UMR2594, F-31326 Castanet-Tolosan, France, 4 Section of Plant Pathology, Swammerdam Institute for Life Sciences, University of Amsterdam, Amsterdam, The Netherlands, 5 Department of Plant Science, Laboratory of Molecular Biology, Wageningen University, Wageningen, The Netherlands

Abstract

Receptor(-like) kinases with Lysin Motif (LysM) domains in their extracellular region play crucial roles during plant interactions with microorganisms; e.g. *Arabidopsis thaliana* CERK1 activates innate immunity upon perception of fungal chitin/chitooligosaccharides, whereas *Medicago truncatula* NFP and LYK3 mediate signalling upon perception of bacterial lipo-chitooligosaccharides, termed Nod factors, during the establishment of mutualism with nitrogen-fixing rhizobia. However, little is still known about the exact activation and signalling mechanisms of MtNFP and MtLYK3. We aimed at investigating putative molecular interactions of MtNFP and MtLYK3 produced in *Nicotiana benthamiana*. Surprisingly, heterologous co-production of these proteins resulted in an induction of defence-like responses, which included defence-related gene expression, accumulation of phenolic compounds, and cell death. Similar defence-like responses were observed upon production of AtCERK1 in *N. benthamiana* leaves. Production of either MtNFP or MtLYK3 alone or their co-production with other unrelated receptor(-like) kinases did not induce cell death in *N. benthamiana*, indicating that a functional interaction between these LysM receptor-like kinases is required for triggering this response. Importantly, structure-function studies revealed that the MtNFP intracellular region, specific features of the MtLYK3 intracellular region (including several putative phosphorylation sites), and MtLYK3 and AtCERK1 kinase activity were indispensable for cell death induction, thereby mimicking the structural requirements of nodulation or chitin-induced signalling. The observed similarity of *N. benthamiana* response to MtNFP and MtLYK3 co-production and AtCERK1 production suggests the existence of parallels between Nod factor-induced and chitin-induced signalling mediated by the respective LysM receptor(-like) kinases. Notably, the conserved structural requirements for MtNFP and MtLYK3 biological activity in *M. truncatula* (nodulation) and in *N. benthamiana* (cell death induction) indicates the relevance of the latter system for studies on these, and potentially other symbiotic LysM receptor-like kinases.

Editor: James Porter, University of North Dakota, United States of America

Funding: This work was supported by the European Community Marie Curie Research Training Network Programme through contract MRTN-CT-2006-035546 "NODPERCEPTION". Work in Toulouse was supported by the French National Research Agency (ANR) through contracts "NodBindsLysM", "SYMPASIGNAL" and "LCOinNONLEGUMES", and has been done as part of the Laboratoire d'Excellence (LABEX) entitled TULIP (ANR-10-LABX-41). FLWT acknowledges the Centre for BioSystems Genomics and Netherlands Genomics Initiative program. RG acknowledges the Dutch Science Organisation NWO (VIDI 864.06.007). The funders had no role in study design, data collection and analysis, decision to publish, or preparation of the manuscript.

Competing Interests: The authors have declared that no competing interests exist.

* E-mail: Th.W.J.Gadella@uva.nl

Introduction

Legumes can establish a mutualism with compatible rhizobia ultimately leading to nodulation, i.e. a formation of specialized symbiotic organs (nodules) in which atmospheric dinitrogen is converted into ammonia by the bacteria in exchange for plant carbohydrates. Nod factors (NFs) play a central role during most *Rhizobium*-legume (RL) symbioses [1]. They are secreted rhizobial signals whose perception by host legume roots is required for root nodule organogenesis, invasion of rhizobia toward a nodule primordium, and accommodation of bacteria inside nodule cells [2–4]. In two model legumes, NF-induced responses during the pre-infection step of RL interaction require *M. truncatula* (*Medicago*)

Nod Factor Perception (*MtNFP*), and *Lotus japonicus* (*Lotus*) *Nod Factor Receptor 1* and *5* (*LjNFR1* and *LjNFR5*) [5–11]. At a later step, *LjNFR1*, *LjNFR5*, *MtNFP*, and an additional *Medicago* gene, *LysM domain-containing Receptor-Like Kinase/Root Hair Curling* (*MtLYK3/MtHCL*), are required for rhizobial infection via so-called infection threads through which the bacteria penetrate nodule primordia [9,12–15]. Additionally, *MtNFP* and *MtLYK3* might co-function during nodule development and/or accommodation of rhizobia inside nodule cells [9,16–18]. Recently demonstrated binding of NF derivatives to LjNFR1 and LjNFR5 confirmed their role as NF receptors [19], whereas the exact mechanism of MtNFP and MtLYK3 activation by compatible NFs remains to be shown.

All four genes encode receptor-like kinases (RLKs) with an extracellular region (ExR) predicted to contain three LysM domains, a transmembrane helix, and a protein kinase domain (KD) within the intracellular region (InR) [5,20–22]. Remarkably, in contrast to MtLYK3 and LjNFR1, which both display kinase activity, MtNFP and LjNFR5 seem to function as pseudokinases that neither show nor rely on the intrinsic kinase activity to signal [9,20–22]. LjNFR5 is hypothesized to form a receptor complex with LjNFR1: a notion consistent with their demonstrated co-functioning during the determination of RL specificity [23]. Similarly, a receptor complex composed of MtNFP and a yet-unidentified LysM-RLK or MtLYK3 is predicted to initiate the pre-infection responses and the infection process, respectively [12,15]. Since mutagenesis studies in *Medicago* have not identified alterations in genes other than *MtNFP* that lead to complete lack of responsiveness to NFs, a function of this additional LysM-RLK in the pre-infection stage is most likely redundant. In addition, MtNFP has been implicated in *Medicago* interactions with pathogens (*Aphanomyces euteiches* and *Colletotrichum trifolii*), and with beneficial arbuscular mycorrhiza (AM) fungi [24–27]. However, it remains to be shown whether MtNFP functions in these processes alone or in co-operation with (an)other RLK(s).

LysM-RLKs in non-legume species also govern plant-microbe interactions. An *MtNFP/LjNFR5* homolog in *Parasponia andersonii*, *PaNFP*, is involved in interactions with *Sinorhizobium* sp. NGR234 and *Rhizophagus irregularis* (formerly *Glomus intraradices*), resulting in nitrogen-fixing and arbuscular mycorrhiza symbiosis, respectively [28]. *Arabidopsis thaliana* (*Arabidopsis*) *LysM-RLK1/CERK1* (*Chitin Elicitor Receptor Kinase 1*) and its ortholog from rice (*Oryza sativa*), OsCERK1, are essential for microbe-associated molecular pattern (MAMP)-triggered immunity. MAMPs are specific molecules conserved in various classes of microorganisms that activate receptor-mediated defence signalling [29–30]. CERK1-mediated innate immunity to fungal and bacterial pathogens is activated upon perception of chitin/chitooligosaccharides (COs), or pepti-doglycan (PGN), respectively [31–35]. In the latter case, both in rice and in *Arabidopsis* PGN binds not to OsCERK1/AtCERK1 but to extracellular LysM domain-containing proteins, termed LYPs or LYMs [35–36]. This in turn is postulated to induce a formation of AtCERK1/AtLYMs receptor complexes, and subsequent signal transduction via the kinase activity of AtCERK1. A similar mechanism operates during COs-induced signalling in rice, involving OsCERK1 and (a) LYP protein(s) [34,36], whereas in *Arabidopsis* COs bind directly to AtCERK1 [37–38]. Therefore, modes of CERK1 activation, even upon perception of the same MAMP, can differ between plant species.

We are interested in NF-induced signalling mediated by MtNFP and MtLYK3, focusing on their postulated interaction *in situ*. However, our attempts to visualize these proteins in *Medicago* root have been unsuccessful, presumably due to stringent regulation of their accumulation (even in the situation of an attempted overproduction). *Nicotiana benthamiana* (*Nicotiana*) has proved to be a useful model for heterologous production and structure-function studies on multiple proteins, providing invaluable insights that guided their subsequent analyses in the respective homologous systems [39]. Therefore, we employed an *Agrobacterium tumefaciens* (*Agrobacterium*)-mediated transient transformation of *Nicotiana* leaves [40], which allowed us to produce both proteins to levels suitable for fluorescence microscopy. Remarkably, we found that heterologous co-production of MtNFP and MtLYK3 resulted in the induction of defence-like responses that are typically observed upon treatment with pathogen-derived molecules. As the apparent (functional) interaction of these LysM-RLKs in *Nicotiana* activated defence-like responses, similar to these mediated by AtCERK1,

our results indicate the existence of parallels between NF-induced and COs-induced signalling.

Materials and Methods

Constructs for Plant Expression

The cDNA of *MtNFP*, *MtLYK3*, and *MtDMI2* (in the latter case the first intron was included in the cDNA), and the genomic sequence of *AtCERK1* were PCR amplified and cloned in a pMON999 vector containing a CaMV 35Sp:: (*sYFP2*, *mCherry*, *3xFLAG* or empty) 35S terminator cassette. The stop codon was removed from the coding sequences during cloning (except when generating untagged *MtNFP* and *MtLYK3* constructs) to allow a translational fusion. Sequences of the primers and linkers are given in Table S1. All point mutations were introduced as described [20]. All constructs were sequenced to verify the correct insert sequence. Constructs generated in pMON999 vector were subsequently recloned into a pBin+ (all *MtNFP* constructs) or pCambia1390 (all *MtLYK3*, *AtCERK1*, and *MtDMI2* constructs) vector using HindIII and SmaI sites. *MtNFP*[ΔInR], *MtNFP-YFP*$_N$, *MtLYK3-YFP*$_N$, *MtLRRII.1-YFP*$_C$, *AtBRI1-YFP*$_C$ (where *YFP*$_N$ or *YFP*$_C$ encode, respectively, the N- and C-terminal part of split YFP used in BiFluorescence Complementation assay) constructs are described [22,41].

Nicotiana Transformations

Agrobacterium tumefaciens GV3101::pMP90 and LBA4404 strains were transformed with the respective constructs via electroporation. The LBA4404 strain was used only in the experiments that compared the effect of different *Agrobacterium* strains on cell death (CD) induction upon MtNFP and MtLYK3 co-production. All results presented in Figures 1–5, Figure S1 and S3, and Tables 1–3 were obtained with *Agrobacterium* GV3101::pMP90 strain. *Agrobacterium*-mediated transformation of *Nicotiana* was performed essentially as described [42], except that *Agrobacterium* cultures were grown in LB medium supplemented with 25 µg/mL of rifampicin and 50 µg/mL of kanamycin. Resuspended cells were incubated at room temperature for at least 1 h before being infiltrated into fully expanded leaves of green house-grown plants using needleless syringes. *Agrobacterium* transformants carrying the respective construct were resuspended in the infiltration medium to desired OD_{600}: all *MtNFP* and *MtNFP*[ΔInR]-*sYFP2* constructs - $OD_{600} = 0.4$; all *MtLYK3* and *AtCERK1* constructs - $OD_{600} = 0.7$; *MtDMI2-sYFP2* - $OD_{600} = 1.0$. Then, they were mixed 1:1 with GV3101::pMP90 transformants carrying: pCambia1390 vector with an empty CaMV 35Sp::35S terminator cassette (for separate expression), a desired *MtNFP* construct or a desired *MtLYK3* construct before being infiltrated into *Nicotiana* leaves. All experiments included mock infiltration with GV3101::pMP90 transformants carrying pCambia1390 vector with an empty CaMV 35Sp::35S terminator cassette, and a positive control (co-expression of WT *MtNFP-FP* and WT *MtLYK3-FP* constructs). Cell death induction upon separate expression or co-expression of each (pair of) constructs was analyzed between 24 and 72 hai in at least three independent experiments, every time using three different plants. In case of no macroscopic symptoms, three leaves were stained with Evans blue to confirm the lack of CD.

To confirm efficient accumulation of MtLRRII.1-YFP$_C$ and AtBRI1-YFP$_C$ fusions in *Nicotiana*, *Agrobacterium* transformants carrying the respective constructs were co-infiltrated at high optical densities (final $OD_{600} = 0.5$) with *Agrobacterium* transformants carrying either *MtNFP-YFP*$_N$ or *MtLYK3-YFP*$_N$ constructs (final $OD_{600} = 0.5$). The observed complementation of YFP fluorescence reported on efficient accumulation, and even unspecific oligomerization of the respective encoded fusions.

Stereoscopic Analysis

Blue light-excitable autofluorescence and far-red chlorophyll autofluorescence in intact *Nicotiana* leaves were imaged using 430/40 excitation and 485/50 emission BP filters, or 480/40 BP excitation and 510 LP emission filters, respectively. Images were captured using CMOS USB DCC1645C camera (THORLabs, Newton NJ, USA) implemented on a Leica MZ FLIII stereoscope. Evans blue staining was performed as described [42]. Leaves were cleared by boiling in acidic lactophenol/ethanol solution (10 g phenol in 10 ml lactic acid, mixed 2:1 with 96% ethanol) until the complete removal of chlorophyll (approximately 3 min per leaf). Ethanol-inextractable autofluorescence was excited with 312 nm wavelength. Images were captured using a Cool Snap CF camera (Photometrix, Tucson AZ, USA).

qRT-PCR Analysis

RNA extraction and qRT-PCR were performed as described [17] except that cDNA was prepared from 500 ng of total RNA (see Table S1 for primer sequences). Two technical replicates from two biological replicates were analyzed and results were collated.

Medicago Transformations

Complementation of *Mtlyk3-1* mutant seedlings was performed as described [20] using *MtLYK3-3xFLAG*, *MtLYK3*[K464A]-*3xFLAG*, and *MtLYK3*[T480A]-*sYFP2* constructs driven by the CaMV 35S promoter. Results were scored as:+(>75% of plants nodulated), reduced (<50% of plants nodulated) or - (0 plants nodulated).

Results

Co-production of MtNFP and MtLYK3 in *Nicotiana* Leaves Induces Cell Death

MtNFP and *MtLYK3* cDNA sequences were fused at their 3′ ends to the sequence encoding a fluorescent protein (FP); either a super yellow fluorescent protein 2 (sYFP2) or mCherry [43–44], and were expressed from a CaMV 35S promoter in *Nicotiana* leaves, where they were delivered by *Agrobacterium*-mediated transformation. These and similar *MtNFP* and *MtLYK3* constructs were shown to complement *Mtnfp* and *Mtlyk3* mutants, respectively [17–18,22], and are therefore suitable for studying the encoded LysM-RLKs. Confocal laser-scanning microscopy analysis demonstrated co-localization of both MtNFP-sYFP2 and MtLYK3-sYFP2 fusions with a plasma membrane (PM) marker (the hypervariable region [HVR] of maize [*Zea mays*] ROP7 fused to the C-terminus of mCherry; [20,22]), hence indicating PM localization of MtNFP and MtLYK3 fusions in *Nicotiana* leaf epidermal cells. Surprisingly, co-infiltration of *Agrobacterium* transformants carrying *MtNFP-FP* or *MtLYK3-FP* constructs, leading to the co-production of the encoded fusions, resulted in collapse and subsequent desiccation of the infiltrated region within 48 hours after infiltration (hai) (Fig. 1A), regardless of the *Agrobacterium* strain used (i.e. GV3101::pMP90 or LBA4404; unpublished data). This cell death (CD) response was not dependent on the tag attached to either protein, since an identical response was observed upon co-production of FP-tagged, 3xFLAG-tagged or untagged MtNFP and MtLYK3 (Fig. 1A, Table 1). Importantly, production of the separate MtNFP or MtLYK3 (fusions) did not induce CD as confirmed with an exclusion dye (Evans blue) staining (Fig. 1A), which reflects compromised membrane permeability attributed with cell death.

To investigate whether a similar CD response could be triggered by heterologous production of other plant RLKs, we analysed the *Nicotiana* response to expression of *Medicago Doesn't Make Infection 2* (*DMI2*; [45]), *MtLRRII.1* [41]), and *Arabidopsis Brassinosteroid Insensitive 1* (*BRI1*; [46]), all driven by the CaMV 35S promoter. Notably, none of these RLKs, alone or in combination with either MtNFP or MtLYK3 fusions, induced CD (Fig. 1B, Table 1), despite being efficiently produced in *Nicotiana* leaves, as confirmed with fluorescence microscopy (see Materials & Methods). Thus, the *Nicotiana* CD response was not a general response to a heterologous production of RLKs but rather a specific response to MtNFP and MtLYK3 co-production.

Production of AtCERK1 also Induces Cell Death in *Nicotiana* Leaves

A rapid tissue collapse at the site of pathogen attack, termed the hypersensitive response (HR), is frequently observed in incompatible plant-pathogen interactions where it is thought to contribute to pathogen restriction and to generate a signal that activates systemic plant defence mechanisms [47–48]. The apparent phenotypic similarity of the CD response to MtNFP and MtLYK3 co-production with the HR elicited by various pathogen-derived components (MAMPs and so-called effectors) [49–50], prompted us to investigate whether co-production of the symbiotic LysM-RLKs might activate defence signalling similar to that mediated by LysM-RLKs functioning in innate immunity. AtCERK1 mediates signalling upon the perception of COs or PGN, although, to our knowledge, CD induction in response to these MAMPs has not been reported so far in any plant species. We therefore investigated the *Nicotiana* response to heterologous production of wild-type (WT) AtCERK1 or its kinase-inactive variant carrying a substitution in the catalytic lysine (Lys 349 in a kinase subdomain II). Both *AtCERK1* and *AtCERK1*[K349E] constructs were generated as fusions to the 5′ end of the *sYFP2* sequence, and their expression in *Nicotiana* leaves was driven by the CaMV 35S promoter.

Notably, heterologous production of AtCERK1-sYFP2 fusion resulted in tissue collapse and desiccation of (nearly) the entire infiltrated region in 20 out of 22 infiltrations 36 hai (Fig. 2A). This CD induction abolished our attempts of precisely characterizing the subcellular localization of AtCERK1 fusion in *Nicotiana* leaf epidermal cells, although we could detect sYFP2 fluorescence at the cell boundary (unpublished data). On the contrary, we observed clear co-localization of AtCERK1[K349E]-sYFP2 fusion with the PM marker using confocal laser-scanning microscopy analysis (Fig. S1). The PM localization of AtCERK1[K349E] fusion in *Nicotiana* leaf epidermal cells is in agreement with the reported subcellular localization of AtCERK1 fluorescent fusion in onion (*Allium cepa*) epidermal cells [31]. Importantly, production of the kinase-inactive variant of AtCERK1 did not result in CD induction, as confirmed with Evans blue staining (Fig. 2B), indicating that biological activity of AtCERK1 in *Nicotiana* leaves was dependent on its kinase activity.

Cell Death Induction Upon MtNFP and MtLYK3 Co-production, and AtCERK1 Production in *Nicotiana* Leaves Requires an Influx of Extracellular Ca^{2+}

An influx of extracellular Ca^{2+} causes an increase in the cytosolic $[Ca^{2+}]$ that is required for MAMP (including COs)-induced activation of a MAPK cascade, ROS production, and gene expression. Thus, Ca^{2+} influx is postulated to occur very early in the plant defence signalling pathway [51–52], possibly immediately upon the activation of the PM-localised MAMP

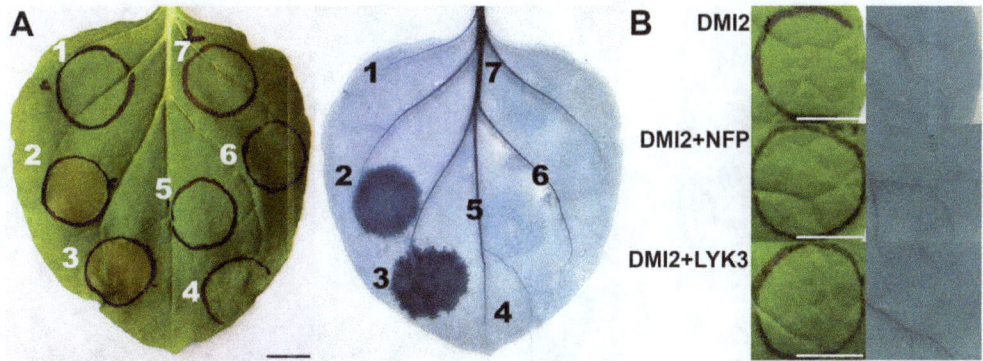

Figure 1. Co-production of MtNFP and MtLYK3 induces cell death in *Nicotiana* leaves. A, The following *MtNFP* and *MtLYK3* constructs were expressed alone or co-expressed in *Nicotiana* leaves: mock infiltration (1); *MtNFP* untagged+*MtLYK3* untagged (2); *MtNFP-sYFP2*+*MtLYK3-sYFP2* (3); *MtLYK3-sYFP2* (4); *MtLYK3* untagged (5); *MtNFP-sYFP2* (6); *MtNFP* untagged (7). Macroscopic observation (left panel) and subsequent Evans blue staining (right panel) are depicted 48 hai. Bar is 1 cm. B, *MtDMI2-sYFP2* construct was expressed alone or co-expressed with either *MtNFP-mCherry* or *MtLYK3-mCherry* construct in *Nicotiana* leaves. Macroscopic observations (left panel) and subsequent Evans blue stainings (right panel) are depicted 48 hai. Bars are 1 cm.

receptors [53] We wanted to know whether an influx of extracellular Ca^{2+} was similarly involved in CD induction upon MtNFP and MtLYK3 co-production or separate production of AtCERK1. To this end, MtNFP-3xFLAG and MtLYK3-3xFLAG fusions or AtCERK1-3xFLAG fusion were (co-)produced in adjacent regions in *Nicotiana* leaves. Twelve hours later, parts of the infiltrated leaf regions were syringe-infiltrated with 5 mM lanthanum chloride (an established inhibitor of the PM-localized calcium channels) or water, and the CD development was monitored between 24 and 72 hours after the first infiltration (with *Agrobacterium*). Notably, in 24 out of 30 leaf regions co-producing MtNFP and MtLYK3 fusions, compromised membrane permeability and tissue collapse were first (i.e. between 36 and 42 hai) localized only (or mostly) outside the lanthanum chloride-treated regions (Fig. 3A). Later on (i.e. 60 hai), 26 out of 30 parts of leaf regions treated with lanthanum chloride showed confluent death of the entire infiltrated region (unpublished data). Similar delay of the CD development was observed 33 hai in 11 out of 21 leaf regions producing AtCERK1 fusion and treated with lanthanum chloride (Fig. 3C). On the contrary, control treatment with water did not affect the development of confluent CD upon (co-)production of MtNFP and MtLYK3 fusions or AtCERK1 fusion (Fig. 3B, D).

Cell Death Upon MtNFP and MtLYK3 Co-production, and AtCERK1 Production in *Nicotiana* Leaves is Associated with an Induction of Defence-like Responses

Subsequently, we investigated whether co-production of MtNFP and MtLYK3 or production of AtCERK1 in *Nicotiana* leaves was associated with an accumulation of phenolic compounds and/or induction of defence-related gene expression, two established hallmarks of plant defence response, including that induced by COs and/or PGN [54–56]. We started by analysing the kinetics of CD development. To this end, *Agrobacterium* transformants carrying *MtNFP-3xFLAG*, *MtLYK3-3xFLAG* or *AtCERK1-3xFLAG* construct were (co-) infiltrated at different time-points in adjacent circles in *Nicotiana* leaves, and CD development was monitored between 24 and 48 hai. In case of co-production of MtNFP and MtLYK3 fusions, macroscopic symptoms of CD were first observed around

Table 1. Cell death induction upon (co-)expression of various RLK-encoding genes in *Nicotiana* leaves.

Construct	Cell death induction		
	Separate expression	Co-expression with *MtNFP-FP* #	Co-expression with *MtLYK3-FP* #
MtNFP–sYFP2	0/12	Not applicable	20/22
MtNFP –3xFLAG	0/9	Not applicable	12/13*
MtNFP	0/9	Not applicable	8/9**
MtLYK3–sYFP2	0/12	20/22	Not applicable
MtLYK3–3xFLAG	0/9	12/13*	Not applicable
MtLYK3	0/9	8/9**	Not applicable
MtDMI2-sYFP2	0/12	0/12	0/12
*MtLRRII.1-YFP*C	0/9	0/9***	0/9***
*AtBRI1- YFP*C	0/7	0/9***	0/9***

– unless stated differently: with *-3xFLAG* (*) untagged (**), or *–YFP*N (***) tagged construct.
Indicated constructs were expressed alone or co-expressed with either *MtNFP* or *MtLYK3* in *Nicotiana* leaves, and the infiltrated regions were marked. Macroscopic symptoms of cell death were scored 48 hai: only infiltrations that resulted in confluent death of (nearly) the entire infiltrated region were scored and are presented as a fraction of total infiltrations performed.

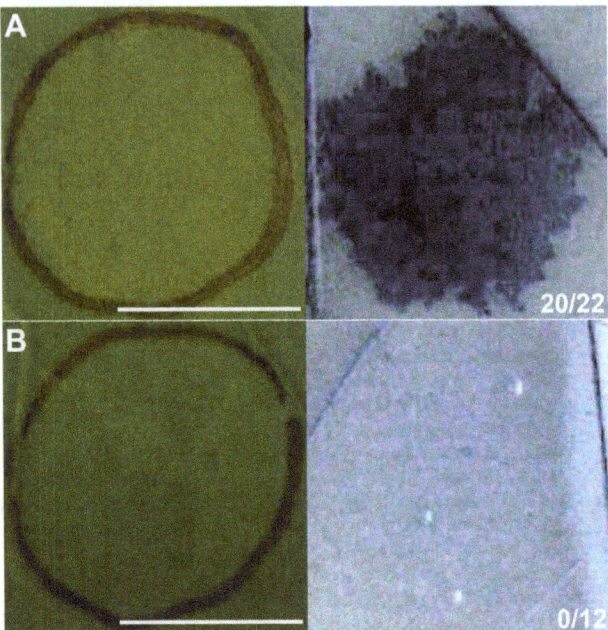

Figure 2. Production of AtCERK1 in *Nicotiana* leaves induces cell death that requires AtCERK1 kinase activity. *AtCERK1-sYFP2* (A) and *AtCERK1*[K349E]-*sYFP2* (B) constructs were expressed in *Nicotiana* leaves. Macroscopic observations (left panel) and subsequent Evans blue stainings (right panel) are depicted 36 hai. Macroscopic symptoms of cell death were scored 36 hai: only infiltrations that resulted in confluent death of (nearly) the entire infiltrated region were scored and are presented (right panel) as a fraction of total infiltrations performed. Bars are 1 cm.

36 hai (Fig. 4A) as a type of flaccidity and the appearance of small patches of collapsed tissue (these were more pronounced on the abaxial side of the leaf). Forty-eight hai, 30 out of 31 infiltrations showed pronounced tissue desiccation of the entire infiltrated region (Fig. 1A). Compromised membrane permeability preceded tissue collapse and often occurred over the entire infiltrated region approximately 33 hai (Fig. 4A). Compromised membrane permeability of leaf regions producing AtCERK1 fusion was observed already approximately 27–30 hai, and pronounced macroscopic symptoms of CD developed 36 hai (Fig. 2A, Fig. 3 C, D).

In addition, co-production of MtNFP-3xFLAG and MtLYK3-3xFLAG fusions resulted in accumulation of blue light-excitable autofluorescence (Fig. 4B) approximately 36 hai. This was not observed upon separate production of either fusion, or upon co-production of MtNFP-3xFLAG and kinase-inactive MtLYK3[G334E]-3xFLAG fusions (unpublished data). Accumulation of ethanol/lactophenol-inextractable and UV-excitable autofluorescence, indicative of phenolic compounds, was detected approximately 36 hai and 30 hai in leaf regions (co-)producing MtNFP-3xFLAG and MtLYK3-3xFLAG fusions or AtCERK1-3xFLAG fusion, respectively (Fig. 4C). Mock infiltration, separate production of MtNFP-3xFLAG or MtLYK3-3xFLAG fusion, production of kinase-inactive AtCERK1[K349E]-3xFLAG fusion or co-production of MtNFP-3xFLAG and kinase-inactive MtLYK3[G334E]-3xFLAG fusions did not result in the accumulation of similar autofluorescence (Fig. 4C).

Subsequently, we investigated induction of defence-related genes expression in *Nicotiana* leaves in response to: separate production and co-production of MtNFP-3xFLAG, MtLYK3-3xFLAG, MtLYK3[G334E]-3xFLAG, and AtCERK1-3xFLAG

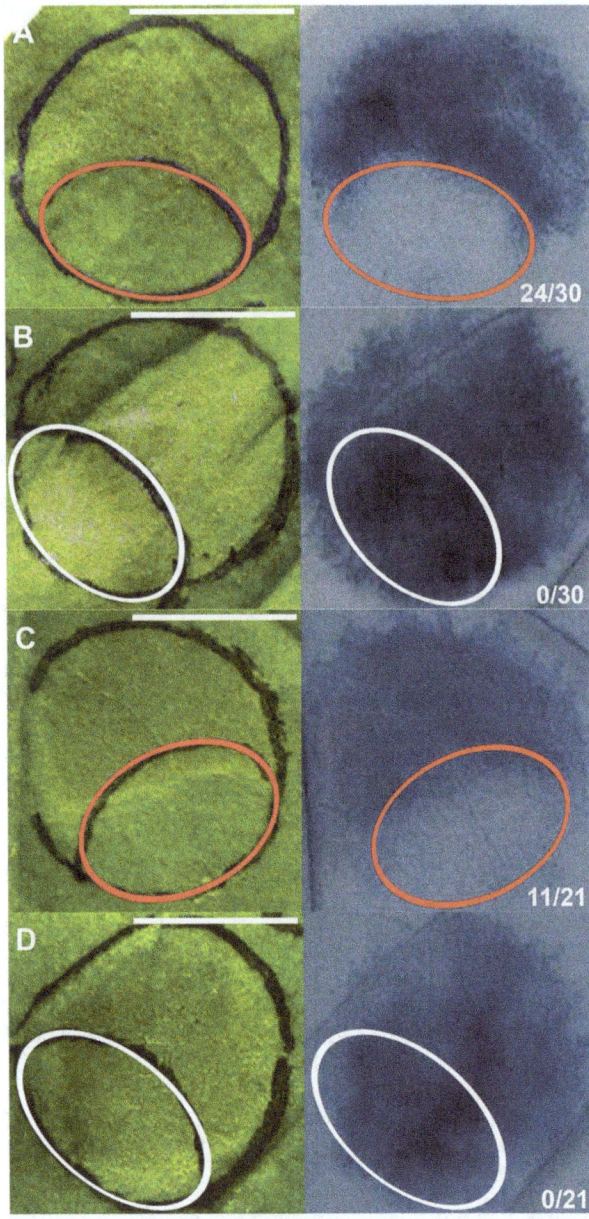

Figure 3. Lanthanum chloride delays the cell death development upon MtNFP and MtLYK3, or AtCERK1 (co-)production. *Agrobacterium* transformants carrying the following constructs were (co-)infiltrated at a final concentration: OD_{600} [*MtNFP-3xFLAG*] = 0.125 and OD_{600} [*MtLYK3-3xFLAG*] = 0.2 (A, B); OD_{600} [*AtCERK1-3xFLAG*] = 0.2 (C, D). Twelve hai parts of the infiltrated regions were syringe-infiltrated with 5 mM lanthanum chloride (circled in red) or water (circled in white). Macroscopic observations (left panel) and subsequent Evans blue stainings (right panel) are depicted 42 hai for leaf regions co-producing MtNFP and MtLYK3 fusions (A, B), and 33 hai for leaf regions producing AtCERK1 fusion (C, D). Cell death development was scored 42 hai (A, B) or 33 hai (C, D): only infiltrations that showed the lack of tissue collapse and no compromised membrane permeability in the lanthanum chloride- or water-treated region were scored and are presented (right panel) as a fraction of total infiltrations performed. Bars are 1 cm.

fusion(s). Induction of: *NbHIN1*– a postulated marker gene for HR [50]; two *PR1* genes, i.e. *NbPR1a acidic* and *NbPR1 basic* [57]; and *NbACRE31*, *NbACRE132*, and *NbCYP71D20*–postulated marker genes for MAMP-triggered immunity [49] was analyzed 24 hai

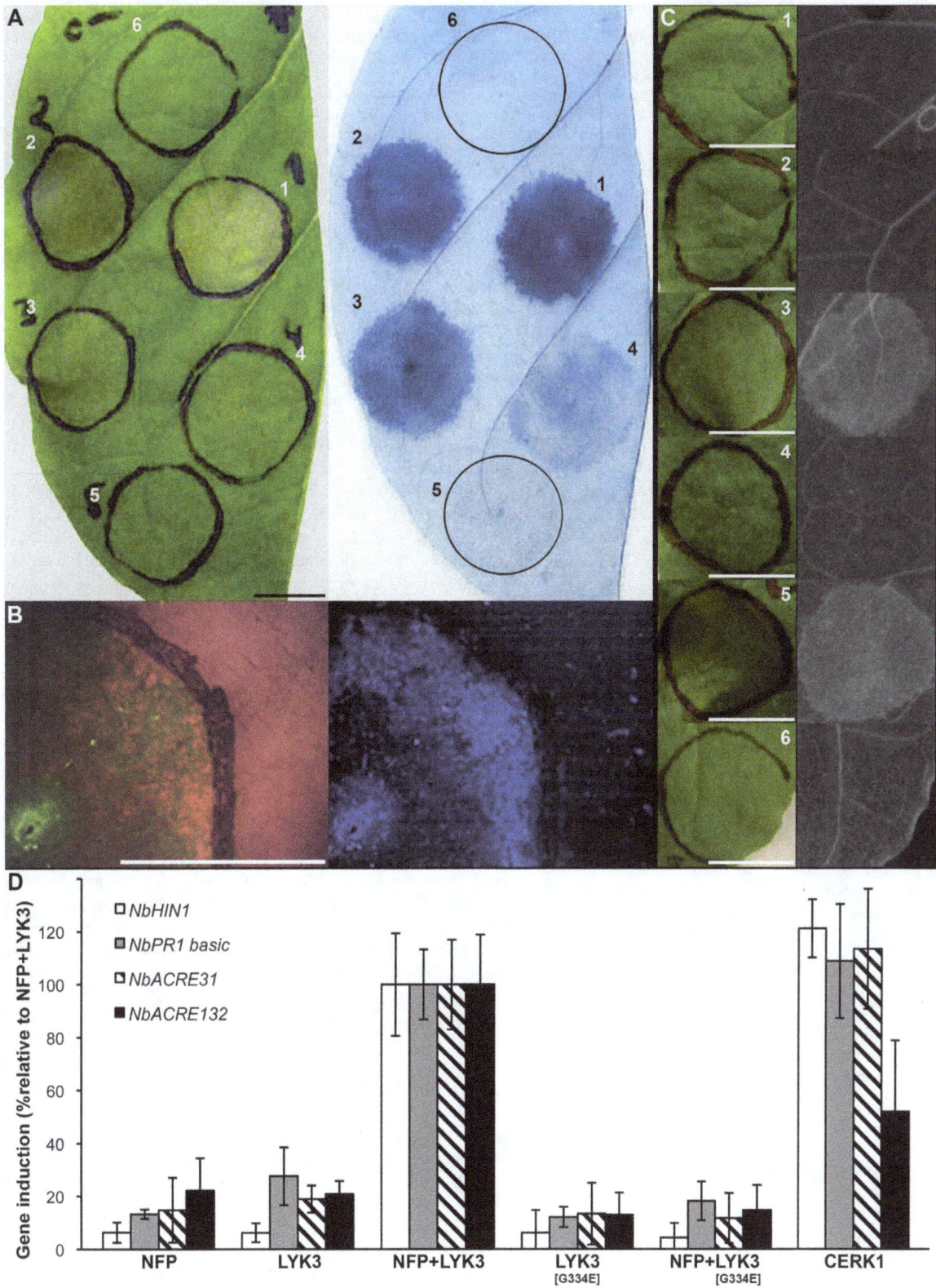

Figure 4. MtNFP and MtLYK3, or AtCERK1 (co-)production in *Nicotiana* leaves induces defence-like responses. A, Kinetics of cell death development in *Nicotiana*. *Agrobacterium* transformants carrying either *MtNFP-3xFLAG* or *MtLYK3-3xFLAG* construct were co-infiltrated into *Nicotiana* leaves at five different time points (1–5). Macroscopic observation (left panel) and subsequent Evans blue staining (right panel) are depicted 42 hai (region 1), 39 hai (region 2), 36 hai (region 3), 33 hai (region 4) and 30 hai (region 5). Mock infiltration (region 6) was done concomitantly with the infiltration of region 1. Bar is 1 cm. B, Changes in leaf autofluorescence upon MtNFP and MtLYK3 co-production. Leaf regions co-producing MtNFP-3xFLAG and MtLYK3-3xFLAG fusions were analyzed between 24 and 48 hai (here depicted 36 hai) using a stereoscope. Note the decrease in chlorophyll content, as indicated by the decrease of far-red autofluorescence of chlorophyll (left panel), and enhanced accumulation of blue light-

excited autofluorescence (right panel) within the infiltrated region. Bar is 1 cm. C, Accumulation of phenolic compounds. The following fusions were (co-)produced in *Nicotiana* leaves: MtNFP-3xFLAG (1); MtLYK3-3xFLAG (2); MtNFP-3xFLAG+MtLYK3-3xFLAG (3); MtNFP-3xFLAG+MtLYK3[G334E]-3xFLAG (4); AtCERK1-3xFLAG (5); or AtCERK1[K349]-3xFLAG (6). Macroscopic observations (left panel) and subsequent UV-excited autofluorescence of ethanol/lactophenol-cleared (right panel) leaf regions are depicted 36 hai (except for 5–30 hai). Bars are 1 cm. D, Induction of *NbHIN1*, *NbPR1 basic*, *NbACRE31*, and *NbACRE132* expression in response to separate production or co-production of: MtNFP-3xFLAG (NFP), MtLYK3-3xFLAG (LYK3), MtLYK3[G334E]-3xFLAG (LYK3[G334E]), and AtCERK1-3xFLAG (CERK1). Leaf samples were collected 24 hai and induction of gene expression was analyzed using qRT-PCR. Histograms represent induction of *NbHIN1* (white columns), *NbPR1 basic* (grey columns), *NbACRE31* (hatched columns), and *NbACRE132* (black columns) normalized by one reference gene, *MtEF1* α. Induction of each gene was normalized to that caused by mock infiltration, and then calculated as % induction relative to the induction observed upon co-production of MtNFP and MtLYK3 fusions. Bars represent standard deviation of the mean. At least two technical replicates from two biological replicates were analyzed.

using quantitative reverse-transcriptase polymerase chain reaction (qRT-PCR). Co-production of MtNFP and MtLYK3 fusions, and separate production of AtCERK1 fusion resulted in an induction of *NbHIN1*, *NbPR1 basic*, *NbACRE31*, and *NbACRE132* gene expression that was significantly higher than that following co-production of MtNFP and MtLYK3[G334E] fusions or separate production of MtNFP, MtLYK3, and MtLYK3[G334E] fusions (Fig. 4D). The *NbPR1a acidic* and *NbCYP71D20* genes did not display significant induction upon (co-) production of any of the protein(s) tested (unpublished data).

Taken together, the indication of a localized accumulation of phenolic compounds and induction of defence-related gene expression suggested that the *Nicotiana* response to MtNFP and MtLYK3 co-production triggered responses that were qualitatively similar to the responses to heterologous production of AtCERK1. In addition, the fact that in both cases the impairment of Ca^{2+} influx delayed the CD development suggested that the apparent functional interaction of two symbiotic LysM-RLKs in *Nicotiana* leaves mimics the action of the MAMP receptor, AtCERK1, and triggers defence-like responses.

Cell Death Induced in *Nicotiana* Leaves Upon MtNFP and MtLYK3 Co-production is a NF-independent Response

Perception of NFs results in triggering host symbiotic programme mediated by MtNFP and/or MtLYK3 [2]. In contrast, co-production of these LysM-RLKs in *Nicotiana* leaves apparently triggered some signalling cascade in the absence of NFs. Therefore, we investigated the effect of NF produced by *Sinorhizobium meliloti*, a microsymbiont of *Medicago*, on this CD response. To this end, *Agrobacterium* transformants carrying either *MtNFP-sYFP2* or *MtLYK3-mCherry* construct were co-infiltrated into *Nicotiana* leaves at varying concentrations (as measured with OD_{600}). Then, purified *Sm*NF at 10^{-7} M (in diluted DMSO) or diluted DMSO alone was applied between 9 and 24 hai to parts of the leaf regions co-producing MtNFP and MtLYK3 fusions, and CD development was monitored between 24 and 72 hours after the first infiltration (with *Agrobacterium*) using Evans blue staining. For all bacterial concentrations and time-points of *Sm*NF/DMSO application tested, compromised membrane permeability in leaf regions co-producing MtNFP and MtLYK3 fusions was observed at similar time irrespective of the *Sm*NF or DMSO treatment (Fig. 5), indicating similar kinetics of CD development. Therefore, we did not obtain evidence for any stimulatory or inhibitory effect of the *Sm*NF on the CD development upon MtNFP and MtLYK3 co-production.

The Intracellular Region of MtNFP and Kinase Activity of MtLYK3 are Required for Cell Death Induction in *Nicotiana* Leaves

The independence of CD induction upon MtNFP and MtLYK3 co-production from the *Sm*NF perception prompted us to compare structural requirements of CD induction and nodulation with

regard to these LysM-RLKs. In case of MtNFP, a recent structure-function study in *Medicago* [22] showed that loss-of-function mutations located in the ExR could be attributed to retention of the mutated protein in the endoplasmic reticulum (ER), whereas most substitutions located in the InR were found not to have an effect on the MtNFP function in nodulation. Therefore, we decided to limit our analysis of MtNFP to three point-mutated variants carrying: Ser 67 Phe (encoded by the *Mtnfp-2* allele), Ser 67 Ala, and Gly 474 Glu substitution; and a truncated variant with almost the entire InR deleted, termed MtNFP[ΔInR] (amino acids: 1–283) (see Table 2). Based on structure-function studies on MtLYK3 and LjNFR1, respectively in *Medicago* and in *Lotus* ([12,20–21], Table 3 in this study), we decided to test the effect of 16 point mutations (listed in Table 3) on MtLYK3 ability to induce CD in *Nicotiana* leaves in the presence of MtNFP. These included: a Pro 87 Ser (encoded by the *Mtlyk3-3* allele) and a Gly 334 Glu (encoded by the *Mtlyk3-1* allele) mutations, and Ala substitutions of Thr 285, Ser 286, Thr 300, Thr 319, Lys 349, Glu 362, Thr 433, Asp 441, Lys 464, Ser 471, Thr 472, Thr 475, Thr 480, and Thr 512. With the exception of the P87S substitution located in the first LysM domain of MtLYK3, all the above mutations are located in the MtLYK3 InR but differ in their effect on MtLYK3 autophosphorylation activity *in vitro* ([20] and Fig. S2; see Table 3). All truncated/mutated variants were prepared as fusions to the N-terminus of sYFP2, and their production and correct PM localization in *Nicotiana* leaf epidermal cells was confirmed, except for two MtNFP variants: MtNFP[S67F]-sYFP2 fusion was retained in the ER, and MtNFP[G474E]-sYFP2 fusion showed a partial PM localization, ([20,22] and Fig. S1). Additionally, we ruled out a possibility that the presence of WT MtNFP-FP or WT MtLYK3-FP fusion might affect stability/localization of the truncated/mutated fusions by confirming their efficient production and PM localization in *Nicotiana* leaf epidermal cells also upon co-production with MtLYK3 or MtNFP fusions (unpublished data).

Subsequently, we analyzed the ability of truncated/mutated MtNFP and MtLYK3 variants to induce CD upon either their separate production or co-production with WT MtLYK3-mCherry or WT MtNFP-mCherry fusion, respectively. In order to compare the CD induction ability of truncated/mutated variants with WT proteins, concomitant co-infiltration with *Agrobacterium* transformants carrying either WT *MtNFP-FP* or WT *MtLYK3-FP* construct was done on every leaf. Development of CD was monitored between 36 and 72 hai, and in case of the absence of or weakly pronounced macroscopic symptoms, the occurrence of CD was further scrutinized with Evans blue staining. None of the truncated/mutated variants was able to induce CD in *Nicotiana* leaves on its own (Table 2, 3). Co-production of MtNFP[S67A]-sYFP2 and MtLYK3 fusions resulted in a confluent death of (nearly) the entire infiltrated region in 6 out of 9 infiltrations, and compromised membrane permeability that could be observed in the entire infiltrated region (Table 2). In contrast, co-production of MtLYK3 fusion with MtNFP[S67F]-sYFP2,

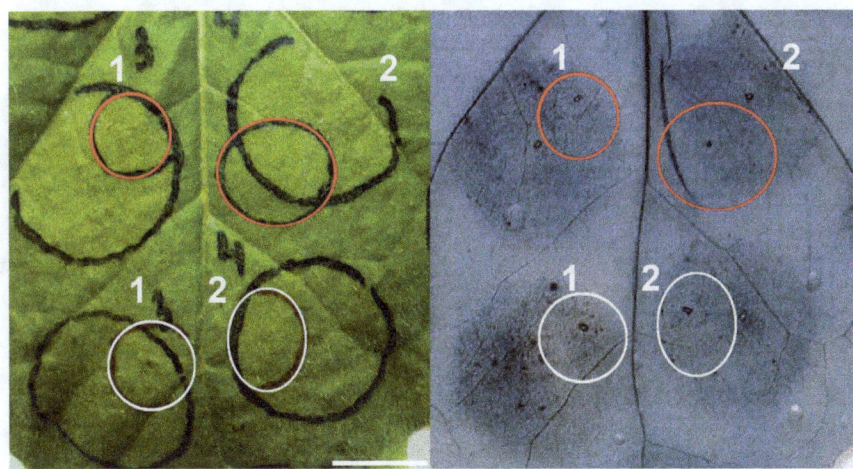

Figure 5. Cell death upon MtNFP and MtLYK3 co-production in *Nicotiana* leaves does not require *Sm*NF. *Agrobacterium* transformants carrying either *MtNFP-3xFLAG* or *MtLYK3-3xFLAG* construct were co-infiltrated into *Nicotiana* leaves at a final concentration: OD_{600} [*MtNFP*] = 0.25 and OD_{600} [*MtLYK3*] = 0.4 (1); OD_{600} [*MtNFP*] = 0.15 and OD_{600} [*MtLYK3*] = 0.25 (2). Twelve hai parts of the transformed regions were syringe-infiltrated with 10^{-7} mM *Sm*NF (circled in red) or DMSO diluted to the same concentration (circled in white). Macroscopic observation (left panel) and Evans blue staining (right panel) are depicted 33 hai. Bar is 1 cm.

MtNFP[G474E]-sYFP2, or MtNFP[ΔInR]-sYFP2 fusion did not induce CD in *Nicotiana* leaves (Table 2). In case of MtLYK3 mutated variants, co-production of MtLYK3[P87S]–sYFP2 and MtNFP fusions induced confluent CD in all infiltrated regions (Table 3). In contrast, co-production of MtNFP fusion with all seven MtLYK3-sYFP2 mutated variants affected for their autophosphorylation activity *in vitro* did not induce CD in *Nicotiana* leaves (Table 3). In case of mutations that do not affect autophosphorylation activity of MtLYK3 kinase, we found that MtLYK3[T285A S286A T300A]-sYFP2 and MtLYK3[S471A]-sYFP2 fusions were as active as WT MtLYK3-sYFP2 fusion for CD induction upon their co-production with MtNFP fusion, whereas MtLYK3[K464A]-sYFP2 fusion induced compromised membrane permeability (but no macroscopic symptoms of cell death) upon co-production with MtNFP fusion (Fig. S3, Table 3). Co-production of MtNFP fusion with MtLYK3[T433A]-sYFP2, MtLYK3[T472A]-sYFP2 or MtLYK3[T512A]-sYFP2 fusion resulted in a confluent death of (nearly) the entire infiltrated region in, respectively, 7 out of 20, 5 out of 11, and 12 out of 20

infiltrations, whereas the remaining leaf regions displayed only (a) small patch(es) of dead tissue (Fig. S3, Table 3).

Taken together, most of the structural requirements regarding the MtNFP and MtLYK3 InR, and the autophosphorylation activity of the MtLYK3 KD, appeared to be identical for biological activity of these LysM-RLKs in both *Medicago* and *Nicotiana*. More specifically, we found out that both nodulation and CD induction displayed the same requirements for 11 out of 15 residues located in the MtLYK3 InR. On the contrary, a single mutation in the MtLYK3 ExR tested (that does not affect the PM localization of the fusion) was found to be crucial for MtLYK3 function in nodulation but not in CD induction. In case of MtNFP, the substitution of Ser 67 similarly abolished (S67F) or did not have an effect (S67A) on MtNFP function in nodulation and CD induction, which seemed to correlate with, respectively, the absence or presence of MtNFP fusion at the PM.

Table 2. Cell death induction activity of MtNFP-sYFP2 truncated/mutated variants in *Nicotiana* leaves.

MtNFP-sYFP2 construct	Subcellular localization[*]	Nodulation activity[*]	Cell death induction	
			Co-expression with *MtLYK3-mCherry*	Separate expression
WT	PM	+	28/30	0/12
S67F (*Mtnfp-2*)	ER	−	0/9	0/9
S67A	PM	+	6/9#	0/9
ΔInR	PM	−	0/20	0/9
G474E	partial PM	−	0/15	0/9

[*]-see [22], PM-plasma membrane, ER-endoplasmic reticulum.
The designated constructs were expressed alone or co-expressed with *MtLYK3-mCherry* construct in *Nicotiana* leaves. Macroscopic symptoms of cell death were scored 48 hai: only infiltrations that resulted in confluent death of (nearly) the entire infiltrated region were scored and are presented as a fraction of total infiltrations performed. # - In the 3 remaining leaf regions, co-expression of *MtNFP*[S67A]-*sYFP2* and *MtLYK3-mCherry* constructs resulted in increased staining with Evans blue in the entire infiltrated region.

Table 3. Cell death induction activity of MtLYK3-sYFP2 mutated variants in *Nicotiana* leaves.

MtLYK3-sYFP2 construct	Auto-phosphorylation activity*	Nodulation activity**	Cell death induction	
			Co-expression with *MtNFP-mCherry*	Separate expression
WT	+	+	28/30	0/12
P87S (Mtlyk3-3)	Not applicable	-	15/15	0/9
T319A	-	-	0/11	0/9
G334E (Mtlyk3-1)	-	-	0/20	0/9
K349A	-	-	0/16	0/9
E362A	-	-	0/15	0/9
D441A	-	-	0/16	0/9
T475A	-	-	0/13	0/9
T480A	-	−(0/24)	0/18	0/9
T285A/S286A/T300A	+	Reduced with T300A	15/16	0/9
T433A	+	Reduced	Reduced 7/20	0/10
K464A	Reduced	Reduced (7/18)	0/12#	0/9
S471A	+	Reduced	9/11	0/9
T472A	+	Reduced	Reduced 5/11	0/9
T512A	Reduced	-	Reduced 12/20	0/9

*-see [20], except for the T480A (Fig. S2), **-see [20], except for the P87S [12], K464A and T480A (this study; number of plants nodulated/number of plants tested). The designated constructs were expressed alone or co-expressed with *MtNFP-mCherry* construct in *Nicotiana* leaves. Macroscopic symptoms of cell death were scored 48 hai: only infiltrations that resulted in confluent death of (nearly) the entire infiltrated region were scored and are presented as a fraction of total infiltrations performed. # - despite the lack of pronounced macroscopic symptoms, the co-expression of *MtLYK3*[K464A]-*sYFP2* and *MtNFP-mCherry* constructs resulted in increased staining with Evans blue in the entire infiltrated region.

Discussion

Co-production of MtNFP and MtLYK3 in *Nicotiana* Induces Defence-like Responses that Resemble *Nicotiana* Responses to AtCERK1 Production

Efficient production of both MtNFP-FP and MtLYK3-FP fusions *in Nicotiana* leaves facilitated characterization of their subcellular localization [20,22] and oligomerization status *in vivo* (manuscript in preparation), and led to the surprising observation of a CD induction (Fig. 1A). This response phenotypically and kinetically (Fig. 4A) resembled the HR elicited in *Nicotiana* spp. by pathogen-derived components [49], or CD induced by (co-) production of certain defence-related proteins, e.g. Pto [58]. Notably, a similar CD response was observed in *Nicotiana* leaves upon production of AtCERK1 (Fig. 2A), a MAMP receptor. Although CD induction was not demonstrated in any plant species in response to COs or PGN [54–56], it has been observed upon deregulation of various stress/defence-related signalling components [59–60], including a mitogen-activated protein kinase kinase (OsMKK4) implicated in COs-induced signalling in rice [61]. In addition, expression of *AtCERK1* from the CaMV 35S promoter (used also in our studies) in *Arabidopsis* was shown to result in a ligand-independent dimerization of AtCERK1 [38]. We hypothesize that overproduction of this LysM-RLK in *Nicotiana* analogously leads to its dimerization and enhanced (deregulated) kinase activity, which in turn is required and sufficient for triggering CD; a response not observed upon ligand-induced activation of AtCERK1 in *Arabidopsis*.

Cell death induction upon MtNFP and MtLYK3 co-production in *Nicotiana* is in agreement with the *Nicotiana* [21] and *Arabidopsis* [19] response to LjNFR1 and LjNFR5 co-production. However, in these studies the associated induction of putative defence-related responses has not been investigated. We here showed that both MtNFP and MtLYK3 co-production and AtCERK1 production in *Nicotiana* leaves triggered local accumulation of phenolic compounds, and a similar induction of expression of 4 out of 6 tested defence-related genes (Fig. 4C, D). We speculate that the two other genes might display different kinetics of induced expression (here analyzed only 24 hai) or might undergo suppression by *Agrobacterium* [62]. Importantly, COs- and/or PGN-induced expression of *PR1*, *ACRE31*, *ACRE132* and a member of a *HIN1* gene family was reported previously [32,52,54], linking these genes to MAMP-induced gene regulation mediated by AtCERK1 in *Arabidopsis* and/or *NbCERK1* in *Nicotiana*. In addition, we found that the lanthanum chloride-induced impairment of a Ca^{2+} influx similarly delayed the CD development upon (co-)production of the LysM-RLKs in our study (Fig. 3A, C). Therefore, we speculate that the signalling triggered upon MtNFP and MtLYK3 co-production in *Nicotiana* mimics AtCERK1-mediated signalling and thereby results in an induction of defence-like responses.

Similarities and Differences between Symbiotic and Defence Signalling Mediated by LysM-RLKs

The similarity between *Nicotiana* response to MtNFP and MtLYK3 co-production and AtCERK1 production suggests a possible overlap in signalling mediated by these LysM-RLKs. Several NF-induced processes, such as: a transient increase of reactive oxygen species (ROS) production; activation of phospholipase C (PLC) and PLD; and prolonged oscillations of (peri)nuclear $[Ca^{2+}]$ are implicated in switching on the symbiotic programme [63–69], whereas a Ca^{2+} influx is postulated to act as a signal for infection thread formation [70]. Interestingly, (CERK1-mediated) COs- and/or PGN-induced responses also involve a Ca^{2+} influx, an elevated ROS production, and PLC activation [33,35,37–38,52–53,71–73]. We speculate that these similar processes might be activated/regulated by related molecular components, hence allowing two *Medicago* LysM-RLKs to activate signalling components present in *Nicotiana* leaf. Remarkably, Nakagawa and associates [74] demonstrated that swapping of the AtCERK1 ExR and a certain part of the AtCERK1 InR for the corresponding regions from LjNFR1 conferred on AtCERK1 a competence, albeit inefficient, for symbiotic signalling during *Lotus-Mesorhizobium loti* interaction. Conversely, our results demonstrate that MtNFP and MtLYK3, when co-produced in *Nicotiana*, are capable of signalling in a similar manner to AtCERK1. We hypothesize that due to the absence of symbiosis-specific "decoders" or "modulators" in *Nicotiana*, MtNFP- and MtLYK3-mediated signalling might be differently interpreted in this species, resulting in the induction of defence-like responses.

Importantly, NF-induced host responses are partially contradictory. On one hand, NFs are postulated to suppress the production in legume roots of salicylic acid and ROS, two potent signals implicated in defence signalling, upon rhizobia perception [63,75]. This differential induction of the symbiotic competence or defence could be regulated quantitatively via NPR1 (*Non-expressor of Pathogenesis-related genes 1*)-mediated signalling [76]. On the other hand, even perception of compatible NFs leads to the induction of (some) defence-related gene expression and phosphorylation of defence-related proteins in the initial stage of symbiosis [7,11,68,74]. An even more striking example of a NF-induced defence-like response comes from *Sesbania rostrata*. NF produced by its microsymbiont, *Azorhizobium caulinodans*, triggers the production of hydrogen peroxide and ethylene that eventually lead to local CD required for the formation of cortical infection pockets [77]. Therefore, our results agree with the hypothesized stimulation of (initial) host's defence responses via LysM-RLKs that mediate NF-induced signalling.

Possible Co-functioning of MtNFP and MtLYK3 in *Nicotiana* and *Medicago*

To explain a possible signalling mechanism employed by the kinase-inactive MtNFP, it has been proposed to form a receptor complex with MtLYK3 and another LYK protein during, respectively, the infection thread growth and pre-infection stage of symbiosis [12,15]. In contrast, it is not known whether MtNFP functions alone during arbuscular mycorrhiza symbiosis [24–26] or resistance towards fungal and oomycete pathogens [27] or requires a similar co-functioning with (an)other RLK(s). We here demonstrated that MtNFP required MtLYK3 to induce CD in *Nicotiana*, and that neither of these LysM-RLKs could be substituted by an unrelated RLK (Fig. 1B, Table 1). We propose that this *Nicotiana* response reflects a functional interaction between these LysM-RLKs. A rather limited heteromerization of MtNFP and a kinase-inactive MtLYK3 variant observed in the PM of *Nicotiana* leaf epidermal cells (manuscript in preparation) does not exclude either hypothesized mechanism: a direct, phosphorylation-dependent physical interaction between MtNFP and WT MtLYK3; or indirect (functional) interaction between MtNFP and MtLYK3 that requires independent activation of different molecular components by either LysM-RLK, and a later convergence of such putative signalling pathways. Interestingly, MtLYK3, but not the homologous [78] AtCERK1, required the presence of MtNFP for CD induction in *Nicotiana* (Fig. 2A), indicating the specific requirement for MtNFP to potentiate the MtLYK3-mediated signalling. This observation agrees with the hypothesized specialization of the LysM-RLKs mediating NF-

induced signalling during the co-evolution of legumes with rhizobia [28].

Cell Death Induction in *Nicotiana* and Nodulation in *Medicago* Share Certain Structural Requirements Regarding MtNFP and MtLYK3

Curiously, CD induction in *Nicotiana* was independent from the NF perception (Fig. 5), and the presence of Pro 87 in the MtLYK3 ExR (Table 3), in contrast to MtNFP and MtLYK3 function in nodulation [12,15]. Further mapping of crucial amino acid residues, and detailed characterization of their exact role in signalling would be required to clarify in the future whether or not nodulation and CD induction indeed hold different structural requirements with regard to the ExRs of these LysM-RLKs. On the contrary, the biological activity of MtNFP in *Nicotiana* was dependent on its PM localization, the presence of its InR, and the Gly 474 (Table 2), thus mimicking the structural requirements of RL symbiosis regarding MtNFP [22], and proteins encoded by *MtNFP* orthologs in *Lotus* [79] and pea (*Pisum sativum*) [5]. The overlap between structural requirements of nodulation and CD induction was even more pronounced with respect to the MtLYK3 InR. Out of 16 residues whose role in nodulation was identified, 9 residues (Thr 319, Gly 334, Lys 349, Glu 362, Thr 433, Asp 441, Thr 472, Thr 475, and Thr 480) were found to be equally important, and 2 residues (Thr 285, Ser 286) were equally dispensable for MtLYK3 biological activity both in *Medicago* and in *Nicotiana* (Table 3). In addition, the K464A and T512A substitutions had a negative effect of MtLYK3 biological activity in both *Nicotiana* and *Medicago*, although this effect was more (K464A) or less (T512A) severely pronounced during CD induction assays than during nodulation (Table 3). Various mutations abolishing MtLYK3 autophosphorylation activity ([20] and Fig. S2) similarly abolished its biological activity in *Medicago* [20] and in *Nicotiana* (Table 3), supporting the hypothesis that autophosphorylation of MtLYK3 is crucial for its signalling function. Importantly, as the role of Thr 480 in nodulation has not been described so far, our results revealed its importance for MtLYK3 function *in vivo*. Notably, the shared structural requirements of nodulation and CD induction were also confirmed with regard to several (putative) phosphorylation sites that do not abolish MtLYK3 autophosphorylation activity *in vitro* (Table 3). Phosphorylation within the InR of a RLK is often required for activation and regulation of its catalytic activity, and for generation of docking sites for (downstream) signalling components [80–82]. The shared importance of three (Thr 433, Thr 472, and Thr 512) of five such phosphorylation sites for MtLYK3 biological activity in both plant species suggests that some of these phosphorylation-dependent functions required for MtLYK3-mediated signalling are conserved in *Nicotiana* leaf.

Demonstrated significant overlap between structural requirements of nodulation and CD induction regarding the MtNFP and MtLYK3 InRs supports our notion of the relevance of the *Nicotiana* system for studies on these, and potentially other (symbiotic) LysM-RLKs. This system presents certain practical advantages over the legume root system, in terms of rapidity and ease of expression of multiple constructs. In view of hypothesized similarities between NF (i.e. lipo-chitooligosaccharide)-induced and COs-induced signalling, analyzing known molecular components/processes involved in the CERK1-mediated signalling [35–38,83] might provide information on the yet-unidentified players implicated in the perception and/or transduction of the NF signal. This would be especially important as still very little is known about the identity of interactors of these symbiotic LysM-RLKs [17–18,21,84]. Possible candidate signalling molecule(s) functioning in co-operation with, or downstream from the LysM-RLKs, and identified in this heterologous system should then be evaluated in legume root in order to confirm their involvement in symbiosis.

Supporting Information

Figure S1 Subcellular localization of various protein fusions in *Nicotiana* leaf epidermal cells. The plasma membrane marker, mCherry-HVR, was co-produced with the designated fusions in *Nicotiana* leaf epidermal cells, and the fluorescence (viewed from abaxial side) was imaged 24 hai using confocal laser scanning microscopy. From left to right: green fluorescence of sYFP2; orange fluorescence of mCherry; superimposition of green, orange, and far-red (chlorophyll) fluorescence with the differential interference contrast (DIC) image. Bars are 20 μm. Note 1: in case of subcellular localization of MtNFP[G474E]-sYFP2 fusion, strong fluorescent puncta (indicated with an arrowhead) at the cell boundary of many cells (sometimes in association with nuclei), and pronounced ER localization (indicated with an arrow) of the fusion were still visible at 48 hai. Nevertheless, some cells showed a more uniform pattern of fluorescence at the cell boundary, and this observation, together with a partial insensitivity of this mutated variant to the PNGaseF treatment [22], indicated that at least some MtNFP[G474E] fusion had reached the PM. Note 2: as all kinase-inactive MtLYK3 variants were produced and correctly localized to the plasma membrane in *Nicotiana* leaf epidermal cells, their lack of biological activity can be attributed to the general abolishment of kinase activity rather than to an individual effect of a particular mutation.

Figure S2 Effect of the Thr 480 Ala substitution on MtLYK3 autophosphorylation activity *in vitro*. The puri-

fied intracellular regions of WT MtLYK3, MtLYK3[G334E], and MtLYK3[T480A], fused to the C terminus of GST, were analyzed for their autophosphorylation activity *in vitro* using radiolabeled ATP (γ-^{32}P ATP) and phosphorimaging (PI). The coomassie blue staining (CB) shows the protein loading.

Figure S3 Various (putative) phosphorylation sites are differentially required for MtLYK3 biological activity in *Nicotiana*. MtLYK3-sYFP2 mutated variants were co-produced with MtNFP-mCherry fusion in *Nicotiana* leaves: MtLYK3[T285A S286A T300A]+MtNFP (1); MtLYK3[T433A]+MtNFP (2); MtLYK3[T512A]+MtNFP (3); MtLYK3[T480A]+MtNFP (4). Macroscopic observation (left panel) and Evans blue staining (right panel) are depicted 48 hai. Bar is 1 cm.

Table S1 Primer and linker sequences.

Materials and Methods S1.

Acknowledgments

We thank Dr. G. Morieri for providing us with purified *Sm*Nod factor, Dr. A. Streng for providing us with *AtCERK1* sequence and for critical reading of the manuscript, and Dr. H. van den Burg for helpful discussions.

Author Contributions

Conceived and designed the experiments: AP-B BL RG JVC TWJG. Performed the experiments: AP-B BL MAK DK-H JVC. Analyzed the data: AP-B BL FLWT RG JVC TWJG. Contributed reagents/materials/analysis tools: AP-B BL MAK DK-H FLWT RG JVC TWJG. Wrote the paper: AP-B BL MAK FLWT RG JVC TWJG.

References

1. Masson-Boivin C, Giraud E, Perret X, Batut J (2009) Establishing nitrogen-fixing symbiosis with legumes: how many rhizobium recipies? Trends Microbiol 17: 458–466.
2. Downie JA (2010) The roles of extracellular proteins, polysaccharides and signals in the interactions of rhizobia with legume roots. FEMS Microbiol Rev 34: 150–170.
3. Gust AA, Willmann R, Desaki Y, Grabherr HM, Nürnberger T (2012) Plant LysM proteins: modules mediating symbiosis and immunity. Trends Plant Sci 17: 495–502.
4. Oldroyd GED (2013) Speak friend, and enter: signalling systems that promote beneficial symbiotic associations in plants. Nature 11: 252–263.
5. Madsen EB, Madsen LH, Radutoiu S, Olbryt M, Rakwalska M, et al. (2003) A receptor kinase gene of the LysM type is involved in legume perception of rhizobial signals. Nature 425: 637–640.
6. Radutoiu S, Madsen LH, Madsen EB, Felle HH, Umehara Y, et al. (2003) Plant recognition of symbiotic bacteria requires two LysM receptor-like kinases. Nature 425: 585–592.
7. El Yahyaoui FE, Küster H, Ben Amor B, Hohnjec N, Pühler A, et al. (2004) Expression profiling in *Medicago truncatula* identifies more than 750 genes differentially expressed during nodulation, including many potential regulators of the symbiotic program. Plant Physiol 136: 3159–3176.
8. Mitra RM, Shaw SL, Long SR (2004) Six non-nodulating plant mutants defective for Nod factor-induced transcriptional changes associated with the legume-rhizobia symbiosis. Proc Natl Acad Sci U S A 101: 10217–10222.
9. Arrighi JF, Barre A, Ben Amor B, Bersoult A, Soriano LC, et al. (2006) The *Medicago truncatula* Lysin [corrected] motif-Receptor-like kinase gene family includes NFP and new nodule-expressed genes. Plant Physiol 142: 265–279.
10. Miwa H, Sun J, Oldroyd GED, Downie JA (2006) Analysis of Nod-factor-induced calcium signalling in root hairs of symbiotically defective mutants of Lotus japonicus. Mol Plant Microbe Interact 19: 914–923.
11. Høgslund N, Radutoiu S, Krusell L, Voroshilova V, Hannah MA, et al. (2009) Dissection of symbiosis and organ development by integrated transcriptome analysis of *Lotus japonicus* mutant and wild-type plants. PLoS ONE 4: e6556.
12. Smit P, Limpens E, Geurts R, Fedorove E, Dolgikh E, et al. (2007) *Medicago* LYK3, an entry receptor in rhizobial nodulation factor signalling. Plant Physiol 145: 183–191.
13. Hayashi T, Banba M, Shimoda Y, Kouchi H, Hayashi M, et al. (2010) A dominant function of CCaMK in intracellular accommodation of bacterial and fungal endosymbionts. Plant J 63: 141–154.
14. Madsen LH, Tirichine L, Jurkiewicz A, Sullivan JT, Heckmann AB, et al. (2010) The molecular network governing nodule organogenesis and infection in the model legume Lotus japonicus. Nature doi: 10.1038/ncomms1009.
15. Bensmihen S, de Billy F, Gough C (2011) Contribution of NFP LysM domains to the recognition of Nod Factors during the *Medicago truncatula/Sinorhizobium meliloti* symbiosis. PLoS ONE 6: e26114.
16. Limpens E, Mirabella R, Fedorova E, Franken C, Franssen H, et al. (2005) Formation of organelle-like N$_2$-fixing symbiosomes in legume root nodules is controlled by DMI2. Proc Natl Acad Sci U S A 102: 10375–10380.
17. Mbengue M, Camut S, de Carvalho-Niebel F, Deslandes L, Froidure S, et al. (2010) The *Medicago truncatula* E3 ubiquitin ligase PUB1 interacts with the LYK3 symbiotic receptor and negatively regulates infection and nodulation. Plant Cell 22: 3474–3488.
18. Haney CH, Riely BK, Tricoli DM, Cook DR, Ehrhardt DW, et al. (2011) Symbiotic rhizobia bacteria trigger a change in localization and dynamics of the *Medicago truncatula* receptor kinase LYK3. Plant Cell 23: 2774–2787.
19. Broghammer A, Krusell L, Blaise M, Sauer J, Sullivan JT, et al. (2012) Legume receptors perceive the rhizobial lipochitin oligosaccharide signal molecules by direct binding. Proc Natl Acad Sci U S A 109: 13859–13864.
20. Klaus-Heisen D, Nurisso A, Pietraszewska-Bogiel A, Mbengue M, Camut S, et al. (2011) Structure-function similarities between a plant receptor-like kinase and the human Interleukin-1 Receptor-Associated Kinase-4. J Biol Chem 286: 11202–11210.
21. Madsen EB, Antolín-Llovera M, Grossmann C, Ye J, Vieweg S, et al. (2011) Autophosphorylation is essential for *in vivo* function of the *Lotus japonicus* Nod Factor Receptor 1 and receptor mediated signalling in cooperation with Nod Factor Receptor 5. Plant J 65: 404–417.
22. Lefebvre B, Klaus-Heisen D, Pietraszewska-Bogiel A, Hervé C, Camut S, et al. (2012) Role of N-glycosylation sites and CXC motifs in trafficking of *Medicago truncatula* Nod Factor Perception protein to the plasma membrane. J Biol Chem 287: 10812–10823.
23. Radutoiu S, Madsen LH, Madsen EB, Jurkiewicz A, Fukai E, et al. (2007) LysM domains mediate lipochitin-oligosaccharide recognition and *Nfr* genes extend the symbiotic host range. EMBO J 26: 3923–3935.

24. Oláh B, Brière C, Bécard G, Dénerié J, Gough C (2005) Nod factors and a diffusible factor from arbuscular mycorrhizal fungi stimulate lateral root formation in *Medicago truncatula* via the DMI1/DMI2 signalling pathway. Plant J 44: 195–207.

25. Maillet F, Poinsot V, André O, Peuch-Pagès V, Haouy A, et al. (2011) Fungal lipochitooligosaccharide symbiotic signals in arbuscular mycorrhiza. Nature 469: 58–64.

26. Czaja LF, Hogekamp C, Lamm P, Maillet F, Andres Martinez E, et al. (2012) Transcriptional responses toward diffusible signals from symbiotic microbes reveal *MtNFP*- and *MtDMI3*-dependent reprogramming of host gene expression by arbuscular mycorrhizal fungal lipochitooligosaccharides. Plant Physiol 159: 1671–1685.

27. Rey T, Nars A, Bonhomme M, Bottin A, Huguet S, et al. (2013) NFP, a LysM protein controlling Nod Factor perception, also intervenes in *Medicago truncatula* resistance to pathogens. New Phytol 198: 875–886.

28. Streng A, Op den Camp R, Bisseling T, Geurts R (2011) Evolutionary origin of *Rhizobium* Nod factor signalling. Plant Signal Behav 6: 1510–1514.

29. Boller T, Felix G (2009) A renaissance of elicitors: perception of microbe-associated molecular patterns and danger signals by pattern-recognition receptors. Annu Rev Plant Biol 60: 379–406.

30. Thomma BPHJ, Nürnberger T, Joosten MHAJ (2011) Of PAMPs and effectors: the blurred PTI-ETI dichotomy. Plant Cell 23: 4–15.

31. Miya A, Albert P, Shinya T, Desaki Y, Ichimura K, et al. (2007) CERK1, a LysM receptor kinase, is essential for chitin elicitor signalling in Arabidopsis. Proc Natl Acad Sci U S A 104: 19613–19618.

32. Wan J, Zhang XC, Neece D, Ramonell KM, Clough S, et al. (2008) A LysM receptor-like kinase plays a critical role in chitin signalling and fungal resistance in Arabidopsis. Plant Cell 20: 471–481.

33. Gimenez-Ibanez S, Hann DR, Ntoukakis V, Petutschnig E, Lipka V, et al. (2009) AvrPtoB targets the LysM receptor kinase CERK1 to promote bacterial virulence on plants. Curr Biol 19: 423–429.

34. Shimizu T, Nakano T, Takamizawa D, Desaki Y, Ishii-Minami N, et al. (2010) Two LysM receptor molecules, CEBiP and OsCERK1, cooperatively regulate chitin elicitor signalling in rice. Plant J 64: 204–214.

35. Willmann R, Lajunen HM, Erbs G, Newman MA, Kolb D, et al. (2011) Arabidopsis lysin-motif proteins LYM1 LYM3 and CERK1 mediate bacterial peptidoglycan sensing and immunity to bacterial infection. Proc Natl Acad Sci U S A 108:19824–19829.

36. Liu B, Li JF, Ao Y, Qu J, Li Z, et al. (2012) Lysin motif-containing proteins LYP4 and LYP6 play dual roles in peptidoglycan and chitin perception in rice innate immunity. Plant Cell 24: 3406–3419.

37. Petutschnig EK, Jones AME, Serazetdinova L, Lipka U, Lipka V (2010) The Lysin Motif receptor-like kinase (LysM-RLK) CERK1 is a major chitin-binding protein in *Arabidopsis thaliana* and subject to chitin-induced phosphorylation. J Biol Chem 285: 28902–28911.

38. Liu T, Liu Z, Song C, Hu Y, Han Z, et al. (2012) Chitin-induced dimerization activates a plant immune receptor. Science 336: 221–234.

39. Oh CS, Martin GB (2010) Effector-triggered immunity mediated by the Pto kinase. Trends Plant Sci 16: 1360–1385.

40. Nguyen HP, Chakravarthy S, Velásquez AC, McLane HL, Zeng L, et al. (2010) Methods to study PAMP-triggered immunity using tomato and *Nicotiana benthamiana*. Mol Plant Microbe Interact 23: 991–999.

41. Lefebvre B, Timmers T, Mbengue M, Moreau S, Hervé C, et al. (2010) A remorin protein interacts with symbiotic receptors and regulates bacterial infection. Proc Natl Acad Sci U S A 107: 2343–2348.

42. van Ooijen G, Mayr G, Kasiem MM, Albrecht M, Cornelissen BJ, et al. (2008) Structure-function analysis of the NB-ARC domain of plant disease resistance proteins. J Exp Bot 59: 1383–1397.

43. Shaner NC, Campbell RE, Steinbach PA, Giepmans BN, Palmer AE, et al. (2004) Improved monomeric red, orange and yellow fluorescent proteins derived from *Discosoma* sp. red fluorescent protein. Nature 12: 1567–1572.

44. Kremers GJ, Goedhart J, van Munster EB, Gadella TW Jr (2006) Cyan and yellow super fluorescent proteins with improved brightness, protein folding, and FRET Förster radius. Biochem 45: 6570–6580.

45. Endre G, Kereszt A, Kevei Z, Mihacea S, Kaló P, et al. (2002) A receptor kinase gene regulating symbiotic nodule development. Nature 417: 962–966.

46. Li J, Chory J (1997) A putative leucine-rich repeat kinase involved in brassinosteroid signal transduction. Cell 90: 929–938.

47. Mur LA, Kenton P, Lloyd AJ, Ougham H, Prats E (2008) The hypersensitive response; the centenary is upon us but how much do we know? J Exp Bot 59: 501–520.

48. Coll NS, Epple P, Dangl JL (2011) Programmed cell death in the plant immune system. Cell Death Differ 18: 1247–1256.

49. Heese A, Hann DR, Gimenez-Ibanez S, Jones AM, He K, et al. (2007) The receptor-like kinase SERK3/BAK1 is a central regulator of innate immunity in plants. Proc Natl Acad Sci U S A 104: 12217–12222.

50. Taguchi F, Suzuki T, Takeuchi K, Inagaki Y, Toyoda K, et al. (2009) Glycosylation of flagellin from *Pseudomonas syringae* pv. *tabaci* 6605 contributes to evasion of host tobacco plant surveillance system. Physiol Mol Plant Pathol 74: 11–17.

51. Ranf S, Eschen-Lippold L, Pecher P, Lee J, Scheel D (2011) Interplay between calcium signalling and early signalling elements during defence responses to microbe- or damage-associated molecular patterns. Plant J 68: 100–113.

52. Segonzac C, Feike D, Gimenez-Ibanez S, Hann DR, Zipfel C, et al. (2011) Hierarchy and roles of pathogen-associated molecular pattern-induced responses in *Nicotiana benthamiana*. Plant Physiol 156: 687–699.

53. Frei dit Frey N, Mbengue M, Kwaaitaal M, Nitsch L, Altenbach D, et al. (2012) Plasma membrane calcium ATPases are important components of receptor-mediated signalling in plant immune responses and development. Plant Physiol 159: 798–809.

54. Gust AA, Biswas R, Lenz HD, Rauhut T, Ranf S, et al. (2007) Bacteria-derived peptidoglycans constitute pathogen-associated molecular patterns triggering innate immunity in Arabidopsis. J Biol Chem 282: 32338–32348.

55. Erbs G, Silipo A, Aslam S, De Castro C, Liparoti V, et al. (2008) Peptidoglycan and muropeptides from pathogens *Agrobacterium* and *Xanthomonas* elicit plant innate immunity: structure and activity. Cell 15: 438–448.

56. Hamel LP, Beaudoin N (2010) Chitooligosaccharide sensing and downstream signalling: contrasted outcomes in pathogenic and beneficial plant-microbe interactions. Planta 232: 787–806.

57. Cornelissen B, Horowitz J, van Kan JA, Goldberg RB, Bol JF (1987) Structure of tobacco genes encoding pathogenesis-related proteins from the PR-1 group. Nucl Acid Res 15: 6799–6811.

58. Mucyn TS, Clemente A, Andriotis VME, Balmuth AL, Oldroyd GED, et al. (2006) The tomato NBARC-LRR protein Prf interacts with Pto kinase *in vivo* to regulate specific plant immunity. Plant Cell 18: 2792–2806.

59. Takahashi Y, Bin Nasir KH, Ito A, Kanzaki H, Matsumura H, et al. (2007) A high-throughput screen of cell-death-inducing factors in *Nicotiana benthamiana* identifies a novel MAPKK that mediates INF1-induced cell death signalling and non-host resistance to *Pseudomonas cichorii*. Plant J 49: 1030–1040.

60. Gao M, Wang X, Wang D, Xu F, Ding X, et al. (2009) Regulation of cell death and innate immunity by two receptor-like kinases in Arabidopsis. Cell 6: 34–44.

61. Kishi-Kaboshi M, Okada K, Kurimoto L, Murakami S, Umezawa T, et al. (2010) A rice fungal MAMP-responsive cascade regulates metabolic flow to antimicrobial metabolite synthesis. Plant J 63: 599–612.

62. Ditt RF, Kerr KF, de Figueiredo P, Delrow J, Comai L, et al. (2006) The *Arabidopsis thaliana* transcriptome in response to *Agrobacterium tumefaciens*. Mol Plant Microbe Interact 19: 665–681.

63. Bueno P, Soto MJ, Rodríguez-Rosales MP, Sanjuan J, Olivares J, et al. (2001) Time-course of lipoxygenase, antioxidant enzyme activities and H_2O_2 accumulation during the early stages of *Rhizobium*-legume symbiosis. New Phytol 152: 91–96.

64. den Hartog M, Musgrave A, Munnik T (2001) Nod factor-induced phosphatidic acid and diacylglycerol pyrophosphate formation: a role for phospholipase C and D in root hair deformation. Plant J 25: 55–65.

65. Charron D, Pingret JL, Chabaud M, Journet EP, Barker DG (2004) Pharmacological evidence that multiple phospholipid signalling pathways link rhizobium nodulation factor perception in *Medicago truncatula* root hairs to intracellular responses, including Ca^{2+} spiking and specific *ENOD* gene expression. Plant Physiol 136: 3582–3593.

66. Lohar DP, Sharopova N, Endre G, Peñuela S, Samac D, et al. (2006) Transcript analysis of early nodulation events in *Medicago truncatula*. Plant Physiol 140: 221–234.

67. Cárdenas L, Martinez A, Sánchez F, Quinto C (2008) Fast, transient and specific intracellular ROS changes in living root hair cells responding to Nod factors (NFs). Plant J 56: 802–813.

68. Serna-Sanz A, Parniske M, Peck SC (2011) Phosphoproteome analysis of *Lotus japonicus* roots reveals shared and distinct components of symbiosis and defense. Mol Plant Microbe Interact 24: 932–937.

69. Sieberer BJ, Chabaud M, Timmers AC, Monin A, Fournier J, et al. (2009) A nuclear-targeted cameleon demonstrates intranuclear Ca^{2+} spiking in *Medicago truncatula* root hairs in response to rhizobial Nodulation factors. Plant Physiol 151: 1197–1206.

70. Shaw SL, Long SR (2003) Nod Factor elicits two separable calcium responses in *Medicago truncatula* root hair cells. Plant Physiol 131: 976–984.

71. den Hartog M, Verhoef N, Munnik T (2003) Nod factor and elicitors activate different phospholipid signalling pathways in suspension-cultured alfalfa cells. Plant Physiol 132: 311–317.

72. Kwaaitaal M, Huisman R, Maintz J, Reinstädler A, Panstruga R (2011) Ionotropic glutamate receptor (iGluR)-like channels mediate MAMP-induced calcium influx in *Arabidopsis thaliana*. Plant J 440: 355–365.

73. Wan J, Tanaka K, Zhang XC, Son GH, Brechenmacher L, et al. (2012) LYK4, a lysine motif receptor-like kinase, is important for chitin signalling and plant innate immunity in Arabidopsis. Plant Physiol 160: 396–406.

74. Nakagawa T, Kaku H, Shimoda Y, Sugiyama A, Shimamura M, et al. (2010) From defense to symbiosis: limited alterations in the kinase domain of LysM receptor-like kinases are crucial for evolution of legume-*Rhizobium* symbiosis. Plant J 65: 169–180.

75. Martinez-Abarca F, Herrera-Cervera JA, Bueno P, Sanjuan J, Bisseling T, et al. (1998) Involvement of salicylic acid in the establishment of the *Rhizobium meliloti*-alfalfa symbiosis. Mol Plant Microbe Interact 11: 153–155.

76. Peleg-Grossman S, Golani Y, Kaye Y, Melamed-Book N, Levine A (2009) NPR1 protein regulates pathogenic and symbiotic interactions between *Rhizobium* and legumes and non-legumes. PLoS ONE 4: e8399.

77. Capoen W, Oldroyd G, Goormachtig S, Holsters M (2010) *Sesbania rostrata*: a case study of natural variation in legume nodulation. New Phytol 186: 340–345.

78. Zhang XC, Cannon SB, Stacey G (2009) Evolutionary genomics of LysM genes in land plants. BMC Evol Biol 9: doi:10.1186/1471-2148-9-183.

79. Murray J, Karas B, Ross L, Brachmann A, Wagg C, et al. (2006) Genetic suppressors of the *Lotus japonicus har1–1* hypernodulation phenotype. Mol Plant Microbe Interact 19: 1082–1091.

80. Wang X, Kota U, He K, Blackburn K, Li J, et al. (2008) Sequential transphosphorylation of the BRI1/BAK1 receptor kinase complex impacts early events in brassinosteroid signalling. Cell 15: 220–235.

81. Lemmon MA, Schlessinger J (2010) Cell signalling by receptor tyrosine kinases. Cell 141: 1117–1134.

82. Oh MH, Wang X, Clouse SD, Huber SC (2012) Deactivation of the *Arabidopsis* Brassinosteroid Insensitive 1 (BRI1) receptor kinase by autophosphorylation within the glycine-rich loop. Proc Natl Acad Sci U S A 109: 327–332.

83. Chen L, Hamada S, Fujiwara M, Zhu T, Thao NP, et al. (2010) The Hop/Sti1-Hsp90 chaperone complex facilitates the maturation and transport of a PAMP receptor in rice innate immunity. Cell Host Microbe 7: 185–196.

84. Ke D, Fang Q, Chen C, Zhu H, Chen T, et al. (2012) The small GTPase ROP6 interacts with NFR5 and is involved in nodule formation in *Lotus japonicus*. Plant Physiol 159: 131–143.

Generation and Characterization of the Western Regional Research Center Brachypodium T-DNA Insertional Mutant Collection

Jennifer N. Bragg[1,2◑], Jiajie Wu[1,2◑¤], Sean P. Gordon[1], Mara E. Guttman[1], Roger Thilmony[1], Gerard R. Lazo[1], Yong Q. Gu[1], John P. Vogel[1]*

1 United States Department of Agriculture- Agriculture Research Service (USDA-ARS), Western Regional Research Center, Albany, California, United States of America,
2 University of California Davis, Davis, California, United States of America

Abstract

The model grass *Brachypodium distachyon* (*Brachypodium*) is an excellent system for studying the basic biology underlying traits relevant to the use of grasses as food, forage and energy crops. To add to the growing collection of *Brachypodium* resources available to plant scientists, we further optimized our *Agrobacterium tumefaciens*-mediated high-efficiency transformation method and generated 8,491 *Brachypodium* T-DNA lines. We used inverse PCR to sequence the DNA flanking the insertion sites in the mutants. Using these flanking sequence tags (FSTs) we were able to assign 7,389 FSTs from 4,402 T-DNA mutants to 5,285 specific insertion sites (ISs) in the *Brachypodium* genome. More than 29% of the assigned ISs are supported by multiple FSTs. T-DNA insertions span the entire genome with an average of 19.3 insertions/Mb. The distribution of T-DNA insertions is non-uniform with a larger number of insertions at the distal ends compared to the centromeric regions of the chromosomes. Insertions are correlated with genic regions, but are biased toward UTRs and non-coding regions within 1 kb of genes over exons and intron regions. More than 1,300 unique genes have been tagged in this population. Information about the Western Regional Research Center *Brachypodium* insertional mutant population is available on a searchable website (http://brachypodium.pw.usda.gov) designed to provide researchers with a means to order T-DNA lines with mutations in genes of interest.

Editor: Samuel P. Hazen, University of Massachusetts Amherst, United States of America

Funding: This work was supported by the Office of Science (BER), US Department of Energy, Interagency Agreement No. DE-AI02-07ER64452 and by USDA CRIS project 5325-21000-013-00. The funders had no role in study design, data collection and analysis, decision to publish, or preparation of the manuscript.

Competing Interests: The authors have declared that no competing interests exist.

* E-mail: john.vogel@ars.usda.gov

◑ These authors contributed equally to this work.

¤ Current address: Shandong Agricultural University, Taian, Shandong, China 271018

Introduction

Brachypodium distachyon (*Brachypodium*) is an annual grass native to the Mediterranean and Middle East and is a member of the Pooideae subfamily [1]. This group also contains cereal and forage grasses including economically important species with complex genomes such as *Triticum aestivum* (bread wheat), *Hordeum vulgare* (barley), and *Avena sativa* (oats). In 2001, *Brachypodium* was proposed as a model system for the study of the Triticeae [2], and in 2005 following a feasibility study from the United States Departments of Energy on the use of plant biomass for the generation of energy and other products, *Brachypodium* was recognized as an attractive model for the improvement of proposed bioenergy feedstocks such as switchgrass and *Miscanthus* [3]. Many aspects of grass biology, such as cell wall composition and architecture [4,5], development, and grain properties, are distinct from dicots. In these cases, a grass such as *Brachypodium*, that possesses all of the biological, physical and genomic attributes required for an experimental system [2,6,7], represents a more relevant model than the dicot model *Arabidopsis*. *Brachypodium*'s compact 272 Mbp diploid genome is similar to rice and sorghum in gene content and gene family structure [8], and the small size, rapid generation time, and simple growth requirements of *Brachypodium* enable high-throughput studies that are not feasible using larger, more demanding species [2,9–11]. Additionally, *Brachypodium* is self-fertile and rarely outcrosses [12] which facilitates breeding homozygous lines for applications that require the maintenance of large numbers of independent genotypes, such as mapping experiments, mutant analysis, and studies of natural diversity [13–17].

Over the last decade, several groups have developed resources that allow *Brachypodium* to serve as powerful functional and structural genomic model system. These resources include highly efficient transformation [18,19] and crossing protocols (http://brachypodium.pw.usda.gov/), bacterial artificial chromosome (BAC) libraries [20–22], genetic markers [12,23], genetic linkage and physical maps [24–26], and a continually increasing germplasm collection [12,27,28]. Publicly accessible databases provide access to the complete genome sequence of *Brachypodium* accession Bd21 [8], array expression data (http://www.brachypodium.org/), and comparative genomics tools (http://brachypodium.pw.usda.gov, http://www.brachybase.org, http://www.phytozome.net, http://www.modelcrop.org, http://mips.

helmholtz-muenchen.de/plant/). The surge of publications in the year following the release of the complete genome sequence testifies to the utility of these *Brachypodium* resources for the study of cereal, forage, and bioenergy grasses in areas as diverse as grass flowering time [29–31], drought response [29], seed dormancy [32], grain development [33,34], iron homeostasis [35], transcription factors [30,36], cell wall composition [37], disease resistance [38], complex genome sequencing [39], and genome evolution [40] (for a review see [9]). Forthcoming sequencing data for more than 50 additional accessions and recombinant inbred lines will serve to increase the utility of *Brachypodium* as an experimental system for the grasses.

While comparative analyses of sequence information can provide educated guesses about gene function, establishing a true link between genes and their biological function requires detailed functional characterization. Sequence indexed insertional mutants are a particularly valuable tool in this context because mutations in a given gene can be identified by simply searching a database. Thus, the large, sequence-indexed T-DNA and transposon-tagged mutant collections available for Arabidopsis and rice are an invaluable resource for such forward genetic studies. In addition to providing loss of function mutants when a T-DNA or transposon lands in a gene, the vectors can be designed to track promoter function and overexpress nearby genes. Gene trapping vectors containing promoterless reporter genes with splice donor and acceptor sites can be used to infer the expression pattern of disrupted genes and to identify promoters with tissue-specific expression patterns [41,42]. Activation tagging vectors contain transcriptional enhancers that cause nearby genes to be overexpressed while maintaining normal expression patterns. Such activation tagged mutants are particularly useful for investigating essential genes in which disruption is lethal, genes with redundant functions where a knockout in one family member fails to produce a phenotype, and the regulation of complex processes in which the activation of a global control gene is required to observe a phenotype [43–45].

A high efficiency transformation method is a prerequisite for the creation of large T-DNA collections. Fortunately, *Brachypodium* is highly amenable to tissue culture and transformation. After conditions were established for the generation of embryogenic callus and production and regeneration of fertile *Brachypodium* plants [46], *Brachypodium* was successfully transformed using a biolistic method [2,47]. Due to complex insertion patterns that often involve many copies of the inserted DNA and local rearrangements, biolistic transformation is not ideal for the generation of mutant populations. *Agrobacterium*-mediated transformation of *Brachypodium* was first reported in 2006 [27]. Subsequently, two groups optimized high efficiency *Agrobacterium*-mediated transformation methods [18,19]. With efficient transformation methods in hand, the generation of *Brachypodium* T-DNA collections was initiated. In 2010, the BrachyTAG collection (http://www.brachytag.org) [48] reported 741 T-DNA lines with flanking sequence tags (FSTs) anchoring insertions sites within the *Brachypodium* genome. These lines contain insertions between 1,500 bp upstream to 500 bp downstream of the coding sequence for 364 *Brachypodium* genes. Based on this data, the authors estimated that at least 51,976 lines would be necessary to obtain a T-DNA insertion in any gene with a 95% probability and that the actual number of lines required to meet this goal is over 100,000. As a means to approach this distant target, we developed a high-throughput pipeline for the production and sequencing of *Brachypodium* T-DNA mutants. Here we present the optimization of this pipeline including a comparison of T-DNA and transposon tagging strategies, further optimization of our transformation

method, and a comparison of several T-DNA vectors. We used these optimized methods to generate 8,491 *Brachypodium* T-DNA mutants and identify 5,285 unique insertion loci for 4,402 of these lines.

Materials and Methods

Plant lines and growth conditions

Brachypodium inbred lines Bd21 and Bd21-3 were compared in initial transformation studies, and Bd21-3 was selected to generate the bulk of the T-DNA mutant population [18]. Plants were grown in a soil mix of 1 part sandy loam, 2 parts sand, 3 parts peat moss, and 3 parts medium grade (#3) vermiculite. A time release fertilizer containing micronutrients (Osmocote Plus 15-9-12, Scotts Co., Marysville, OH) was added at the time of planting. Plants were grown in both greenhouses and growth chambers. Growth chambers conditions were 20 hr light : 4 hr dark photoperiod, cool-white fluorescent lighting at a level of 150 $\mu Em^{-2} s^{-1}$, and temperatures of 24°C during the day and 18°C at night. Greenhouse conditions were no shading, 24°C in the day and 18°C at night with the day length extended to 16 hours by supplemental lighting. T_1 seeds were harvested from senesced T_0 plants after they were completely dried (typical yield ranged between 50–150 T_1 seeds per plant). When additional seeds were required for particular lines, 6–12 T_1 seeds were sown and the harvested seeds were collected in bulk. If large quantities of seeds are required, plants can be grown under short day conditions and then vernalized or moved to long day conditions. In this case, over 1,000 seeds can be obtained from an individual plant.

T-DNA constructs

Several constructs were used in this study. The previously described construct pOL001 [27] was used as a benchmark for the evaluation of the new constructs and for the initial production of T-DNA lines. pOL001 contains a *HptII* gene under control of the CaMV 35S promoter for selection of transgenic callus and a GUS reporter gene under control of the maize ubiquitin promoter (**Fig. 1A**).

Ac-Ds and En-I(Spm) transposon systems were tested using the constructs Ac-DsATag-Bar_gosGFP [49] (**Fig. 2A**) and pdSpm-R [50]. These constructs are single vector transposon tagging constructs and contain both the transposase and the corresponding mobile element. We built seven additional constructs (the pIJ vectors) starting from pCAMBIA vector backbones (http://www.cambia.org/) (**Fig. 1A**). The first constructs (pIJ, pIJB) were used to compare the transformation efficiency of hygromycin and BASTA selections. The pIJ2LB and pIJB2LB constructs were used to determine if the presence of two left border sequences improved the efficiency of FST generation by decreasing the transfer of vector DNA beyond the left border [51]. The final vectors were modified from pIJ2LB to create derivatives designed for gene trapping (pIJ2LBP and pIJ2LBP2) and activation tagging (pIJ2LBA). The construction of these vectors is described below, and primers used in their construction are listed in **Table S1**. Sequencing was performed on the T-DNA regions of all constructs to ensure that no mutations were introduced during PCR and to confirm proper orientation after ligation of digestion fragments.

The pUbi-BASK vector (courtesy of Jim Thomson) was the source of the maize ubiquitin promoter with the ubiquitin 5′ intron used in the constructs below. To generate pUbi-BASK the pAHC20 vector [52] had the internal EcoRI restriction site within the maize ubiquitin promoter removed using site directed mutagenesis. The *BAR* coding sequence was excised from

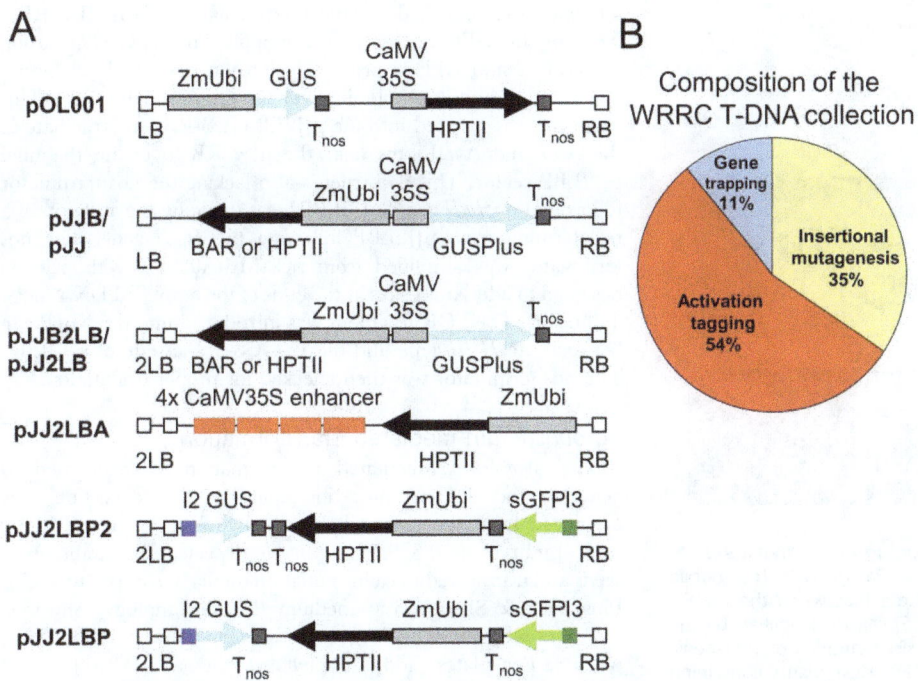

Figure 1. T-DNA constructs used to generate the USDA-ARS-WRRC T-DNA collection. A. Illustrations of the T-DNA regions of pOL001 and the seven new constructs used to generate the majority of the population are shown. White boxes: T-DNA left border (LB or 2LB) and right border sequences (RB). Light gray boxes: *Z. mays* ubiquitin promoter with intron (ZmUbi) and cauliflower mosaic virus 35S promoter (CaMV 35S). Dark grey boxes: nos terminator sequences (T$_{nos}$). Red boxes: cauliflower mosaic virus 35S transcriptional enhancer sequences (4× CaMV35S enhancer). Black arrows: hygromycin phosphotransferase selection gene (*HptII*) or phosphinothricin acetyltransferase selection gene (*BAR*). Blue arrows and boxes: ß-glucuronidase reporter genes (*GUS* and *GUSPlus*) with a rice tubulin intron containing splice donor and acceptor sites (I2). Green arrow and box: green fluorescent protein reporter gene (*sGFP*) with a rice tubulin intron containing splice donor and acceptor sites (I3). **B.** Pie chart showing the composition of the T-DNA population. The yellow represents lines made with the insertional mutagenesis vectors including pOL001, pJJH, pJJB, pJJ2LB, and pJJB2LB. The blue represents lines made with the gene trap vectors pJJ2LBP and pJJ2LBP2, and the red represents lines made with the activation tagging vector pJJ2LBA.

pAHC20 with BamHI and KpnI and a BamHI/AscI/SpeI/KpnI (BASK) multiple cloning site was inserted in its place.

pJJB. This vector has a phosphinothricin acetyl transferase selection (*BAR*) gene under the control of the maize ubiquitin promoter. The *BAR* gene sequence was amplified from pMNRTT224nb (courtesy of Venkatesan Sundaresan) using the BARBamHIF and BARKpnIR primers. The PCR product was digested with BamHI and KpnI and inserted into the corresponding sites of the pUbi-BASK vector directly after the maize ubiquitin promoter with the ubiquitin 5′ intron and preceding the nos terminator. The Ubi-BAR-nos fragment was then inserted into the HindIII and EcoRI sites of the pCAMBIA0305.2 vector (http://www.cambia.org/). The resulting construct also has a GUSPlus reporter gene under control of the CaMV 35S promoter and a kanamycin resistance gene for selection in bacteria.

pJJB2LB. To generate this vector the pCAMBIA0305.2 left border (LB) was replaced with the double left border sequence (2LB) from pL3 [53]. This 741 base pair region consists of a nopaline LB and its flanking regions followed by an octopine LB and its flanking regions. The 2LB sequence was amplified with primers designed to introduce a SacII site adjacent to the octopine border and an AseI site adjacent to the nopaline border (pL3-2LB-F1 and pL3-2LB-R1). The 2LB PCR fragment was introduced between the AseI and SacII sites of pCAMBIA0305.2. Then, the Ubi-BAR-nos sequence (as for pJJB) was introduced between the HindIII and EcoRI sites.

pJJ. This vector has the hygromycin phosphotransferase selection gene *HptII* under control of the maize ubiquitin

promoter. The maize ubiquitin promoter sequence with the ubiquitin 5′ intron was amplified from the pUbi-BASK vector using the primers UbiHEF and UbiOLR, and *HptII* was amplified from pGHyg [18] using the primers HygOLF and HygEcoRIR. The PCR products were digested with BamHI, and the resulting fragments ligated and used as a template for PCR with the UbiHEF and HygEcoRIR primers. This PCR product was digested with EcoRV and EcoRI and introduced into the SmaI and EcoRI sites of pCAMBIA0380. This construct lacks a nos terminator following the *HptII* sequence.

pJJ2LB. To build this construct, the Ubi-Bar-nos cassette was removed from pJJB2LB by digestion with EcoRI and HindIII and replaced with the Ubi-Hyg fragment from pJJ. This fragment is also lacking the nos terminator following the *HptII* gene.

pJJ2LBA. This construct contains a CaMV 4x35S activation tagging cassette and hygromycin selection under control of the maize ubiquitin promoter. The CaMV 4x35S enhancer fragment was derived from Ac-DsATag-BAR_gosGFP [49], in which it was flanked by two EcoRI restriction sites and contained one internal EcoRI. Ac-DsATag-BAR_gosGFP was digested with HindIII to release a fragment containing the CaMV 4x35S region and the Ubi-BAR-nos cassette, the ends of the fragment were blunted using mung bean nuclease and the fragment was then inserted into pCR4-TOPO. The ~1.7 kb CaMV 4x35S enhancer fragment was released by partial digestion with EcoRI and introduced into the EcoRI site of pJJ2LB. Due to the repetitive nature of the CaMV 4x35S region, we were only able to sequence through three copies of the enhancer sequence after cloning into pJJ2LBA.

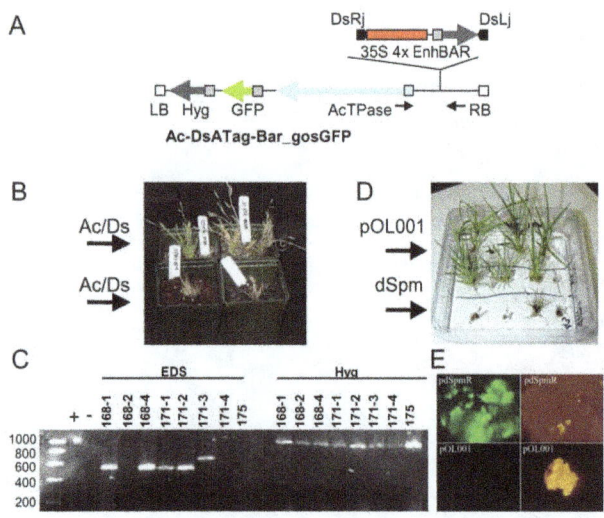

Figure 2. Evaluation of transposon tagging in *Brachypodium*. A. Diagram of the T-DNA region of Ac-DsATag-Bar_gosGFP. The mobile element is depicted on the upper line. Note that when the mobile element excises the empty donor site (EDS) can be amplified by the flanking PCR primers, solid black arrows. **B.** Examples of transgenic plants created using Ac-DsATag-Bar_gosGFP. Most plants containing the transposable elements died before flowering or failed to set seed. **C.** PCR using primers that flank the Ds mobile element. An EDS produces a band while an intact donor site will not produce a band. Note that five of eight lines contain an EDS indicating that Ds was excised. The hygromycin resistance gene was amplified from all lines indicating that they contained at least part of the T-DNA. **D.** Examples of transgenic plants created using the pdSpm-R tagging construct and the control vector pOL001. Most plants containing the transposable elements died when very small. **E.** Calluses transformed with the pdSpm-R tagging construct were fluorescent green due to the expression of GFP contained on the T-DNA. By contrast, calluses transformed with a control plasmid, pOL001, were not fluorescent. For both constructs, pictures on the left are taken under UV light through a GFP filter and the pictures on the right are under white light.

However, enzymatic digestion confirmed the insert was of the expected size for four enhancer copies. This construct does not have a nos terminator following the *HptII* gene, but still transforms efficiently.

pIJ2LBP and pIJ2LBP2. These constructs were modified from pIJ2LB. Both contain hygromycin selection driven by the maize ubiquitin promoter and two gene trap reporters, a promoterless GUS gene (I2-GUS-T_n) near the 2LB and a promoterless GFP gene (T_n-sGFP-I3) near the RB. The gene trap cassettes were derived from the pGA2717 vector and contain rice tubulin intron sequences with splice donor and acceptor sites before the reporter gene sequences [41]. To make pIJ2LBP, the constitutive GUSPlus reporter and the RB were first removed from pIJ2LB by digestion with SphI. Primers (RB-f2 and RB-r2) were designed to amplify the pCAMBIA0305.2 right border and to introduce SpeI, AvrII and AseI restriction sites just before the RB and an SpeI site just following the RB sequence. This PCR fragment was introduced into the SpeI sites of pIJ2LB lacking the GUSPlus and original RB to create the intermediate pIJ2LB-RB2. Next, the I2-GUS-T_n cassette was released from pGA2717 by digestion with HpaI and KpnI, the ends were blunted, and the fragment cloned into the TOPO Blunt vector (Life Technologies, Grand Island, NY). The I2-GUS-T_n cassette was released from TOPO Blunt with EcoRI and ligated into pIJ2LB-RB2. Clones with the insert oriented with the I2 intron adjacent to the 2LB

sequence were selected for the intermediate pIJ2LB-GUS-RB2. Next, the I3-sGFP-T_n cassette was amplified from pGA2717 using the GFP-f2 and GFP-r2 primers to introduce an AscI site before the I3 intron and an AvrII site after the nos terminator (T_n). This fragment was inserted into the pIJ2LB-GUS-RB2 intermediate at the AscI and AvrII sites near the new RB to create the final pIJ2LBP vector. This construct is also lacking the nos terminator following the *HptII* gene. pIJ2LBP2 was made by the addition of a nos terminator to pIJ2LBP following the *HptII* gene. The nos terminator was amplified from pCAMBIA0305.2 with primers designed to add XmaI sites at the ends of the amplified DNA (nos-f and nos-r). The PCR fragment was introduced into the XmaI site between the *HptII* gene and the I2-GUS-T_n cassette of pIJ2LBP. The nos terminator was then checked for proper orientation.

Agrobacterium-mediated Transformation

The *Agrobacterium*-mediated transformation protocol used to generate the T-DNA insertion lines in this collection was optimized from the protocol described in Vogel and Hill 2007 [18]. Embryos (<0.3–0.7 mm) were dissected from immature seeds and transferred to callus initiation media (CIM, per L: 4.43 g Linsmaier & Skoog basal medium (Phytotechnology, Shwanee Mission, KS #L689), 30 g sucrose, 1 ml 0.6 mg/ml $CuSO_4$, pH 5.8. For plates, add 2 g phytagel (Sigma #P-8169). After autoclaving, add 0.5 ml of 5 mg/ml 2,4-D stock solution.) Following 3–4 weeks incubation in the dark at 28°C, embryogenic callus was subcultured onto fresh CIM plates. A second subculture was performed after two more weeks. The calluses from the second subculture were grown for one week before being used for transformation. On the day of transformation, calluses were bathed for 5 minutes in a suspension of *Agrobacterium* strain AGL1 containing the desired vector for transformation [54] ($OD_{600} = 0.6$) prepared in liquid CIM containing 200 µM 2,4-D and 0.1% Synperonic PE/F68 (Sigma #81112, formerly Pluronic F68). After removing as much of the *Agrobacterium* suspension as possible, the calluses were transferred to petri dishes containing a piece of sterile filter paper for co-cultivation for 3 days in the dark at 22°C. Note that co-cultivation under desiccating conditions is critical to the success of the transformation protocol. Next the callus pieces were moved to CIM plates containing 150 mg/L timentin and the appropriate selective agent to kill untransformed plant tissue - either 40 mg/L hygromycin B (Phytotechnology H397) or 60 mg/L DL-Phosphinothricin (Phytotechnology P679) - and incubated in the dark at 28°C for 1 week. Healthy sectors of hygromycin resistant transgenic callus were subcultured to fresh CIM plates one time for an additional two weeks of selection. BASTA selected callus was subjected to two additional rounds of subculture for an additional four weeks of selection. Note it is not necessary to obtain a callus with only healthy transgenic tissue because even small pieces of healthy callus surrounded by dead and dying callus will produce plantlets efficiently. Between 3 weeks (hygromycin selection) to 5 weeks (BASTA selection) after co-cultivation, calluses were transferred to regeneration media (per L: 4.43 g Linsmaier & Skoog (LS) basal medium, 30 g maltose, 2 g phytagel, pH 5.8; after autoclaving, 1.0 ml of sterile 0.2 mg/ml kinetin stock solution was added) containing 150 mg/L timentin and the appropriate selective agent. Plates were incubated in the light (cool-white fluorescent lighting at a level of 65 µEm^{-2} s^{-1} with a 16 hr light : 8 hr dark cycle) at 28°C. Callus pieces began to turn green and shoots appeared between 2–4 weeks. Individual plantlets were moved to tissue culture boxes (we used sundae cups made for food service applications from Solo Corporation, Lake Forest, IL Cat. # SOL-TS5 (cups) and SOL-DL-100 (dome lids)) containing MS sucrose medium (per L: 4.42 g Murashige & Skoog

(MS) basal medium with vitamins (Phytotechnology M519), 30 g sucrose, and 2 g phytagel, pH 5.7) and incubated in the light (cool-white fluorescent lighting at a level of 65 $\mu Em^{-2} s^{-1}$ with a 16 hr light : 8 hr dark cycle) at 28°C. After plantlets had formed roots and were approximately 2–5 cm tall, they were transplanted to soil and placed in a growth chamber for flowering (20 hr light, 4 hr dark, 24°C during the day and 18°C at night, cool-white fluorescent lighting at a level of 150 $\mu Em^{-2} s^{-1}$). In this case, vernalization was not required to induce rapid flowering. Alternatively, to promote rapid flowering in plantlets moved directly to greenhouse conditions (no shading, 24°C in the day and 18°C at night with supplemental lighting to extend daylength to 16 h), plants were vernalized in tissue culture boxes or in soil under light (continuous cool-white fluorescent lighting, 4 $\mu Em^{-2} s^{-1}$) at 4°C for 2–4 weeks depending on the season.

PCR of empty donor sites in Ac/Ds lines

PCR was performed on DNA extracted from leaves of Ac-DsATag-Bar_gosGFP T-DNA lines. Primers HygBamHIF (5′ttggatccatgaaaaagcctgaactcacc3′) and HygKpnIR (5′ttggtaccc-tatttcttttgccctcgg3′) were used to amplify the *HptII* gene. Locations of the primers R13pMOg22 (5′ggaaacgacaatctgatctctagg 3′) and Ac-promRev (5′ctcagtggttatggatgggagttg3′) that were used to identify the empty donor sites are shown as small black arrows in **Fig. 2A**.

DNA Extraction

DNA extraction was based on the method described by Shiaman Chao and Daryl Somers (http://maswheat.ucdavis. edu). To prepare tissue, two young leaves (approximately 3 inches) from T_0 plants were sampled into the wells of a 96-well plate (E&K Scienfitic, EK-22280). Glass or chrome steel beads (http://www.biospec.com) were added to the wells (either one 6.3 mm and two to three 3.5 mm glass beads or two 3.2 mm chrome steel beads per well), and the plates were lyophilized overnight. Directly after lyophilization, the plates were covered with sealing mats (E&K Scienfitic, EK-80080), and tissue was ground in a Retsch MM301 Ball Mill at 30 cycles/sec for 1 min. The grinding was repeated 5 times, for a total of 5 min. Before opening plates, the ground tissue was centrifuged at 4000 rpm for 20 min. at 4°C. To isolate DNA, 800 µl of extraction buffer (0.1 M Tris-HCl pH 7.5, 0.05 M EDTA pH 8.0, 1.25% SDS) preheated to 65°C was added to each well. Plates were sealed with a new sealing mat and shaken thoroughly. Chrome steel beads were removed prior to adding extraction buffer, but glass beads were left in the wells. To prevent contamination from neighboring wells, a flat weight was placed on top of the sealing mat and secured with several rubber bands. Plates were incubated at 65°C for 0.5–1 hour, with mixing every 5–10 min. Plates were transferred to ice for 15 min. before centrifuging at 4000 rpm for 5 min. at 4°C. Next, 400 µl cold 6M ammonium acetate was added to the wells, and the plates inverted several times and placed on ice for 15 min. Plates were then centrifuged at 4,000 rpm for 15 min. at 4°C to collect the precipitated proteins and plant tissue, and 900 µl of the supernatant was transferred into another plate containing 540 µl of isopropanol in each well. Plates were mixed thoroughly and placed on ice or at −20°C for 30 min. to precipitate DNA. Plates were then centrifuged at 4,000 rpm for 30 min. at 4°C, the supernatant decanted, and the plate inverted on a paper towel for a few seconds. Pellets were washed with 1 ml of 70% ethanol and centrifuged at 4,000 rpm for 20 min at 4°C. After removing the supernatant, the pellets were dried in the hood overnight and resuspended in 125 µl TE buffer (10 mM Tris-Cl pH 7.5, 1 mM EDTA pH 8.0). To dissolve the DNA, the samples

were placed at 4°C overnight and then transferred to a 96-well microtiter plate. Estimated yield is 10–20 ng/µl.

Inverse PCR

Restriction enzyme digestions were performed in 15 µl reactions containing 75–150 ng (7.5 µl) of DNA, 0.2 µl enzyme, and 1× NEB #1 buffer. Samples were incubated at 37°C for 3 hours, then at 65°C or 80°C (depending on enzymes) for 20 min. The enzymes used to digest individual T-DNA constructs at the right or left borders are listed in **Table 1**. Digestion products were purified by precipitation with 95% ethanol and 3M sodium acetate and resuspended in 10 µl of TE buffer (pH 8.0). Ligations were performed by adding 0.125 µl T4 DNA ligase, 1.0 µl of 10× ligation buffer, and 3.875 µl H_2O to the digested DNA and incubating for 16 hr at 16°C. PCR reactions (10 µl) were prepared for the ligation products as follows: 2 µl 5× Go Taq buffer (NEB), 1 µl dNTPs (2.5 mM), 0.2 µl Primer1 (10 µM), 0.2 µl Primer2 (10 µM), 0.5 µl DMSO (100%), 0.05 µl Go Taq, 2 µl DNA (ligation), and 4.05 µl H_2O. The PCR program used was 95°C for 5 min. and 94°C 30 sec, followed by 35 cycles (94°C 30 sec, 60°C 30 sec, 72°C 1 min. and 45 sec), and the program ended with 72°C 10 min. Some primer pairs needed 2–3 additional cycles to improve amplifications. If the PCR did not work well, another round of PCR was performed using nested primers and 0.5 µl of the first round PCR product as a template.

FST Sequencing

Unconsumed dNTPs and primers were removed from PCR products using ExoSAP-IT (Affymetrix) per manufacturer's instructions. Sequencing was performed in either 96 (10 µl reactions) or 384 (5 µl reactions) well plates with the BigDye Terminator v3.1 sequencing kit (Applied Biosystems) using the following program: 98°C 5 min, followed by 39 cycles (96°C 10 sec, 50°C 10 sec, 60°C 4 min), and holding at 4°C. The sequencing products were precipitated with 95% ethanol and 3M sodium acetate, washed with 70% ethanol, dried in a hood (hours to overnight). The pellets were dissolved in 8.5 µl (for a 96 well plate) or 5 ul (for a 384-well plate) of Hi-Di Formamide (Applied Biosystems), and the plates stored at −20°C until use. Before loading the sequencer, the samples were denatured at 96°C for 3 min. Primer locations and sequences are listed in **Table 2**.

GUS staining of JJ2LBP and JJ2LBP2 lines

Stem, leaf and floral tissue samples were collected from young *Brachypodium* plants into microcentrifuge tubes. GUS staining solution (0.1 M sodium phosphate pH 7.0, 0.5 mM potassium ferrocyanide, 0.5 mM potassium ferricyanide, 0.5% v/v Triton X-100, 0.15% w/v X-Gluc) was added directly to the tubes, and samples were vacuum infiltrated for 5 min before placing at 37°C in the dark overnight. The GUS staining solution was removed with a pipet and 95% EtOH was added to remove any chlorophyll that might mask the blue staining and to fix the tissue.

Results

Optimization of transformation

We performed a series of experiments to improve transformation efficiency, defined as the number of transgenic plantlets regenerated per number of callus pieces co-cultivated with *Agrobacterium*, and decrease the labor required per transgenic plant produced. The following modifications were found to improve transformation efficiency. The production of high quality embryogenic callus was increased by the addition of copper sulfate at a final concentration of 0.6 mg/L to the callus initiation

Table 1. Restriction enzymes and primers used for Inverse PCR.

Construct	Digestion	Border	PCR primers and annealing temperatures	Cycles*	Sequencing primer
pOL001	BfaI	LB	001-LB-F8R8, 62°C	38	001-LB-R6
			001-LB-F8R9, 57°C	38*	
			001-LB-F6R8, 60°C	35	
	BfaI	RB	001-SP3, 62°C	35	001-SP4
			001-RB-R6, 62°C	35	
	HhaI	LB	001-LB-F9R7, 60°C	35	001-LB-R9
	HhaI	RB	001-SP3, 62°C	35	001-SP4
			001-RB-R6, 62°C	35	
	TaqI	RB	001-SP3, 62°C	35	001-SP4
			001-RB-R6, 62°C	35	
pJJ2LB	HpyCH4IV	LB	pJJ2LB -LB-F4R2, 60°C	35	UH2LB-LB-R5
	HpyCH4IV	RB	pJJ2LB -RB-F1R1, 53°C	35	UH2LB-RB-F3
	BfaI	LB	pJJ2LB -LB-F1R2, 60°C	35	UH2LB-LB-R5
	TaqI	RB	pJJ2LB -RB-F1R1, 53°C	35	UH2LB-RB-F3
pJJ2LBP2	BfaI	LB	pJJ2LBP2-LB-F3R3, 60°C	35*	PJJ2LBP2-LB-R1
			pJJ2LBP2-LB-F3R1, 60°C	35	PJJ2LBP2-LB-R2
	BfaI	RB	pJJ2LBP2-RB-F1R1, 63°C	35	PJJ2LBP2-RB-F3
	HpyCH4IV	LB	pJJ2LBP2-LB-F5R5, 50°C	35	PJJ2LBP2-LB-R3
	HpyCH4IV	RB	pJJ2LBP2-RB-F2R3, 50°C	37	PJJ2LBP2-RB-F3
	TaqI	RB	pJJ2LBP2-RB-F2R3, 50°C	37	PJJ2LBP2-RB-F3
pJJ2LBA	HpyCH4IV	LB	pJJ2LBA-LB-F2R1, 58°C	35	PJJ2LBA-LB-R2
			pJJ2LBA-LB-F1R1, 58°C	35	
	HpyCH4IV	RB	pJJ2LBA-RB-F1R1, 60°C	35	PJJ2LBA-RB-F3
	HpaII	RB	pJJ2LBA-RB-F1R1, 60°C	35	PJJ2LBA-RB-F3

*indicates PCR primers requiring a second round of PCR.

media [19]. Callus pieces were moved to media containing the selective agent (hygromycin or Basta) directly after the 3 day co-cultivation step, rather than after a 7 day recovery on media lacking the selective agent. This modification permitted us to eliminate one of the subculture steps after co-cultivation decreasing labor and supplies and accelerating the recovery of transgenic plants. After rooting, the regenerated plantlets were placed at 4°C for two to three weeks before moving them into soil to cue early flowering in the greenhouse and bypass the need for growing the plants under long days in a growth chamber. *Brachypodium* accessions Bd21-3 and Bd21 were compared as the source of embryogenic callus. Transformation efficiency was similar for the two lines when a microscope was used to identify embryogenic callus during subculture. However, Bd21-3 was selected for the production of our insertional mutant population, because it forms a more strongly yellow colored callus with organized structures that eased selection of the correct callus and increased transformation efficiency when subculture was performed without the aid of a microscope.

Optimization of the transformation vector was responsible for the greatest improvement in transformation efficiency. We conducted experiments to compare the recovery of transgenic plants using hygromycin or Basta selection under the control of three different promoters (**Table 3**). The pOL001 vector previously was used to produce transgenic *Brachypodium* with transformation efficiency up to 41% [18]. This construct contains hygromycin selection under the control of the CaMV 35S promoter and was used as a benchmark in all transformation optimization experiments. Three additional hygromycin selection vectors and three vectors designed with Basta selection were compared with pOL001. The pGA2717 rice transformation vector [41] contains hygromycin selection driven by the rice tubulin promoter. Two additional vectors, pJJ and pJJ2LB, were constructed by placing hygromycin selection under control of the maize ubiquitin promoter and inserted into pCAMBIA0305.2 (http://www.cambia.org/). The pJJ vector utilizes the single left border (LB) from pCAMBIA0305.2, but in the pJJ2LB vector, the pCAMBIA0305.2 LB was replaced with a double left border from the pL3 vector [53]. The pJJB and pJJB2LB vectors were constructed similarly, but have the maize ubiquitin promoter driving Basta selection. The final vector, pSMAb801 has the CaMV 35S promoter directing expression of the Basta selection gene [55].

Transformation efficiency was evaluated for these seven vectors in multiple experiments (**Table 3**). Our results show that transgenic plants can be recovered using either hygromycin or Basta as the selective agent, however, vectors containing hygromycin selection, with the exception of vector pGA2717, were much more efficient (averages 22.9 to 55.8%) than those employing Basta selection (averages 2.2 to 8.3%). The hygromycin-selected pGA2717 vector yielded the lowest transformation efficiency observed, 0.7%, possibly due to the function of the rice tubulin promoter. However, since we did not sequence the vector we cannot rule out the possibility that there was something wrong

Table 2. Sequences and locations of primers used for Inverse PCR.

Name	Sequence (5'-3')	Location
pOL001-LB-F6	GACCCGGTCGTGCCCCTCT	8548-8566
pOL001-LB-R6	TTAAAAACGTCCGCAATGTGTTATTAAG	8484-8457
pOL001-LB-F8	TGCCTGCAGTGCAGCGTGACC	8531-8551
pOL001-LB-R8	TTCAGTACATTAAAAACGTCCGCAATGTG	8493-8465
pOL001-LB-R9	GATAAGCTGTCAAACATGAGAATTCAG	8515-8489
pOL001-SP3	CGTCATCGGCGGGGGTCATAAC	14003-14024
pOL001-SP4	TCTCCGCTCATGATCAGATTGTCG	14039-14062
pOL001-RB-R6	AAGCACATACGTCAGAAACCATTATTGCG	13915-13887
pJJ2LB -LB-F1	CTCGCTAAACTCCCCAATGTCAAG	901-924
pJJ2LB -LB-F4	ACAGCGGGCAGTTCGGTTTCA	824-844
pJJ2LB -LB-R2	CTCGTCCGAGGGCAAAGAAATAGG	152-129
pJJ2LB -LB-R5	TCCTGTGTGAAATTGTTATCCGC	103-81
pJJ2LB -RB-F1	ACGCGATAGAAAACAAAATATAGC	5664-5687
pJJ2LB -RB-R1	GCGGGACTCTAATCATAAAAACC	5626-5648
pJJ2LB -RB-F3	GCGCAAACTAGGATAAATTATCG	5688-5710
pJJ2LBP2-LB-F3	CTCTACACCACGCCGAACACCTG	793-815
pJJ2LBP2-LB-R3	CCTGTGTGAAATTGTTATCCGCTCAC	102-77
pJJ2LBP2-LB-R1	CACAACATACGAGCCGGAAGCATA	69-46
pJJ2LBP2-LB-F5	GTGGCTAATTACATGACTAACTTGG	182-206
pJJ2LBP2-LB-R5	ATAGCACCGTGGTAGTAAGAATG	147-169
pJJ2LBP2-RB-F1	GCTCCTCGCCCTTGCTCACCAT	6221-6242
pJJ2LBP2-RB-R1	TGTTGCCGGTCTTGCGATGATTATC	5460-5436
pJJ2LBP2-RB-F3	GCAGTGAATTAACATAGCAGAGAA	6355-6378
pJJ2LBP2-RB-F2	ACAAACAAGAAATGGCAGTGAAT	6341-6363
pJJ2LBP2-RB-R3	TCTAACCAACTTGTTTATTGCTAATG	6333-6308
pJJ2LBA-LB-F1	ACAGGAAACAGCTATGACATGATTACGA	98-125
pJJ2LBA-LB-R1	ACACAACATACGAGCCGGAAGCATA	46-70
pJJ2LBA-LB-R2	GAGCCGGAAGCATAAAGTGTAAAG	36-59
pJJ2LBA-RB-F1	TCGAGAGGGGTCCAGAGGCA	4053-4072
pJJ2LBA-RB-R1	CTGTCGGCATCCAGAAATTGCG	4001-4022
pJJ2LBA-RB-F3	AGATGCCGTGCCGTCTGCT	4080-4098

Table 3. T-DNA constructs evaluated for transformation efficiency.

Construct	Average Efficiency*	Promoter	Selection	LB copies
pOL001	22.9	CaMV 35S	hygromycin	1
pJJ	55.8	Maize ubiquitin	hygromycin	1
pJJ2LB	36.5	Maize ubiquitin	hygromycin	2
pGA2717	0.7	Rice tubulin	hygromycin	1
pJJB	3.0	Maize ubiquitin	Basta	1
pJJB2LB	8.3	Maize ubiquitin	Basta	2
pSMAb801	2.2	CaMV35S	Basta	1

*Efficiency is calculated as the percentage of callus pieces co-cultivated with *Agrobacterium* that go on to produce transgenic plants.

the plants tested were positive for GUS reporter gene expression and greater than 90% of the regenerated plantlets survived to set T_1 seed (data not shown). Using our optimized method and vectors, transformation efficiencies averaged 42% during the production of the mutant population and efficiencies of 50–75% were achieved for individual experiments.

Generation of T-DNA lines

The bulk (82.6%) of the T-DNA population was generated from three constructs pJJ2LB, pJJ2LBA and pJJ2LBP2 (**Fig. 1**). All three constructs were designed with two left borders to try to limit the transfer of vector DNA beyond the left border [51]. Constructs pJJ2LB and pJJ2LBP2 can only affect gene function by insertion into coding or regulatory regions. In addition, pJJ2LBP2 can function as a gene trap to identify adjacent promoters because it contains reporterless GUS and GFP genes with multiple splice acceptors at the left and right borders respectively [41]. In addition to disrupting gene function by insertion, the pJJ2LBA can act as an activation tag that causes the overexpression of nearby genes because it contains four copies of the CaMV 35S enhancer sequence adjacent to the LB [43,56]. Overall, 8,491 fertile lines were produced that comprise the WRRC *Brachypodium* T-DNA insertional mutant population described in this report (**Fig. 1B and Table S2**). Since this is an ongoing project, additional lines continue to be produced and readers are directed to the project website (http://brachypodium.pw.usda.gov/) for the most up to date totals.

Evaluation of transposon tagging in *Brachypodium*

To determine if transposon tagging could be used to generate insertional mutants in *Brachypodium* efficiently, we tested two transposon systems (**Fig. 2**) that have been used previously for large-scale mutagenesis in rice and Arabidopsis, Ac/Ds and EnSpm [57,58]. *Agrobacterium*-mediated transformation was used to deliver T-DNAs containing the transposon constructs. Both T-DNAs contain hygromycin selection for plant transformation, a GFP reporter gene, and a transposase. In addition, each construct contains a mobile element harboring four copies of the CaMV 35S enhancer sequence for activation tagging. The average transformation efficiency of Ac-DsATag-Bar_gosGFP (Fig. 1A) [49] over six transformations was 27.2%, but most plants either died before setting seed or produced non-viable seed (**Fig. 2B**). Only 5.9% (7 of 119) of the Ac/DsAtag-Bar_gosGFP T_0 transgenic plants survived to set T_1 seed. PCR was performed on genomic DNA extracted from eight Ac/DsAtag-Bar_gosGFP T_0 plants (**Fig. 2C**).

with the construct. Of the plantlets regenerated on hygromycin, 83.5% survived to produce T_1 seed whereas only 47.1% of the plantlets regenerated on Basta survived (data not shown). Furthermore, the strong selection of hygromycin permitted transfer of callus to regeneration media after three weeks of selection rather than five, accelerating the time to recovery of transgenic plants compared to Basta selection. The addition of a second left border did not affect plant generation or survival (data not shown).

Additional experiments compared transformation efficiency for pOL001 and pJJ2LB. Hygromycin selection is driven by different promoters in these two vectors (CaMV 35S in pOL001 and maize ubiquitin in pJJ2LB). In six side by side transformation experiments 1,037 pieces of callus were co-cultivated with *Agrobacterium* carrying pOL001 and 991 pieces were co-cultivated with *Agrobacterium* carrying pJJ2LB. Transformation efficiency was significantly higher for pJJ2LB containing the maize ubiquitin promoter (48%) than for pOL001 containing the CaMV 35S promoter (32.2%). For both of these constructs, more than 95% of

All samples tested positive for the presence of the *HptII* gene indicating the T-DNA was present in the plants. Transposition of the Ds element leaves behind an empty donor site that can be detected when PCR is performed with primers located outside of the Ds region (**Fig. 2A**). If the Ds element remains in place, the distance between the primers is too large to amplify a fragment. Five of the 8 plants tested for an empty donor site yielded a band with a size indicating that the Ds element had moved (**Fig. 2C**) and none of these lines set seed. In the case of line 171-3, the larger PCR fragment suggests that the Ds excision was incomplete. Line 175, one of the lines in which the Ds element did not move, is also a line that produced T_1 seed. We conclude the Ac/Ds transposon functions in *Brachypodium*, but is lethal, possibly because it is too active. Similarly, transformations with a construct containing dSpm (pdSpm-R) [50] yielded 22 plantlets, all but one of which died while still very small (**Fig. 2D**). However, all callus pieces transformed with pdSpm-R displayed GFP fluorescence indicating that they were transformed with the construct (**Fig. 2E**). Although there is potential for optimization of transposon systems for the production of mutants in *Brachypodium*, we chose to focus on T-DNA tagging to generate insertional mutants.

Expression of ß–glucuronidase (GUS) from the pJJ2LBP and pJJ2LBP2 gene trap vectors

To assess the function of the gene trap vectors, 235 lines containing pJJ2LBP and 500 lines containing pJJ2LBP2 lines were assayed for GUS activity as an indicator that the promoterless GUS reporter gene was being expressed by a *Brachypodium* promoter. Since the T_1 generation examined was segregating for the transgenes, we first used PCR to identify transgenic T_1 plants by amplifying the *HptII* gene contained on the T-DNA (data not shown). Leaf, stem, and floral samples for each transgenic plant were placed in a GUS staining solution and incubated overnight. Upon visual examination of cleared tissue, 53 (7.2%) of the lines showed GUS expression in at least one tissue type (**Fig. 3A, B, C**). Seven lines showed blue staining in all three tissues sampled (**Fig. 3B**), 14 lines in vegetative tissues only (leaf and stem), and 22 lines in floral tissue only (**Fig. 3C**). The remaining 10 positive lines had GUS expression in flowers and either stem or leaf tissue. Seeds for 95 lines were germinated on MS media containing hygromycin and tested for GUS expression in roots, but blue staining was not detected in the roots (data not shown). Using the vector from which we derived our gene trap cassettes, pGA2717, Ryu et. al. observed GUS staining in 4.8% of 3,140 rice lines [41], a value similar to what we observed in *Brachypodium*. Another study of 8,200 rice lines using a similar vector, pTAG8, reported GUS staining in vegetative tissue for 11% of the lines tested and in reproductive tissue for 22% of the lines tested [42]. These results demonstrate that the GUS reporter derived from the pGA2717 rice gene trap vector functions in *Brachypodium* and suggest there is room for optimization of gene trap vectors for *Brachypodium*.

Phenotypes observed in T-DNA insertional mutant lines

T_1 seeds were planted for nearly 2,000 mutant lines for which insertions had been identified within or near genes (see below for details on insertion sites). The T_1 plants represent the first generation segregating for T-DNA insertions and provided the opportunity to survey for mutant phenotypes and to bulk seeds. Phenotypes were noted in approximately 5% of the lines planted. Classes of phenotypes included: size mutants such as tall, short, or dwarf plants (35 lines); early and late flowering time mutants (28 lines); pigmentation mutants such as albinos, variegated leaves, and mottled leaves (25 lines); weak or sickly plants (5 lines); mutants with altered morphology, including curling stems and

Figure 3. Phenotypes observed in the T-DNA population. A–C. GUS staining of lines created with pJJ2LBP and pJJ2LBP2 vectors. Leaf, stem and floral tissue (arranged left to right) collected from plants transformed with gene trap vectors were stained to detect GUS expression. Ruler divisions are in millimeters. **A.** Blue color was not detected in line JJ5896 indicating an absence of GUS expression. **B.** GUS expression is indicated by the blue color in all three tissue types for line JJ6405. **C.** For line JJ6135 expression was primarily detected in the developing embryo indicated by the black arrow. **D–I.** Images highlighting a selection of mutant phenotypes observed in the T-DNA population. **D.** Wild-type Bd21-3 plant. **E.** Flowering time variation illustrated by a late flowering mutant in the front right of the pot. **F.** Morphology variation represented by a mutant with curling stems and leaves. **G.** Size differences illustrated by the row of short plants on the right. **H.** Increased segmentation, branching and delayed flowering time (left) compared to the wild-type plant (right). **I.** Pigmentation mutants.

leaves or branching variations (8 lines); and fertility mutants (3 lines). A few examples of these phenotypes are shown in **Figure 3**.

Generating flanking sequence tags using inverse PCR

Inverse PCR (IPCR) of genomic DNA from individual transgenic plants was used to obtain sequences adjacent to the T-DNA insertions in mutant lines. Sequences that match the *Brachypodium* genome serve as flanking sequence tags (FSTs) and define the locations of the T-DNA insertion sites (ISs). The IPCR method (**Fig. 4**) depends on restriction enzyme digestion and the ligation of a digestion site within the T-DNA to the nearest genomic digestion site for the same enzyme [57,58]. To determine the best means of recovering FSTs, we performed sequencing reactions from both the LB and the RB of the T-DNA. Multiple enzymes and sequencing primers were also tested. Primers oriented out of the T-DNA (designated T primers) yield sequences directly adjacent to the T-DNA and are directed into the flanking genomic region. Primers oriented into the T-DNA (designated RE primers) result in sequences starting at the genomic restriction enzyme site and are directed toward the T-DNA (**Fig. 4**).

Initial studies tested the efficiency of recovering FSTs after digestion with the enzyme HpyCH4IV. This enzyme was chosen because it is located near both of the T-DNA borders, and therefore, one digestion and ligation reaction could be used for four sequencing reactions. Sequencing from the LB using T primers (LB-T) returned FSTs for 232 of 567 lines tested (40.9%), and reactions from the LB using RE primers (LB-RE) produced FSTs for only 189 of the same 567 lines (33.3%). Similar results were observed when sequencing from the RB. Reactions using T primers (RB-T) returned FSTs for 178 of 378 lines tested (47.1%) and reactions using the RE primers (RB-RE) reactions produced FSTs for 172 of 474 lines tested (36.3%). These results indicate that at both borders, sequencing directly from the T-DNA into the genomic sequence using T primers was more efficient at generating FSTs than sequencing from the genomic restriction site back toward the T-DNA using RE primers. In all four sets of reactions described above (LB-T, LB-RE, RB-T, and RB-RE), the majority of the sequences recovered contained vector sequences (51–61%), and 7–15% of the sequences failed to match any known sequence or were not readable due to low quality scores. A comparison of early sequencing reactions from the pJJ vector

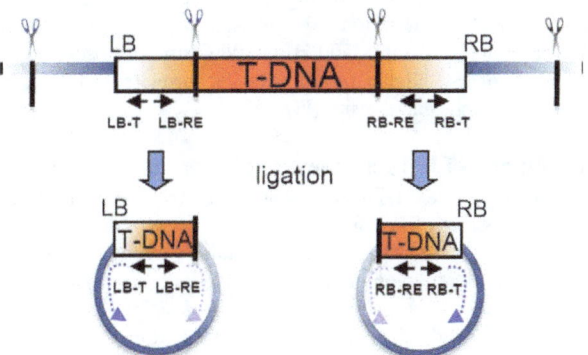

Figure 4. IPCR strategy for obtaining T-DNA flanking sequences. The diagram illustrates the IPCR strategy used to obtain sequences flanking the T-DNA insertion sites (ISs). Restriction enzymes (shown as scissors) with recognition sites (black lines) near the T-DNA border sequences (LB and RB) were used to digest DNA from T-DNA insertion lines. Enzymes cut both within the T-DNA and genomic sequence, and ligations were performed to circularize purified digestion products. PCR of the ligation products was performed using T-DNA specific primers and sequencing was performed with nested primers. The orientations and locations of primers with respect to restriction sites are shown as black arrows. Primers located within the T-DNA directed toward the junction with genomic DNA are designated T primers (LB-T and RB-T). Sequencing reactions using these primers return genomic sequence directly adjacent to the IS (dark blue dotted arrows). Primers directed into the T-DNA are designated RE primers (LB-RE and RB-RE). After enzyme digestion and ligation, sequencing reactions using these primers return genomic sequence starting from the closest restriction enzyme recognition site within the *Brachypodium* genomic sequence and directed toward the T-DNA insertion (light blue dotted arrows).

derivatives showed that the addition of a second LB did not improve the efficiency of FST recovery (data not shown).

In an effort to increase the efficiency of recovering FSTs, IPCR was conducted using two enzymes adjacent to the RB of the pJJ vectors. The two enzymes, HpyCH4IV and HpaII, were used in RB-T reactions for 852 T-DNA lines. HpyCH4IV digestion resulted in FSTs for 350 lines (41.1%) and HpaII digestion returned FSTs for 344 lines (40.4%). In combination, FSTs for 433 lines (50.8%) were obtained from the IPCR reactions after digestion with the two enzymes. The two enzyme approach increased the efficiency of obtaining FSTs from the RB approximately 25% over the single enzyme approach. When LB-T reactions were included for the HpyCH4IV digestion, the number of lines with FSTs reached 531 (62.3%), further increasing the efficiency by more than 20%.

In a separate set of tests, FST recovery was evaluated when IPCR reactions were performed after digestion with three different enzymes. Using BfaI in LB reactions, we obtained FSTs for 38.6% of the lines tested. In HpyCH4IV reactions from the LB and RB, we recovered FSTs for 55.3% of the lines. Together, these two enzymes yielded FSTs for 66.4% of the lines tested. Addition of IPCR reactions from the RB using a third enzyme, TaqI, only increased the total to 67.4% of the lines tested. As a result of these experiments, we decided to use the two borders, two enzymes approach to obtain FSTs for our T-DNA lines.

Identifying flanking sequence tags and assigning insertion sites

Data from 17,637 inverse PCR sequencing reactions for 7,145 T-DNA lines were compared to the *Brachypodium distachyon* genome assembly v1.0 using BLASTn (**Table 4**). To maximize the detection of T-DNA insertion loci, we assigned an e-value cutoff of 10^{-3}. Using this criterion, 7,389 sequences (41.9%) matched the *Brachypodium* genome. These sequences represent 4,402 (61.6%) of the lines analyzed. The remaining 10,248 sequences that failed to match the *Brachypodium* genome were vector sequences, sequences with no identified matches, or poor quality sequences. The top scoring BLAST hit for each FST was used for subsequent analysis. The only exceptions were FSTs that exactly matched more than one location in the genome (discussed below). The average FST length was 195 bases and the median length 142 bases. We defined the base of the FST closest to the T-DNA to be the insertion site (IS). However, since we often did not sequence across the junction, the actual IS may differ slightly. Multiple sequencing reactions were performed for many of the T-DNA lines using T and RE primers that anneal near the RB or LB. Since independent reactions may return different sequences for a single line, a particular line can have FSTs assigned to more than one location in the genome. This is not surprising because the average number of T-DNA insertions per line is ~2 [18]. When a single line has multiple FSTs, they are distinguished by the addition of a numerical suffix to the name of the T-DNA line. For example, four sequencing reactions were performed for the T-DNA line JJ3, yielding four FSTs. The FSTs were designated as JJ3.0, JJ3.1, JJ3.2, and JJ3.3. The JJ3.0 FST is located on chromosome 1, and the other three FSTs are assigned to chromosome 2. When a T-DNA line has more than one FST located on the same chromosome within a 1 kb range, the FSTs were treated as a single IS. For example, JJ3.1 and JJ3.2 are both located on Bd2 separated by only 284 bases, and therefore are counted as one T-DNA IS. Using this classification, the 7,389 FSTs in this collection represent 5,285 distinct ISs. Of these sites, 1,538 (29.1%) ISs are supported by more than one FST (in 1,501 lines) (**Tables 5 and 6**).

Table 4. Flanking sequence tags (FSTs) generated for T-DNA lines.

Sequences	Number of sequences	Percentage of total sequences
Generated	17,637	100.0
BLASTed vs *B. distachyon* genome	17,599	99.9
Matches to *B. distachyon* genome	7,389	42.0
Unique insertion sites (IS)	5,285	30.0
Lines	**Number of lines**	**Percentage of lines**
Sequenced	7,145	100.0
FST assigned	4,402	61.6
No FST assigned	2,743	38.4
FST assigned to sterile lines	251	5.7

In the majority of the T-DNA lines with assigned FSTs, one (82.4%) or two (15.4%) ISs were identified (**Table 5**), and fewer than 3.0% of the lines were assigned three or more ISs. In cases where the T-DNA has inserted into a repetitive sequence, the BLAST search returned multiple hits with equal scores and prevented assignment of an unambiguous IS. We observed this for 0.6% of the lines (28 lines) with FSTs. There are two primary locations in the genome where these insertions mapped. One line (IJ4195) returned a sequence that was assigned to 29 loci near the centromere of chromosome Bd4 in a region that contains *Brachypodium* centromeric retroelements, and 27 lines gave sequences that were assigned to 649 loci in the first 215 kb of the short arm of Bd5 in a region encoding 26S ribosomal RNA genes. To simplify our analyses we only used one IS for each FST. For the majority of the T-DNA lines, the single FST assigned from the top scoring BLAST hit for each sequencing reaction is displayed on the USDA-ARS-WRRC T-DNA website (described in a later section). However, for the insertions in repetitive regions, each of the potential genomic locations for the equal scoring BLAST hits is displayed.

Distribution of ISs within the *Brachypodium* genome

The distribution of ISs across the five *Brachypodium* chromosomes was analyzed by plotting the number of insertions within 500 kb windows moving across the length of each chromosome starting from the beginning of the short arm to the end of the long arm (**Fig. 5**). Insertions span the entire length of all five chromosomes. For chromosomes 1, 2, 3, and 5, the number of insertion sites detected per chromosome is generally proportional

to the chromosome length and the distribution ranges from 19.2 to 20.9 ISs/Mb with an average of 20.2 ISs/Mb (**Table 7**). Fewer ISs were detected on chromosome 4 relative to its size (15.9 ISs/Mb) than were observed for the other chromosomes and results in an average over the entire genome of 19.3 ISs/Mb. Overall, the number of genes/Mb is directly proportional to the length of the chromosome. Thus, the lower number of ISs/Mb on chromosome 4 may be partly attributable to the lower number of genes/Mb on this chromosome (**Table 7**). A non-uniform distribution of ISs similar to that reported for T-DNA insertions in rice [57–59], *Arabidopsis* [60], and the BrachyTAG [48] collection also is observed in the WRRC population. In general, the distal ends of the chromosomes have a greater density of insertions, and fewer insertions are detected near the centromeric regions (**Fig. 5**). The *Brachypodium* v1.0 annotation [8] was used to plot the number of genes present in the same 500 kb windows used to visualize the distribution of ISs. Higher numbers of ISs correlate well with the regions of higher gene density (**Fig. 5**).

Distributions of FSTs in genic and intergenic regions

To analyze the T-DNA distribution between genic and intergenic regions (**Fig. 6 and Table 8**), we compared the assigned IS loci to the v1.0 annotation reported by the International Brachypodium Initiative [8]. The report identified 25,532 protein coding genes in the *Brachypodium* genome with an average gene length of 3,336 bases, including exons, introns, and UTRs. This represents 31.3% of the 272 Mb genome. In our population, 28.4% of the ISs (1,499) reside within genes, a value slightly lower than the percentage of the genome assigned to genes. To more precisely describe the ISs we further categorized

Table 5. Number of T-DNA integration events detected per line.

IS detected/line	Number of lines	Percentage of lines with FSTs
1	3,629	82.4
2	678	15.4
3	84	1.9
4	8	0.2
5	2	<0.1
6	1	<0.1
Total	**4,402**	**100**

Table 6. Number of FSTs supporting individual ISs.

FST/IS	Number if ISs	Percentage of ISs
1	3,747	70.9
2	1,109	21.0
3	315	6.0
4	95	1.8
5	15	0.3
6	4	<0.1
Total	**5,285**	**100**

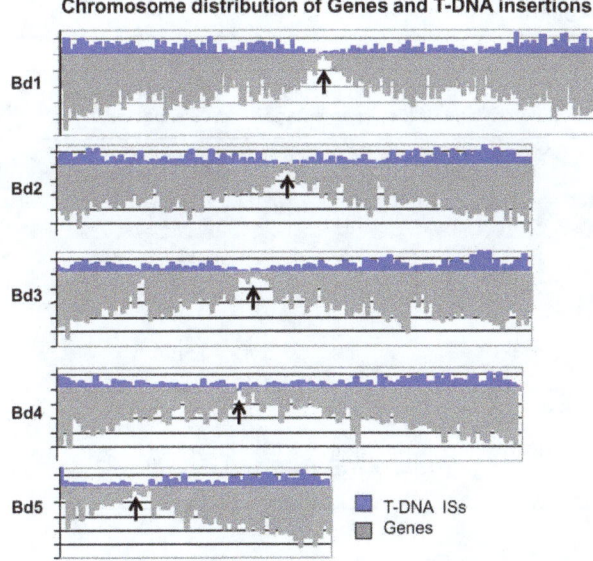

Chromosome distribution of Genes and T-DNA insertions

Bd1

Bd2

Bd3

Bd4

Bd5

■ T-DNA ISs
■ Genes

Figure 5. Distribution of Genes and T-DNA insertions along the chromosomes. T-DNA insertions (blue) and Brachypodium genes (gray) are plotted in 500 kb windows moving across the length of each chromosome. Each division on the Y-axis represents 20 insertions sites or genes, respectively. Chromosome numbers are listed to the left. Insertions span the length of all five chromosomes and are positively correlated with genic regions. A greater density of insertions is observed near the distal ends of the chromosomes than is observed near the centromeres (approximate locations indicated by black arrows).

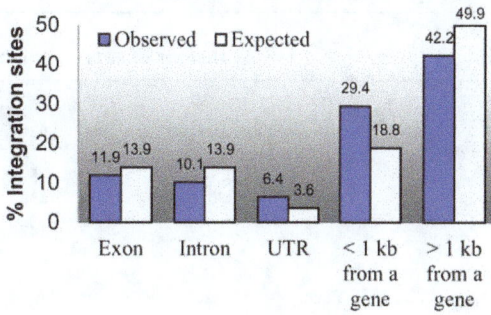

Distribution of integration sites between genic and intergenic regions

Figure 6. Distribution of insertion sites between genic and intergenic regions. A comparison of the observed and expected percentages of T-DNA insertions is illustrated for each insertion class. Insertions in UTRs and within 1 kb of a gene are observed to be substantially higher than expected. Intergenic insertions >1 kb from a gene are lower than expected.

insertions into genes as insertions into exons, introns, and UTRs. We estimate the percentage of the genome represented by each of these classes as 13.9% exon sequences, 13.9% introns and 3.6% UTRs. It should be noted that UTRs are significantly underestimated because sequences were only annotated as UTRs if there was EST support for the UTR. Using these estimates, we observed a lower number of T-DNA insertions in exons (11.9%) and in introns (10.1%) compared to what would be expected from the percentages of the genome represented by these features. This is in accordance with the lower than expected proportion of T-DNA insertions observed for whole genes. In contrast, the 6.4% (338) of the T-DNA insertions assigned to UTRs is nearly twice the amount expected (3.6%) based on the percentage of the genome assigned to UTRs. In our population, T-DNA insertions slightly favored 3' UTRs (191, 3.7%) over 5' UTRs (147, 2.6%). We were also interested in identifying insertions near genes, because sequences adjacent to genes often encode elements critical for

the regulation of gene expression. To do this, we defined an IS to be near a gene if it was located within 1 kb of the 5' or 3' end of a gene model included in v1.0 annotation of the *Brachypodium* genome. The 2,000 bases flanking each of the 25,532 *Brachypodium* genes represent 18.8% of the genome. Of the ISs in intergenic space, 29.4% (1,556) are present within 1 kb of gene sequences. Similar to what is observed for UTRs, the integration of T-DNAs near genes is significantly higher than the proportion expected, but in this case, T-DNA insertions were more often observed 5' of genes (931, 59.2%) than 3' of genes (625, 40.8%).

Online resources for accessing the USDA-ARS Western Regional Research Center (WRRC) T-DNA collection

The goal of this project was to add to the growing collection of genomic resources available for *Brachypodium* by creating a large collection of sequence-indexed T-DNA lines. Mutant lines are available to interested researchers through a link from the *Brachypodium* resource page of the Genomics and Gene Discovery Research Unit at the USDA-ARS, WRRC (http://brachypodium.pw.usda.gov/). The WRRC T-DNA website (http://brachypodium.pw.usda.gov/TDNA/) includes a GBrowse window for visualization of FSTs in the context of adjacent genomic features and a window for BLAST searches against the regions adjacent to T-DNA insertion sites. In addition, an Excel table listing FST details (**Table S2**) and FASTA file containing the genomic regions flanking T-DNA ISs (**File S1**) are available for download. Instructions for ordering lines are provided. In

Table 7. Chromosome distribution of FSTs, ISs, and genes.

Chromosome	Mb	# FSTs		FST/Mb	ISs	ISs/Mb	genes/Mb
Bd1	74.8	2,201		29.4	1,568	20.9	99.8
Bd2	59.3	1,738		29.3	1,234	20.8	97.6
Bd3	59.9	1,601		26.7	1,148	19.2	94.1
Bd4	48.6	1,092		22.5	774	15.9	86.1
Bd5	28.3	757		26.7	561	19.8	85.9
Total	**271**	**7,389**	Avg.	**26.9**	**5,285**	**19.3**	**92.7**

Table 8. Distribution of FSTs and ISs between genic and intergenic regions.

Insertion class	Number of FSTs	Number of ISs	Percentage of ISs	Expected percentage	Difference from expected
Insertions in genes (intron, exon, UTR)	2,025	1,499	28.4	31.3	−2.9
Exon	855	629	11.9	13.9	−2.0
Intron	705	532	10.1	13.9	−3.8
UTR	465	338	6.4	3.6	2.8
5′ UTR	192	147	2.8	-	-
3′ UTR	273	191	3.6	-	-
Intergenic insertions <1 kb from a gene	2,203	1,556	29.4	18.8	10.6
5′ of gene	1327	931	17.6	-	-
3′ of gene	876	625	11.8	-	-
Intergenic >1 kb from a gene	3,161	2,230	42.2	49.9	−7.7
Total	7,389	5,285			

addition, the WRRC T-DNA collection can be viewed as a track in the http://www.brachypodium.org/ Gbrowse window.

Discussion

By modifying our transformation protocol we were able to significantly increase transformation efficiency and reduce the length of time necessary to generate transgenic plants. Specifically, we eliminated the recovery step where callus was placed on callus inducing media without hygromycin for a week after co-cultivation. This reduced the transformation time by one week and eliminated the labor and materials required for one transfer. In studies evaluating different transformation vectors, we found hygromycin selection to be more efficient than BASTA for the production of transgenic plants. After co-cultivation, callus selected using BASTA required two subculture steps prior to regeneration, whereas higher transformation efficiencies were achieved from hygromycin-selected callus transferred to regeneration media after only one subculture. Our vector comparisons also demonstrated that the promoter driving the selectable marker greatly affected transformation efficiency with the maize ubiquitin promoter demonstrating the highest efficiency among the promoters tested. Using our optimized protocol we achieved an average transformation efficiency of 42%. Significantly, this high efficiency was achieved in a production setting where calluses were transferred and discarded on a set timetable to minimize labor and space required per transgenic line produced. These improvements provide considerable time and cost savings when transformations are conducted on a large scale.

Using our optimized transformation method, we generated 8,491 fertile T-DNA lines making the WRRC collection the largest collection of T-DNA lines in any grass with the exception of rice. To increase the utility of this collection, we used inverse PCR to sequence the DNA flanking the insertion sites. Our initial experiments focused on optimizing the IPCR method. We found 20–26% higher FST recovery when we performed sequencing reactions directly from the T-DNA into the genomic sequence compared to reactions from the genomic restriction site back toward the T-DNA. Adding sequencing reactions generated from a second enzyme digestion near the RB increased recovery of FSTs by approximately 25%, and we increased FST recovery by 20% when we included reactions from the LB. However, adding a third enzyme at the RB only produced a marginal increase in FST

identification, and vectors with two LBs did not improve FST recovery.

In total, we identified 5,285 specific insertion sites in the *Brachypodium* genome. We successfully identified T-DNA insertions

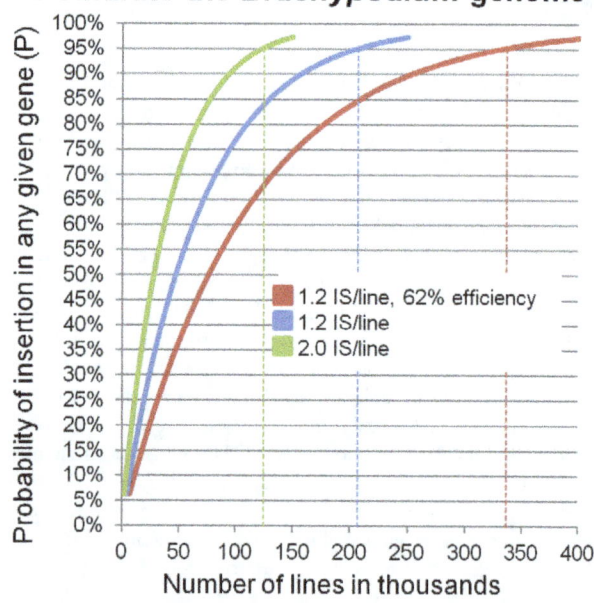

Figure 7. Estimated numbers of T-DNA mutants needed to saturate the *Brachypodium* genome. The number of T-DNA mutants needed for saturation mutagenesis of the *Brachypodium* genome is estimated using the formula $\{P = 1 - (1 - [x/g])^n\}$. P is the probability of finding an insertion in a particular gene; g is the genome size in kb; x is the average gene length in kb; n is the number of T-DNA insertions needed. The red line shows the number of lines estimated using our current sequencing efficiency of identifying FSTs in 62% of the lines tested with an average of 1.2 IS/line. The blue and green lines illustrate estimates based on identifying FSTs in 100% of the lines with an average of 1.2 IS/line or 2.0 IS/line, respectively. Dashed lines indicate the number of lines needed for P = 0.95 for each efficiency of identifying insertion sites.

in 62% of the lines that we sequenced and found an average of 1.2 insertions sites per line. These sites represent 1,499 insertions in genes and another 2,203 insertions within 1 kb of a gene. This latter class of insertions may alter gene expression if they lie within regulatory regions or if the T-DNA is an activation tagging construct. The WRRC collection comprises mutants with insertions in or near (within 1 kb) 8.8% (2,245) of the annotated *Brachypodium* genes and represents a significant addition to existing collections of available *Brachypodium* insertional mutants. However, the number of lines is far from saturation. The formula $\{P = 1 - (1 - [x/g])^n\}$ [61] can be used to calculate the probability (P) of finding an insertion in a particular gene ("g" is the genome size in kb, "x" is the average gene length in kb, and "n" is the number of T-DNA insertions needed). Applying the above formula to *Brachypodium* assuming our current 62% efficiency of retrieving a useful FST from a line and an average of 1.2 T-DNA insertions detected per line, we calculate that a collection of 76,000 lines would have a 50% probability of containing an insertion in any average 3.3 kb *Brachypodium* gene (**Fig. 7**). To reach P = 0.95, a collection on the order of 329,000 lines would be necessary. Due to redundancy of hits, the first T-DNA insertions will hit the highest diversity of genes. Thus, even the smaller P = 0.5 collection would have great utility. These are large, yet not unrealistic, numbers. This estimation assumes random integration, but because T-DNA insertions are preferentially identified near genes, fewer lines should be needed to reach saturation. Furthermore, the utility of the existing collections can be increased through improvements in detection of insertions missed in the first sequencing attempts (**Fig. 7**). Current efforts are focused on identifying insertion sites in the 38% of the lines of the WRRC collection lacking FSTs and increasing the average number of T-DNA insertions detected to approach the expected number of 2 per line (estimated at 9,000 additional insertion sites).

The insertion sites are accessible through the project website (http://brachypodium.pw.usda.gov/) where they can be searched by BLAST, Gbrowse or downloaded as a table. Lines are freely available to anyone in the scientific community. The WRRC T-DNA collection represents a significant and growing resource for plant science research.

Supporting Information

Table S1   Cloning primer sequences.

Table S2   Excel file containing a complete list of FST sequences with their genomic locations and sequencing information.

File S1   Text file containing a list of all FST sequences in FASTA format.

Acknowledgments

We would like to thank Andy Pereira and Venkatesan Sundaresan for transposon constructs. We thank Jeannie Lin and Jim Thomson for the pUbi-BASK vector. We thank Todd Mockler and Henry Priest for incorporating our FST data into Brachybase.

Author Contributions

Conceived and designed the experiments: JNB JW YQG JPV. Performed the experiments: JNB JW SPG MAG RLT. Analyzed the data: JNB JW SPG MAG RLT GRL YQG JPV. Contributed reagents/materials/analysis tools: JNB JW SPG MAG RLT GRL YQG JPV. Wrote the paper: JNB JW SPG MAG RLT GRL YQG JPV.

References

1. Catalán P, Müller J, Hasterok R, Jenkins G, Mur LAJ, et al. (2012) Evolution and taxonomic split of the model grass *Brachypodium distachyon*. Annals of Botany. Available:http://www.ncbi.nlm.nih.gov/pubmed/22213013. Accessed 26 January 2012.

2. Draper J, Mur LA, Jenkins G, Ghosh-Biswas GC, Bablak P, et al. (2001) *Brachypodium distachyon*. A new model system for functional genomics in grasses. Plant Physiol 127: 1539–1555.

3. (2006) Breaking the Biological Barriers to Cellulosic Ethanol: A Joint Research Agenda, US. Department of Energy, Office of Science and Office of Energy Efficiency Available at: http://genomicsgtl.energy.gov/biofuels/.

4. Carpita NC (1996) Structure and biogenesis of the cell walls of grasses. Annu Rev Plant Physiol Plant Mol Biol 47: 445–476. doi:10.1146/annurev.arplant.47.1.445.

5. Vogel J (2008) Unique aspects of the grass cell wall. Curr Opin Plant Biol 11: 301–307. doi:10.1016/j.pbi.2008.03.002.

6. Bevan MW, Garvin DF, Vogel JP (2010) *Brachypodium distachyon* genomics for sustainable food and fuel production. Curr Opin Biotechnol 21: 211–217. doi:10.1016/j.copbio.2010.03.006.

7. Garvin D, Gu Y, Hasterok R, Hazen SP, Jenkins G, et al. (2008) Development of genetic and genomic research resources for *Brachypodium distachyon*, a new model system for grass crop research. The Plant Genome 48: 69–84.

8. International *Brachypodium* Initiative. (2010) Genome sequencing and analysis of the model grass *Brachypodium distachyon*. Nature 463: 763–768. doi:10.1038/nature08747.

9. Brkljacic J, Grotewold E, Scholl R, Mockler T, Garvin DF, et al. (2011) *Brachypodium* as a model for the grasses: today and the future. Plant Physiol 157: 3–13. doi:10.1104/pp.111.179531.

10. Bragg J, Tyler L, Vogel JP (2010) *Brachypodium distachyon*, a model for bioenergy crops. Handbook of Bioenergy Crop Plants. CRC Press.

11. Mur LAJ, Allainguillaume J, Catalán P, Hasterok R, Jenkins G, et al. (2011) Exploiting the *Brachypodium* Tool Box in cereal and grass research. New Phytol 191: 334–347. doi:10.1111/j.1469-8137.2011.03748.x.

12. Vogel JP, Tuna M, Budak H, Huo N, Gu YQ, et al. (2009) Development of SSR markers and analysis of diversity in Turkish populations of *Brachypodium distachyon*. BMC Plant Biol 9: 88. doi:10.1186/1471-2229-9-88.

13. Wiebe K, Harris NS, Faris JD, Clarke JM, Knox RE, et al. (2010) Targeted mapping of Cdu1, a major locus regulating grain cadmium concentration in durum wheat (*Triticum turgidum* L. var *durum*). Theor Appl Genet 121: 1047–1058. doi:10.1007/s00122-010-1370-1.

14. Rosewarne GM, Singh RP, Huerta-Espino J, Herrera-Foessel SA, Forrest KL, et al. (2012) Analysis of leaf and stripe rust severities reveals pathotype changes and multiple minor QTLs associated with resistance in an Avocet×Pastor wheat population. Theor Appl Genet 124: 1283–1294. doi: 10.1007/s00122-012-1786-x

15. Bouteillé M, Rolland G, Balsera C, Loudet O, Muller B (2012) Disentangling the intertwined genetic bases of root and shoot growth in Arabidopsis. PLoS ONE 7: e32319. doi:10.1371/journal.pone.0032319.

16. Brown AN, Lauter N, Vera DL, McLaughlin-Large KA, Steele TM, et al. (2011) QTL mapping and candidate gene analysis of telomere length control factors in maize (*Zea mays* L.). G3 (Bethesda) 1: 437–450. doi:10.1534/g3.111.000703.

17. Liu L, Stein A, Wittkop B, Sarvari P, Li J, et al. (2012) A knockout mutation in the lignin biosynthesis gene *CCR1* explains a major QTL for acid detergent lignin content in *Brassica napus* seeds. Theor Appl Genet 124: 1573–1586. doi: 10.1007/s00122-012-1811-0.

18. Vogel J, Hill T (2008) High-efficiency *Agrobacterium*-mediated transformation of *Brachypodium distachyon* inbred line Bd21-3. Plant Cell Rep 27: 471–478. doi:10.1007/s00299-007-0472-y.

19. Vain P, Worland B, Thole V, McKenzie N, Alves SC, et al. (2008) *Agrobacterium*-mediated transformation of the temperate grass *Brachypodium distachyon* (genotype Bd21) for T-DNA insertional mutagenesis. Plant Biotechnol J 6: 236–245. doi:10.1111/j.1467-7652.2007.00308.x.

20. Huo N, Gu YQ, Lazo GR, Vogel JP, Coleman-Derr D, et al. (2006) Construction and characterization of two BAC libraries from *Brachypodium distachyon*, a new model for grass genomics. Genome 49: 1099–1108. doi:10.1139/g06-087.

21. Huo N, Lazo GR, Vogel JP, You FM, Ma Y, et al. (2008) The nuclear genome of *Brachypodium distachyon*: analysis of BAC end sequences. Funct Integr Genomics 8: 135–147. doi:10.1007/s10142-007-0062-7.

22. Hasterok R, Marasek A, Donnison IS, Armstead I, Thomas A, et al. (2006) Alignment of the genomes of *Brachypodium distachyon* and temperate cereals and grasses using bacterial artificial chromosome landing with fluorescence in situ hybridization. Genetics 173: 349–362. doi:10.1534/genetics.105.049726.

23. Sonah H, Deshmukh RK, Sharma A, Singh VP, Gupta DK, et al. (2011) Genome-wide distribution and organization of microsatellites in plants: an

insight into marker development in *Brachypodium*. PLoS ONE 6: e21298. doi:10.1371/journal.pone.0021298.

24. Garvin DF, McKenzie N, Vogel JP, Mockler TC, Blankenheim ZJ, et al. (2010) An SSR-based genetic linkage map of the model grass *Brachypodium distachyon*. Genome 53: 1–13. doi:10.1139/g09-079.

25. Gu YQ, Ma Y, Huo N, Vogel JP, You FM, et al. (2009) A BAC-based physical map of *Brachypodium distachyon* and its comparative analysis with rice and wheat. BMC Genomics 10: 496. doi:10.1186/1471-2164-10-496.

26. Huo N, Garvin DF, You FM, McMahon S, Luo M-C, et al. (2011) Comparison of a high-density genetic linkage map to genome features in the model grass *Brachypodium distachyon*. Theor Appl Genet 123: 455–464. doi:10.1007/s00122-011-1598-4.

27. Vogel J, Garvin D, Leong O, Hayden D (2006) *Agrobacterium*-mediated transformation and inbred line development in the model grass *Brachypodium distachyon*. Plant Cell Tissue Organ Culture 85: 199–211.

28. Filiz E, Ozdemir BS, Budak F, Vogel JP, Tuna M, et al. (2009) Molecular, morphological, and cytological analysis of diverse *Brachypodium distachyon* inbred lines. Genome 52: 876–890. doi:10.1139/g09-062.

29. Higgins JA, Bailey PC, Laurie DA (2010) Comparative genomics of flowering time pathways using *Brachypodium distachyon* as a model for the temperate grasses. PLoS ONE 5: e10065. doi:10.1371/journal.pone.0010065.

30. Cao S, Kumimoto RW, Siriwardana CL, Risinger JR, Holt BF 3rd (2011) Identification and characterization of NF-Y transcription factor families in the monocot model plant *Brachypodium distachyon*. PLoS ONE 6: e21805. doi:10.1371/journal.pone.0021805.

31. Faricelli ME, Valárik M, Dubcovsky J (2010) Control of flowering time and spike development in cereals: the earliness per se *Eps-1* region in wheat, rice, and *Brachypodium*. Funct Integr Genomics 10: 293–306. doi:10.1007/s10142-009-0146-7.

32. Barrero JM, Jacobsen JV, Talbot MJ, White RG, Swain SM, et al. (2012) Grain dormancy and light quality effects on germination in the model grass *Brachypodium distachyon*. New Phytol 193: 376–386. doi:10.1111/j.1469-8137.2011.03938.x.

33. Guillon F, Larré C, Petipas F, Berger A, Moussawi J, et al. (2012) A comprehensive overview of grain development in *Brachypodium distachyon* variety Bd21. J Exp Bot 63: 739–755. doi:10.1093/jxb/err298.

34. Guillon F, Bouchet B, Jamme F, Robert P, Quéméner B, et al. (2011) *Brachypodium distachyon* grain: characterization of endosperm cell walls. J Exp Bot 62: 1001–1015. doi:10.1093/jxb/erq332.

35. Yordem BK, Conte SS, Ma JF, Yokosho K, Vasques KA, et al. (2011) *Brachypodium distachyon* as a new model system for understanding iron homeostasis in grasses: phylogenetic and expression analysis of Yellow Stripe-Like (YSL) transporters. Ann Bot 108: 821–833. doi:10.1093/aob/mcr200.

36. Mochida K, Yoshida T, Sakurai T, Yamaguchi-Shinozaki K, Shinozaki K, et al. (2011) In silico analysis of transcription factor repertoires and prediction of stress-responsive transcription factors from six major gramineae plants. DNA Res 18: 321–332. doi:10.1093/dnares/dsr019.

37. Christensen U, Alonso-Simon A, Scheller HV, Willats WGT, Harholt J (2010) Characterization of the primary cell walls of seedlings of *Brachypodium distachyon*-a potential model plant for temperate grasses. Phytochemistry 71: 62–69. doi:10.1016/j.phytochem.2009.09.019.

38. Peraldi A, Beccari G, Steed A, Nicholson P (2011) *Brachypodium distachyon*: a new pathosystem to study Fusarium head blight and other Fusarium diseases of wheat. BMC Plant Biol 11: 100. doi:10.1186/1471-2229-11-100.

39. Berkman PJ, Skarshewski A, Manoli S, Lorenc MT, Stiller J, et al. (2012) Sequencing wheat chromosome arm 7BS delimits the 7BS/4AL translocation and reveals homoeologous gene conservation. Theor Appl Genet 124: 423–432. doi:10.1007/s00122-011-1717-2.

40. Massa AN, Wanjugi H, Deal KR, O'Brien K, You FM, et al. (2011) Gene space dynamics during the evolution of *Aegilops tauschii*, *Brachypodium distachyon*, *Oryza sativa*, and *Sorghum bicolor* genomes. Mol Biol Evol 28: 2537–2547. doi:10.1093/molbev/msr080.

41. Ryu C-H, You J-H, Kang H-G, Hur J, Kim Y-H, et al. (2004) Generation of T-DNA tagging lines with a bidirectional gene trap vector and the establishment of an insertion-site database. Plant Mol Biol 54: 489–502. doi:10.1023/B:PLAN.0000038257.93381.05.

42. Yu S-M, Ko S-S, Hong C-Y, Sun H-J, Hsing Y-I, et al. (2007) Global functional analyses of rice promoters by genomics approaches. Plant Molecular Biology 65: 417–425. doi:10.1007/s11103-007-9232-1.

43. Weigel D, Ahn JH, Blázquez MA, Borevitz JO, Christensen SK, et al. (2000) Activation tagging in Arabidopsis. Plant Physiol 122: 1003–1014. doi:10.1104/pp.122.4.1003.

44. Jeon J, Lee S, Jung K, Jun S, Jeong D, et al. (2000) T-DNA insertional mutagenesis for functional genomics in rice. Plant Journal 22: 561–570.

45. Wan S, Wu J, Zhang Z, Sun X, Lv Y, et al. (2008) Activation tagging, an efficient tool for functional analysis of the rice genome. Plant Molecular Biology 69: 69–80. doi:10.1007/s11103-008-9406-5.

46. Bablak P, Draper J, Davey M, Lynch P (1995) Plant regeneration and micropropagation of *Brachypodium distachyon*. Plant Cell, Tissue and Organ Cult 42: 97–107.

47. Christiansen P, Andersen CH, Didion T, Folling M, Nielsen KK (2005) A rapid and efficient transformation protocol for the grass *Brachypodium distachyon*. Plant Cell Rep 23: 751–758. doi:10.1007/s00299-004-0889-5.

48. Thole V, Worland B, Wright J, Bevan MW, Vain P (2010) Distribution and characterization of more than 1000 T-DNA tags in the genome of *Brachypodium distachyon* community standard line Bd21. Plant Biotechnol J 8: 734–747. doi:10.1111/j.1467-7652.2010.00518.x.

49. Trijatmiko K (2005) Activation tagging using the En-I and Ac-Ds maize transposon systems in rice. In: Comparative analysis of drought resistance genes in Arabidopsis and rice. Wageningen, The Netherlands: Wageningen University.

50. Kumar CS, Wing RA, Sundaresan V (2005) Efficient insertional mutagenesis in rice using the maize En/Spm elements. Plant J 44: 879–892. doi:10.1111/j.1365-313X.2005.02570.x.

51. Thole V, Worland B, Snape JW, Vain P (2007) The pCLEAN dual binary vector system for *Agrobacterium*-mediated plant transformation. Plant Physiol 145: 1211–1219. doi:10.1104/pp.107.108563.

52. Christensen AH, Quail PH (1996) Ubiquitin promoter-based vectors for high-level expression of selectable and/or screenable marker genes in monocotyledonous plants. Transgenic Research 5: 213–218. doi:10.1007/BF01969712.

53. Podevin N, De Buck S, De Wilde C, Depicker A (2006) Insights into recognition of the T-DNA border repeats as termination sites for T-strand synthesis by *Agrobacterium tumefaciens*. Transgenic Res 15: 557–571. doi:10.1007/s11248-006-9003-9.

54. Lazo GR, Stein PA, Ludwig RA (1991) A DNA transformation-competent Arabidopsis genomic library in *Agrobacterium*. Biotechnology (NY) 9: 963–967.

55. Mori M, Tomita C, Sugimoto K, Hasegawa M, Hayashi N, et al. (2007) Isolation and molecular characterization of a Spotted leaf 18 mutant by modified activation-tagging in rice. Plant Mol Biol 63: 847–860. doi:10.1007/s11103-006-9130-y.

56. Jeong D-H, An S, Kang H-G, Moon S, Han J-J, et al. (2002) T-DNA insertional mutagenesis for activation tagging in rice. Plant Physiol 130: 1636–1644. doi:10.1104/pp.014357.

57. Jeong D-H, An S, Park S, Kang H-G, Park G-G, et al. (2006) Generation of a flanking sequence-tag database for activation-tagging lines in japonica rice. Plant J 45: 123–132. doi:10.1111/j.1365-313X.2005.02610.x.

58. An S, Park S, Jeong D-H, Lee D-Y, Kang H-G, et al. (2003) Generation and analysis of end sequence database for T-DNA tagging lines in rice. Plant Physiol 133: 2040–2047. doi:10.1104/pp.103.030478.

59. Zhang J, Guo D, Chang Y, You C, Li X, et al. (2007) Non-random distribution of T-DNA insertions at various levels of the genome hierarchy as revealed by analyzing 13 804 T-DNA flanking sequences from an enhancer-trap mutant library. Plant J 49: 947–959. doi:10.1111/j.1365-313X.2006.03001.x.

60. Alonso JM, Stepanova AN, Leisse TJ, Kim CJ, Chen H, et al. (2003) Genome-wide insertional mutagenesis of *Arabidopsis thaliana*. Science 301: 653–657. doi:10.1126/science.1086391.

61. Krysan PJ, Young JC, Sussman MR (1999) T-DNA as an insertional mutagen in Arabidopsis. Plant Cell 11: 2283–2290. doi:10.1105/tpc.11.12.2283.

Development of an Agrobacterium-Mediated Stable Transformation Method for the Sensitive Plant *Mimosa pudica*

Hiroaki Mano[1], Tomomi Fujii[2], Naomi Sumikawa[1¤a], Yuji Hiwatashi[1,2¤b], Mitsuyasu Hasebe[1,2]*

1 Division of Evolutionary Biology, National Institute for Basic Biology, Okazaki, Japan, **2** School of Life Science, Graduate University for Advanced Studies, Okazaki, Japan

Abstract

The sensitive plant *Mimosa pudica* has long attracted the interest of researchers due to its spectacular leaf movements in response to touch or other external stimuli. Although various aspects of this seismonastic movement have been elucidated by histological, physiological, biochemical, and behavioral approaches, the lack of reverse genetic tools has hampered the investigation of molecular mechanisms involved in these processes. To overcome this obstacle, we developed an efficient genetic transformation method for *M. pudica* mediated by *Agrobacterium tumefaciens* (Agrobacterium). We found that the cotyledonary node explant is suitable for Agrobacterium-mediated transformation because of its high frequency of shoot formation, which was most efficiently induced on medium containing 0.5 µg/ml of a synthetic cytokinin, 6-benzylaminopurine (BAP). Transformation efficiency of cotyledonary node cells was improved from almost 0 to 30.8 positive signals arising from the intron-sGFP reporter gene by using Agrobacterium carrying a super-binary vector pSB111 and stabilizing the pH of the co-cultivation medium with 2-(*N*-morpholino)ethanesulfonic acid (MES) buffer. Furthermore, treatment of the explants with the detergent Silwet L-77 prior to co-cultivation led to a two-fold increase in the number of transformed shoot buds. Rooting of the regenerated shoots was efficiently induced by cultivation on irrigated vermiculite. The entire procedure for generating transgenic plants achieved a transformation frequency of 18.8%, which is comparable to frequencies obtained for other recalcitrant legumes, such as soybean (*Glycine max*) and pea (*Pisum sativum*). The transgene was stably integrated into the host genome and was inherited across generations, without affecting the seismonastic or nyctinastic movements of the plants. This transformation method thus provides an effective genetic tool for studying genes involved in *M. pudica* movements.

Editor: Boris Alexander Vinatzer, Virginia Tech, United States of America

Funding: This work was supported in part by Grants-in-Aid for scientific research from Ministry of Education, Culture, Sports, Science and Technology and Japan Society for the Promotion of Science, Japan. https://www.jsps.go.jp/english/e-grants/. The funders had no role in study design, data collection and analysis, decision to publish, or preparation of the manuscript.

Competing Interests: The authors have declared that no competing interests exist.

* E-mail: mhasebe@nibb.ac.jp

¤a Current address: Electronics-Inspired Interdisciplinary Research Institute, Toyohashi University of Technology, Toyohashi, Japan
¤b Current address: National Plant Phenomics Centre, Institute of Biological, Environmental and Rural Sciences, Aberystwyth University, Aberystwyth, United Kingdom

Introduction

Being fixed in the soil, rooted plants have evolved a variety of strategies to survive stressful environments. Despite lacking muscular and nervous systems, which play pivotal roles in animal motility, certain plant species have acquired the ability to undergo rapid leaf movements in response to external stimuli [1]. The compound leaves of the leguminous species *Mimosa pudica* exhibit seismonastic movement within seconds [2] of being touched or subjected to other types of stimulation [3]. This rapid movement has been suggested to reduce predation risks [4] by scaring away predators [1], decreasing the visibility of the leaves [1], or exposing protective thorns that are usually obscured behind the leaves [5]. The physiological mechanisms underlying seismonastic movement have been studied extensively since the 19th century [6]. This movement is caused by a loss of turgor pressure in one half of the pulvinus (extensor [7]), which is located at the base of each primary petiole, pinna, and pinnule (leaflet) [6]. Nuclear magnetic resonance (NMR) imaging demonstrated that water is translocated from the extensor half to the other half (flexor [7]) of the pulvinus during the movement [8]. At the cellular level, individual "motor cells" in the extensor half of the pulvinus shrink following outflow of intracellular water [9,10], which is accompanied by a large efflux of K^+ and Cl^- ions [11,12,13,14]. These rapid movements of water and ions are difficult to explain by a simple diffusion model [12,15], suggesting that special mechanisms, such as solute-water co-transporters or contractile proteins, are involved in this process [15]. Pharmacological and cytological studies indicate that fragmentation of the actin cytoskeleton [16,17], dephosphorylation of its tyrosine residues [17,18], and changes in Ca^{2+} level [10,19] in pulvinar motor cells participate in the movement. The seismonastic reaction can be propagated over a distance by an electrical action potential [20], which is likely transmitted through the protoxylem [20,21] and the phloem [22]. Chemical substance(s) also contribute to the long-range transmission of the movement [23] and several candidate substances were identified by chemical analysis and bioassays [24,25]. Mechanoreceptor cells in *M. pudica* have long been enigmatic; however, a recent study identified such cells on the tertiary pulvinus [26].

Despite many advances in our understanding of the physiology of seismonastic movements in *M. pudica*, the genetic mechanisms underlying this phenomenon remain to be unraveled, due to the lack of reverse genetic tools for this species. Until now, there was no technique for introducing desired genes into this plant's genome. Agrobacterium-mediated genetic transformation is widely used to generate transgenic plants [27] and is a well-established technique in model legumes such as *Medicago truncatula* [28] and *Lotus japonicus* [29]. However, transformation of other "recalcitrant" legumes, including *M. pudica*, remains challenging, because of the low frequency of shoot formation *in vitro* and the difficulty in transferring genes to cells that are capable of forming shoots [30,31,32].

In the present study, we developed an efficient Agrobacterium-mediated transformation method in *M. pudica*. To overcome the obstacles described above, we examined the shoot formation frequency of several types of explants and selected the cotyledonary node explant, which formed shoots at the highest frequency among the explants tested, as the starting material. We found that a super-binary vector, pSB111 [33], which exhibits improved transformation efficiencies due to the presence of additional virulence genes in the vector backbone [34], increases the number of transformed cells on the cotyledonary node. Furthermore, we demonstrated that controlling the pH during co-cultivation is required for efficient transformation. We thus established an effective transformation method for *M. pudica* that can be used to conduct reverse-genetic studies on the seismonastic movements of this plant.

Materials and Methods

Construction of T-DNA vectors and preparation of Agrobacterium cells

A DNA fragment containing the coding sequence of synthetic green fluorescent protein (*sGFP*) [35] was PCR amplified from pUGW4 [36] using a pair of primers (5′-AAAGT CGACT CGTGA GCAAG GGCGA GGAG-3′ and 5′-TTGAG CTCTT ACTTG TACAG CTCGT CCATG C-3′) and subcloned into the pCR-Blunt II-TOPO vector (Life Technologies, Carlsbad, USA). A DNA fragment containing the first intron of the castor bean *CAT-1* gene [37] was amplified from pIG121-Hm [38] with primers (5′-CTAAG CTTCG CAAGA CCCTT CCTC-3′ and 5′-ATTTC ACGGG TTGGG GTTTC TACAG GACG T-3′), digested with *Sal*I and *Xba*I, and inserted into the *Sal*I/*Xba*I site of the pCR-Blunt II-sGFP construct. Then a DNA fragment containing the intron-sGFP region was excised by digestion with *Sac*I and *Xba*I, and inserted into the *Sac*I/*Xba*I site of pIG121-Hm to produce the pIF121-Hm vector, in which the coding sequence of the beta-glucuronidase gene (*uidA*) [39] was replaced by that of *sGFP*. pIF121-Hm was then introduced into four different *Agrobacterium tumefaciens* (Agrobacterium) strains (AGL1, GV2260, EHA101, and LBA4404 [40]) by electroporation. A super-binary vector, pSB111-GFP, was prepared according to the method of Komari *et al.* [33], with modifications described below. A DNA fragment spanning the intron-*sGFP* sequence and the hygromycin phosphotransferase gene (*hpt*) was amplified from pIF121-Hm with primers (5′-GCAAC GCAAT TAATG TGAGT TAGCT C-3′ and 5′-GGGCT CGAGA GGGAA GAAAG CGAAA GGAG-3′), digested with *Hind*III and *Xho*I, and inserted into the *Hind*III/*Xho*I site of an intermediate vector, pSB11. The resultant construct, pSB11-GFP, was introduced into LBA4404 harboring

pSB1 by electroporation and the pSB111-GFP vector was then produced by homologous recombination between pSB1 and pSB11-GFP in Agrobacterium. Agrobacterium cells harboring pIF121-Hm or pSB111-GFP were selected on LB medium containing 50 μg/ml hygromycin B (Life Technologies) and stored as glycerol stocks at −80°C.

Preparation of cultivation media

Germination medium (GM) consisted of half-strength basal MS salts (1/2 MS; Wako, Osaka, Japan) and 0.2% (w/v) gellan gum (Phytagel; Sigma-Aldrich, St. Louis, USA) at pH 5.8. Shoot induction medium (SIM) consisted of 1/2 MS, 2% (w/v) sucrose, 1× Gamborg's vitamins (Sigma-Aldrich), 0.5 μg/ml 6-benzylaminopurine (BAP; Sigma-Aldrich), and 0.3% gellan gum at pH 5.8. Selection medium (SEM) was prepared by supplementing SIM with 15 μg/ml hygromycin B (Sigma-Aldrich) and 150 μg/ml cefotaxime sodium salt (Sanofi K.K., Tokyo, Japan). Co-cultivation medium (COM) consisted of 1/2 MS, 2% sucrose, 1× Gamborg's vitamins, 0.5 μg/ml BAP, and 0.1% (w/v) 2-(*N*-morpholino)ethanesulfonic acid (MES; Dojindo Laboratories, Kamimashiki-gun, Japan) at pH 6.1 or other values as indicated in the text. Each cultivation medium was prepared as follows: MS basal salts, sucrose, vitamins, BAP, 1-naphthaleneacetic acid (NAA; Sigma-Aldrich), and MES were dissolved in water and the pH was adjusted with KOH or HCl. Then the medium was combined with gellan gum and sterilized by autoclaving at 120°C for 20 min, or alternatively, by filtration through a 0.22-μm PES PLUS membrane (Asahi Glass, Tokyo, Japan) or 0.45-μm PVDF membrane (Millex HV; Merck-Millipore, Billerica, USA). Hygromycin B, cefotaxime, D-glucose, and acetosyringone were added after autoclaving. Cultivation media with minor modifications, for example those with different concentrations of phytohormones, were prepared in a similar manner.

Sterilization of seeds

M. pudica "WASE (an early flowering accession)" seeds were purchased from Sakata Seed (Yokohama, Japan). Approximately 400 seeds in a 50-ml conical tube were washed briefly in 20 ml of 70% ethanol, and put under vacuum (-0.8 MPa) for 10 min in another 20 ml of 70% ethanol. Then the seeds were transferred to 20 ml of 50% commercial bleach (TOPVALU Kitchen Bleach; Aeon, Chiba, Japan) containing NaClO, NaOH, and alkylamineoxide, at concentrations not disclosed by the company, put under vacuum for 10 min, and then washed in another 20 ml of 50% bleach for 30 min with reciprocal shaking at 120 rpm. The seeds were rinsed with sterilized hot water (60°C) at least five times and soaked in 18 ml of hot water (60°C) for 10 min to remove seed coat waxes. After the addition of 2 ml of Plant Preservative Mixture (PPM; Plant Cell Technology, Washington DC, USA), the seeds were put under vacuum for 10 min and then shaken reciprocally at 120 rpm for 30 min. The seeds were placed in a 6×6 array in a Plant Box (a plastic cultivation box with dimensions of 60×60×100 mm; Asahi Glass, Tokyo, Japan) containing 80 ml of GM, and germinated at 25°C for 54 to 60 hours in the dark. Seedlings with hypocotyls of 3 to 8 mm in length were used for subsequent experiments.

Preparation of explants and optimization of shoot induction conditions

Explants were prepared under a dissecting microscope in a laminar flow cabinet. Seedlings were dissected on three sheets of filter paper wetted with COM in a petri dish. After the seed coat was removed with forceps, the primary root and cotyledons were separated from the remaining part of the seedling using a surgical blade (No. 11; Feather, Osaka, Japan) (Figure 1). The epicotyl was cut off from the remaining part to produce the cotyledonary node explant and the associated hypocotyl. Sixteen explants were placed on SIM (25 ml in a 90×20 mm dish) or its derivatives containing different phytohormone concentrations. The explants were cultured at 25°C under 12-hour light (12L; with a light intensity of 120–180 µmol m^{-2} s^{-1})/12-hour dark (12D) cycles, and the medium was changed every 2 weeks. The number of shoots equal to or longer than 2 mm was counted on each explant after 4 or 6 weeks of cultivation.

Transformation

An aliquot of the Agrobacterium stocks was streaked on solid LB medium containing 50 µg/ml hygromycin B and cultured at 30°C for 48 to 60 hours. A single colony was inoculated into 5 ml of liquid LB medium containing 25 µg/ml hygromycin B and precultured at 28°C for 24 hours with rotatory shaking at 180 rpm. After large aggregates were removed by gravity settling, the liquid Agrobacterium culture was again inoculated into 40 ml of fresh liquid LB medium containing 25 µg/ml hygromycin B in a 200 ml baffled flask at a concentration of $OD_{600} = 0.15$. Then the culture was incubated at 28°C with rotatory shaking at 180 rpm for around 4 hours, until OD_{600} reached 0.6. The Agrobacterium

cells were harvested by centrifugation at 5,000×g for 10 min at 25°C, resuspended in 20 ml of COM, centrifuged again, and resuspended in COM at a concentration of $OD_{600} = 0.3$. Finally, the Agrobacterium suspension was supplemented with 40 µg/ml acetosyringone, 0.2% (w/v) D-glucose, and, in some cases, 0.03% (v/v) Silwet L-77.

Twenty of the cotyledonary node explants were soaked in 10 ml of the Agrobacterium suspension with (or without) 0.03% Silwet L-77 in a glass test tube (16.5×165 mm). In some cases, the explants were sonicated with a Branson Sonifier 150 (Branson Ultrasonics, Danbury, USA) with three pulses of 5-s duration at the maximum output power (14 W). The explants were maintained under normal pressure or vacuum (−0.8 MPa) for 10 min and then collected with a tea strainer. The explants were transferred to a plastic dish (90×15 mm) containing 10 ml of the Agrobacterium suspension without Silwet L-77. Alternatively, the explants were directly transferred to the plastic dish without undergoing sonication, vacuum, or Silwet L-77 treatment. Then the dish was sealed with Parafilm (Bemis, Neenah, USA) and cultured for 3 days at 25°C in the dark. To monitor pH changes in co-cultivation medium, 200 µl of the medium was sampled at each time point and the pH was measured using a compact pH meter (Twin pH AS-212; As One, Osaka, Japan). After the co-cultivation, the explants were transferred to SEM (25 ml in a 90×15 mm dish) with forceps and

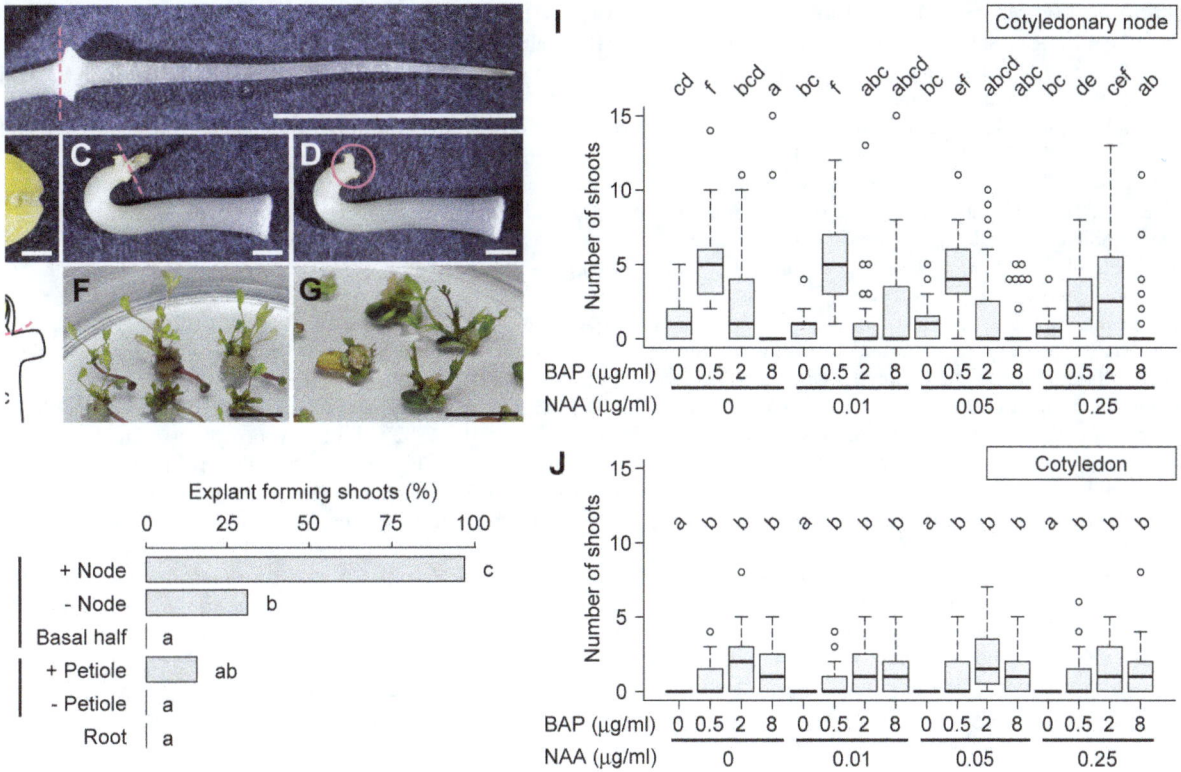

Figure 1. Shoot formation from *M. pudica* explants. A–E. Preparation of explants. A 2-day-old seedling cultured in the dark (A) was divided into the root, the cotyledons with petiole (B), and the remaining part (C). The epicotyl containing the shoot apex was then removed from the remaining part (C) to prepare the cotyledonary node explant (D) as illustrated in (E). Dashed lines in (A), (C), and (E) indicate the cutting positions. The circle in (D) indicates the position of the cotyledonary node. SA, shoot apex; Hc, hypocotyl. F, G. Shoot formation from the cotyledonary node (F) and petiolate cotyledon (G) explants after 4 and 6 weeks of cultivation in the presence of 0.5 µg/ml BAP, respectively. H. Comparison of the frequency of explants forming shoots after 4 weeks of cultivation with 0.5 µg/ml of BAP (n = 32). I, J. Effects of BAP and NAA on shoot formation from cotyledonary node (I) and petiolate cotyledon (J) explants after 4 and 6 weeks of cultivation, respectively. The distribution of the number of shoots formed per explant is shown as box-and-whisker plots (n = 32). Lower and upper whiskers indicate the range of values within 1.5 times the interquartile range from the box and circles indicate outliers. Significant differences were observed between two groups that do not share the same lowercase letter [P<0.05 by Fisher's exact test with Holm's P-value adjustment (H) or Steel-Dwass test (I, J)]. Scale bars, 1 cm (A, F, G), 1 mm (B–D).

continued to be cultured at 25°C under 12L (with a light intensity of 120–180 µmol m^{-2} s^{-1})/12D cycles, with the medium changed every 5 days. After 10 days of selection on SEM, GFP-positive signals located at the cotyledonary node region of each explant (Figure 1D) were counted visually under a fluorescence dissecting microscope SZX16 (Olympus, Tokyo, Japan) equipped with a SZX2-FGFPHQ filter. The number of GFP-positive signals was based on the number of spatially discrete spots, which were predominantly attributable to individual GFP-expressing cells but also included small clusters of cells. The number of explants possessing GFP-positive shoot buds was similarly counted after 30 days of selection. In this experiment, the shoot buds entirely consisting of GFP-positive cells were counted as GFP-positive buds, while chimeric buds containing only some GFP-positive cells were excluded.

Each explant was further cultured on SEM until a GFP-positive callus grew up to 2 mm in length. The GFP-positive callus was surgically excised from the explant and trimmed from GFP-negative tissue. The excised callus was cultured on SEM for an additional 5 days and then cultured on SIM (25 ml in a 90×20 mm dish), with the medium changed every 5 days. After the initiation of shoot elongation, the callus was transferred to a Plant Box containing 80 ml of SIM and continued to be cultured, with the medium changed every 10 days, until the shoots developed at least two compound leaves.

Root induction and whole plant formation

Vermiculite (Fujimi Engei, Shizuoka, Japan) was poured into a Plant Box to a depth of approximately 3 cm, sterilized by autoclaving, and then irrigated with sterilized water, cultivation medium, or phytohormone solution. Cultivation media and water solidified with 0.3% and 1.5% Phytagel, respectively, were also prepared (80 ml per Plant Box). Different concentrations of Phytagel were used for cultivation media and water due to difficulties in solidifying media at low salt concentrations. Regenerated shoots of 2 to 3 cm in length and containing two or more compound leaves were cut with dissecting scissors and placed on vermiculite or gellan gum medium in a 3×3 array for each box. The shoots were kept at 25°C under 12L (with a light intensity of 120–180 µmol m^{-2} s^{-1})/12D cycles without changing the medium and the number of shoots forming any length of root was counted every 7 days. Twenty-seven shoots arising from three independent explants (n = 9 each) were examined for each experimental condition.

Once the root length of a regenerated plantlet reached 3 cm in total, the plantlet was transferred to a soft plastic pot (75×65 mm) containing an equal volume of granulated culture soil (Nippi Engei Baido 1; Nihon Hiryo, Tokyo, Japan) and vermiculite. It was cultured at 27°C under 14L (with a light intensity of 50–120 µmol m^{-2} s^{-1})/10D cycles for approximately 1 month and then transferred to a larger pot (120×100 mm). Liquid nutrient (Hyponex high grade; Hyponex Japan, Osaka, Japan) was occasionally supplied after the inflorescences became visible. Each inflorescence was enclosed in a small plastic bag 1 day before flowering and self-pollinated by rubbing it gently on the day of flowering. Collected seeds were stored at room temperature in a desiccator.

Southern blot analysis

For genomic DNA extraction, immature leaves were sampled before leaflet opening. The leaves (100–200 mg in fresh weight) were frozen in liquid nitrogen and crushed to a fine powder with a mortar and pestle. The specimen was transferred to a 50 ml conical tube, combined with 20 ml of 2× CTAB buffer [2% (w/v)

hexadecyltrimethylammonium bromide (CTAB), 1.4 M NaCl, 20 mM EDTA, 100 mM Tris-HCl (pH 8.0)] that had been heated to just below boiling point, and vortexed immediately. Then the sample was supplemented with 0.1% (v/v) 2-mercaptoethanol and incubated at 60°C for 1 hour with reciprocal shaking at 80 rpm. After the addition of 20 ml of chloroform, the sample was mixed on a rotator for 10 min and centrifuged at 10,000×g for 30 min at 25°C. The upper aqueous phase was transferred to a new tube, supplemented with 1/10 volume of 10% (w/v) CTAB containing 0.7 M NaCl, and re-extracted with chloroform. The sample solution was combined with an equal volume of 2-propanol and centrifuged at 10,000×g for 30 min at 25°C. The precipitation was rinsed with 5 ml of 70% ethanol, air-dried for 10 min, and resuspended in 400 µl of TE [10 mM Tris-HCl (pH 8.0), 1 mM EDTA] containing 0.1 mg/ml RNase A. The sample was incubated at 37°C for 1 hour with reciprocal shaking at 80 rpm, then supplemented with 1 mg/ml proteinase K, and further incubated at 56°C for 30 min with shaking. Genomic DNA in the extract was purified with Genomic-tip 100/G (Qiagen, Venlo, Netherlands) according to the manufacturer's instructions.

Genomic DNA fragments digested with *Eco*RI (5 µg per lane) were electrophoresed in 0.7% SeaKem GTG agarose (Takara Bio, Otsu, Japan) in 1× TAE buffer. Then the DNA fragments were transferred to a Hybond N+ membrane (GE Healthcare, Little Chalfont, UK) using the VacuGene XL Vacuum Blotting System (GE Healthcare). Southern hybridization was performed using an AlkPhos Direct Labeling and Detection System with CDP-Star (GE Healthcare). The DNA probe for *sGFP* was prepared by PCR with primers (5′-ATGGT GAGCA AGGGC GAGGA GC-3′ and 5′-TTACT TGTAC AGCTC GTCCA TGCC-3′) and the pSB111-GFP vector was used as template. Hybridization and subsequent primary washes were performed at 55°C and 65°C, respectively. Hybridization signals were detected using a LAS-3000 Mini luminescent image analyzer (Fujifilm, Tokyo, Japan).

Results

Optimization of shoot induction conditions

We examined shoot formation from several explants derived from 2-day-old *M. pudica* seedlings (Figure 1). After a 4-week cultivation in the presence of 0.5 µg/ml 6-benzylaminopurine (BAP), 97% (31 of 32) of the hypocotyls associated with the cotyledonary node (Figure 1D), hereafter referred to as cotyledonary node explants, formed shoots around the node (Figure 1F, H). Lower frequencies of shoot formation were observed from isolated cotyledons with petioles (5 of 32) and hypocotyls cut just beneath the node (10 of 32), and no shoot formation occurred from the basal halves of hypocotyls, roots, or cotyledons that lacked petioles (Figure 1H). These results indicate that tissues in and around the cotyledonary nodes of *M. pudica* have the ability to form shoots, as do those of other leguminous species [30,31,32].

We optimized the concentrations of two kinds of phytohormones, the cytokinin BAP and the auxin 1-naphthaleneacetic acid (NAA), both of which affect the number of shoots formed on the cotyledonary node in other leguminous species [41,42,43]. Shoots were most efficiently induced on the cotyledonary node in the presence of 0.5 µg/ml BAP and no NAA, which resulted in 5.2±0.5 (mean ± SE) shoots per explant after 4 weeks of cultivation (n = 32; Figure 1I). Shoot formation frequency of petiolate cotyledon explants was also examined at various phytohormone concentrations, but only 2.1±0.4 shoots per explant or fewer were induced, even after a longer cultivation period (n = 32; 6 weeks; Figure 1J). Based on these observations, the cotyledonary node was selected as the target tissue for

Agrobacterium infection, and explants were cultured in medium supplemented with 0.5 μg/ml BAP in subsequent experiments.

Agrobacterium-mediated transformation of cotyledonary nodes

For the Agrobacterium-mediated transformation of cotyledonary node cells, we prepared two kinds of binary vectors: a conventional binary vector, pIF121-Hm, and a super-binary vector, pSB111-GFP, which possesses additional virulence genes [33,34]. The T-DNA region of each vector carries an intron-sGFP reporter gene (Figure 2A), which can be used to selectively visualize transformed cells in living plant tissues, but does not label Agrobacterium cells [37].

Agrobacterium tumefaciens infection is triggered by the transcriptional activation of its virulence genes [44] in response to phenolic compounds such as acetosyringone [45], monosaccharides [46,47], and acidic pH values [48]. We thus examined the effects of supplementing the co-cultivation medium with acetosyringone, D-glucose, and MES buffer adjusted to pH 5.8, alone or in combination, on transformation efficiency (Figure 2B). The number of GFP-positive signals in the cotyledonary node increased in the presence of both acetosyringone and MES (Figure 2B). Although the addition of glucose further increased the transformation efficiency of explants treated with acetosyringone and MES buffer, the increase was not significant (Figure 2B). However, since the addition of glucose did potentially increase the transformation efficiency, we used all three compounds in the subsequent experiments. A comparison of the two binary vectors and Agrobacterium strains demonstrated that the transformation efficiency was higher when pSB111-GFP was combined with Agrobacterium strain LBA4404 than when pIF121-Hm was combined with any of four different Agrobacterium strains (Figure 2C). These results suggest that the addition of acetosyringone and of a buffer capable of maintaining an acidic pH enhance the transformation efficiency of *M. pudica*, as does the use of a super-binary vector.

We further assessed the effect of pH on Agrobacterium infection. As reported previously [49], the addition of MES to cultivation media reduced the amount of pH changes during autoclaving (Figure S1A). However, the smaller change in pH after autoclaving was not the direct cause of the improved infection efficiency, because the cotyledonary nodes of explants cultured on filtration-sterilized, non-buffered medium had almost no positive signals, as did those cultured on autoclaved, non-buffered medium (Figure S1B). Monitoring the pH of the medium during the co-cultivation period revealed that the pH in the non-buffered medium containing Agrobacterium and the explants dropped below 4.7 within the first three hours (Figure 3A). A similar decrease in pH was observed in the non-buffered medium containing only Agrobacterium, but not in the medium alone or in medium containing only explants (Figure 3A), suggesting that the conspicuous acidification of the co-cultivation medium was mainly caused by Agrobacterium. The addition of MES buffer relieved, but did not completely prevent, the excessive acidification and kept the pH of the medium above 5.0 for at least 9 hours during co-cultivation (Figure 3A). The addition of MES also improved the transformation efficiency when using a solid co-cultivation medium, but to a lesser extent than the liquid co-cultivation medium (Figure S1C). Optimization of the initial pH value demonstrated that transformation was most efficient in liquid co-cultivation medium adjusted to pH 6.1 (Figure 2D). This value was higher than those reported for the maximum induction of virulence genes in octopine-type Agrobacterium strains (pH 5.2

to 5.3) [48,50], and possibly counterbalanced the pH decrease during co-cultivation (Figure 3B).

To further improve the transformation efficiency, we examined the effects of sonication [51] and vacuum infiltration [52] prior to the co-cultivation period. We also gauged the effects of transiently adding a detergent, Silwet L-77 [53,54], to the Agrobacterium suspension at 0.03% (v/v) during the sonication and/or vacuum treatments. Compared to the control experiment, none of the treatments, individually or combined, significantly altered the number of GFP-positive signals after 10 days of selection (Figure 2E). On the other hand, the number of explants forming GFP-positive shoot buds after 30 days varied depending on the treatments (Figure 2F). A significant, two-fold increase was observed in the explants treated only with Silwet L-77 (Figure 2F), suggesting that the detergent facilitates Agrobacterium infection of cells that are capable of forming shoots, which are possibly situated deep inside the cotyledonary node. The additional use of sonication and/or vacuum in combination with the Silwet L-77 treatment reduced the emergence of GFP-positive shoot buds (Figure 2F), possibly due to the increased damage of cells at the cotyledonary node.

Taken together, the transformation efficiency of the cotyledonary node of *M. pudica* was drastically improved by three different factors: the use of the super-binary vector, the addition of MES buffer to the co-cultivation medium, and transient treatment with Silwet L-77 before co-cultivation.

Root induction and whole plant formation

After 1 month or longer of selection with hygromycin B, transformed cells in the cotyledonary node formed GFP-positive calluses with shoot buds (Figure 2I–J). These calluses were surgically isolated from the explants and continued to be cultured on SIM for further shoot development (Figure 2K–M). Well-developed shoots possessing at least two compound leaves (Figure 4A, B) were used in a root induction experiment in which three nutrient conditions (water, 1/2 MS, or 1/2 MS containing sucrose and vitamins) and two supporting materials (gellan gum or vermiculite) were tested. For both supporting materials, higher root induction efficiencies were obtained with water than with the MS-based media (Figure 4E), suggesting that poor nutrient conditions favored rooting. Vermiculite increased root induction efficiencies to a greater extent than did gellan gum (Figure 4E), possibly due to the improved permeability to air [55]. Roots were most efficiently induced by vermiculite supplied with water, which resulted in rooting of 81% (22 of 27) of the regenerated shoots after 21 days of cultivation (Figure 4E). This efficiency, together with the fact that the transformed shoots can readily be multiplied by vegetative propagation on SIM, ensures the root induction on practically all transformed shoots. We also examined the effects of three auxins, NAA, indole-3-acetic acid (IAA), and indole-3-butyric acid (IBA), all of which were used for root induction in various plants [43,56,57,58]. None of these compounds, however, improved the root induction efficiency of our system any further, when used at a concentration of 0.5 μg/ml (Figure 4E). The resultant plantlets were transferred to soil after their roots reached 3 cm in total length (Figure 4C, D) and their establishment in the soil was confirmed by further cultivation.

Using the optimized conditions described above, we evaluated the transformation efficiency of *M. pudica* throughout the entire procedure. A total of 160 cotyledonary node explants were subjected to the Agrobacterium-mediated transformation, and monitored for 12 months after co-cultivation (Figure 5A). Sixty-three percent (101 of 160) of the explants formed GFP-positive calluses during selection and more than a half of them (57 of 101)

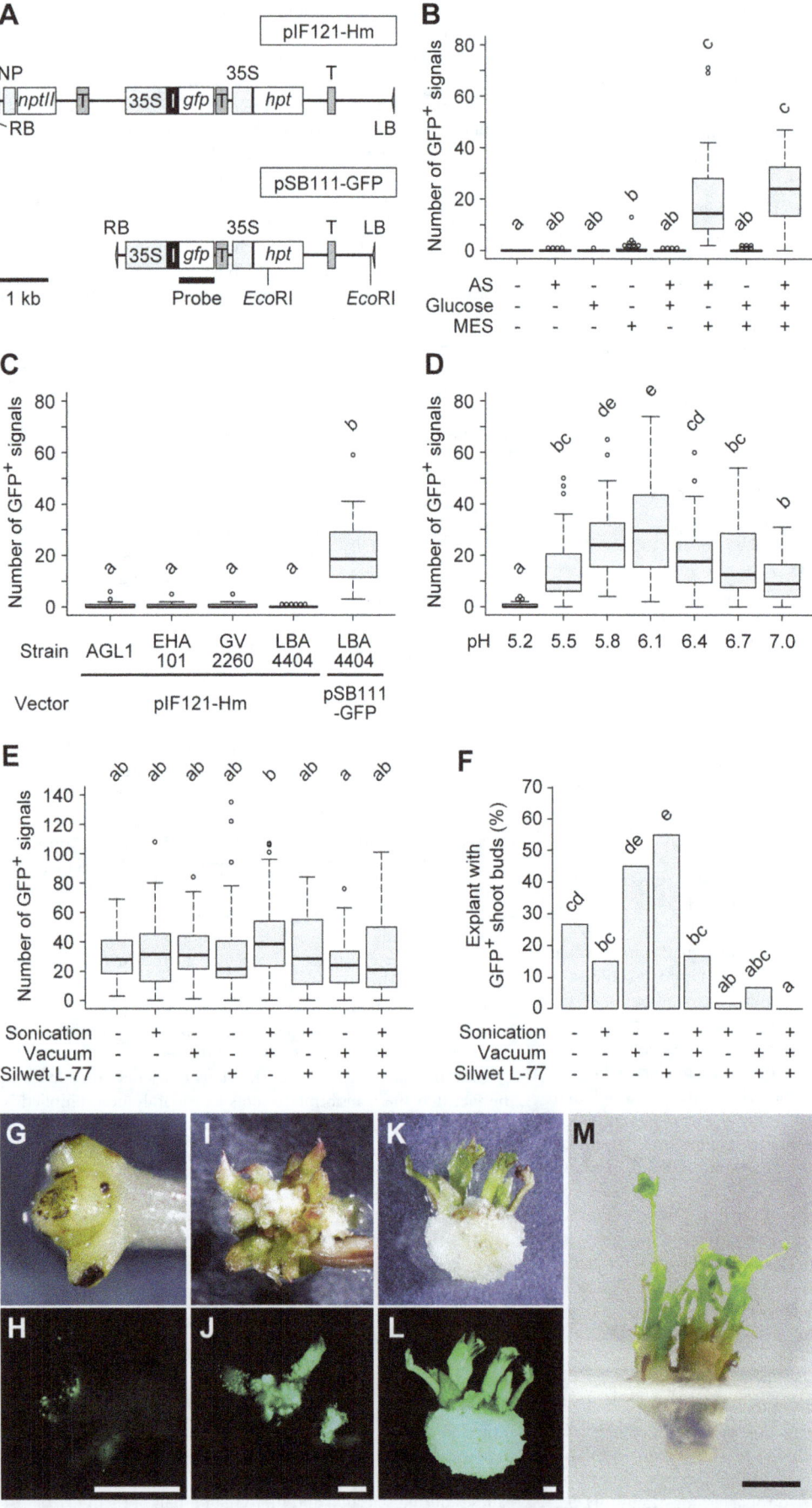

Figure 2. Optimization of Agrobacterium infection conditions. A. Schematic representations of the T-DNA regions of the binary vectors. The position of a probe used for Southern blot analysis and the *Eco*RI cutting sites are shown for pSB111-GFP. RB, right border [72]; LB, left border [72]; 35S, CaMV 35S promoter [73]; NP, *NOS* promoter [74]; T, *NOS* terminator [74]; I, first intron of the castor bean *CAT-1* gene [37]; *nptII, neomycin phosphotransferase II* gene [75]; *hpt, hygromycin phosphotransferase* gene [76]; and *gfp, sGFP* gene [35]. B. Effects of co-cultivation in the presence of acetosyringone (AS), D-glucose, and MES buffer on transformation efficiency (n = 40). The pH of each medium was adjusted to 5.8 before autoclaving. C. Effects of Agrobacterium strains and vectors on transformation efficiency (n = 40). Co-cultivation medium containing acetosyringone, glucose, and MES (pH 5.8) was used. D. Effects of initial pH of co-cultivation media (n = 60). E, F. Effects of sonication, vacuum infiltration, and Silwet L-77 detergent treatments prior to co-cultivation (n = 60). The number of GFP-positive signals on the cotyledonary node of each explant was counted after 10 days of selection (B–E). The frequency of explants possessing GFP-positive shoot buds was counted after 30 days of selection (F). Significant differences were observed between two groups that do not share the same lowercase letter [P<0.05 by the Steel-Dwass test (B–E) or Fisher's exact test with Holm's P-value adjustment (F)]. G–J. Bright-field (G, I) or green fluorescent (H, J) images of cotyledonary node explants after 10 (G, H) or 30 (I, J) days of selection. K, L. Bright-field (K) or green fluorescent (L) images of an isolated GFP-positive callus after 51 days of selection. M. After an additional 21 days of cultivation (a total of 72 days), the same callus shown in (K) formed multiple shoots with compound leaves. Scale bars, 1 mm (G–L), 1 cm (M).

initiated shoot elongation on SIM. Forty-two of the 57 shoots developed two or more compound leaves and 30 of these successfully rooted and became established in the soil. These results demonstrated that 18.8% of the explants (30 of 160) produced at least one independent line of T_0 plants (Figure 5A). The number of transgenic T_0 plants continued to increase even after 12 months of cultivation (Figure 5A), suggesting that the efficiency would further increase with time. On the other hand, four independent T_0 plants (derived from 2.5% of the explants) became established in as little as 4 months (Figure 5A), enabling us to recover their T_1 progeny within a total of 8 months (Figure 5B).

Molecular and biological analyses of transgenic plants

We performed a genomic Southern blot analysis on regenerated T_0 plants using the sGFP sequence as a probe (Figure 2A). Among 13 independent lines tested, approximately two-thirds of transgenic plants (9 of 13) possessed a single T-DNA insertion, while the others (4 of 13) had two insertions (Figure 6). This simple pattern of insertion represents an advantage of the present method over particle bombardment, which is used to transform other leguminous species [30,31], but which often results in complex patterns of DNA insertions [59,60].

Transmission of the transgene to T_1 progeny was confirmed in all lines tested (n = 10) by observing the GFP fluorescence

(Figure 5C). In most cases (9 of 10), the segregation ratio of the GFP fluorescence in a selfed T_1 progeny was in good agreement with that expected from the number of T-DNA insertions (Table S1; 3:1 and 15:1 for one and two T-DNA insertions, respectively). These results provide further evidence for the simplicity of T-DNA insertion patterns produced by the present method and also indicate the non-chimeric nature of each T_0 plant. Transmission of the transgene to T_2 progeny was also confirmed for one line (Figure 5D), demonstrating the stable transmission of transgenes across generations.

Finally, the transgenic plants were examined for their ability to undergo characteristic movements. All of the T_0 (n = 70) and T_1 (n = 10) plants showed both seismonastic movement in response to touch (Video S1) and nyctinastic movement (data not shown), suggesting that the transformation procedure presented here does not impair these movements. In sum, the present study provides a genetic tool to investigate the molecular mechanisms underlying the intriguing movements of *M. pudica*.

Discussion

In this study, we developed a robust protocol for the genetic transformation of *M. pudica*. A key improvement for the successful transformation of *M. pudica* was the use of MES buffer to maintain

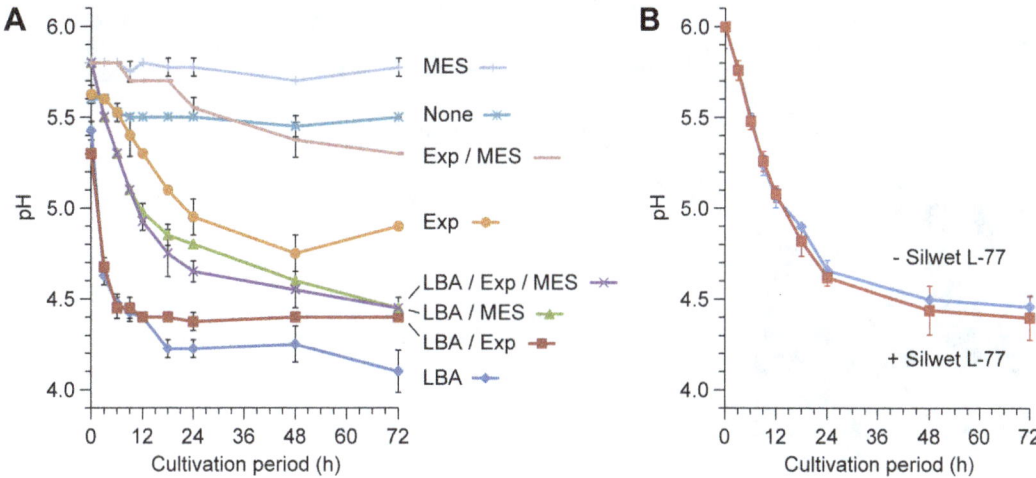

Figure 3. Changes in medium pH during co-cultivation. A. Effects of Agrobacterium strain LBA4404 harboring pSB111-GFP (LBA), cotyledonary node explants (Exp), and MES buffer (MES) on medium pH (n = 4). The co-cultivation media were initially adjusted to pH 5.8 and sterilized by filtration to circumvent the pH decrease caused by autoclaving. In the absence of MES buffer, pH values had already declined in the time it took to prepare the Agrobacterium suspension in co-cultivation medium (~30 minutes). B. Changes in pH of MES-buffered medium initially adjusted to pH 6.1 (n = 5). The medium was sterilized by autoclaving and then used for co-cultivation of Agrobacterium and explants. The effect of Silwet L-77 treatment prior to co-cultivation was also examined. Data are the means ± SD. The pH of the medium was measured at 0, 3, 6, 9, 12, 18, 24, 48, and 72 hours after the initiation of cultivation.

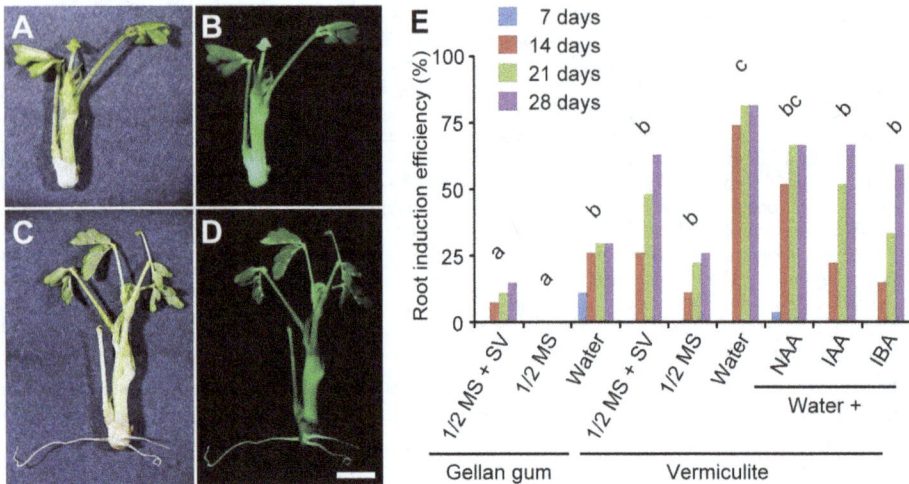

Figure 4. Optimization of root induction conditions. A–D. Bright-field (A, C) and green fluorescent (B, D) images of a transformed shoot before (A, B) and after (C, D) 14 days of cultivation in irrigated vermiculite. Scale bar, 5 mm. E. Comparison of root induction conditions (n = 27). A statistical analysis was conducted using the frequencies at 14 days of cultivation. Significant differences (P<0.05 by Fisher's exact test with Holm's P-value adjustment) were observed between two groups that do not share the same lowercase letter. 1/2 MS, half-strength MS salts; SV, 2% sucrose and 1× Gamborg's vitamins.

the pH during co-cultivation. Although the pH-dependent activation of Agrobacterium virulence genes was previously demonstrated [48,50] and several studies emphasized the importance of buffering agents in co-cultivation media [61,62], the requirement to stabilize the pH with buffering agents seems to depend on the transformation system being used. For example,

only one-quarter (17 of 67) of the transformation methods given in a protocol book [27] that covers a wide range of plant species and transformation systems describes the use of buffer reagents during Agrobacterium preparation and/or co-cultivation. This variability may be due to differences in other conditions that possibly affected pH stability, such as the composition of co-cultivation media, the

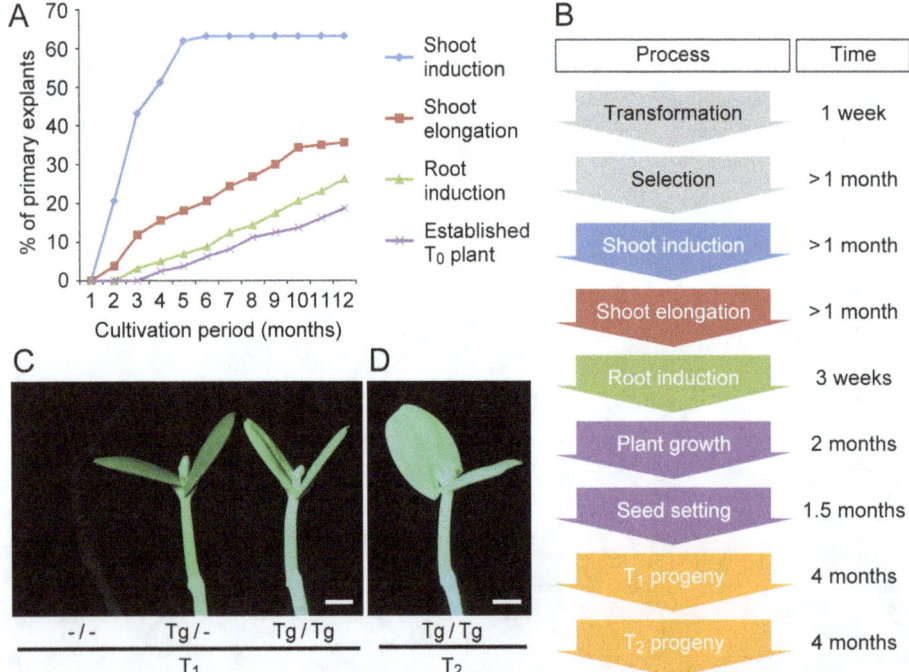

Figure 5. Establishment of transgenic plants. A. Time course of establishment of transgenic T$_0$ plants from cotyledonary node explants (n = 160). Each line indicates the frequency of Agrobacterium-infected explants that reached the process indicated in (B). B. Schematic representation of the entire transformation procedure. C. Green fluorescent images of T$_1$ seedlings of a single T-DNA insertion line (#1 in Figure 6). The zygosity of T$_1$ progeny [non-transformant (−/−), hemizygote (Tg/−) or homozygote (Tg/Tg)] produced by self-crossing of a T$_0$ plant could be determined based on fluorescence intensity of the plantlet. D. Green fluorescent image of homozygous T$_2$ seedling produced by self-crossing of a homozygous T$_1$ plant. Scale bar, 2 mm.

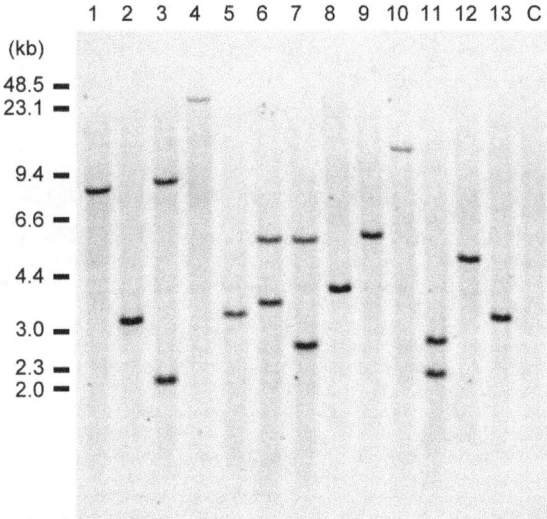

Figure 6. Genomic Southern blot analysis of T₀ plants. Genomic DNAs of 13 independent T₀ plants (lines #1-13) and a non-transformed plant (C) were analyzed by *Eco*RI digestion and detection of the *sGFP* sequence. The size of several bands is shorter than the minimal length expected from the intact T-DNA sequence (3.0 kb; Figure 2A), suggesting that the T-DNA sequence had undergone truncation and/or rearrangement.

plant species and tissues being used, the Agrobacterium strains, and procedures used to prepare them. Although a previous study reported that MES reduces the transformation efficiency [63], the present study indicates that controlling the pH with buffering agents may improve the efficiency of Agrobacterium-mediated transformation. The direction of pH change also varied with the transformation system being used; the pH dropped to below 4.5 in a trumpet lily (*Lilium* x *formolongi*) system [62] and the present study, whereas it rose to 7.2 in a Tepary bean (*Phaseolus acutifolius*) system [61]. These observations, together with the finding that the optimal initial pH of the co-cultivation medium (pH 6.1) differed from that reported for the maximal activation of the virulence genes in other octopine-type strains (pH 5.2 to 5.3) [48,50], indicate the importance of pH optimization for each transformation system.

The transformation efficiency of the present study (18.8%) is comparable to efficiencies obtained for extensively studied, recalcitrant legumes such as soybean (16.4%) [64] and pea (13.5%) [65]. This level of transformation efficiency is sufficient for conventional transgenic analyses that introduce a limited number of foreign DNAs of interest. On the other hand, further improvement of the method may be needed for high-throughput genetic screenings, such as insertional mutagenesis [66], activation tagging [67], and the FOX hunting system [68], which rely on a large number of transgenic plants. One possible approach for improvement would be to increase the Agrobacterium infection efficiency with thiol compounds, which are effective for the transformation of soybean [69,70]. Another approach would be to increase the frequency of shoot formation from the transformed calluses, because only 30% (30 of 101) of the calluses produced well-developed shoots, even after a long cultivation period (Figure 5A). Further optimization of cultivation conditions, such as temperature, lighting, nutritional composition, and phytohormones, would increase the transformation efficiency and/or accelerate shoot formation.

In this study, the transgenic plants were recovered via a combination of hygromycin selection and fluorescence-based visual selection, in which GFP-positive transgenic calluses were surgically isolated from surrounding non-transgenic regions. Compared to hygromycin selection alone, this dual selection system facilitates and accelerates the establishment of transgenic T₀ plants that consist entirely of transformed cells. On the other hand, our preliminary observation indicated that 70% (21 of 30) of the regenerated shoots (>5 mm) exhibited GFP fluorescence after 60 days of cultivation on SEM. This result suggests that transformants can also be recovered using antibiotic selection alone, although further investigation is required to evaluate the recovery rate of transgenic plants under these conditions.

Despite the recent development of new reverse genetic tools, such as virus-induced gene silencing (VIGS) [71], Agrobacterium-mediated transformation still plays a pivotal role in plant biology research. In the present study, we establish a method whereby this invaluable genetic technique may be applied to *M. pudica*, a classic model organism in plant physiology.

Supporting Information

Figure S1 Effects of MES buffer on transformation efficiency. A. Changes in pH of co-cultivation media after autoclaving. Data are the means ± SD (n = 3). A diagonal line is shown for clear visualization of the pH changes from initial values. B. Comparison of sterilization methods of co-cultivation media in the presence or absence of 0.1% MES buffer (n = 20). Each co-cultivation medium was adjusted to pH 5.8 before sterilization and supplemented with both acetosyringone and glucose. C. Comparison of liquid and solid co-cultivation media in the presence or absence of 0.1% MES buffer (n = 20). The pH of each co-cultivation medium was adjusted to 5.8 before autoclaving. Gellan gum (0.3%) was used to solidify the solid co-cultivation media. The number of GFP-positive signals on the cotyledonary node of each explant was counted after 10 days of selection. Significant differences (P<0.05 by the Steel-Dwass test) were observed between two groups that do not share the same lowercase letter (B, C). D, E. Comparison of sterilization methods of co-cultivation medium optimized for transformation (n = 80). Co-cultivation medium containing acetosyringone, glucose, and MES buffer (pH 6.1) and the treatment with Silwet L-77 prior to co-cultivation were used in this experiment. No significant differences were observed in either the number of GFP-positive signals after 10 days of selection (D; by the Mann-Whitney U-test) or the frequency of explants possessing GFP-positive shoot buds after 30 days of selection (E; by Fisher's exact test).

Table S1 Segregation of GFP expression in selfed T₁ progeny.

Video S1 Movie of the seismonastic movement of transgenic *M. pudica*. A homozygous T₁ seedling (10 days old) is shown. Green and red signals represent the GFP fluorescence and the autofluorescence of chloroplasts, respectively. This movie was taken using a SZX16 microscope equipped with a SZX2-FGFP long-pass filter and coupled to a DP71 digital camera (Olympus).

Acknowledgments

We thank the Japan Tobacco Plant Innovation Center for providing pSB11 and LBA4404 harboring pSB1. We also thank the Functional

Genomics Facility and the Model Plant Research Facility of National Institute for Basic Biology for technical assistance.

References

1. Braam J (2005) In touch: plant responses to mechanical stimuli. New Phytol 165: 373–389.
2. Volkov AG, Foster JC, Ashby TA, Walker RK, Johnson JA, et al. (2010) *Mimosa pudica*: Electrical and mechanical stimulation of plant movements. Plant Cell Environ 33: 163–173.
3. Roblin G (1979) *Mimosa pudica*: A model for the study of the excitability in plants. Biol Rev 54: 135–153.
4. Jensen EL, Dill LM, Cahill Jr JF (2011) Applying behavioral-ecological theory to plant defense: light-dependent movement in *Mimosa pudica* suggests a trade-off between predation risk and energetic reward. Am Nat 177: 377–381.
5. Eisner T (1981) Leaf folding in a sensitive plant: A defensive thorn-exposure mechanism? Proc Natl Acad Sci U S A 78: 402–404.
6. Weintraub M (1952) Leaf movements in *Mimosa pudica* L. New Phytol 50: 357–382.
7. Satter RL, Galston AW (1981) Mechanisms of control of leaf movements. Annu Rev Plant Physiol 32: 83–110.
8. Tamiya T, Miyazaki T, Ishikawa H, Iriguchi N, Maki T, et al. (1988) Movement of water in conjunction with plant movement visualized by NMR imaging. J Biochem 104: 5–8.
9. Fleurat-Lessard P, Frangne N, Maeshima M, Ratajczak R, Bonnemain JL, et al. (1997) Increased expression of vacuolar aquaporin and H⁺-ATPase related to motor cell function in *Mimosa pudica* L. Plant Physiol 114: 827–834.
10. Yao H, Xu Q, Yuan M (2008) Actin dynamics mediates the changes of calcium level during the pulvinus movement of *Mimosa pudica*. Plant Signal Behav 3: 954–960.
11. Allen RD (1969) Mechanism of the seismonastic reaction in Mimosa pudica. Plant Physiol 44: 1101–1107.
12. Kumon K, Suda S (1984) Ionic fluxes from pulvinar cells during the rapid movement of *Mimosa pudica* L. Plant Cell Physiol 25: 975–979.
13. Samejima M, Sibaoka T (1980) Changes in the extracellular ion concentration in the main pulvinus of *Mimosa pudica* during rapid movement and recovery. Plant Cell Physiol 21: 467–479.
14. Toriyama H (1955) Observational and experimental studies of sensitive plants VI. The migration of potassium in the primary pulvinus. Cytologia (Tokyo) 20: 367–377.
15. Morillon R, Liénard D, Chrispeels MJ, Lassalles JP (2001) Rapid movements of plants organs require solute-water cotransporters or contractile proteins. Plant Physiol 127: 720–723.
16. Fleurat-Lessard P, Roblin G, Bonmort J, Besse C (1988) Effects of colchicine, vinblastine, cytochalasin B and phalloidin on the seismonastic movement of *Mimosa pudica* leaf and on motor cell ultrastructure. J Exp Bot 39: 209–221.
17. Kanzawa N, Hoshino Y, Chiba M, Hoshino D, Kobayashi H, et al. (2006) Change in the actin cytoskeleton during seismonastic movement of *Mimosa pudica*. Plant Cell Physiol 47: 531–539.
18. Kameyama K, Kishi Y, Yoshimura M, Kanzawa N, Sameshima M, et al. (2000) Tyrosine phosphorylation in plant bending. Nature 407: 37.
19. Turnquist HM, Allen NS, Jaffe MJ (1993) A pharmacological study of calcium flux mechanisms in the tannin vacuoles of *Mimosa pudica* L. motor cells. Protoplasma 176: 91–99.
20. Sibaoka T (1962) Excitable cells in Mimosa. Science 137: 226.
21. Samejima M, Sibaoka T (1983) Identification of the excitable cells in the petiole of *Mimosa pudica* by intracellular injection of procion yellow. Plant Cell Physiol 24: 33–39.
22. Fromm J, Eschrich W (1988) Transport processes in stimulated and non-stimulated leaves of *Mimosa pudica* - II. Energesis and transmission of seismic stimulations. Trees (Berl West) 2: 18–24.
23. Ricca U (1916) Soluzione d'un problema di fisiologia - La propagazione di stimolo nella "Mimosa". Nuovo G Bot Ital 23: 51–170.
24. Schildknecht H (1983) Turgorins, hormones of the endogenous daily rhythms of higher organized plants - Detection, isolation, structure, synthesis, and activity. Angew Chem Int Edi Engl 22: 695–710.
25. Ueda M, Yamamura S (1999) The chemistry of leaf-movement in *Mimosa pudica* L. Tetrahedron 55: 10937–10948.
26. Visnovitz T, Világi I, Varró P, Kristóf Z (2007) Mechanoreceptor cells on the tertiary pulvini of *Mimosa pudica* L. Plant Signal Behav 2: 462–466.
27. Wang K (2006) *Agrobacterium* Protocols, 2nd ed. Methods Mol Biol 343. Totowa: Humana Press. 512 p.
28. Frugoli J, Harris J (2001) *Medicago truncatula* on the move! Plant Cell 13: 458–463.
29. Udvardi MK, Tabata S, Parniske M, Stougaard J (2005) *Lotus japonicus*: legume research in the fast lane. Trends Plant Sci 10: 222–228.
30. Chandra A, Pental D (2003) Regeneration and genetic transformation of grain legumes: An overview. Curr Sci 84: 381–387.
31. Eapen S (2008) Advances in development of transgenic pulse crops. Biotechnol Adv 26: 162–168.
32. Somers DA, Samac DA, Olhoft PM (2003) Recent advances in legume transformation. Plant Physiol 131: 892–899.
33. Komari T, Hiei Y, Saito Y, Murai N, Kumashiro T (1996) Vectors carrying two separate T-DNAs for co-transformation of higher plants mediated by *Agrobacterium tumefaciens* and segregation of transformants free from selection markers. Plant J 10: 165–174.
34. Komari T, Takakura Y, Ueki J, Kato N, Ishida Y, et al. (2006) Binary vectors and super-binary vectors. Methods Mol Biol 343: 15–41.
35. Chiu W, Niwa Y, Zeng W, Hirano T, Kobayashi H, et al. (1996) Engineered GFP as a vital reporter in plants. Curr Biol 6: 325–330.
36. Nakagawa T, Kurose T, Hino T, Tanaka K, Kawamukai M, et al. (2007) Development of series of gateway binary vectors, pGWBs, for realizing efficient construction of fusion genes for plant transformation. J Biosci Bioeng 104: 34–41.
37. Ohta S, Mita S, Hattori T, Nakamura K (1990) Construction and expression in tobacco of a β-glucuronidase (GUS) reporter gene containing an intron within the coding sequence. Plant Cell Physiol 31: 805–813.
38. Akama K, Shiraishi H, Ohta S, Nakamura K, Okada K, et al. (1992) Efficient transformation of *Arabidopsis thaliana*: comparison of the efficiencies with various organs, plant ecotypes and *Agrobacterium* strains. Plant Cell Rep 12: 7–11.
39. Jefferson RA, Kavanagh TA, Bevan MW (1987) GUS fusions: β-glucuronidase as a sensitive and versatile gene fusion marker in higher plants. EMBO J 6: 3901–3907.
40. Hellens R, Mullineaux P, Klee H (2000) A guide to *Agrobacterium* binary Ti vectors. Trends Plant Sci 5: 446–451.
41. Geetha N, Venkatachalam P, Prakash V, Sita GL (1998) High frequency induction of multiple shoots and plant regeneration from seedling explants of pigeonpea (*Cajanus cajan* L.). Curr Sci 75: 1036–1041.
42. Gulati A, Jaiwal PK (1994) Plant regeneration from cotyledonary node explants of mungbean (*Vigna radiata* (L.) Wilczek). Plant Cell Rep 13: 523–527.
43. Venkatachalam P, Jayabalan N (1997) Effect of auxins and cytokinins on efficient plant regeneration and multiple-shoot formation from cotyledons and cotyledonary-node explants of groundnut (*Arachis hypogaea* L.) by in vitro culture technology. Appl Biochem Biotechnol 67: 237–247.
44. Gelvin SB (2012) Traversing the cell: *Agrobacterium* T-DNA's journey to the host genome. Front Plant Sci 3: 52.
45. Stachel SE, Messens E, Van Montagu M, Zambryski P (1985) Identification of the signal molecules produced by wounded plant cells that activate T-DNA transfer in *Agrobacterium tumefaciens*. Nature 318: 624–629.
46. Cangelosi GA, Ankenbauer RG, Nester EW (1990) Sugars induce the *Agrobacterium* virulence genes through a periplasmic binding protein and a transmembrane signal protein. Proc Natl Acad Sci U S A 87: 6708–6712.
47. Shimoda N, Toyoda-Yamamoto A, Nagamine J, Usami S, Katayama M, et al. (1990) Control of expression of *Agrobacterium vir* genes by synergistic actions of phenolic signal molecules and monosaccharides. Proc Natl Acad Sci U S A 87: 6684–6688.
48. Stachel SE, Nester EW, Zambryski PC (1986) A plant cell factor induces *Agrobacterium tumefaciens vir* gene expression. Proc Natl Acad Sci U S A 83: 379–383.
49. De Klerk GJ, Hanecakova J, Jásik J (2008) Effect of medium-pH and MES on adventitious root formation from stem disks of apple. Plant Cell Tiss Organ Cult 95: 285–292.
50. Turk SCHJ, Melchers LS, den Dulk-Ras H, Regensburg-Tuïnk AJG, Hooykaas PJJ (1991) Environmental conditions differentially affect *vir* gene induction in different *Agrobacterium* strains. Role of the VirA sensor protein. Plant Mol Biol 16: 1051–1059.
51. Trick HN, Finer JJ (1997) SAAT: sonication-assisted *Agrobacterium*-mediated transformation. Transgenic Res 6: 329–336.
52. Bechtold N, Ellis J, Pelletier G (1993) *In planta Agrobacterium* mediated gene transfer by infiltration of adult *Arabidopsis thaliana* plants. C R Acad Sci III 316: 1194–1199.
53. Clough SJ, Bent AF (1998) Floral dip: a simplified method for *Agrobacterium*-mediated transformation of *Arabidopsis thaliana*. Plant J 16: 735–743.
54. Wu H, Sparks C, Amoah B, Jones HD (2003) Factors influencing successful *Agrobacterium*-mediated genetic transformation of wheat. Plant Cell Rep 21: 659–668.
55. Jay-Allemand C, Capelli P, Cornu D (1992) Root development of in vitro hybrid walnut microcuttings in a vermiculite-containing gelrite medium. Sci Hortic (Amsterdam) 51: 335–342.
56. De Klerk GJ, Ter Brugge J, Marinova S (1997) Effectiveness of indoleacetic acid, indolebutyric acid and naphthaleneacetic acid during adventitious root formation in vitro in *Malus* 'Jork 9'. Plant Cell Tiss Organ Cult 49: 39–44.
57. Kollárová K, Henselová M, Lišková D (2005) Effect of auxins and plant oligosaccharides on root formation and elongation growth of mung bean hypocotyls. Plant Growth Regul 46: 1–9.

Author Contributions

Conceived and designed the experiments: HM YH MH. Performed the experiments: HM TF NS. Analyzed the data: HM. Wrote the paper: HM YH MH.

58. Rout GR (2006) Effect of auxins on adventitious root development from single node cuttings of *Camellia sinensis* (L.) Kuntze and associated biochemical changes. Plant Growth Regul 48: 111–117.

59. Dai S, Zheng P, Marmey P, Zhang S, Tian W, et al. (2001) Comparative analysis of transgenic rice plants obtained by *Agrobacterium*-mediated transformation and particle bombardment. Mol Breed 7: 25–33.

60. Travella S, Ross SM, Harden J, Everett C, Snape JW, et al. (2005) A comparison of transgenic barley lines produced by particle bombardment and *Agrobacterium*-mediated techniques. Plant Cell Rep 23: 780–789.

61. De Clercq J, Zambre M, Van Montagu M, Dillen W, Angenon G (2002) An optimized *Agrobacterium*-mediated transformation procedure for *Phaseolus acutifolius* A. Gray. Plant Cell Rep 21: 333–340.

62. Ogaki M, Furuichi Y, Kuroda K, Chin DP, Ogawa Y, et al. (2008) Importance of co-cultivation medium pH for successful *Agrobacterium*-mediated transformation of *Lilium* x *formolongi*. Plant Cell Rep 27: 699–705.

63. Becker J, Vogel T, Iqbal J, Nagi W (1994) *Agrobacterium*-mediated transformation of *Phaseolus vulgaris*. Adaptation of some conditions. Ann Rep Bean Improv Coop 37: 127–128.

64. Olhoft PM, Flagel LE, Donovan CM, Somers DA (2003) Efficient soybean transformation using hygromycin B selection in the cotyledonary-node method. Planta 216: 723–735.

65. Grant J, Cooper P (2006) Peas (*Pisum sativum* L.). Methods Mol Biol 343: 337–345.

66. Krysan PJ, Young JC, Sussman MR (1999) T-DNA as an insertional mutagen in Arabidopsis. Plant Cell 11: 2283–2290.

67. Walden R, Fritze K, Hayashi H, Miklashevichs E, Harling H, et al. (1994) Activation tagging: a means of isolating genes implicated as playing a role in plant growth and development. Plant Mol Biol 26: 1521–1528.

68. Ichikawa T, Nakazawa M, Kawashima M, Iizumi H, Kuroda H, et al. (2006) The FOX hunting system: an alternative gain-of-function gene hunting technique. Plant J 48: 974–985.

69. Olhoft PM, Lin K, Galbraith J, Nielsen NC, Somers DA (2001) The role of thiol compounds in increasing *Agrobacterium*-mediated transformation of soybean cotyledonary-node cells. Plant Cell Rep 20: 731–737.

70. Olhoft PM, Somers DA (2001) L-Cysteine increases *Agrobacterium*-mediated T-DNA delivery into soybean cotyledonary-node cells. Plant Cell Rep 20: 706–711.

71. Lu R, Martin-Hernandez AM, Peart JR, Malcuit I, Baulcombe DC (2003) Virus-induced gene silencing in plants. Methods 30: 296–303.

72. Wang K, Genetello C, Van Montagu M, Zambryski PC (1987) Sequence context of the T-DNA border repeat element determines its relative activity during T-DNA transfer to plant cells. Mol Gen Genet 210: 338–346.

73. Odell JT, Nagy F, Chua NH (1985) Identification of DNA sequences required for activity of the cauliflower mosaic virus 35S promoter. Nature 313: 810–812.

74. Depicker A, Stachel S, Dhaese P, Zambryski P, Goodman HM (1982) Nopaline synthase: transcript mapping and DNA sequence. J Mol Appl Genet 1: 561–573.

75. Bevan MW, Flavell RB, Chilton MD (1983) A chimaeric antibiotic resistance gene as a selectable marker for plant cell transformation. Nature 304: 184–187.

76. Waldron C, Murphy EB, Roberts JL, Gustafson GD, Armour SL, et al. (1985) Resistance to hygromycin B - A new marker for plant transformation studies. Plant Mol Biol 5: 103–108.

Overexpression of the *AtSHI* Gene in Poinsettia, *Euphorbia pulcherrima*, Results in Compact Plants

M. Ashraful Islam[1], Henrik Lütken[2], Sissel Haugslien[1], Dag-Ragnar Blystad[1], Sissel Torre[3], Jakub Rolcik[4], Søren K Rasmussen[2], Jorunn E Olsen[3], Jihong Liu Clarke[1]*

1 Bioforsk - Norwegian Institute for Agricultural and Environmental Research, Ås, Norway, **2** Department of Plant and Environmental Sciences, Faculty of Science, University of Copenhagen, Frederiksberg, Denmark, **3** Department of Plant and Environmental Sciences, Norwegian University of Life Sciences, Ås, Norway, **4** Palacky University, Olomouc, Czech Republic

Abstract

Euphorbia pulcherrima, poinsettia, is a non-food and non-feed vegetatively propagated ornamental plant. Appropriate plant height is one of the most important traits in poinsettia production and is commonly achieved by application of chemical growth retardants. To produce compact poinsettia plants with desirable height and reduce the utilization of growth retardants, the *Arabidopsis SHORT INTERNODE* (*AtSHI*) gene controlled by the cauliflower mosaic virus *35S* promoter was introduced into poinsettia by *Agrobacterium*-mediated transformation. Three independent transgenic lines were produced and stable integration of transgene was verified by PCR and Southern blot analysis. Reduced plant height (21–52%) and internode lengths (31–49%) were obtained in the transgenic lines compared to control plants. This correlates positively with the *AtSHI* transcript levels, with the highest levels in the most dwarfed transgenic line (TL1). The indole-3-acetic acid (IAA) content appeared lower (11–31% reduction) in the transgenic lines compared to the wild type (WT) controls, with the lowest level (31% reduction) in TL1. Total internode numbers, bract numbers and bract area were significantly reduced in all transgenic lines in comparison with the WT controls. Only TL1 showed significantly lower plant diameter, total leaf area and total dry weight, whereas none of the *AtSHI* expressing lines showed altered timing of flower initiation, cyathia abscission or bract necrosis. This study demonstrated that introduction of the *AtSHI* gene into poinsettia by genetic engineering can be an effective approach in controlling plant height without negatively affecting flowering time. This can help to reduce or avoid the use of toxic growth retardants of environmental and human health concern. This is the first report that *AtSHI* gene was overexpressed in poinsettia and transgenic poinsettia plants with compact growth were produced.

Editor: Turgay Unver, Cankiri Karatekin University, Turkey

Funding: This research was supported by the Danish grant "Joint Proof-of-Concept Fund" to Professor Søren K Rasmussen and the Research Council of Norway grant KMB 199398/I10 to Dr Jihong Liu Clarke. The funders had no role in study design, data collection and analysis, decision to publish, or preparation of the manuscript.

Competing Interests: The authors have declared that no competing interests exist.

* E-mail: jihong.liu-clarke@bioforsk.no

Introduction

The ornamental industry is one of the fastest growing industries worldwide, especially in Japan and China. Global production of ornamental potted plants and cut flowers comprises about 50 billion €, corresponding to an estimated global consumption between 100 and 150 billion € [1,2]. The market for cut flowers and potted ornamental plants is not only determined by producers' choices but also by a continuously growing demand for novelties and high quality [3,4]. Compaction of plants is one of the most important traits in many ornamental potted plants, e.g. poinsettia [2].

Euphorbia pulcherrima Willd. Ex Klotzsch, poinsettia, is a non-food, non-feed and vegetatively propagated ornamental plant, known as a contemporary symbol of Christmas in many parts of the world [5]. It is a short day plant and flowering is initiated when the day length is shorter than a critical length [6]. Generally, poinsettia has an elongated natural growth habit. Dwarf characteristics can be obtained either by directly using dwarf cultivars or by grafting cultivars on dwarf rootstocks [7]. Similarly, spraying with growth retardants such as CCC (chlormequat chloride) or alar (dimethylaminosuccinamic acid), that among others inhibit the biosynthesis of the plant hormone gibberellin (GA), results in compact ornamental potted plants [8]. However, growth retardants are expensive, time consuming to apply and have negative impact on human health as well as the environment. Moreover, the growth regulators will likely be banned in EU countries in the near future [9–11].

In the poinsettia industry, alternative strategies like manipulation of temperature, light quality and light duration have previously been tested to control elongation growth of poinsettias [12–15]. In northern areas short term diurnal temperature drops, obtained by opening vents in the morning, are commonly used to reduce shoot elongation. However, in warmer periods and warmer areas of the world this is not applicable. Furthermore, in poinsettia phytoplasma is introduced to induce free-branching and this can also result in compact growth [16]. However, the phytoplasma is lost upon exposure of the plant materials to heat treatment as well as meristematic and somatic embryogenesis tissue culture, which is commonly used to obtain disease free plants by removing pathogens such as the poinsettia mosaic virus (PnMV) [16].

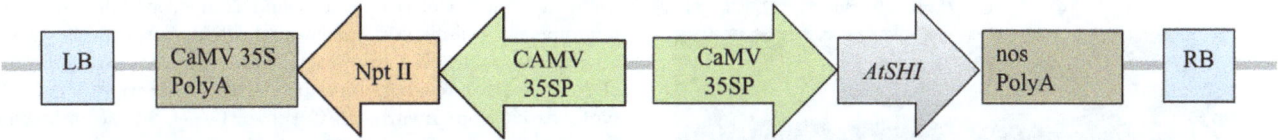

Figure 1. Gene construct: pKanIntron-35S-*SHI* used for *Agrobacterium*-mediated transformation of poinsettia.

Genetic engineering is increasingly adopted as an important alternative to conventional breeding [1,17]. Recently, it was shown that introduction of the *Arabidopsis thaliana SHORT INTERNODES (SHI)* gene into *Kalanchoë* and *Populus* resulted in dwarfed growth without any morphological abnormalities [18,19]. However, in *Populus* the dwarfing effect on the stem was only very weak, although the internode and petiole lengths were significantly reduced. On the other hand, overexpression of *GA2-oxidase* genes, which control GA inactivation, resulted in dwarfed plants with delayed flowering time in *Solanum* and *Arabidopsis* [20,21]. Also, antisense silencing of the GA biosynthesis gene *GA20-oxidase* resulted in smaller leaves, delayed flowering time and reduced fertility in *Arabidopsis* [8]. Overexpression of the *Arabidopsis* GA signalling gene *GIBBERELLIN INSENSITIVE (GAI)* in apple and *Chrysanthemum* reduced plant height, but was correlated with reduced rooting ability and delayed flowering in the respective species [7,22]. In light of these observations, introduction of the *AtSHI* gene to poinsettia might be highly interesting as a means to control elongation growth without introducing undesired morphological or developmental changes. *SHI* gene is a plant specific transcription factor belonging to the *SHI* gene family, and it has been identified in different plants species like tomato, rice, soybean and *Medicago truncatula* [23]. The *Arabidopsis SHI* gene family consists of 10 members; *SHI, STYLISH 1 (STY1)* and *STY2*, *LATERAL ROOT PRIMORDIUM 1 (LRP1)* and *SHI-RELATED SEQUENCE 3 to 8 (SRS3 to SRS8)* [18,23–25]. These corresponding proteins have two highly conserved regions, a RING-like zinc finger motif positioned in the N-terminal end and an IGGH domain of unknown function in the C-terminal part of the protein [23,25,26].

All genes could be amplified from all tissues of *Arabidopsis*, except *SRS8*, indicating it might be a pseudogene [25]. STY1 is the closest paralog of SHI having two identical domains in the N and C terminal. The *SHI* family genes have both diverse and redundant functions in plant growth and are involved in shoot apical region development as well as flower and leaf development [23,25,26]. In these respects *SHI/STY*-related genes appear important in gynoecium development, vascular formation and organ identity in floral whorls two and three [25]. It has been documented that *SHI* and *STY* are expressed in the apical region of the developing gynoecium [25,27–29]. Overexpression of *SRS7* conferred dwarfed growth in *Arabidopsis* with small and curled leaves, anther dehiscence and abnormal floral development [30].

SHI family members act as DNA binding transcription activators and may act on plant growth and development by affecting phytohormones like auxin and GA, which among other things control shoot elongation in response to different stimuli [31–35]. Plants overexpressing *SHI* are dwarfed, but a normal, more elongated phenotype can be restored by application of auxin [27,28]. Also, the *YUCCA4 (YUC4)* auxin biosynthesis gene is induced by the SHI/STY family proteins. However, SHI and STY appear to differ slightly in function as shown by lower affinity of SHI than STY1 to the *YUC4* promoter in the yeast two-hybrid system [36]. Application of indole-3-acetic acid (IAA) to the apical meristem has been shown to increase biosynthesis of bioactive GA

[37–39]. *SHI* overexpressing *Arabidopsis* plants show increased levels of the inactive GA_{34} compared to wild type plants [24]. Furthermore, in *Brassica SHI*-related genes have been identified as negative regulators of GA-induced cell division [40].

In this study, we report for the first time successful use of genetic engineering as a tool to control elongation growth in poinsettia, which is among the economically most important potted ornamental plants worldwide. Compact growth was obtained by overexpressing the *AtSHI* gene by using a recently developed *Agrobacterium*-mediated transformation method for poinsettia [41]. Apart from the desired dwarfed growth habit, no developmental abnormalities were scored and flowering time was unaffected.

Materials and Methods

Plant Materials

Poinsettia (*Euphorbia pulcherrima* Willd. ex Klotzsch) cv. Millenium cuttings were grown in the greenhouse under 16 h photoperiod provided by high pressure sodium (HPS) lamps (400W, GAN 4-550, Gavita, Superagro, Andebu, Norway) at $21\pm2°C$ with an average relative air humidity (RH) $70\pm5\%$. For *Agrobacterium tumefaciens*-mediated transformation 5–15 mm long internode stem explants were excised from 8–10 weeks old poinsettia plants. The explants were surface sterilized with 70% ethanol (1 min), 1% NaOCl (10 min) and then rinsed thoroughly three times with sterile deionized and autoclaved water for 3, 10 and 20 min. After sterilization, stem segments (1–1.5 mm thickness) were excised and utilized for *Agrobacterium* transformation.

Transformation, Selection and Plant Regeneration

Plasmid vector pAt35S:SHI was constructed and introduced into *Agrobacterium tumefaciens* strain GV3850 as previously described in details by Lütken et al. [19]. A brief schematic drawing of the *SHI* gene expression cassette is shown in Figure 1. The *Agrobacterium* culture and subsequent transformation were carried out basically as described by Clarke et al. [41]. The stem segments were inoculated in the *Agrobacterium* suspension for 10 min with gentle shaking. The stem segments were then blotted on sterile filter paper and transferred to callus induction medium (CIM) (MS medium supplemented with 0.2 mg l^{-1} BAP, 0.2 mg l^{-1} CPA and 30 g l^{-1} sucrose) for co-cultivation up to 72 h in the dark at 24°C. After co-cultivation, the explants were transferred to the CIM medium with antibiotic selection containing 10 mg l^{-1} kanamycin and 400 mg l^{-1} claforan (Aventis Pharma Ltd, Norway) for about 10 days. The embryogenic calli were subsequently transferred after every three weeks to somatic embryo induction medium (SEIM) (MS medium contains 0.3 mg l^{-1} NAA, 0.15 mg l^{-1} 2ip and 30 g l^{-1} sucrose) supplemented with antibiotics 25 mg l^{-1} kanamycin and 400 mg l^{-1} claforan. Shoots and plantlets which derived from the somatic embryos were transferred to root induction (RI) medium (1/2 strength MS with 20 g l^{-1} sucrose) with or without hormones (1 mg l^{-1} IAA or IBA). The pH was 5.7–5.8 in all MS media. Regenerated plants were transferred to soil and grown in a greenhouse once the root system was developed.

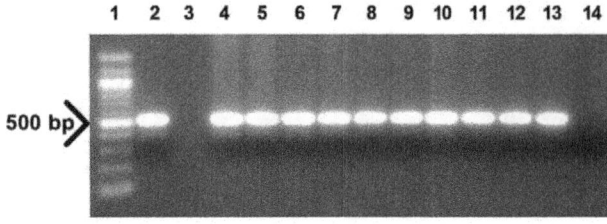

Figure 2. PCR analysis of poinsettia transformed with the *AtSHI* **gene (genomic DNA was extracted from leaves).** Lane 1:100 bp marker, lane 2: plasmid control, lane 3: blank, lanes 4–11: eight plants from independent transgenic lines 1 (TL1) (individuals 1–8), lane 12: TL2, lane 13: TL3 and lane 14: WT control line. The arrow indicates the 500 bp band.

Light conditions provided by fluorescent tubes (Philips Master TL-D Super 58W/840, Eindhoven, The Netherlands) for *in vitro* cultures were 23 μmol m^{-2} s^{-1} for callus induction and 30 μmol m^{-2} s^{-1} for the SEIM and RI media, respectively, under a 16 h photoperiod at 24°C.

PCR Analysis

Genomic DNA was isolated from the putative transgenic poinsettia plants using the DNeasy Plant Mini Kit (Qiagen GmbH, Hilden, Germany) according to the manufacturer's instructions. PCR reactions (20 μl) were performed using 100 ng DNA, 0.2 μM of each primer, and 2xHotStarTaq Mastermix and supplied water from HotStarTaq® Plus Master Mix Kit (Qiagen GmbH, Hilden, Germany). Primer sequences used to amplify a fragment of approximately 500 bp for the *AtSHI* gene (At5g66350) were 5′-ACTCTAACGCTGACGGTGGA-3′ (forward) and 5′-TGCTGACCGGTAGAAAGCTG-3′ (reverse). PCR amplification was performed in a C1000™ thermal cycler (BIO-RAD, Singapore) using the following conditions: 15 min at 95°C (1 cycle) followed by 30 s at 95°C, 45 s at 56°C, and 1 min at 72°C (35 cycles) with a final extension of 7 min at 72°C (1 cycle). PCR products were analysed by ultraviolet light after electrophoresis on 0.8% (W/V) agarose gels.

Southern Blot Analysis

Total genomic DNA was isolated from the leaves of the WT control plants and the PCR positive transgenic poinsettia lines using the DNeasy Plant Maxi Kit (Qiagen GmbH, Hilden, Germany). Southern blot analysis was performed according to Sambrook et al. [42]. Ten micrograms of genomic DNA were digested with *Hind*III for 5 h and separated on 1% (W/V) TBE agarose gel overnight at 37 V, and subsequently transferred onto Genescreen Plus(™) Hybridization Transfer membrane (NEF 988001 PK, Boston, MA, USA). The 500 bp PCR product representing the coding region of the *AtSHI* gene as described above was used as a probe for hybridization. Membranes were hybridized with the ^{32}P-labelled probe overnight at 65°C. After hybridization the membranes were washed and then exposed to film as described by Clarke et al. [41].

RNA Isolation and cDNA Synthesis

Young leaves of transgenic poinsettia and control plants were harvested for RNA extraction with E.Z.N.A Plant RNA Mini Kit (Omega bio-tek, Norcross, GA, USA) according to the manufacturer's instructions. RNAs were subsequently treated with Turbo DNA-*free*™ kit (Ambion Inc., Austin, TX, USA) to eliminate DNA contamination. RNA quality and quantity were evaluated using agarose gel electrophoresis and Nanodrop 2000 Spectrophotometer (Wilmington, Delaware, USA), respectively. Two micrograms total RNA from each sample was used to synthesize cDNA in a 20 μl reaction using the cDNA SuperScript® VILO™ synthesis kit from Invitrogen (Carlsbad, CA, USA) according to the manufacturer's instructions.

Real-Time Quantitative PCR

Real- time quantitative PCR analysis was performed in a 25 μl reaction volume, using 2.5 μl twentyfold diluted cDNA as a template with 12.5 μl of 1x Power SYBR® green PCR master mix (Applied Biosystems, Warrington, UK) and 0.4 μM of each primer. Primers were designed using Primer3 online software and sequences are listed in Table 1.

Reactions were conducted in an Applied Biosystems 7900HT Real-time PCR system combined with SDS 2.3 version software (Applied Biosystems, Singapore). The PCR conditions were as follows: 50°C for 2 min and 95°C for 10 min, followed by 40 cycles of 95°C for 15 s, 60°C for 1 min. Dissociation curve analysis was carried out to verify the specificity of the PCR amplification. PCR efficiencies were calculated from a dilution series of genomic DNA of transgenic plants for each primer pair of target gene and endogenous gene by following the equation $E = 10^{-1/\text{slope}}$. There were three replications for all samples including the control samples except the TL2, and a no template control was included for each primer pair. The transcript level was estimated by threshold cycle (C_t) values of each sample. The differences of C_t values between the endogenous α–tubulin and the target gene were normalized using the formula $2^{-\Delta\Delta Ct}$ [43]. The relative transcript levels (fold-differences) of the genes were converted to scale $\log_{10}$ values.

Phenotypical Analysis of Transgenic Lines (*AtSHI*) Compared to Control Plants

The independent transgenic lines, TL1–6 from TL1, TL2, TL3 and control plants were vegetatively propagated. After root formation, the plants were potted in 13 cm plastic pot filled with *Sphagnum* peat (Veksttorv, Ullensaker Almenning, Nordkisa, Norway). The plants were kept at 21±2°C, 70±5% RH in the greenhouse under a 16 h photoperiod (8 h darkness) at an irradiance of 150±25 μmol m^{-2} s^{-1} provided by HPS lamps (400W, GAN 4–550). The plants were pinched over 3–4 leaves allowing three shoots to grow per plant. Four plants from each transgenic line including WT control plants were transferred to growth chambers (controlled environment) without any natural light. Light was provided by General Electric company, Fairfield, CT, USA at an irradiance of 100±20 μmol m^{-2} s^{-1} (measured by LI-COR Quantum/Radiometer/Photometer, Model LI-250, Lincoln, Nebraska, USA) under a 10 h photoperiod for flowering. The temperature was 21°C and the RH was 70±5%.

The plants were watered daily during the growth period with commercial nutrient solutions. The side shoots developed during the growth period were removed and counted and only three shoots were allowed to grow per plant. The length of these three shoots was measured from the base of the stem to the shoot apical meristem every fourth day until flowering. At the end of the experimental period (after flowering) petiole length of four mature leaves was measured. The number of leaves and bracts (namely the transition leaves which formed red color more than 40%) were counted and the average internode lengths were calculated by dividing the shoot height by the number of leaves and bracts. Relative chlorophyll content was measured from the middle leaf of the three side shoots on each plant by a chlorophyll content meter (Model CL-01, Hansatech instruments, Norfolk, England). Leaf

Table 1. Primers for real-time PCR expression analysis of *AtSHI* in poinsettia.

Target gene	Forward primer (5'-3')	Reverse primer (5'-3')	Product size
AtSHI	AGCTATGGCAACACCCAAAC	ATCCAGCCTTTGTTGCTGTT	71
α-tubulin	TGGAGCTCTCTTTGCTTCAA	CCAACAAAGCTGCATAGCAA	81

and bract areas were measured by an area meter (Model 3100 area meter, LI-COR, Lincoln, Nebraska, USA). Specific leaf and bract area (SLA and SBA, respectively) were each determined from area/dry weight. After recording the fresh weight, the dry weight was recorded after drying at 65°C until a constant mass was reached. The number of days until visible bracts and cyathia was counted. At the selling stage plants were moved to a postharvest test room to compare differences in cyathia abscission and bract necrosis. The climate in the test room was 21°C, 30–40% RH and an irradiance of 10 μmol m^{-2} s^{-1} was provided 12 h daily by fluorescent tubes (Philips Master TL-D 58W/830).

Three replicate experiments in growth chambers were carried out in which phenotypic observations were performed, samples for auxin analysis collected and postharvest quality tested. In addition to the growth chamber experiments, two to four plants from each line were grown in the greenhouse with conditions as described above for evaluation of the morphological performance of transgenic plants under long day conditions (16 h). Both short and long day treatment experiments were conducted during November 2011 through January 2012.

Auxin Measurements

In the growth chamber experiment, elongating shoot tips were harvested after three weeks of starting short day conditions and immediately placed in liquid N_2. The samples were freeze-dried using a freeze dryer machine (Heto Holten A/S, Allerød, Denmark). For each genotype, 3 replicate samples, each containing 3 shoot tips (elongating parts of the stem) were used for auxin analysis. The frozen plant materials were ground in a mortar and extracted for 5 min with 1 ml cold phosphate buffer (50 mM, pH 7.0) containing 0.2% sodium diethyldithiocarbomate. ^{15}N and ^{2}H$_5$-labeled internal standards of indole -3-acetic acid (IAA) and IAA metabolites were obtained from OlChemlm (Olomouc, Czech Republic). The measurements were performed in duplicate. The samples were processed according to Pěnčík et al. [44] using high performance liquid chromatography (HPLC) coupled to a tandem triple-quadrupole mass spectrometer (MS/MS).

Statistical Analysis

Different growth parameters of transgenic and WT control plants were subjected to analysis of variance (General Linear Model procedure) and Tukey's pair wise comparison test (p≤0.05) using Minitab version 16 (Minitab Inc., State College, PA, USA).

Results

PCR Screening and Southern Blot Analysis of Transgenic Poinsettia Plants

To generate stable compact growth transgenic poinsettia plants by overexpressing the *AtSHI* gene, *Agrobacterium*-mediated transformation experiments were carried out using stem segment explants. After selection and regeneration through somatic embryogenesis, regenerated poinsettia plants were obtained and established in the greenhouse. Using *AtSHI* specific primers,

genomic DNA of putative transgenic poinsettia plants was screened by PCR analyses for presence of *AtSHI*. Three independent transgenic lines were confirmed, one plant was selected for each line except the transgenic line one (TL1) of which eight plants from the same clone were used (Figure 2). Transgenic lines were further analyzed by Southern blot hybridization. Results of Southern blot analysis confirmed the stable integration of transgene into the poinsettia genome (Figure 3). Of the three transgenic lines analyzed, line TL1 (with three individual transgenic plants) showed single copy of the transgene whereas TL2 and TL3 with one transgenic plant each showed two copies of the transgene integration (lanes 2–4, 5 and 6 in Figure 3). Lane 7 is the WT negative control, whereas lane 1 is the positive control.

AtSHI Expression in the Transgenic Lines

AtSHI transcript levels were analysed by real-time quantitative PCR in *AtSHI*-expressing transgenic poinsettia lines and the WT controls. Three transgenic lines (TL) were analyzed. Of these, two plants (TL1–4 and TL1–6) were from TL1, while TL2 and TL3 with one plant each were included for the quantitative real time PCR. *AtSHI* transcript levels varied among the transgenic lines as shown in Figure 4. The highest relative levels of transcript were found in TL1–6, whereas the lowest expression was found in TL3.

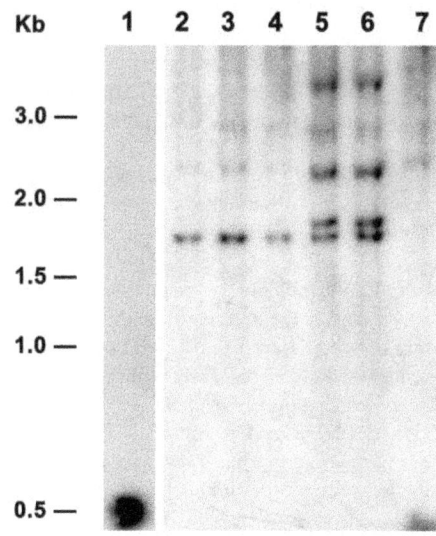

Figure 3. Southern blot analysis of PCR positive transgenic poinsettia lines overexpressing the *AtSHI* gene. *Hind*III-digested total genomic DNAs were hybridized with a *AtSHI* probe (the 500 bp PCR product). Lane 1: positive control, the 500 bp PCR product; lane 2–4: TL1–3, TL1–4, TL1–6 of the transgenic line one (TL1); lane 5: TL2; lane 6: TL3, lane 7: WT control.

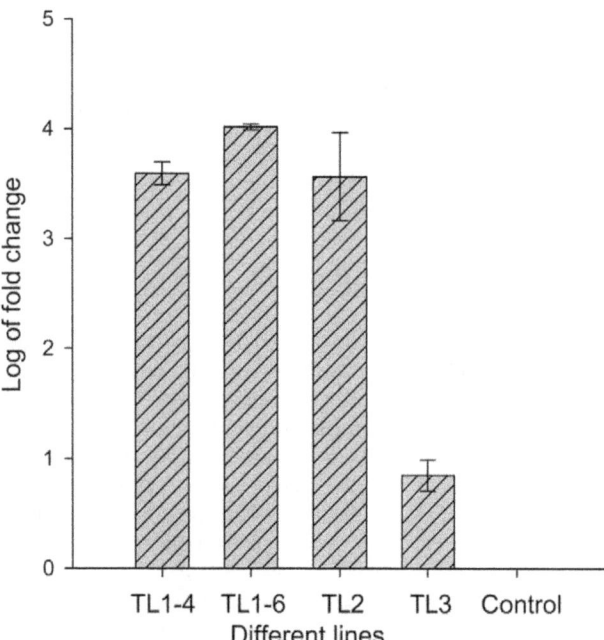

Figure 4. Quantitative real-time PCR analysis of *AtSHI* transgene in different transgenic lines of poinsettia. Two microgram total RNAs from transgenic poinsettia lines and the endogenous control, α-tubulin gene were used for synthesize cDNAs prior the real-time qPCR analysis. Values are means of three technical replications except TL2 (n = 2). Data were analyzed using the $2^{-\triangle\triangle C_T}$ method and represented as $\log_{10}$ values. Vertical bars represent the ± SE (standard error).

Reduced Growth Elongation without Change in Flowering Time and Keeping Quality in *AtSHI* Expressing Transgenic Lines

Over-expression of *AtSHI* in poinsettia significantly reduced plant height compared to WT control plants. When grown under short day (SD) conditions, the TL1 transgenic line showed the strongest height reduction response (52%) whereas the TL2 and TL3 plants were reduced in height by 49% and 30%, respectively, compared to WT control plants (Figures 5A, 6A). Under long day (LD) conditions, the height reduction was 25%, 21% and 23% in TL1, TL2 and TL3, respectively, compared to control plants (Figure 5B). No differences were observed in leaf color or leaf shape, and there was no significant difference in petiole length (Figure 6 B, C).

As investigated under SD conditions, overexpression of the *AtSHI* gene significantly reduced the internode lengths and the internode number compared to WT control plants (Figure 7A, B). On average, internode lengths were significantly reduced by 49%, 41% and 31% and the internode number by 32%, 41% and 33% for TL1, TL2 and TL3, respectively. The transgenic plants had significantly lower bract number and reduced bract area as shown in Figure 7C and D. The average bract number was reduced by 44%, 50% and 40% and the bract area by 68%, 62% and 47% for TL1, TL2 and TL3, respectively. In TL1 the total dry weight of stems, leaves and bracts, specific leaf area, total leaf area and shoot diameter differed significantly from the WT control plants (Table 2). No significant difference in relative chlorophyll content was found between the transgenic and control plants (Table 2).

Bract formation started after four weeks when plants were kept under SD conditions and visible cyathia were observed after 5

weeks. No significant difference in time to initiation of flowering was observed between the transgenic and control plants. The development of bract color formation appeared a bit faster in control plants compared to transgenic plants. Bract necrosis was not observed and no significant differences in cyathia abscission or keeping quality between the transgenic plants and control plants were observed (data not shown). In the postharvest room, the cyathia abscission started after three weeks and one week later about 90% of the bracts were abscised from all plants (data not shown).

Reduced Height and Internode Lengths Correlate with Reduced IAA Levels

To investigate the mechanism of the reduced plant height and internode lengths in *AtSHI* overexpressed plants, the levels of IAA were investigated in shoot tips of transgenic poinsettia lines as well as the WT control plants. In the transgenic plants the IAA levels showed a reduction of 11–31% compared to the WT control plants as shown in Table 3. The lowest IAA levels (31% less) were recorded in the transgenic line TL1.

Discussion

In this study we have demonstrated that the over-expression of the *AtSHI* gene is an efficient tool to reduce plant height in the economically highly important ornamental plant poinsettia. PCR analysis verified the *AtSHI*-expressing transgenic poinsettia lines, while Southern blot analysis further confirmed the transgene integration into the poinsettia genome. With the vegetatively propagated nature of poinsettia, the desired compact growth characteristic in the transgenic poinsettia lines over-expression of the *AtSHI* gene will be maintained and propagated by cuttings, a clear advantage over sexually propagated plants. Recently, a dwarfing effect of this gene was shown also in *Kalanchoë* and *Populus* [18,19]. *SHI* gene has among others been suggested to act to control contents of plant hormones involved in control of shoot elongation [27,31,36,45].

AtSHI Expression in Poinsettia Results in Reduced Internode Elongation

AtSHI gene expression resulted in a significant reduction of shoot height compared with the control plants under SD as well as LD conditions (Figures 5, 6A). The strongest shoot length reductions were observed in TL1, which compared to control plants showed 52% and 25% reduction under SD and LD, respectively. In all three transgenic lines, the reduction in shoot lengths was more pronounced under SD compared to LD. This could be ascribed to a generally more vigorous height growth under the higher light sum of the LDs of 16 h. Under LD plants were vegetatively growing whereas under SD conditions shoot elongation ceased due to floral induction. The overexpression of *AtSHI* in poinsettia is comparable with observations made in *Populus*, *Kalanchoë* and *Arabidopsis* [18,19] where it also significantly reduced the height. In this study, the largest reduction in internode length in transgenic poinsettia was 49% compared to WT control plants.

In our current study, the number of internodes was reduced in the *AtSHI* overexpressed transgenic poinsettia plants. Such information is available from neither *Kalanchoë* nor *Arabidopsis*, which is a rosette plant and does not have elongated internodes, but the result is in contrast to the results from *Populus* [18,19]. On the other hand, the petiole length was reduced in the *Populus* [18,19]. In poinsettia, bract number and bract area were reduced significantly in the transgenic plants compared to the control

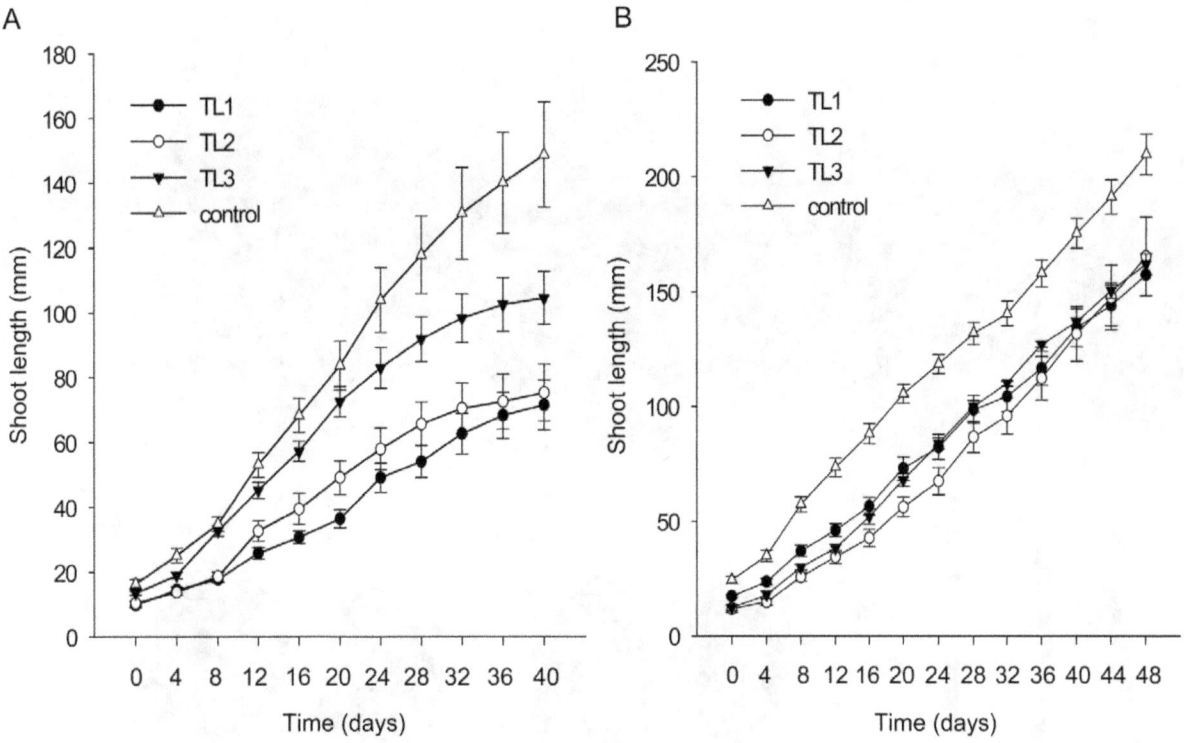

Figure 5. Height comparison among the different transgenic lines (TL) of *AtSHI* **overexpressing poinsettia and untransformed control plants grown under short day (10 h) (A) and long day conditions (16 h) (B).** Vertical bars represent the ± SE (standard error), n = 11–12 and 6–12 in A and B, respectively.

plants. A reduction of 47 to 67% in bract area was observed in the different transgenic lines (Figure 7D). The reduced bract size may have impact on the ornamental value of poinsettia. About 41% higher total leaf area and 8% lower specific leaf area were observed in the transgenic line TL1 compared to control plants (Table 2). Leaf colour and alterations were visually observed where the transgenic lines did not show any differences from the control plants. In Figure 6C, both serrate lobed and non-serrate lobed leaves are present. We have observed both types of leaves in the same plants of three transgenic lines and the control plants. Thus, there was no difference in leaf shape of transgenic and control plants. Fridborg et al. [24] observed darker green leaves in *Arabidopsis*, whereas Lütken et al. [19] did not observe any differences in leaf colour in *Kalanchoë* but *AtSHI* was over-expressed in both plants. The overexpression of *AtSHI* caused pleiotropic changes during the developmental stages of *Arabidopsis*. The reason for differences of phenotypic characteristics might be due to the different habits of growth and flowering stage of *Arabidopsis* and *Kalanchoë* as well as poinsettia. Life span and vegetative stage are very short in *Arabidopsis* compared to *Kalanchoë* or poinsettia. No significant difference in relative chlorophyll content was observed among the transgenic lines and control plants. This might suggest that the *AtSHI* overexpression does not reduce photosynthesis in plants. Rather, the relative chlorophyll mean value was higher in the transgenic lines compared to the WT control plants. However, 56% total dry weight in TL1 was significantly lower compared to the WT controls. About 23% reduction in shoot diameter in TL1 was observed compared to the control plants. This result is similar to *SHI* overexpressing *Kalanchoë*, but in contrast to *Arabidopsis* [19,24]. There was no significant difference of branching among the transgenic *AtSHI* poinsettias and control plants (data not shown). This is in contrast to the *shi* mutant in *Arabidopsis*, where more branches were observed in *SHI* overexpressing plants [24].

Reduced Elongation Growth Correlates with Reduced IAA and Higher Expression Levels of *AtSHI*

The reduced stem extension of the plants expressing the *AtSHI* gene was correlated with reduced endogenous levels of IAA levels (Tables 2, 3), with the lowest levels in the shortest lines (TL1). SHI/STY family members regulate plant development. In *Arabidopsis* the *STY1* interacts with the promoter of the auxin biosynthesis gene *YUC4* and induces its transcription [36,46]. YUC family proteins act as rate limiting enzymes of the tryptophan-dependent auxin biosynthesis pathways [47,48]. In *Arabidopsis* it was shown that the SHI/STY family controls the developmental process through regulation of auxin biosynthesis [27]. Our results with reduced height and reduced IAA levels in poinsettia overexpressing *AtSHI* are similar to those obtained in a previous study of the moss *Physcomitrella patens* [27]. Two genes of *PpSHI* reduced elongation and reduced endogenous auxin levels in this moss [45]. The *Arabidopsis* auxin mutant (*ettin-1*) is affected by *SHI/STY* family mutants [27,49]. It is also reported that elongation involves cell division and cell elongation due to cell wall modification activated by auxin among others [50]. Elongation is also well known to be related to GA levels or sensitivity to GA [20,51,52]. However, it has been reported that IAA promotes the biosynthesis of active gibberellin (GA) and that auxin transport inhibitors reduce the active GA content in pea and *Arabidopsis* [37,38,53,54]. This reduction of shoot length and internode length is also consistent with results obtained in the moss *P. patens* [45]. The transgenic moss lines showed reduced levels of auxin presumably related to the overexpression of *PpSHI*.

Figure 6. Transgenic *AtSHI* overexpressing poinsettia plants (A), petioles (B) and leaves (C). In each figure: from the left different transgenic lines; TL1, TL2, TL3 and non-transformed control plants are shown. The plants were grown at $21\pm2°C$, a 10 h photoperiod at an irradiance of 100 ± 20 µmol m^{-2} s^{-1}.

In *AtSHI* overexpressing lines of *Populus*, the internode number and the concentration of cytokinin were increased. The height and number of shoot apical meristems (SAM) was reduced due to reduced content of cytokinins [55]. The auxin level was also lower in transgenic plant tissue. This supports the *Arabidopsis* data where auxin levels and biosynthesis were reduced by the introduction of the *SHI* family member gene *sty1* [27,36]. In contrast, *AtSHI* expressing *Populus* did not show any change in auxin concentration, but the cytokinin levels were decreased [18]. No significant difference was observed in the petiole length in poinsettia and a similar result was obtained in *Arabidopsis*.

The *AtSHI* transcript levels correlated well with the observed phenotype. Plants that contained relatively high levels of the *AtSHI*

transcript were severely dwarfed, whereas less dwarfed plants contained lower transcript levels (Figures. 4, 5A).

Flowering Time and Keeping Quality are Not Affected in the *AtSHI* Overexpressing Transgenic Poinsettia Plants

In the ornamental industry, the most important feature is flowering. The attractive part of poinsettia is the bract colour (formation of red anthocyanin pigment in the transition leaves) formation, which was observed at nearly the same time in both control and transgenic plants. The completion of bract colour formation was a little bit faster in control plants compared to transgenic plants. This is quite similar to previous studies, where no difference in flowering time was found in *sty1–2* and wild type

Table 2. Comparison of growth parameters among transgenic (T) lines of poinsettia overexpressing the *AtSHI* gene, TL1, TL2, TL3 and control plants under short day conditions of a 10 h photoperiod in a controlled environment.

	Line TL1	Line TL2	Line TL3	Control
Height (mm)	71.5±7.7 b	75.3±8.8 b	104.5±8.2 b	148.8±16.3 a
Relative chlorophyll content	27.7±1.0 a	28.0±2.1 a	25.5±2.1 a	24.6±2.2 a
Shoot diameter (mm)	10.2±0.6 b	12.0±1.1 a	13.9±0.3 a	13.2±0.3 a
Total leaf area (cm^2)	199.8±17.3 c	247.6±32.3 bc	385.1±29.0 a	341.0±27.0 ab
Specific leaf area (cm^2 g^{-1})	283.3±3.7 a	271.9±6.6 ab	265.1±5.8 b	262.7±5.0 b
Total dry weight (g)	1.2±0.1b	1.8±0.3 ab	2.7±0.3 a	2.7±0.3 a

Different letters within a parameter indicate significant differences (Tukey's test at p<0.05). n = 11–12. Mean value ± SE (standard error) are given.

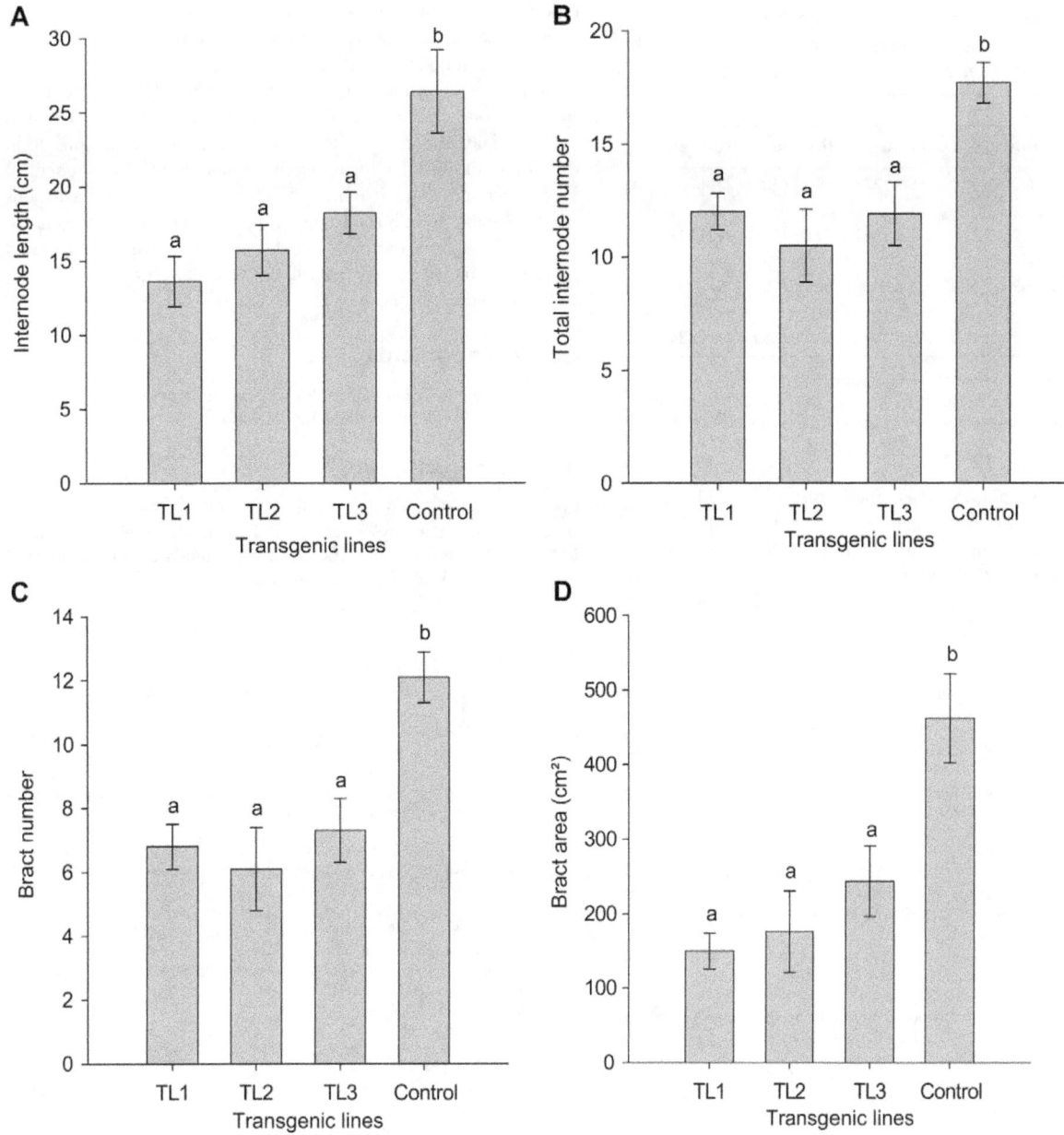

Figure 7. Effects of *AtSHI* overexpression on internode length (A), total internode number (sum of bracts and leaves) (B), bract number (C) and bract area (D) of different transgenic lines and control plants of poinsettia. Plants were grown under short day conditions of a 10 h photoperiod, n = 11–12. Mean values with different letters are significantly different based on ANOVA followed by a Tukey's test at p≤0.05. Vertical bars represent the ± SE (standard error).

Arabidopsis plants [46]. Multiple inputs like photoperiod, light quality and GA converge to regulate flowering [56,57]. GA promotes flowering in some LD plants like *Arabidopsis* and *Lolium* and inhibits flowering in SD plants such as rice [58,59]. However, GA's involvement in floral initiation is complex and varies from species to species [60,61]. The bract formation was visible after four weeks when plants were kept under SD conditions. Cyathia abscission was observed in the postharvest room under standard conditions. The cyathia abscission started after three weeks and at the end of 4 weeks about 90% bracts were abscised in both transgenic and WT control plants (data not shown). Furthermore, no negative effects were seen on cyathia abscission or bract necrosis or the postharvest quality due to *AtSHI* overexpression in

poinsettia. In *Arabidopsis* a negative effect of *SHI* overexpression is late flowering, but this was not observed in either *Kalanchoë* or in our study of poinsettia [19].

Conclusions

The economic importance of poinsettia provides a driving force to improve this important ornamental plant by using genetic engineering. With respect to control of plant morphology, this method is time saving, convenient and environmentally friendly compared to conventional breeding and application of hazardous chemical growth retardants. We have here demonstrated that compact poinsettia plants without delay in flowering or change in keeping quality can be obtained by using ectopic *AtSHI* expression

Table 3. Endogenous levels of auxin and their metabolites (pmol g^{-1} DW) in transgenic lines and control plants of poinsettia grown under a short photoperiod of 10 h.

Line	IAA	IAA-Asp	IAA-Glu	Total auxin
Transgenic line TL1	2004.8±14.2 a	169.9±23 a	99.6±7.5 a	2274.3±35.3 a
Transgenic line TL2	2575±104 a	240.3±29.7 a	89.4±4.0 ab	2905±127 a
Transgenic line TL3	2290±371 a	171.8±6.3 a	62.63±4.6 b	2524±375 a
WT control	2892±328 a	204.3±10 a	98.19±9.59 a	3194±340 a

Different letters within a parameter indicate significant differences (Tukey's test at $p \leq 0.05$). Mean value ± SE (standard error) are given; Three shoots of one plant were treated as one replicate and three separate replicates were analyzed.

as a tool. For poinsettia, dwarf characteristics with good keeping quality are required for high ornamental and market value. Dwarf plants are also convenient for handling and transportation compared to more elongated plants, and need less space in expensive production facilities. Although there is generally,

especially in Europe, a negative attitude towards genetic engineering of food crops which are consumed by humans, the attitude might be less negative for plants grown for other purposes such as non-food, non-feed ornamental plants. Hence, the use of *Agrobacterium*-mediated transformation to introduce *AtSHI* into commercially grown poinsettia cultivars can be very promising in the poinsettia industry, being environmentally friendly, beneficial to the economy and to human health by avoiding hazardous effects of plant growth retardant application. Due to the vegetative propagation nature of poinsettia, the acquired dwarfing effect and other desirable characteristics will be stably inherited in vegetative cuttings used in propagation.

Acknowledgments

Thanks are due to Erling Fløistad for his practical support. We are grateful for Dr Nicholas Clarke for linguistic correction.

Author Contributions

Conceived and designed the experiments: MAI HL DRB ST SKR JEO JLC. Performed the experiments: MAI HL SH JR. Analyzed the data: MAI HL DRB ST JR SKR JEO JLC. Contributed reagents/materials/analysis tools: MAI HL JR. Wrote the paper: MAI HL DRB ST SKR JEO JLC.

References

1. Chandler S, Tanaka Y (2007) Genetic modification in floriculture. Crit Rev Plant Sci 26: 169–197.
2. Lütken H, Clarke JL, Müller R (2012) Genetic engineering and sustainable production of ornamentals: current status and future directions. Plant Cell Rep 31: 1141–11573.
3. Ascough G, Erwin J, Van Staden J (2008) Biotechnology and ornamental horticulture. South Afr J Bot 74: 357.
4. Müller R (2011) Physiology and genetics of plant quality improvement. PhD Thesis, University of Copenhagen.
5. Ecke IP, Faust JE, Higgins A, Williams J (2004) The Ecke Poinsettia Manual. Ball Publishing, Illinois. 1–40.
6. Ecke P, Matkin O, Hartley DE (1990) The Poinsettia Manual (3rd edn.). Paul Ecke Poinsettias, Encinitas, USA: 1–267.
7. Zhu L, Li X, Welander M (2008) Overexpression of the *Arabidopsis gai* gene in apple significantly reduces plant size. Plant Cell Rep 27: 289–296.
8. Coles JP, Phillips AL, Croker SJ, Garcia-Lepe R, Lewis MJ, et al. (1999) Modification of gibberellin production and plant development in Arabidopsis by sense and antisense expression of gibberellin 20-oxidase genes. Plant J 17: 547–556.
9. Rademacher W (2000) Growth retardants: effects on gibberellin biosynthesis and other metabolic pathways. Ann Rev Plant Biol 51: 501–531.
10. De Castro VL, Goes KP, Chiorato SH (2004) Developmental toxicity potential of paclobutrazol in the rat. Int J Env Health Res 14: 371–380.
11. Sørensen MT, Danielsen V (2006) Effects of the plant growth regulator, chlormequat, on mammalian fertility. International J Androl 29: 129–133.
12. Islam MA, Kuwar G, Clarke JL, Blystad DR, Gislerød HR, et al. (2012) Artificial light from light emitting diodes (LEDs) with a high portion of blue light results in shorter poinsettias compared to high pressure sodium (HPS) lamps. Sci Hortic 147: 136–143.
13. Clifford SC, Runkle ES, Langton FA, Mead A, Foster SA, et al. (2004) Height control of poinsettia using photoselective filters. HortScience 39: 383–387.
14. Moe R, Fjeld T, Mortensen LM (1992) Stem elongation and keeping quality in poinsettia (*Euphorbia pulcherrima* Willd.) as affected by temperature and supplementary lighting. Sci Hortic 50: 127–136.
15. Mata DA, Botto JF (2009) Manipulation of light environment to produce high-quality poinsettia plants. HortScience 44: 702–706.
16. Lee M, Klopmeyer M, Bartoszyk IM, Gundersen-Rindal DE, Chou TS, et al. (1997) Phytoplasma induced free-branching in commercial poinsettia cultivars. Nature Biotech 15: 178–182.
17. Potera C (2007) Blooming biotechnology. Nature Biotech 25: 963–970.
18. Zawaski C, Kadmiel M, Ma C, Gai Y, Jiang X, et al. (2011) *Short internodes* like genes regulate shoot growth and xylem proliferation in *Populus*. New Phytol 191: 678–691.
19. Lütken H, Jensen LS, Topp SH, Mibus H, Müller R, et al. (2010) Production of compact plants by overexpression of *AtSHI* in the ornamental *Kalanchoë*. Plant Biotech J 8: 211–222.
20. Dijkstra C, Adams E, Bhattacharya A, Page A, Anthony P, et al. (2008) Overexpression of a gibberellin 2-oxidase gene from *Phaseolus coccineus* L. enhances

gibberellin inactivation and induces dwarfism in *Solanum* species. Plant Cell Rep 27: 463–470.
21. Curtis IS, Hanada A, Yamaguchi S, Kamiya Y (2005) Modification of plant architecture through the expression of GA 2-oxidase under the control of an estrogen inducible promoter in *Arabidopsis thaliana* L. Planta 222: 957–967.
22. Petty L, Thomas B, Jackson S, Harberd N (2001) Manipulating the gibberellin response to reduce plant height in *Chrysanthemum morifolium*. Acta Hort 560: 87–90.
23. Fridborg I, Kuusk S, Robertson M, Sundberg E (2001) The *Arabidopsis* protein SHI represses gibberellin responses in Arabidopsis and barley. Plant Physiol 127: 937.
24. Fridborg I, Kuusk S, Moritz T, Sundberg E (1999) The *Arabidopsis* dwarf mutant shi exhibits reduced gibberellin responses conferred by overexpression of a new putative zinc finger protein. The Plant Cell 11: 1019–1032.
25. Kuusk S, Sohlberg JJ, Eklund DM, Sundberg E (2006) Functionally redundant SHI family genes regulate *Arabidopsis gynoecium* development in a dose dependent manner. The Plant J 47: 99–111.
26. Kuusk S, Sohlberg JJ, Long JA, Fridborg I, Sundberg E (2002) *STY1* and *STY2* promote the formation of apical tissues during *Arabidopsis gynoecium* development. Development 129: 4707–4717.
27. Sohlberg JJ, Myrenås M, Kuusk S, Lagercrantz U, Kowalczyk M, et al. (2006) *STY1* regulates auxin homeostasis and affects apical–basal patterning of the *Arabidopsis gynoecium*. The Plant J 47: 112–123.
28. Ståldal V, Sohlberg JJ, Eklund DM, Ljung K, Sundberg E (2008) Auxin can act independently of CRC, LUG, SEU, SPT and STY1 in style development but not apical basal patterning of the *Arabidopsis gynoecium*. New Phytol 180: 798–808.
29. Smyth DR, Bowman JL, Meyerowitz EM (1990) Early flower development in *Arabidopsis*. The Plant Cell 2: 755–767.
30. Kim SG, Lee S, Kim YS, Yun DJ, Woo JC, et al. (2010) Activation tagging of an *Arabidopsis SHI-RELATED SEQUENCE* gene produces abnormal anther dehiscence and floral development. Plant Mol Biol 74: 337–351.
31. Fujita T, Sakaguchi H, Hiwatashi Y, Wagstaff SJ, Ito M, et al. (2008) Convergent evolution of shoots in land plants: lack of auxin polar transport in moss shoots. Evol Dev 10: 176–186.
32. Stavang JA, Lindgård B, Erntsen A, Lid SE, Moe R, et al. (2005) Thermoperiodic stem elongation involves transcriptional regulation of gibberellin deactivation in pea. Plant Physiol 138: 2344–2353.
33. Stavang JA, Gallego-Bartolomé J, Gómez MD, Yoshida S, Asami T, et al. (2009) Hormonal regulation of temperature-induced growth in *Arabidopsis*. The Plant J 60: 589–601.
34. Yamaguchi S (2008) Gibberellin metabolism and its regulation. Annu Rev Plant Biol 59: 225–251.
35. Zhao Y (2010) Auxin biosynthesis and its role in plant development. Annual Rev Plant Biol 61: 49–64.
36. Eklund DM, Ståldal V, Valsecchi I, Cierlik I, Eriksson C, et al. (2010) The *Arabidopsis thaliana* STYLISH1 protein acts as a transcriptional activator regulating auxin biosynthesis. The Plant Cell 22: 349–363.
37. Ross JJ, O'neill DP, Smith JJ, Kerckhoffs LHJ, Elliott RC (2000) Evidence that auxin promotes gibberellin A$_1$ biosynthesis in pea. The Plant J 21: 547–552.

38. Frigerio M, Alabadí D, Pérez-Gómez J, García-Cárcel L, Phillips AL, et al. (2006) Transcriptional regulation of gibberellin metabolism genes by auxin signaling in *Arabidopsis*. Plant Physiol 142: 553–563.

39. Wolbang CM, Chandler PM, Smith JJ, Ross JJ (2004) Auxin from the developing inflorescence is required for the biosynthesis of active gibberellins in barley stems. Plant Physiol 134: 769–776.

40. Hong JK, Kim JS, Kim JA, Lee SI, Lim MH, et al. (2010) Identification and characterization of *SHI* family genes from *Brassica rapa* L. ssp. *pekinensis*. Genes Genom 32: 309–317.

41. Clarke JL, Spetz C, Haugslien S, Xing S, Dees MW, et al. (2008) *Agrobacterium tumefaciens*-mediated transformation of poinsettia, *Euphorbia pulcherrima*, with virus-derived hairpin RNA constructs confers resistance to *Poinsettia mosaic virus*. Plant cell Rep 27: 1027–1038.

42. Sambrook J, Fritsch E, Maniatis T (1989) Molecular Cloning: A Laboratory Manual, 2nd edn.: Cold Spring Harbor Laboratory Press, Cold Spring Harbor, NY.

43. Livak KJ, Schmittgen TD (2001) Analysis of relative gene expression data using real-time quantitative PCR and the $2^{-\Delta\Delta C_T}$ method. Methods 25: 402–408.

44. Pěnčík A, Rolčík J, Novák O, Magnus V, Barták P, et al. (2009) Isolation of novel indole-3-acetic acid conjugates by immunoaffinity extraction. Talanta 80: 651–655.

45. Eklund DM, Thelander M, Landberg K, Ståldal V, Nilsson A, et al. (2010) Homologues of the *Arabidopsis thaliana SHI/STY/LRP1* genes control auxin biosynthesis and affect growth and development in the moss *Physcomitrella patens*. Development 137: 1275.

46. Ståldal V, Cierlik I, Chen S, Landberg K, Baylis T, et al. (2012) The *Arabidopsis thaliana* transcriptional activator *STYLISH1* regulates genes affecting stamen development, cell expansion and timing of flowering. Plant Mol Biol 78: 545–559.

47. Cheng Y, Dai X, Zhao Y (2006) Auxin biosynthesis by the YUCCA flavin monooxygenases controls the formation of floral organs and vascular tissues in *Arabidopsis*. Genes Develop 20: 1790–1799.

48. Zhao Y, Christensen SK, Fankhauser C, Cashman JR, Cohen JD, et al. (2001) A role for flavin monooxygenase-like enzymes in auxin biosynthesis. Sci 291: 306–309.

49. Nemhauser JL, Feldman LJ, Zambryski PC (2000) Auxin and *ETTIN* in *Arabidopsis* gynoecium morphogenesis. Development 127: 3877–3888.

50. Taiz L, Zeiger E (2010) Plant Physiology (5th edition), Chapter 19 Auxin: The Growth Hormone. Sinauer Associates Inc., Publishers, Sunderland, Massachusetts USA, 545–578.

51. Fleet CM, Sun T (2005) A DELLAcate balance: the role of gibberellin in plant morphogenesis. Current Opi Plant Biol 8: 77–85.

52. Sakamoto T, Morinaka Y, Ishiyama K, Kobayashi M, Itoh H, et al. (2003) Genetic manipulation of gibberellin metabolism in transgenic rice. Nature Biotech 21: 909–913.

53. O'Neill DP, Ross JJ (2002) Auxin regulation of the gibberellin pathway in pea. Plant Physiol 130: 1974–1982.

54. Ross J (1998) Effects of auxin transport inhibitors on gibberellins in pea. J Plant Growth Reg 17: 141–146.

55. Werner T, Motyka V, Laucou V, Smets R, Van Onckelen H, et al. (2003) Cytokinin-deficient transgenic *Arabidopsis* plants show multiple developmental alterations indicating opposite functions of cytokinins in the regulation of shoot and root meristem activity. The Plant Cell 15: 2532–2550.

56. Simpson GG, Dean C (2002) *Arabidopsis*, the Rosetta stone of flowering time? Science 296: 285–289.

57. Mouradov A, Cremer F, Coupland G (2002) Control of flowering time: interacting pathways as a basis for diversity. The Plant Cell Online 14: S111–S130.

58. Eriksson S, Böhlenius H, Moritz T, Nilsson O (2006) GA$_4$ is the active gibberellin in the regulation of LEAFY transcription and *Arabidopsis* floral initiation. The Plant Cell Online 18: 2172–2181.

59. Izawa T, Oikawa T, Sugiyama N, Tanisaka T, Yano M, et al. (2002) Phytochrome mediates the external light signal to repress FT orthologs in photoperiodic flowering of rice. Genes and Develop 16: 2006–2020.

60. Koorneef M, Elgersma A, Hanhart C, Loenen-Martinet EP, Rijn L, et al. (1985) A gibberellin insensitive mutant *of Arabidopsis thaliana*. Physiol Plant 65: 33–39.

61. Mutasa-Göttgens E, Hedden P (2009) Gibberellin as a factor in floral regulatory networks. J Expt Bot 60: 1979–1989.

17

Temperature Effects on Bacterial Phytochrome

Ibrahim Njimona, Rui Yang, Tilman Lamparter*

Karlsruhe Institute of Technology KIT, Botanical Institute, Karlsruhe, Germany

Abstract

Bacteriophytochromes (BphPs) are light-sensing regulatory proteins encoded in photosynthetic and non-photosynthetic bacteria. This protein class incorporate bilin as their chromophore, with majority of them bearing a light- regulated His kinase or His kinase related module in the C-terminal. We studied the His kinase actives in the temperature range of 5°C to 40°C on two BphPs, Agp1 from *Agrobacterium tumefaciens* and Cph1 from cyanobacterium *Synechocystis* PCC 6803. As reported, the phosphorylation activities of the far red (FR) irradiated form of the holoprotein is stronger than that of the red (R) irradiated form in both phytochromes. We observed for the apoprotein and FR irradiated holoprotein of Agp1 an increase in the phosphorylation activities from 5°C to 25°C and a decrease from 25°C to 40°C. At 5°C the activities of the apoprotein were significantly lower than those of the FR irradiated holoprotein, which was opposite at 40°C. A similar temperature pattern was observed for Cph1, but the maximum of the apoprotein was at 20°C while the maximum of the FR irradiated holoprotein was at 10°C. At 40°C, prolonged R irradiation leads to an irreversible bleaching of Cph1, an effect which depends on the C-terminal His kinase module. A more prominent and reversible temperature effect on spectral properties of Agp1, mediated by the His kinase, has been reported before. His kinases in phytochromes could therefore share similar temperature characteristics. We also found that phytochrome B mutants of *Arabidopsis* have reduced hypocotyl growth at 37°C in darkness, suggesting that this phytochrome senses the temperature or mediates signal transduction of temperature effects.

Editor: Enamul Huq, University of Texas at Austin, United States of America

Funding: The work was funded by the Deutsche Forschungsgemeinschaft, La 799/8-1. The authors acknowledge support by Deutsche Forschungsgemeinschaft and Open Access Publishing Fund of Karlsruhe Institute of Technology. The funders had no role in study design, data collection and analysis, decision to publish, or preparation of the manuscript.

Competing Interests: The authors have declared that no competing interests exist.

* Email: tilman.lamparter@kit.edu

Introduction

Histidine kinases (HKs) are sensory homodimeric proteins which commonly consist of an N-terminal sensory module and a C-terminal effector module. The C-terminal effector module is composed of the dimerization/histidine phosphotransfer (DHp) and catalytic/ATP-binding (CA) domains. HKs are known to catalyze three distinct phosphotransfer reactions: autophosphorylation of the histidine substrate within the DHp domain, and transphosphorylation as well as dephosphorylation of the cognate response regulator (RR). These proteins are essential signal carriers in plants and microorganisms [1,2]. Bacteria and fungi respond to transient environments through transmembrane integrated HKs, which act in concert with their RRs to elicit necessary adaptive responses that are critical for their survival and virulence [3,4]. The HKs and RRs have evolved as two-component signal transduction systems (TCS), thereby stimulation of the HK leads to autophosphorylation at a conserved histidine residue which initiates a signaling cascade [5]. In the prototypical two-component pathway, the phosphoryl group is transferred from the HK directly to the RR. In the case of hybrid HKs, the phosphoryl group is first transferred intramolecularly to an aspartate in the receiver domain of the same polypeptide. A histidine transferase then shuttles the phosphoryl group to a soluble RR. The net difference between phosphorylation and dephosphorylation of RR accounts for the quick biological response either by altering the transcriptional level of specific downstream target genes or by direct modulation of molecular motors [6–8]. Because of wide prevalence in bacteria and fungi and the absence in humans, TCSs have been considered attractive targets for the development of new potential drugs to control bacterial and fungal infections [9].

Phytochromes are widely distributed photochromic photoreceptors found in bacteria, plants and fungi. These bilin proteins are sensitive in the red and far red region of the visible spectrum [10]. The N-terminal photosensory chromophore (PCM) module is composed of a PAS (Period/Arnt/Single minded), GAF (cGMP phosphodiesterases/Adenylate cyclase/FhlA) and PHY (phytochrome) domain. In fungal and bacterial phytochromes including Agp1 from *Agrobacterium tumefaciens*, a biliverdin chromophore is covalently attached to a conserved Cys at the N-terminal of the PAS domain [11,12], whereas a conserved Cys in the GAF domain is used by plant- and various cyanobacterial phytochromes including Cph1 from *Synechocystis* PCC 6803 [13–16] that use phytochromobilin and phycocyanobilin as chromophore, respectively [17]. Phytochromes are synthesized in the red absorbing form (Pr) which upon absorption of red light photoconverts to the far red absorbing form (Pfr). Pfr converts back to Pr upon irradiation with far red. Besides photoconversion, Pfr to Pr and Pr to Pfr dark conversion have been found for different types of phytochromes. Photoconversion is coupled to protein structural changes [18–21] which result in a modulation of signal transduction.

The C-terminus of plant, fungal and most bacterial phytochromes shows significant homology to histidine kinases [22]. The substrate histidine in the H-box of the protein is highly conserved in bacterial and fungal phytochromes, but missing in plant phytochromes with the exception of monocotyledonous PhyA [23–25]. Bacterial and fungal phytochromes act as bona fide HKs in which photoconversion alters their kinase activity [26]. In cyanobacterial Cph1 and in Agp1 from *A. tumefaciens*, the bacterial phytochromes used in the present study, autophosphorylation is strong in the Pr and weaker in the Pfr form [18,26,27].

Crystal X-ray diffraction and cryo-EM studies show that bacterial phytochromes are homodimeric proteins in which the two subunits are arranged in a parallel manner [28,29]. The GAF domain forms most of the contacts with the chromophore. The PAS and GAF domains are connected in a knotted structure with the PHY domain forming a tongue-like structure that folds back on the GAF domain and the chromophore, providing a short connection between chromophore and HK [13,30–32]. The PHY domain is shown to stabilize the Pfr form of Cph1 and Agp1 [13,19] and is connected to the C-terminal HK by a long helix [28]. Intramolecular signal transduction is probably mediated through this connection.

In Agp1, the autophosphorylation shows an unexpected temperature dependency. The strongest phosphorylation signal was obtained at 25°C. With increasing temperature the phosphorylation decreased and was almost undetactable at 40°C, although the cells are still growing at this temperature. Spectral properties of Agp1 are affected by the temperature in a way that had not yet been described for other phytochromes: at 40°C, continuous red irradiation leads to photoconversion of Pfr into another bleached form termed Prx. This temperature effect is dependent on the presence of the HK [33]. These findings led to the suggestion that Agp1 might act as a temperature sensor. A similar loss of phosphorylation at 40°C has been described for the VirA HK of *A. tumefaciens*, which is responsible for activation of virulence genes [34].

In the present study, we extended our studies on Agp1 to a broader temperature range and investigated HK temperature effects of Cph1, a well characterized cyanobacterial phytochrome. Whereas the temperature optimum of Agp1 was between 20°C and 25°C, the Cph1 holoprotein had a temperature optimum at 10°C. At elevated temperature, prolonged irradiation leads to an irreversible bleaching of Cph1, an effect again dependent on the HK.

Materials and Methods

Protein expression, purification and assembly

All expression vectors used in these studies encode protein with a C-terminal hexahistidine tag for Ni^{2+} affinity purification. The vectors are based on the pQE12 (Qiagen) expression vector. The cloning of pAG1 and pF10, expression vectors (for full-length Agp1 and Cph1), is described in [27,35] and the cloning of expression vectors for the truncated proteins lacking the HK is described in [36,37]. *Escherichia coli* XL1-blue cells with the desired expression vector were grown in 2 L lysogeny broth medium with 0.3 μM final concentration of ampicillin and tetracycline at 37°C until an optical density at 600 nm (OD_{600}) of 0.6 was reached. Specific protein expression was induced by adding isopropyl-1-thiol-ß-D-galactopyranoside to a final concentration of 50 μM. After 40 hours incubation at 18°C, the OD_{600} reached 2.3. At this point, the cells were collected and centrifuged at 5000 g for 10 min and the pellet was suspended in 20 mL of extraction buffer (300 mM NaCl, 50 mM Tris/Cl, 5 mM EDTA, pH 7.8) with freshly added 2 mM (final concentration) ß-mercaptoethanole. The cells were lysed with a French Pressure Cell (Aminco) with two passages at 1380 bar. The lysed cells were centrifuged at 20 000 g for 30 min. The soluble proteins in the supernatant were precipitated with 50% ammonium sulfate, and the pellet dissolved in dilute imidazole buffer (300 mM NaCl, 50 mM Tris/HCl, 10 mM imidazole, pH 7.8). The sample was centrifuged again and the supernatant loaded to a 3 cm×5 cm Ni^{2+} affinity chromatography column (Qiagen, Hilden Germany) that was equilibrated with dilute imidazole buffer. The column was then washed with this buffer until non-bound proteins were washed out. The apoprotein was eluted with concentrated imidazole buffer (300 mM NaCl, 50 mM Tris/Cl, 250 mM imidazole, pH 7.8). The eluted protein was pooled and precipitated with 50% ammonium sulfate. The resulting pellet was suspended in extraction buffer and centrifuged again. The supernatant was used for later steps. All purification steps were done at 4°C. The apoprotein concentration was estimated by absorption at 280 nm [27]. For details on the expression and purification of apo-Agp1 and apo-Cph1 see previous publications [18,27]. For chromophore assembly, 50 μM of DTT and biliverdin (Frontier Scientific) or phycocyanobilin [35] to a final concentration of 60 μM were added to 20 μM protein. The sample was incubated in darkness at 20°C until assembly was complete as monitored by UV/visible photometry. Excess chromophore was separated from the holoprotein using NAP-10 desalting columns (GE Healthcare) as described by the manufacturer. The final buffer for spectral assays was 300 mM NaCl, 50 mM Tris/Cl, 5 mM EDTA, pH 7.8. The pH, adjusted at 4°C, varies with the temperature and was 7.0 at 40°C. Therefore, control measurements were undertaken to test spectra and phosphorylation at 20°C between pH 7 and 8. We found no pH effects on these features that could explain the changes obtained upon temperature change.

Spectral assays

Absorption spectra were measured in a JASCO V-550 photometer with temperature control unit and a custom-built computer-controlled irradiation device. The scan speed was set to 1000 nm min^{-1}; scans were measured between 900 nm and 250 nm. For photoconversion of Cph1, 730 nm far red and 655 nm red light emitting diodes were used. Far red (500 μmol $m^{-2} s^{-1}$) was typically given for 120 s. Red (400 μmol $m^{-2} s^{-1}$) was either given for 60 s of for prolonged times as indicated in the text.

Kinase assays at different temperatures

For autophosphorylation, an earlier protocol [18,33] was adopted. Phosphorylation experiments of the holoproteins were performed under blue-green safelight using 505 nm light emitting diodes or in darkness. The protein concentration was 12 μM. Each sample contained 5 μl Agp1 or Cph1, irradiated with either 655 nm red light from a light emitting diode (20 μmol $m^{-2} s^{-1}$) or with 780 nm far red from a light emitting diode (80 μmol $m^{-2} s^{-1}$). The irradiation time was always 2 min, and the temperature during this irradiation was 25°C. Directly after irradiation, 15 μl phosphorylation buffer (final concentrations 25 mM Tris/HCl, 5 mM $MgCl_2$, 4 mM β-ME, 0.2 mM EDTA pH 7.8, 50 mM KCl, 5% ethylene glycol, 0.45 μM (50 MBq/ml) $γ^{-32}P$ ATP, pH 7.8) were added to each sample. After mixing, the samples were immediately transferred to a thermal block, which was pre-set to a given temperature. The incubation time was 30 min. Thereafter, the phosphorylation reaction was stopped by adding 10 μl of loading buffer (30% glycerol, 6% SDS, 300 mM

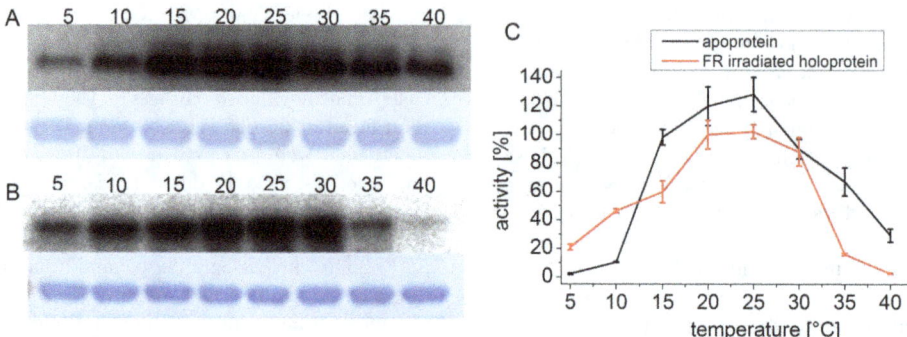

Figure 1. Autophosphorylation of Agp1 at various temperatures. Autoradiogram (above) and Coomassie-stained blot (below) of Agp1 apoprotein (A) and FR irradiated holoprotein (B). The incubation temperature during the phosphorylation assay in °C is given above each lane. (C) Mean phosphorylation intensities of three experiments ± SE as shown in (A) and (B) are plotted over temperature. The 100% value corresponds to the mean signal of the holoprotein at 25°C.

DTT, 0.01% bromphenol blue, 240 mM Tris/HCl, pH 6.7) to each reaction mixture. Then, 10 µl of each sample were loaded an SDS-PAGE gel (10% acrylamide in separating gel). After electrophoresis, the protein was transferred onto a PVDF membrane (Millipore) with a Trans-Blot semi-dry blot apparatus (Bio-Rad). The dried membrane was exposed to a phosphoimager plate (Fuji) for about 5 min, followed by quantification using the fluorescent image analyzer FLA 2000 (Fuji), and integrated analysis software. The protein on the PVDF membrane was stained with Simply Blue safestain (Invitrogen). Apoprotein phosphorylation experiments were performed in the same manner as for the holoprotein except that the protein was not irradiated and that all incubation steps were under artificial light. Direct comparisons of apo- and holoprotein Agp1 at 25°C revealed a 1.2±0.1 apoprotein/holoprotein (FR irradiated) ratio. This ratio was used to normalize the temperature dependent activities of holo- and apoprotein in Agp1, which were probed in different sets of experiments. The kinase activity of Cph1 was assayed in a similar manner. The ratio of apoprotein/holoprotein (Pr) phosphorylation at 20°C was 0.97±0.1.

Arabidopsis hypocotyl growth

Arabidopsis thalinia phytochrome B mutants N6217 (Col ecotype) and N6218 (Ws ecotype) were obtained from the European Arabidopsis Stock Center Nottingham and propagated in the Botanical Garden. Seeds were surface sterilized for 1 min in 70% EtOH, for 7 min in 1:1 (v/v) sodium hypochlorite, rinsed 4 times with sterile H_2O, and placed on growth medium consisting of 1/2 Murashige and Skoog salts [38], 0.05% MES, 0.8% (w/v)

agar without sucrose. The plates were incubated for 2 days at 4°C in darkness. Germination was induced by applying 2 hours white light at 23°C. Thereafter, seeds were taken back to darkness for 22 h at 23°C. Germinated seeds were then incubated at different temperatures as given in the text for 48 h on vertical petri dishes. All treatments are done in green safelight or darkness. Thereafter images were taken and hypocotyl lengths were measured using Image J.

Results

The previous measurements on Agp1 were performed at temperatures of 25°C and above [33]. In order to find the temperature optimum of phosphorylation, we expanded the temperature range for phosphorylation studies. As shown in Fig. 1, the temperature optimum for the apoprotein and the holoprotein is at 20°C to 25°C. As before, the autophosphorylation of the apoprotein at 40°C was still around 20% of the activity at 25°C, but negligible in the holoprotein. Autophosphorylation of the holoprotein at 5°C was still reasonably high, around 20% of the 25°C level. Thus, the order between holo and apoprotein is reverted at low temperature.

A decrease of phosphorylation activity with increasing temperature has also been described for other HK-like VirA and DesK [34,39]. We therefore asked whether other bacterial phytochromes follow the same pattern. We performed similar experiments with the cyanobacterial phytochrome Cph1 from *Synechocystis* PCC 6803. Apo- and holoprotein were again investigated in separate experiments to keep the number of samples in each experiment in

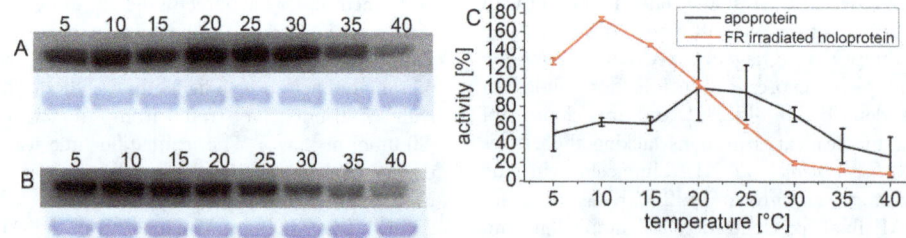

Figure 2. Autophosphorylation of Cph1 at various temperatures. Autoradiogram (above) and Coomassie-stained blot (below) of Cph1 apoprotein (A) and far red light irradiated holoprotein (B). Holoprotein was irradiated with FR and incubated with γ-^{32}P ATP in darkness at the indicated temperatures. (C) Mean phosphorylation intensities of three experiments ± SE as shown in (A) and (B) are plotted over temperature. The 100% value corresponds to the mean signal of the holoprotein at 20°C.

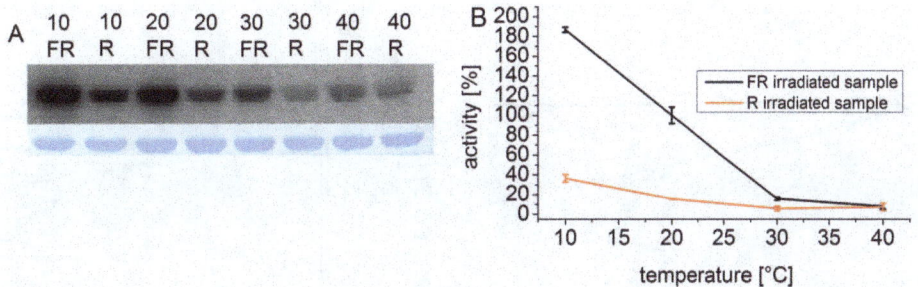

Figure 3. Autophosphorylation of Cph1 at various temperatures. (A) Autoradiogram (above) and Coomassie-stained blot (below) of Cph1 holoprotein. Samples were irradiated either with FR or R as indicated and incubated with $\gamma^{-32}P$ ATP in darkness at the indicated temperatures (°C). (B) Mean phosphorylation intensities of three experiments $\pm$ SE as shown in (A) are plotted over temperature. The 100% value corresponds to the mean signal of the FR irradiated sample at 20°C. as shown in (A).

a range that can be handled. In a third series, the phosphorylation of the holoprotein in the Pr and Pfr forms were compared.

The temperature maximum of the Cph1 holoprotein in the Pr form was at 10°C, and at 5°C the activity was still 80% of the highest value (Fig. 2). At 40°C, the activity was less than 5%. The temperature optimum of the apoprotein was at 20–25°C. The chromophore has thus also a significant impact on the temperature behavior of Cph1 phosphorylation. At 10°C, the activity of the holoprotein is approximately 2.5 fold higher than that of the apoprotein (Fig. 2). With increasing temperature, the ratio between far-red- and red-irradiated holoprotein samples (predom-

inately Pr and Pfr, respectively) diminished (Fig. 3), as in Agp1. Under our measuring conditions this ratio was 5:1 at 10°C, but almost 1:1 at 40°C.

Spectral properties

At 35°C and 40°C, Agp1 acquires unusual spectral properties. Continuous irradiation results in the conversion of Pfr into a new spectral species with a decreased extinction coefficient, termed Prx, which might have a deprotonated chromophore. The effect is only found in the presence of the HK [33]. Here, we tested the effect of continuous irradiation on Cph1 at different temperatures.

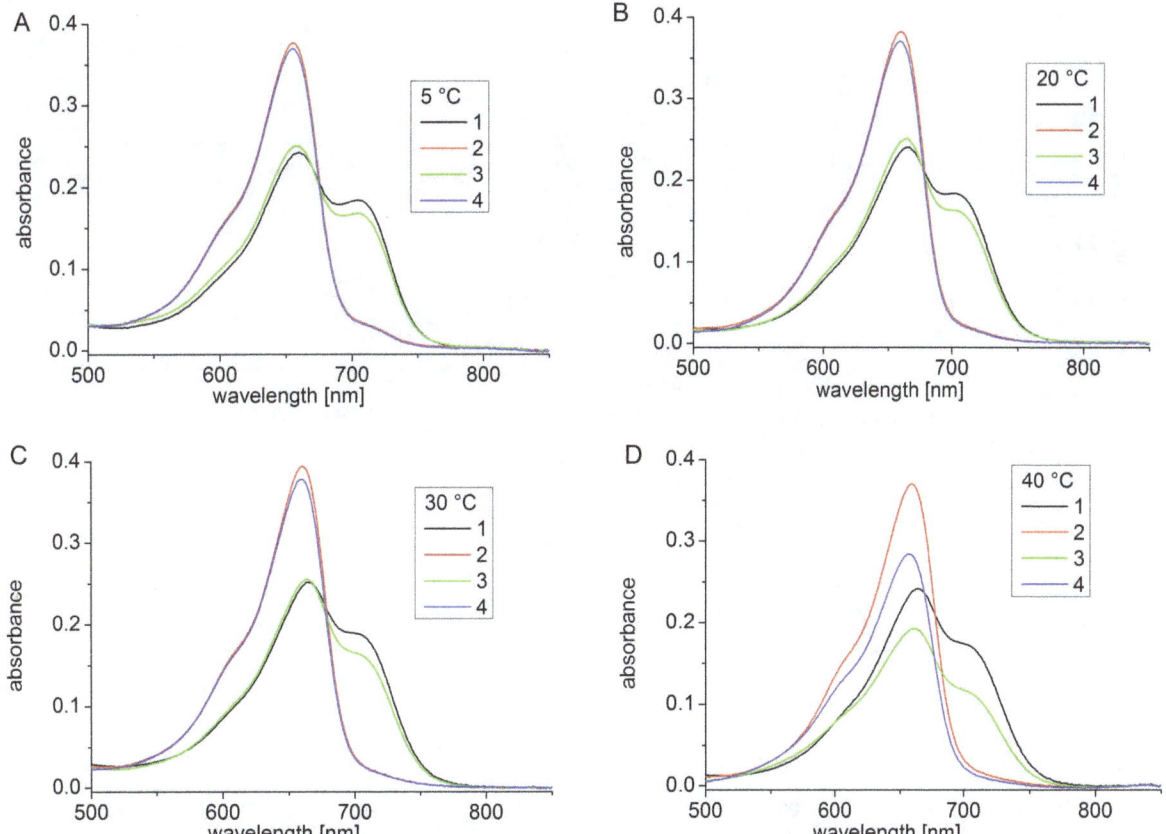

Figure 4. Spectra of full length Cph1 at 5°C (A), 20°C (B), 30°C (C) and 40°C (D). Shown are measurements after temperature adaptation and 2 min R irradiation (1, black lines), after a subsequent FR irradiation (2, red lines), a long (2 h) R irradiation (3, green lines) and a subsequent FR irradiation (4, blue lines). The R light intensity was 400 μmol m^{-2} s^{-1}.

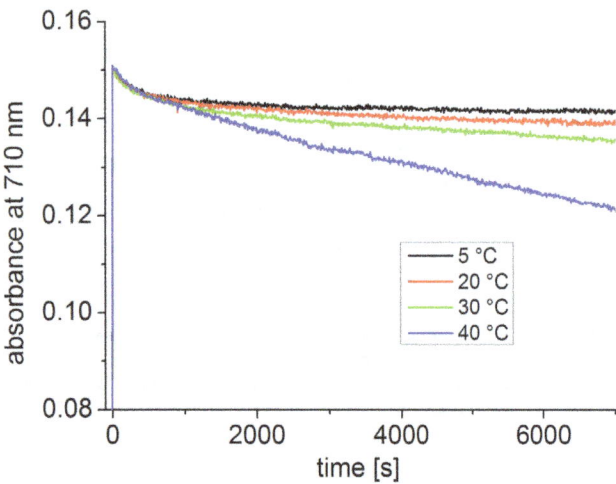

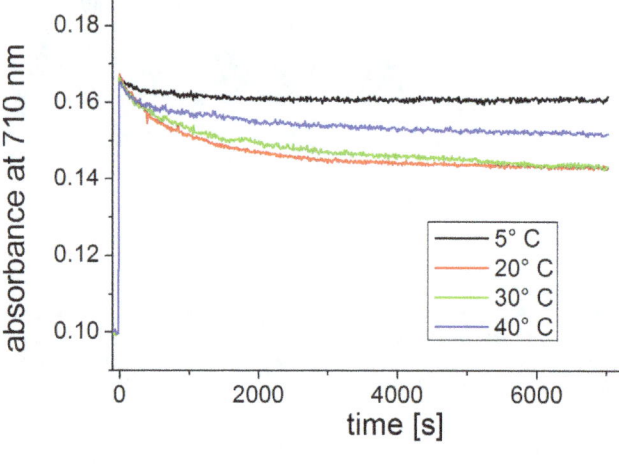

Figure 5. Time drive measurements with full length Cph1. Absorbance at 710 nm (the absorbance maximum of Pfr) was measured continuously; at t = 0 s, R (400 μmol m^{-2} s^{-1}) was switched on. The protein concentration of this sample differed slightly from that in Fig. 4.

Figure 7. Time drive measurements with PCM of Cph1. Absorbance at 710 nm (the absorbance maximum of Pfr) was measured continuously; at t = 0 s, R (400 μmol m^{-2} s^{-1}) was switched on. The protein concentration of this sample differed from that in Fig. 6.

Spectra were taken after FR and R irradiations at 5°C, 20°C, 30°C and 40°C. In a typical setting, Cph1 reference spectra were first taken after 2 min R and after 2 min FR. The samples were then irradiated for 2 h with R. One spectrum was recorded after this treatment and a final one after a 2 min FR irradiation. Spectra

and absorbance changes as measured during the 2 h R irradiation period are shown in Figures 4 and 5 for the full length protein. For the full length protein we observed a slight (less than 5%) loss of absorbance at 710 nm upon continued irradiation at 5°C, 20°C, and 30°C (Fig. 5) that affects both Pr and Pfr (Fig. 4). This effect results probably from bleaching of a sensitive subpopulation of

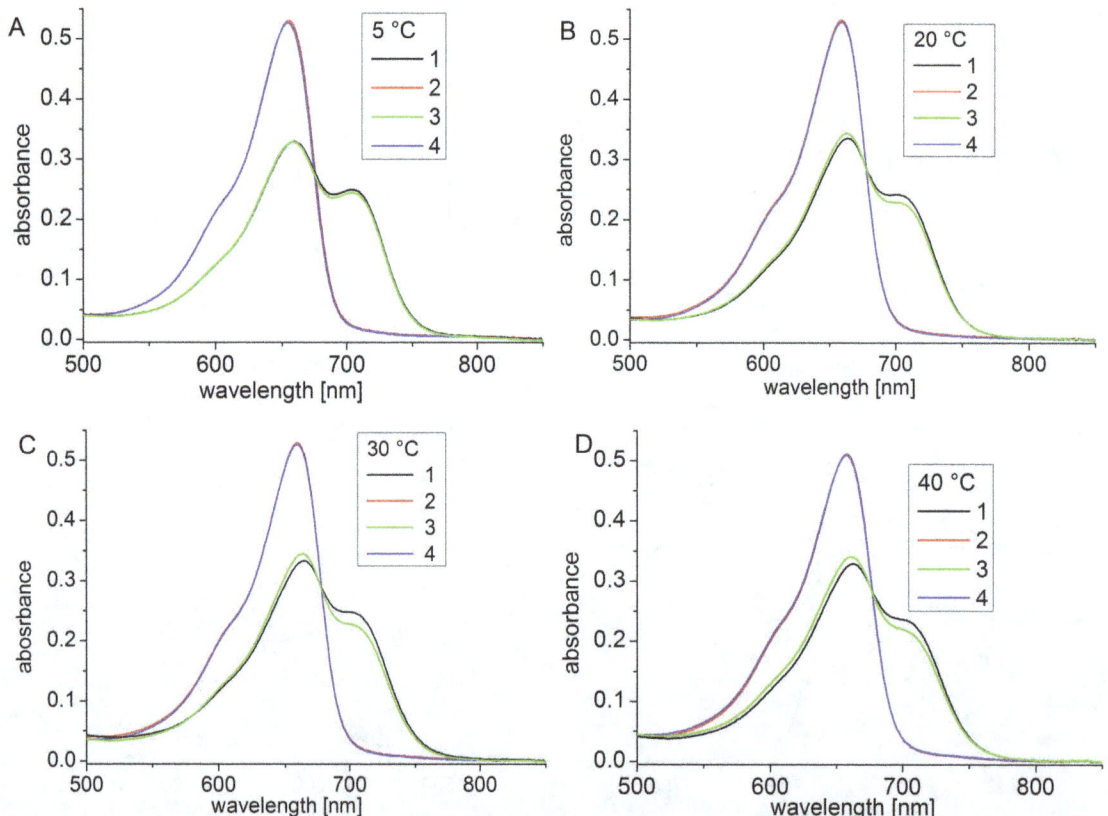

Figure 6. Spectra of the N-terminal PCM of Cph1 at 5°C (A), 20°C (B), 30°C (C) and 40°C (D). Shown are spectra after temperature adaptation and 2 min R irradiation (1, black lines), after a subsequent FR irradiation (2, red lines), a long (2 h) R irradiation (3, green lines) and a subsequent FR irradiation (4, blue lines). The R light intensity was 400 μmol m^{-2} s^{-1}.

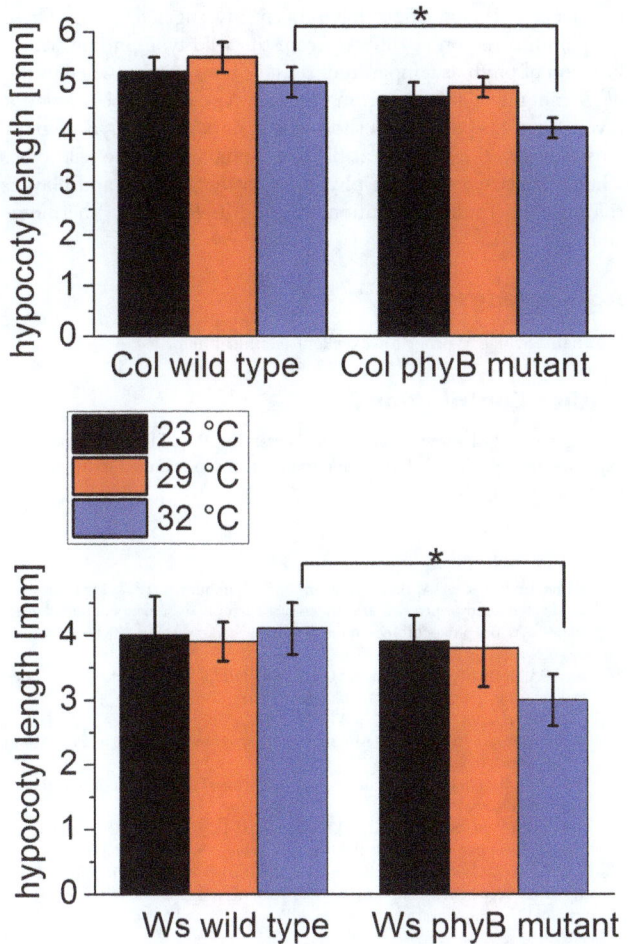

Figure 8. Hypocotyl lengths of Arabidopsis wild type and mutants grown at different temperature in darkness. Mean values ± SE.

Cph1 under the high light intensities (400 µmol m^{-2} s^{-1}). Heterogeneity of Cph1 has been demonstrated by various methods [40,41]. When the samples were irradiated continuously with red light, the absorbance at 710 nm decreased steadily. At 40°C, prolonged irradiation resulted in a continuous loss of absorbance in the range of the Pfr absorbance maximum, which reached approximately 20% after 2 h (Fig. 5). This loss affected both Pr and Pfr absorbance (Fig. 4D). There were no indications for denaturation of the protein during these experiments: denaturation is accompanied by aggregation and increase of scattering, which was not found. When irradiated Cph1 was brought back from 40°C to 20°C, the absorbance did not recover (data not shown). Opposed to Agp1, the temperature effect on the spectra is therefore not reversible. Spectra and absorbance changes of the N-terminal PCM fragment are shown in Figs. 6 and 7, respectively. The spectra recorded at 40°C did not significantly differ from those at the other temperatures. There was a weak bleaching effect at all temperatures which was strongest at 30°C (Fig. 7). Under continuous illumination with red light, the absorbance at 710 nm decreased slightly during the initial hour but did not change thereafter (Figs. 6 and 7). The temperature effect is thus dependent on the presence of the His kinase.

In vivo temperature effect

In order to see whether there could be a temperature effect of Agp1 in *A. tumefaciens*, we performed growth and swimming plate assays of wild type and knockout mutants but we could not find an effect that would be significantly different between the wild type and the *agp1*- mutant (Njimona, Rottwinkel, Lamparter, unpublished results). The biological effects of Cph1 or Agp1 are not known [42]. For this reason an *in vivo* effect of temperature mediated by these phytochromes might be hard to find. We thus performed a temperature experiment with seedlings of *Arabidopsis thaliana*, the model plant for phytochrome responses. The hypocotyl growth of seedlings is inhibited by light, mediated by phytochromes and cryptochromes [43]. We compared hypocotyl lengths of seedlings grown at different temperatures between two different *phyB* mutants and the corresponding wild types Columbia (Col) and Wassilevskia (Ws). In order to avoid interference with light, we performed these growth experiments in darkness only. In both wild type seedlings, no significant difference was found between hypocotyls grown at 23°C, 29°C or 32°C, whereas both *phyB* mutants had shorter hypocotyls at 32°C when compared with the corresponding wild type (Fig. 8). This difference was significant with an error probability of <5%. The differences between wild type and mutant at other temperatures were not significant. The fact that two independent mutants had a similar phenotype indicates that the result is not related to a second site mutation. Phytochrome B is thus required for proper hypocotyl growth at elevated temperature in darkness.

Discussion

In the ecosystem, organisms have to withstand to fluctuations in temperature on a daily basis. Previous studies have shown that photoreceptors may have a unique role in reacting to environmental temperature change [44]. For phytochromes a thermometer function has not been described. However, most phytochromes have a C-terminal His kinase or His kinase related region which could function as a temperature sensor. Here, we investigated the effect of temperature on the HK activities of two model bacteriophytochromes, Agp1 and Cph1 and on the hypocotyl growth of *Arabidopsis*.

In several cases, HKs act as temperature sensors, either as their sole function, exemplified by Hik33 of *Synechocystis* PCC 6803 [45] or DesK from *Bacillus subtilis*, [46] or combined with sensing chemical factors as in VirA of *Agrobacterium tumefaciens* [34]. All proteins shows high autophosphorylation activity at 25°C and low activity at 37°C. Recently, Mironov et al. suggested a dual function of Hik33 as red light regulation of cold responses in *Synechocystis* PCC6803 [47,48]. In the bacteriophytochrome Agp1, a temperature increase above 25°C results in a reduction of HK activity. An even more pronounced effect with an optimum at 10°C was found for the Cph1 holoprotein from *Synechocystis*. It might well be that a decrease of kinase activity with increasing temperature is universal for bacterial phytochromes and other HKs. Both Agp1 and Cph1 also have in common that elevated temperature changed their spectral properties in a HK dependent manner. Phytochrome fragments with PAS and GAF domains but missing PHY domains have photoproduct spectra [18,19,49] similar to those of full length Agp1 and Cph1 at elevated temperature ([33] and present work). Elevated temperature could result in partial unfolding of the PHY domain, mediated through the HK that is linked to the C-terminus of the PHY domain. This could lead to disturbance of the chromophore hydrogen bonding network and deprotonation of the chromophore. A third feature common to Cph1 and Agp1 is the combined effect of chromophore abundance and temperature on HK activity. In both proteins the apoprotein kinase activity was stronger than that of the holoprotein at elevated temperature, but weaker in the low temperature range.

Several links between phytochromes and temperature effects have been reported. The phytochrome from *P. aeruginosa* seems to be involved in stress responses and quorum sensing; heat tolerance was impaired in a phytochrome knockout mutant but not in the wild type [50]. In seed plants, phytochromes are important for temperature adaptations [51], with seed germination providing a classic example [52,53]. In *A. thaliana*, which has five phytochromes (phyA to phyE), the effect of temperature on seed germination has been investigated in wild-type plants and in double and triple *phyA*, *phyB*, and *phyE* mutants. The germination/temperature pattern differed from the wild-type pattern in most mutants (Heschel et al. 2007). The *HFR1* and *PIF4* transcription factors play a central role in temperature adaptation of *A. thaliana*. In *hfr1*, *pif4*, and *phyB* mutants temperature effects

on hypocotyl elongation have been investigated [54]. These temperature responses differed from the wild type, in line with a function of phyB as temperature sensor. The question whether the *phyB* mutant is different from the wild type in darkness was not investigated [54]. Our data showed that dark grown *phyB* mutants have a reduced hypocotyl at 32°C as compared to the wild type. These results suggests that phyB acts in the Pr form and plays a role in temperature adaptation; it could itself act as a thermometer.

Acknowledgments

We thank Sybille Wörner for excellent technical help.

Author Contributions

Conceived and designed the experiments: IN RY TL. Performed the experiments: IN RY. Analyzed the data: IN RY. Contributed to the writing of the manuscript: IN TL.

References

1. Gao R, Stock AM (2009) Biological Insights from Structures of two-component Proteins. Annual Review of Microbiology 63: 133–154.
2. Mougel C, Zhulin IB (2001) CHASE: an extracellular sensing domain common to transmembrane receptors from prokaryotes, lower eukaryotes and plants. Trends Biochem Sci 26: 582–584.
3. Falke JJ, Bass RB, Butler SL, Chervitz SA, Danielson MA (1997) The two-component signaling pathway of bacterial chemotaxis: a molecular view of signal transduction by receptors, kinases, and adaptation enzymes. Annu Rev Cell Dev Biol 13: 457–512.
4. Mizuno T (1998) His-Asp phosphotransfer signal transduction. J Biochem 123: 555–563.
5. West AH, Stock AM (2001) Histidine kinases and response regulator proteins in two-component signaling systems. Trends Biochem Sci 26: 369–376.
6. Russo FD, Silhavy TJ (1993) The essential tension: opposed reactions in bacterial two-component regulatory systems. Trends Microbiol 1: 306–310.
7. Stock AM, Robinson VL, Goudreau PN (2000) Two-component signal transduction. Annu Rev Biochem 69: 183–215.
8. Vescovi EG, Sciara MI, Castelli ME (2010) Two component systems in the spatial program of bacteria. Current Opinion in Microbiology 13: 210–218.
9. Gotoh Y, Eguchi Y, Watanabe T, Okamoto S, Doi A, et al. (2010) Two-component signal transduction as potential drug targets in pathogenic bacteria. Curr Opin Microbiol 13: 232–239.
10. Rockwell NC, Su YS, Lagarias JC (2006) Phytochrome structure and signaling mechanisms. Annu Rev Plant Biol 57: 837–858.
11. Lamparter T, Carrascal M, Michael N, Martinez E, Rottwinkel G, et al. (2004) The biliverdin chromophore binds covalently to a conserved cysteine residue in the N-terminus of *Agrobacterium* phytochrome Agp1. Biochemistry 43: 3659–3669.
12. Blumenstein A, Vienken K, Tasler R, Purschwitz J, Veith D, et al. (2005) The *Aspergillus nidulans* phytochrome FphA represses sexual development in red light. Curr Biol 15: 1833–1838.
13. Essen LO, Mailliet J, Hughes J (2008) The structure of a complete phytochrome sensory module in the Pr ground state. Proc Natl Acad Sci U S A 105: 14709–14714.
14. Fry KT, Mumford FE (1971) Isolation and partial characterization of a chromophore-peptide fragment from pepsin digests of phytochrome. Biochemical and Biophysical research communications 45: 1466–1473.
15. Wu SH, Lagarias JC (2000) Defining the bilin lyase domain: lessons from the extended phytochrome superfamily. Biochemistry 39: 13487–13495.
16. Lamparter T, Esteban B, Hughes J (2001) Phytochrome Cph1 from the cyanobacterium *Synechocystis* PCC6803: purification, assembly, and quaternary structure. Eur J Biochem 268: 4720–4730.
17. Lamparter T (2004) Evolution of cyanobacterial and plant phytochromes. FEBS Lett 573: 1–5.
18. Esteban B, Carrascal M, Abian J, Lamparter T (2005) Light-induced conformational changes of cyanobacterial phytochrome Cph1 probed by limited proteolysis and autophosphorylation. Biochemistry 44: 450–461.
19. Noack S, Michael N, Rosen R, Lamparter T (2007) Protein conformational changes of *Agrobacterium* phytochrome Agp1 during chromophore assembly and photoconversion. Biochemistry 46: 4164–4176.
20. Heyes DJ, Khara B, Sakuma M, Hardman SJO, O'Cualain R, et al. (2012) Ultrafast Red Light Activation of Synechocystis Phytochrome Cph1 Triggers Major Structural Change to Form the Pfr Signalling-Competent State. Plos One 7.
21. Takala H, Bjorling A, Berntsson O, Lehtivuori H, Niebling S, et al. (2014) Signal amplification and transduction in phytochrome photosensors. Nature 509: 245–248.
22. Schneider-Poetsch HA, Braun B, Marx S, Schaumburg A (1991) Phytochromes and bacterial sensor proteins are related by structural and functional homologies. Hypothesis on phytochrome- mediated signal- transduction. FEBS Lett 281: 245–249.
23. Kay SA, Keith B, Shinozaki K, Chua N-H (1989) The sequence of the rice phytochrome gene. Nucleic Acids Res 17: 2865–2866.
24. Cowl JS, Hartley N, Xie DX, Whitelam GC, Murphy GP, et al. (1994) The PHYC gene of Arabidopsis. Absence of the third intron found in PHYA and PHYB. Plant Physiol 106: 813–814.
25. Quail PH (1997) The phytochromes: A biochemical mechanism of signaling in sight? BioEssays 19: 571–579.
26. Yeh KC, Wu SH, Murphy JT, Lagarias JC (1997) A cyanobacterial phytochrome two-component light sensory system. Science 277: 1505–1508.
27. Lamparter T, Michael N, Mittmann F, Esteban B (2002) Phytochrome from *Agrobacterium tumefaciens* has unusual spectral properties and reveals an N-terminal chromophore attachment site. Proc Natl Acad Sci U S A 99: 11628–11633.
28. Li H, Zhang JR, Vierstra RD, Li HL (2010) Quaternary organization of a phytochrome dimer as revealed by cryoelectron microscopy. Proc Natl Acad Sci U S A 107: 10872–10877.
29. Scheerer P, Michael N, Park JH, Nagano S, Choe HW, et al. (2010) Light-induced conformational changes of the chromophore and the protein in phytochromes: bacterial phytochromes as model systems. ChemPhysChem 11: 1090–1105.
30. Wagner JR, Brunzelle JS, Forest KT, Vierstra RD (2005) A light-sensing knot revealed by the structure of the chromophore-binding domain of phytochrome. Nature 438: 325–331.
31. Yang X, Kuk J, Moffat K (2008) Crystal structure of Pseudomonas aeruginosa bacteriophytochrome: photoconversion and signal transduction. Proc Natl Acad Sci U S A 105: 14715–14720.
32. Burgie ES, Bussell AN, Walker JM, Dubiel K, Vierstra RD (2014) Crystal structure of the photosensing module from a red/far-red light-absorbing plant phytochrome. Proc Natl Acad Sci U S A.
33. Njimona I, Lamparter T (2011) Temperature effects on Agrobacterium phytochrome Agp1. Plos One 6: e25977.
34. Jin SG, Song YN, Deng WY, Gordon MP, Nester EW (1993) The regulatory VirA-protein of *Agrobacterium tumefaciens* does not function at elevated temperatures. Journal of Bacteriology 175: 6830–6835.
35. Lamparter T, Mittmann F, Gärtner W, Börner T, Hartmann E, et al. (1997) Characterization of recombinant phytochrome from the cyanobacterium *Synechocystis*. Proc Natl Acad Sci U S A 94: 11792–11797.
36. Inomata K, Hammam MAS, Kinoshita H, Murata Y, Khawn H, et al. (2005) Sterically locked synthetic bilin derivatives and phytochrome Agp1 from *Agrobacterium tumefaciens* form photoinsensitive Pr- and Pfr-like adducts. J Biol Chem 280: 24491–24497.
37. Strauss HM, Schmieder P, Hughes J (2005) Light-dependent dimerisation in the N-terminal sensory module of cyanobacterial phytochrome 1. FEBS Lett 579: 3970–3974.
38. Murashige T, Skoog F (1962) A Revised Medium for Rapid Growth and Bio Assays with Tobacco Tissue Cultures. Pphysiol Plant 15: 473–497.
39. Albanesi D, Martin M, Trajtenberg F, Mansilla MC, Haouz A, et al. (2009) Structural plasticity and catalysis regulation of a thermosensor histidine kinase. Proc Natl Acad Sci U S A 106: 16185–16190.
40. Kim PW, Rockwell NC, Martin SS, Lagarias JC, Larsen DS (2014) Dynamic inhomogeneity in the Photodynamics of Cyanobacterial Phytochrome Cph1. Biochemistry 53: 2818–2826.

41. Song C, Psakis G, Lang C, Mailliet J, Gärtner W, et al. (2011) Two ground state isoforms and a chromophore D-ring photoflip triggering extensive intramolecular changes in a canonical phytochrome. Proc Natl Acad Sci U S A 108: 3842–3847.

42. Hübschmann T, Yamamoto H, Gieler T, Murata N, Börner T (2005) Red and far-red light alter the transcript profile in the cyanobacterium Synechocystis sp. PCC 6803: impact of cyanobacterial phytochromes. FEBS Lett 579: 1613–1618.

43. Poppe C, Sweere U, Drumm-Herrel H, Schäfer E (1998) The blue light receptor cryptochrome 1 can act independently of phytochrome A and B in Arabidopsis thaliana. Plant J 16: 465–471.

44. Blazquez MA, Ahn JH, Weigel D (2003) A thermosensory pathway controlling flowering time in Arabidopsis thaliana. Nat Genet 33: 168–171.

45. Suzuki I, Kanesaki Y, Mikami K, Kanehisa M, Murata N (2001) Cold-regulated genes under control of the cold sensor Hik33 in Synechocystis. Molecular Microbiology 40: 235–244.

46. Albanesi D, Martin M, Trajtenberg F, Mansilla MC, Haouz A, et al. (2009) Structural plasticity and catalysis regulation of a thermosensor histidine kinase. Proc Natl Acad Sci U S A 106: 16185–16190.

47. Mironov KS, Sidorov RA, Kreslavskii VD, Bedbenov VS, Tsydendambaev VD, et al. (2014) Cold-induced gene expression and x3 fatty acid unsaturation is controlled by red light in Synechocystis. Journal of Photochemistry and Photobiology B: Biology.

48. Mironov KS, Sidorov RA, Trofimova MS, Bedbenov VS, Tsydendambaev VD, et al. (2012) Light-dependent cold-induced fatty acid unsaturation, changes in membrane fluidity, and alterations in gene expression in Synechocystis. BBA Bioenergetics 1817: 1352–1359.

49. Karniol B, Wagner JR, Walker JM, Vierstra RD (2005) Phylogenetic analysis of the phytochrome superfamily reveals distinct microbial subfamilies of photoreceptors. Biochem J 392: 103–116.

50. Barkovits K, Harms A, Benkartek C, Smart JL, Frankenberg-Dinkel N (2008) Expression of the phytochrome operon in Pseudomonas aeruginosa is dependent on the alternative sigma factor RpoS. FEMS Microbiol Lett 280: 160–168.

51. Franklin KA (2009) Light and temperature signal crosstalk in plant development. Current Opinion in Plant Biology 12: 63–68.

52. Payne PI, Dyer TA (1972) Phytochrome and temperature relations in Lactuca sativa L. Grand Rapids seed germination after thermo-dormancy. Nat New Biol 235: 145–147.

53. Fielding A, Kristie DN, Dearman P (1992) The temperature dependence of Pfr action governs the upper temperature. Photochem Photobiol 56: 623–627.

54. Foreman J, Johansson H, Hornitschek P, Josse EM, Fankhauser C, et al. (2011) Light receptor action is critical for maintaining plant biomass at warm ambient temperatures. Plant Journal 65: 441–452.

18

Detection of the Virulent Form of AVR3a from *Phytophthora infestans* following Artificial Evolution of Potato Resistance Gene *R3a*

Sean Chapman[1,9], Laura J. Stevens[1,2,3,9], Petra C. Boevink[1,3], Stefan Engelhardt[2,3], Colin J. Alexander[4], Brian Harrower[1], Nicolas Champouret[5], Kara McGeachy[1], Pauline S. M. Van Weymers[1,2,3], Xinwei Chen[1,3], Paul R. J. Birch[1,2,3], Ingo Hein[1,3]*

1 Cell and Molecular Sciences, James Hutton Institute, Invergowrie-Dundee, United Kingdom, 2 Division of Plant Sciences, University of Dundee at James Hutton Institute, Invergowrie-Dundee, United Kingdom, 3 Dundee Effector Consortium, Invergowrie-Dundee, United Kingdom, 4 Biomathematics and Statistics Scotland, Invergowrie-Dundee, United Kingdom, 5 J.R. Simplot Company, Simplot Plant Sciences, Boise, Idaho, United States of America

Abstract

Engineering resistance genes to gain effector recognition is emerging as an important step in attaining broad, durable resistance. We engineered potato resistance gene *R3a* to gain recognition of the virulent AVR3a^{EM} effector form of *Phytophthora infestans*. Random mutagenesis, gene shuffling and site-directed mutagenesis of *R3a* were conducted to produce R3a* variants with gain of recognition towards AVR3a^{EM}. Programmed cell death following gain of recognition was enhanced in iterative rounds of artificial evolution and neared levels observed for recognition of AVR3a^{KI} by R3a. We demonstrated that R3a*-mediated recognition responses, like for R3a, are dependent on SGT1 and HSP90. In addition, this gain of response is associated with re-localisation of R3a* variants from the cytoplasm to late endosomes when co-expressed with either AVR3a^{KI} or AVR3a^{EM} a mechanism that was previously only seen for R3a upon co-infiltration with AVR3a^{KI}. Similarly, AVR3a^{EM} specifically re-localised to the same vesicles upon recognition by R3a* variants, but not with R3a. R3a and R3a* provide resistance to *P. infestans* isolates expressing AVR3a^{KI} but not those homozygous for AVR3a^{EM}.

Editor: Frederik Börnke, Leibniz-Institute for Vegetable and Ornamental Crops, Germany

Funding: LS is supported by The James Hutton Institute and the University of Dundee through a joint PhD student bursary. PSMVW is supported by the U.S. Department of Agriculture through National Institute of Food and Agriculture (NIFA) project 2011-68004-30154 (http://www.csrees.usda.gov/). NC is employed by J.R. Simplot Company, Simplot Plant Sciences, 5369 West Irving Street, Boise, ID 83706, USA. This work was funded by the Rural & Environment Science & Analytical Services (RESAS) Division of the Scottish Government (http://www.scotland.gov.uk/Topics/Research/About/EBAR/research-providers) and the Biotechnology and Biological Sciences Research Council (BBSRC) (http://www.bbsrc.ac.uk/home/home.aspx) through the joint projects CRF/2009/SCRI/SOP & BB/H018441/1 (IH), BB/K018299/1 and RESAS funded work package 6.4. The funders had no role in study design, data collection and analysis, decision to publish, or preparation of the manuscript.

* Email: Ingo.Hein@hutton.ac.uk

9 These authors contributed equally to this work.

Introduction

In a process known as effector triggered immunity (ETI), plant disease resistance (*R*) genes can facilitate immunity to phylogenetically diverse and unrelated pathogens that express cognate effector molecules [1]. Effectors that are recognised by *R* genes, either directly or indirectly, and provoke successful plant defences are genetically defined as avirulence (*Avr*) genes. The best described group of plant *R* gene products contains a nucleotide-binding (NB) domain and leucine-rich repeats (LRRs), collectively known as NB-LRRs [2]. NB-LRRs are strictly regulated by plants as, upon their activation, many elicit programmed cell death (PCD) as part of the hypersensitive response (HR) which prevents further spread of disease in plant tissues [3]. Together with effectors, *R* genes are at the forefront of host/pathogen co-evolution [4]. NB-LRRs are one of the largest gene families in

plants and more than 750 members have recently been described in potato [5–6]. The organisation of many NB-LRRs into physically-linked clusters is providing insight into their evolution, which can involve duplication followed by diversification.

In agriculture, successful deployment of *R* genes to control important diseases in crop plants has so far been hampered by the ability of pathogens often to rapidly circumvent detection by the host plant's innate immune system. Advances in studying pathogen effector diversity coupled with the ability to engineer *R* genes offers the opportunity to develop more durable resistances that specifically target essential effectors and known variants [7–8].

The *Phytophthora infestans* effector AVR3a is an essential effector for this pathogen to cause late blight on potato. Stable silencing of *Avr3a* in the *P. infestans* isolate 88069 significantly reduces infection in susceptible *Solanum tuberosum* (potato) cv. Bintje and in the model solanaceous plant species *Nicotiana* benthamiana [9–10]. Two forms of AVR3a are prevalent in current *P. infestans* isolates and differ in only two amino acids within the mature protein; AVR3aE^{80}M^{103} (AVR3a^{EM}) and AVR3aK^{80}I^{103} (AVR3a^{KI}) [11–12]. AVR3a^{KI} elicits ETI upon recognition by the potato resistance protein R3a, a member of the coiled-coil (CC) NB-LRR gene family [13]. This response is evaded by AVR3a^{EM} which consequently is free to promote virulence [9,11,13]. However, a weak R3a-dependent response to AVR3a^{EM} can be observed under UV light [14]. The mechanism of R3a-mediated recognition of AVR3a has been investigated recently [15]. Upon activation by AVR3a^{KI}, both the effector protein and R3a rapidly re-localize from the host cytoplasm to late endosomes, components of the endocytic pathway, which is thought to be a prerequisite for subsequent HR development. The un-recognised AVR3a^{EM} form of the effector does not cause this re-localisation. There is no evidence of direct interaction between AVR3a^{KI} and R3a, but bimolecular fluorescence complementation (BiFC) assays reveal that the two proteins are in close proximity at late endosomes [15].

Artificial evolution has previously been used to alter the recognition specificity of the potato CC-NB-LRR resistance protein Rx to gain recognition of different strains of *Potato virus X* (PVX) and a distantly related virus, *Poplar mosaic virus* (PopMV) [16–17]. Random mutagenesis, screening for beneficial mutations and designed amalgamation of these mutations has been used to generate transgenic plants of the model species *N. benthamiana* that are resistant to previously virulent strains. DNA shuffling, also known as directed evolution, was first developed in the early 1990 s and has since been used to generate a wide variety of novel genes and proteins [18]. DNA shuffling has previously been used in the functional analysis of the resistance gene *Pto* [19]. In this study, the LRR of *R3a* has been subjected to error-prone PCR and iterative rounds of DNA shuffling to identify gain-of-recognition variants (R3a*) through functional screening in *N. benthamiana*. R3a* gene products with varying degrees of AVR3a^{EM} recognition were generated in three rounds of DNA shuffling and a subsequent round of site-directed mutagenesis. The best-performing clones from each round were taken forward for further analysis and compared to wild-type R3a. R3a* variants demonstrated significantly improved gain-of-recognition towards AVR3a^{EM} that manifested itself as a gain of re-localisation to late endosomes, but not yet as a gain of resistance.

Materials and Methods

Construction of plasmid vectors

The plasmid pBinPlus.R3a [13] was amplified with the primer pairs R3a-5-Asc/R3a-1564-Bam-M and R3a-1564-Bam-P/R3a-3-Not (Table S1) and the *Asc*I/*Bam*HI and *Bam*HI/*Not*I digested amplification products cloned in a three way ligation in to *Asc*I/*Not*I digested binary vector pGRAB [20] to produce pGRAB.35S::R3a. A derivative of the former plasmid, pGRAB.-R3a::R3a, was produced in which the *Cauliflower mosaic virus* 35S (35S) promoter sequence was replaced with the *R3a* promoter sequence. This derivative was produced by digestion of pBin-Plus.R3a with *Pme*I and *Xho*I, and insertion of the released fragment, containing 2358 bases upstream of the translational start site, in to pGRAB.35S::R3a that had been treated with *Bsp*EI, T4 DNA polymerase and *Xho*I in order.

Mutagenesis and DNA shuffling

Shuffling of 2283 bp of the R3a LRR region was performed as described by Stemmer [18]. First round PCRs were carried out with the primer pair R3a-1564-Bam-P/R3a-3-Not in the presence of 0.5 mM MnCl$_2$ as described by Leung et al. [21] with *Taq* DNA polymerase (Roche, Mannheim, Germany). Following DNaseI treatment, fragments of circa 500 bp were reassembled through forty rounds of primer-less thermo-cycling. Gel-purified products of circa 2.3 kb were amplified through thirty cycles of PCR and, following gel-purification, digested with *Bam*HI and *Not*I prior to cloning in to pGRAB.35S::R3a digested with the same enzymes. The ligated population was amplified in *E. coli* resulting in a population with a complexity of circa 125 K, a vector background of circa 14% and a base mutation rate of 0.43%. Aliquots of plasmid, gel-purified on account of plasmid instability, were transformed in to *A. tumefaciens* cells. In the second and third rounds *Pfu*Ultra II Fusion HS DNA polymerase (Stratagene, La Jolla, CA, USA) was used in thermo-cycling reactions to produce shuffled populations with lower mutation rates, less than 0.05%. The second and third round populations had complexities of circa 200 K and 150 K, respectively, and both had a vector background of circa 10%.

Site-directed mutagenesis

A mixture of two templates, the wild-type gene and clone Rd1-2 from the first round of shuffling containing the Q931R codon substitution, was used in PCR amplifications with the primer pairs R3a-1564-Bam-P/R3a-1740W-M, R3a-1740W-P/R3a-1841W-M, R3a-1841W-P/R3a-2743W-M, R3a-2743W-P/R3a-3028-M. The products of the primary PCRs were used in an overlap PCR reaction with the flanking primer pair R3a-1564-Bam-P/R3a-3028-M. The product of the secondary PCR reaction was digested with *Bam*HI and *Aat*II and cloned between the same unique sites of pGRAB.R3a::R3a. A population of circa 50K clones, with a vector background of 1% and random base mutation rate of 0.05% was produced.

Plant growth conditions

N. benthamiana plants were grown in a glasshouse with a 16 h day period at 22°C and an 8 h night period at 18°C. Supplementary lighting was provided below 200 W m^{-2} and screening above 450 W m^{-2}.

Screening of mutated R3a clones

DNA populations prepared from *E. coli* were transformed in to *Agrobacterium tumefaciens* strain AGL1 [22] carrying the helper plasmids pSoup and pBBR1MCS1.VirG$_{N54D}$ [23] for screening. *Agrobacterium* cultures grown from single transformed colonies were co-infiltrated with cultures of *Agrobacterium* transformed with pGRAB.35S::AVR3a^{EM} according to the method of Engelhardt et al. [15] in to *N. benthamiana* leaves with each of the components at the same final OD$_{600}$ of 0.1–0.5. Reference mixtures were infiltrated in to opposing half-leaves and between two and seven days post infiltration leaves were inspected to assess visible symptoms and plant auto-fluorescence under day-light and 365 nm illumination from a Blak-Ray lamp (UVP, Upland, CA, USA), respectively.

Symptom scoring

Circa five week old plants were used for symptom scoring with two adjacent, expanded leaves, both circa 90 mm in length, being used for infiltrations. Symptoms were scored on an arbitrary scale for nine days after infiltration. Symptom scores were plotted; the areas under the curve determined and mean scores calculated by dividing by the duration. A linear mixed modelling approach was adopted using GenStat for Windows, 16th edition (VSN International Ltd., Hemel Hempstead, UK). The data were analysed in two stages: first, the stability of the phenotypes over repeated experiments was examined by fitting a model with experiment as a fixed effect. Infiltration mixture, leaf age, position of infiltration site on leaf, experiment and their interactions were set as fixed effects with plant and leaf within an individual plant as the random effects. This allowed for terms in the model which specifically tested for significant interactions with experiment. Lack of any significant interaction with the infiltration mixtures would provide evidence that the relative responses of the phenotypes were consistent. Having determined that the infiltration mixtures behaved consistently over the experiments, the second stage fitted a model with experiment as a random effect with plant and leaf within plant nested below this. Multiple comparison tests then examined the differences in response amongst the infiltration mixtures.

Virus-induced gene silencing (VIGS) of *SGT1* and *HSP90*

Tobacco rattle virus (TRV)-induced gene silencing in *N. benthamiana* was performed as described previously [14]. *Agrobacterium* cultures transformed with the binary TRV RNA1 construct, pBINTRA6, or the TRV RNA2 vector constructs PTV00, PTV:*eGFP*, PTV:*HSP90* or PTV:*SGT1* were re-suspended to $OD_{600} = 0.5$ for the RNA1 construct and $OD_{600} = 1.0$ for the RNA2 constructs. Re-suspended RNA1 and RNA2 cultures were mixed in a 1:1 ratio and infiltrated into non-cotyledonous leaves of *N. benthamiana* plants at the 5-leaf stage. For each of the biological replicates, six plants per treatment were used and six plants were used as non-TRV controls. Three weeks after treatment with the VIGS constructs, plants were infiltrated with culture mixtures ($OD_{600} = 0.5$) designed to express R3a, Rd2-1, Rd3-1 or Rd4-1 and AVR3a^{KI} or AVR3a^{EM}. HRs were scored at 6dpi.

Confocal laser scanning microscopy

Imaging was performed on a Leica TCS-SP2 AOBS microscope (Leica Microsystems) using HCX APO L, 40x/0.8, and 63x/0.9 water dipping lenses or a Zeiss 710 using a Plan APO 40x/1.0 water dipping lens. Images were collected using line by line sequential scanning. The optimal pinhole diameter and the same gain levels were used within experiments. YFP and CFP were imaged using 514 nm and 405 nm excitation, respectively, and emissions were collected between 520–563 nm and 455–490 nm, respectively. Photoshop CS5.1 software (Adobe Systems) was used for post-acquisition image processing.

Agrobacterium tumefaciens transient assays (ATTAs)

Functional *Agrobacterium tumefaciens* transient assays (ATTAs) were carried out in *N. benthamiana*. Cultures carrying pGRAB.-R3a::R3a, pGRAB.R3a::Rd2-1, pGRAB.R3a::Rd3-1, pGRAB.-R3a::Rd4-1 or pGRAB empty vector were re-suspended as described before to OD600 = 0.1 for each construct. Each of the five resuspensions was infiltrated into separate areas of leaves. Four leaves on each of sixteen plants were infiltrated in each replicate. Two days post infiltration, leaves were detached and infiltration sites inoculated with *AVR3a^{KI}* homozygous *P. infestans* isolate 7804.b or *AVR3a^{EM}* homozygous isolate 88069. Leaves were incubated in transparent sealed boxes at 100% humidity in a cool room and covered for the first 12 hours. Lesion sizes were measured up to 15dpi.

Production of transgenic potato plants

R3a wild-type gene and the three modified versions Rd2-1, Rd3-1 and Rd4-1 were cloned under *R3a* native regulatory elements in a pCambia-based binary vector with kanamycin resistance as a selectable marker, using standard restriction enzyme methods to create pSIM2093, pSIM3027, pSIM3028 and pSIM3029 respectively. *Rpi-vnt1* under its native regulatory elements was cloned in to pSIM401 to generate pSIM1620. The binary vectors were then transformed into *Agrobacterium* strains AGL1 and LBA4404 for plant transformation. Ranger Russet was transformed as described in Duan *et al.* [24]. For each construct twenty to thirty lines were regenerated with kanamycin selection. These were tested for late blight resistance and assessed at 7dpi.

Results

Identification of R3a mutants with enhanced AVR3a^{EM} recognition

A binary vector, pGRAB.35S::R3a, containing the R3a open reading frame (Accession number AY849382.1) under the control of *Cauliflower mosaic virus* 35S promoter and terminator sequences was produced to allow specific mutation and shuffling of the R3a LRR domain. In this plasmid a unique *Bam*HI site was introduced silently at nucleotide position 1567 in the ORF, ninety nucleotides upstream of the sequence encoding the LRR domain as denoted by Huang *et al.* [13]. Following mutagenesis and DNA shuffling of the LRR region, regenerated recombinant clones were screened for enhanced AVR3a^{EM} recognition through *Agrobacterium*-mediated transient expression in co-infiltrations with binaries expressing the virulent elicitor, AVR3a^{EM}. Clones were screened for enhanced recognition through comparison of visible symptoms of PCD and induced auto-fluorescence with reference to the responses produced by the wild-type R3a gene and AVR3a^{EM}. Clones with putatively improved recognition isolated from primary screens were screened again to confirm the phenotypic improvement and to eliminate auto-activators. In the first cycle, screening of approximately three thousand clones identified eleven R3a* clones with improved phenotypes. The eleven clones from the first round are referred to as Rd1-1 to Rd1-11 and contained, in addition to synonymous changes, base changes that resulted in between one and four amino acid substitutions (Fig. 1a; Table S2). Three of the clones contained single amino acid substitutions (Rd1-1 [K920E], Rd1-2 [Q931R], Rd1-3 [R618Q]) identifying these changes as being responsible for enhanced recognition. Interestingly, the single amino acid substitution K920E in R3a was recently also identified in a complementary study by Segretin *et al.* [25] as a substitution that enhances recognition towards AVR3a^{EM}. The largest numbers of amino acid substitutions were found in LRRs #3 and #15 and included those found in the three clones with single substitutions (Fig. 1a).

A second round of shuffling was carried out to combine beneficial changes and remove deleterious substitutions using the eleven isolated clones, Rd1-1 to Rd1-11, and the wild-type gene as starting material. Screening of approximately six hundred recombinant clones identified four which gave responses comparable to or greater than the best performing clone from the first round, Rd1-1 [K920E]. All of these R3a* clones from the second round, Rd2-1 to Rd2-4, contained the amino acid substitution E620D in LRR #3 in addition to at least one other coding change (Fig. 1a; Table S2).

As recognition of AVR3a^{EM} improved and gave stronger responses it became more difficult to discriminate the differences in responses between modified clones with AVR3a^{EM} and also in comparison to the response of the wild-type *R3a* gene with AVR3a^{KI}. Therefore, a third round of shuffling, using the clones from the first and second rounds of shuffling and the wild-type

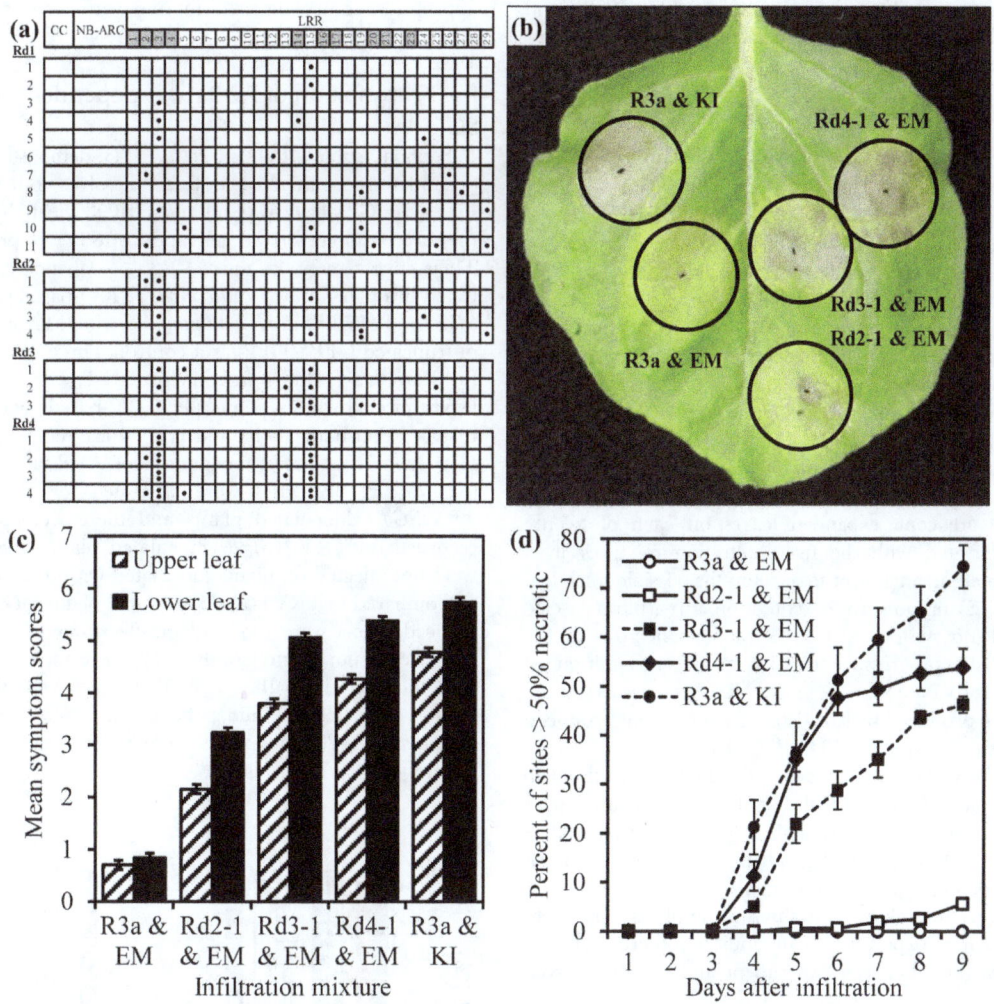

Figure 1. Four rounds of mutagenesis and shuffling identified R3a mutants with enhanced recognition of AVR3a^EM and disease responses (R3a*). (a) Schematic showing locations of non-synonymous mutations found in the LRRs of R3a* clones isolated from the four rounds of mutagenesis and shuffling (Rd1 to Rd4). LRRs containing amino acids under diversifying selection are shaded above. (b) Representative *N. benthamiana* leaf showing responses of best performing clones from second, third and fourth rounds (Rd2-1, Rd3-1 & Rd4-1) to AVR3a^EM (EM), compared to responses of wild-type R3a to AVR3a^KI (KI) and AVR3a^EM five days after co-infiltration with resistance genes under transcriptional control of *R3a* promoter. (c) Mean disease scores from the four experiments, each of nine days duration, for different infiltration mixtures in upper (hatched) and lower (solid) paired leaves. Error bars show +/− standard error. (d) Time-course of percentage of sites showing necrosis development, greater than 50% necrosis of individual infiltrated sites, for the five infiltrated mixtures. Mean percentages of the four experiments. Each experiment includes data for 40 infiltration sites (upper and lower leaves combined) and error bars show +/− standard errors.

gene as starting material, was performed. However, the shuffled LRR domains were cloned in to a binary vector, pGRAB.-R3a::R3a, containing the *R3a* gene or *R3a** variants under the wild-type *R3a* promoter, rather than the strong 35S promoter. The purpose of this was to protract the timing of the cell death responses and facilitate the discrimination of differences (Fig. 1b). Screening of approximately 300 clones from this third round population identified three R3a* clones, Rd3-1 to Rd3-3, that gave responses greater than the best performing clone from the second round of shuffling, Rd2-1 [T585A, E620D]. All of these clones contained the E620D change found in the second round clones and a number of other amino acid changes including at least one change in LRR #15. A previously conducted comparison of functional R3a with three paralogous, non-functional resistance gene analogues revealed 13 positions under diversification [13]. These positions, with the exception of one in the CC-domain, are located in LRRs 1 to 4 and 14 to 23.

Intriguingly, the majority of mutations present in the clones from the second and third rounds are within these LRRs (Fig. 1a).

One limitation of shuffling is in recombining beneficial mutations that are in close proximity due to the limited frequency of crossing-over. Therefore, as multiple mutations in LRRs #3 and #15 had been found to enhance AVR3a^EM recognition, an alternate approach was adopted. A population of site-directed mutants was produced using degenerate oligonucleotides that encoded different pairs of amino acids at positions 585 (T/A), 618 (R/Q), 620 (E/D), 918 (R/G), 920 (K/E), 923 (D/G) and 931 (Q/R) with the potential for 128 different permutations. Screening of approximately 400 clones from this population with the best performing clone from the third round of shuffling (Rd3-1 [E620D, L668P, Q931R]) as a reference, identified eight clones that produced responses with AVR3a^EM comparable to or greater than the reference. Five of these clones were found to have identical sequences. Representative unique clones, Rd4-1 to Rd4-4, contain the amino acid substitutions R618Q, E620D, K920E and Q931R, but lacked either of the designed substitutions R918G or D923G (Table S2).

Iterative rounds of shuffling have progressively improved AVR3a^{EM} recognition by R3a* variants, producing faster and stronger PCD responses upon co-infiltration

The enhanced recognition of AVR3a^{EM} was assessed more accurately in single leaf comparisons. The R3a* constructs Rd2-1, Rd3-1 and Rd4-1 were transiently expressed under the control of the native *R3a* promoter with AVR3a^{EM} and the responses compared with those produced by the wild-type gene with AVR3a^{EM} and AVR3a^{KI}. The best performing clone from the first round was not included in this analysis on account of the relatively poor response produced when it was under the control of the *R3a* promoter.

Symptom development on *N. benthamiana* was monitored for 9 days after infiltration in four independent experiments. In each experiment two adjacent, expanded leaves on each of twenty plants were infiltrated with the five infiltration mixtures in a circularly permuted arrangement to account for possible intra-leaf position effects. Symptoms were scored on an arbitrary scale ranging from 0 (no symptoms) to 10 (complete necrosis of the infiltrated area). A progressive increase in the recognition of AVR3a^{EM} was observed for the three R3a* clones with the necrotic response produced by Rd4-1 being close to that produced by the wild-type gene with AVR3a^{KI} (Fig. 1b).

A mixed model with experiment as a fixed effect was fitted to test for consistent responses of the infiltration mixtures over repeated experiments. The experiments showed significant differences in mean response (p = 0.001), but there was no significant interaction (p = 0.306) between experiment and infiltration mixture indicating that the relative responses of the phenotypes were stable over repeated experiments. In addition, there were no significant interactions between experiment and the other fixed effects.

Since the phenotypes were determined to be stable, a second analysis of the data with experiment as a random effect was now fitted. This showed highly significant effects (p<0.001) from infiltration mixture and leaf age (upper vs lower leaf). Further, there was a highly significant interaction (p<0.001) between infiltration mixture and leaf age because the combination of R3a and AVR3a^{EM} did not show a difference in mean scores between younger and older leaves, in contrast to the other combinations which showed significant differences (Fig. 1c). Position of infiltration site on leaf was non-significant and all other interactions of fixed effect were also non-significant. Multiple comparisons using Bonferroni correction with an experiment-wise significance level of 5% showed significant differences between the mean symptom scores for all infiltration mixtures within either younger or older leaves with a comparison-wise significance level of 0.0011 (Table S3). Ordering of the responses was the same for both younger and older leaves with R3a and AVR3a^{KI}> Rd4-1 & AVR3a^{EM}> Rd3-1 & AVR3a^{EM}> Rd2-1 & AVR3a^{EM}> R3a and AVR3a^{EM}. While the clone Rd2-1 from the second round gave symptoms when co-expressed with AVR3a^{EM}, it rarely produced an HR phenotype as shown in figure 1d, which shows the proportion of sites with more than 50% of the infiltrated area necrotic. Despite improved recognition of AVR3a^{EM} by the R3a* variants relative to wild-type R3a, their recognition of AVR3a^{KI} was not impaired (data not shown), indicating that this specificity had not been significantly attenuated. Further, when expressed from the strong 35S promoter in the absence of AVR3a^{EM} or AVR3a^{KI} none of the clones produced necrosis, indicating that they maintained

appropriate control mechanisms and were not auto-activators (Fig. S1).

R3a* recognition of Avr3a^{EM} is dependent on HSP90 and SGT1

Previous studies by Bos *et al.* [14] demonstrated that R3a-dependent recognition of AVR3a^{KI} involves both SGT1 (suppressor of the G2 allele of *skp1*) and HSP90 (heat shock protein 90) that are required for the activation of other R proteins [26–27]. Their involvement in the AVR3a^{EM}-dependent responses was tested through *Tobacco rattle virus* (TRV)-based gene silencing of *SGT1* and *HSP90* in *N. benthamiana* with TRV-based expression of truncated GFP (eGFP) as a control. Three biological replicates for R3a, Rd2-1, Rd3-1 and one for Rd4-1, with infiltrations in two leaves of each of six plants per TRV-based silencing construct, revealed that both SGT1 and HSP90 are required to mediate an HR upon R3a*-based recognition of AVR3a^{KI} and AVR3a^{EM} (Fig. 2, Fig. S2). HRs were abolished for all infiltrations on TRV:*SGT1* inoculated plants and there were almost no HRs recorded on TRV:*HSP90* inoculated plants (Fig. 2). The HRs were not affected on plants inoculated with TRV:*eGFP*.

Compared to TRV:*eGFP* inoculated plants, *SGT1* and *HSP90* silenced plants were morphologically stunted, a phenotype that had been reported previously [14]. Nevertheless, upon infection with the bacterial pathogen *Erwinia amylovora* that produces a SGT1 and HSP90 independent non-host response in *N. benthamiana* [28], all plants were able to mount the expected HR response (Fig. 2, Fig. S2).

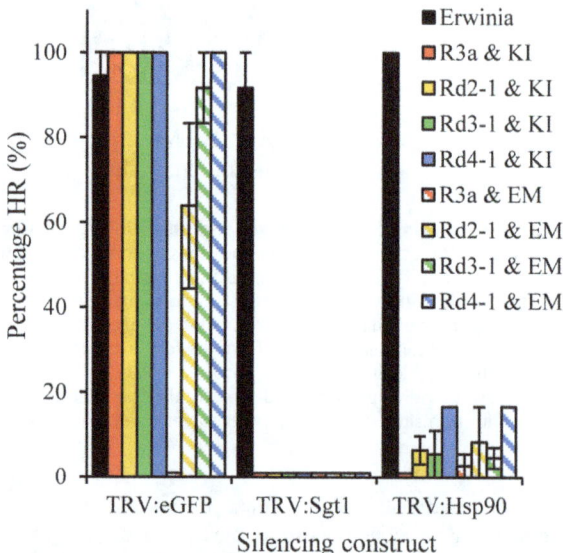

Figure 2. HR responses resulting from R3a* recognition of AVR3a^{EM} and AVR3a^{KI}, like those caused by wild-type R3a recognition of AVR3a^{KI}, are dependent on SGT1 and HSP90. SGT1- and HSP90-silenced plants were produced using TRV-based vectors. These plants and control plants inoculated with TRV:*eGFP* were infiltrated with different combinations of *Agrobacterium* cultures designed to express R3a, R3a* variants, AVR3a^{KI} (KI) or AVR3a^{EM} (EM). The percentage of sites (N = 12) showing HR responses six days after infiltration was recorded. The graph shows the mean percentages from three independent experiments with the exception that the dependence on HSP90 of Rd4-1 responses was only tested in a single experiment. The non-host bacterial pathogen *Erwinia amylovora* was used as a control for an SGT1- and HSP90-independnet HR response. Error bars show +/− standard error. Zero values have been transformed to 1% to facilitate their observation.

In a gain of mechanism, R3a* variants re-localize to late endosomes upon co-infiltration with Avr3a^{KI} or Avr3a^{EM}

In a previous study, Engelhardt et al. [15] demonstrated that, upon recognition of AVR3a^{KI} but not AVR3a^{EM}, wild-type R3a re-localizes from the host cytoplasm to specific late endosomes that can be labelled with the cyan fluorescent protein marker PS1-CFP [29]. This re-localization was found to be a pre-requisite for subsequent HR development for untagged R3a co-expressed with AVR3a^{KI} [15]. To study if R3a* variants with enhanced recognition of AVR3a^{EM} had gained the capacity to re-localize upon detection of AVR3a^{EM} and continued to exhibit this phenotype following detection of AVR3a^{KI}, N-terminal fusions of R3a* variants Rd2-1, Rd3-1 and Rd4-1 with yellow fluorescent protein (YFP) were generated as described previously with expression of these constructs driven by a 35S promoter. Western-blot analysis of protein extracts from inoculated tissue demonstrated the integrity of the fusion proteins (Fig. S3). As demonstrated for YFP-R3a wild-type fusions by Engelhardt et al. [15], YFP-R3a* fusions did not elicit HRs alone or in the presence of AVR3a^{KI} or AVR3a^{EM}, probably due to steric hindrance of the signalling domains of R3a (data not shown).

As anticipated, following transient expression in N. benthamiana, all YFP-R3a/R3a* fusions when expressed by themselves displayed cytoplasmic localizations (Fig. S4). In accord with the observations described by Engelhardt et al. [15], the localisation of YFP-R3a remained cytoplasmic upon co-infiltration with AVR3a^{EM} (Fig. 3), but changed to fast moving, PS1-CFP labelled vesicles, following recognition of AVR3a^{KI}. The YFP-R3a* fusions of Rd2-1, Rd3-1 and Rd4-1 proteins maintained this mechanistically characteristic re-localisation following co-expression with AVR3a^{KI} (Fig. 3). However, in contrast to YFP-R3a, all selected YFP-R3a* variants also displayed highly reproducible re-localization to PS1-CFP labelled vesicles after the perception of AVR3a^{EM} (Fig. 3; Fig S5).

AVR3a^{KI} and AVR3a^{EM} re-localize to endosomes upon co-infiltration with R3a* variants but not, in the case of AVR3a^{EM}, with wild-type R3a

As shown by Engelhardt et al. [15] AVR3a^{KI}, but not AVR3a^{EM}, also re-localizes from the cytoplasm to endosomes upon co-expression with R3a. This was demonstrated by N-terminal fusions of AVR3a^{KI} and AVR3a^{EM} to green fluorescent protein as well as by BiFC, also known as split-YFP, assays [9,15,30]. The latter revealed that wild-type R3a and AVR3a^{KI} are found in close proximity at PS1-CFP labelled vesicles [15].

To investigate if the vesicular co-association of R3a and AVR3a^{KI} was extended to the R3a* variants, BiFC was used to analyse and localize protein–protein interactions in planta. As described previously for wild-type R3a, the N-terminal portion of YFP, YN, was fused to the N-terminal end of the R3a* variants Rd2-1, Rd3-1 and Rd4-1. The constructs used to express the C-terminal portion of YFP, YC, fused to AVR3a^{KI} and AVR3a^{EM} were as described previously [15] with all constructs being transiently expressed in N. benthamiana from the 35S promoter.

In accord with previous findings [15], co-expression of YN-R3a with YC-AVR3a^{KI} gave strong YFP fluorescence, whereas co-expression with YC-AVR3a^{EM} did not give detectable YFP fluorescence (Fig. 4). Like the YN fusion to wild-type R3a, all the YN-R3a* fusions when co-expressed with AVR3a^{KI} gave strong, punctate, YFP signals (Fig. 4), but unlike the wild-type fusion also gave YFP fluorescence signals at PS1-CFP labelled vesicles when co-expressed with AVR3a^{EM} (Fig. 4; Fig. S6). This indicates that AVR3a^{EM} is also within close proximity of the re-localized R3a*

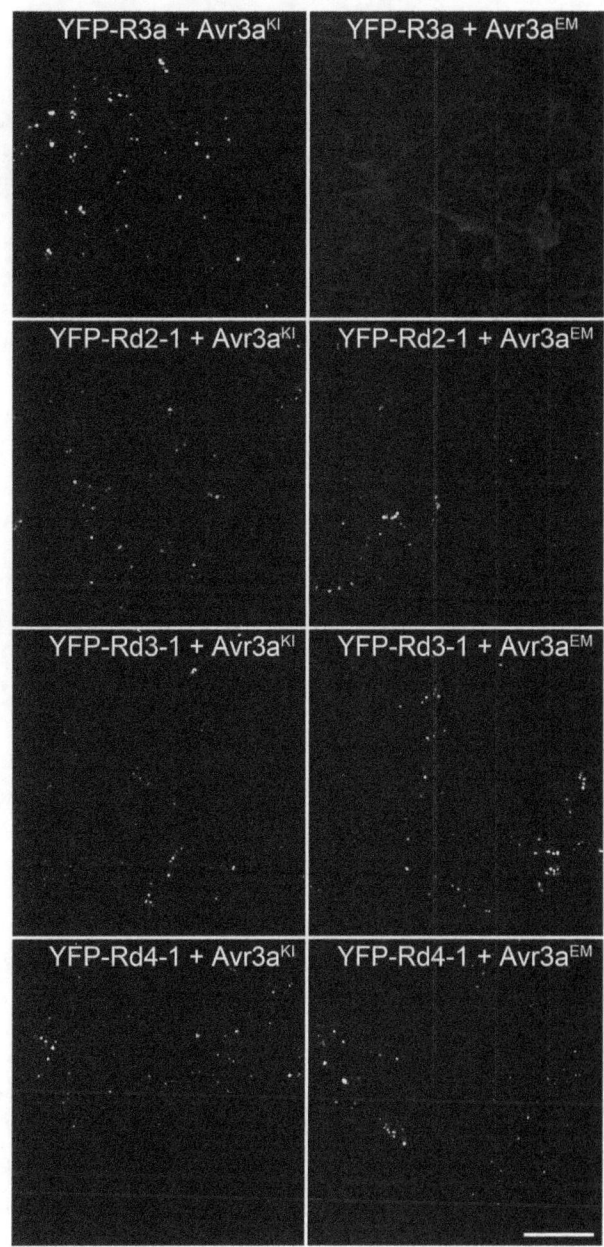

Figure 3. YFP fusions to R3a* variants re-localize to vesicles after the perception of both of AVR3a^{KI} and AVR3a^{EM}, whereas YFP-R3a remains cytoplasmic in the presence of AVR3a^{EM}. Two days after infiltration of mixtures of Agrobacterium cultures designed to express AVR3a^{KI}, AVR3a^{EM}, YFP-R3a or YFP fusions to the R3a* variants, infiltrated N. benthamiana leaf tissue was examined under a confocal laser scanning microscope. Scale bar = 50 μm.

gene products. Thus, in line with the gain of recognition of AVR3a^{EM} by the R3a* variants and subsequent necrosis responses, the R3a* variants and AVR3a^{EM} show the same mechanistic re-localization as observed for R3a and AVR3a^{KI}.

R3a* variants maintain resistance towards AVR3a^{KI}-expressing P. infestans isolates but have not gained resistance towards AVR3a^{EM} homozygous isolates

To evaluate if R3a* variants with gain of AVR3a^{EM} recognition and re-localisation mechanism yield effective disease resistance,

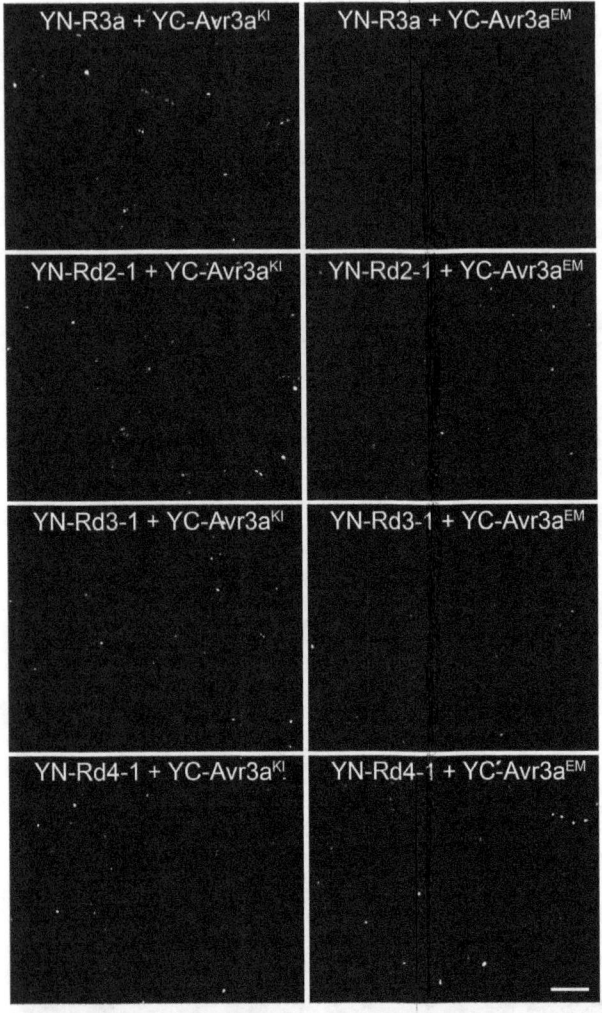

Figure 4. Both YC-AVR3a^{KI} and YC-AVR3a^{EM} when co-expressed with YN-R3a* fusions give vesicle associated YFP fluorescence like YC-AVR3a^{KI} and YN-R3a, whereas YC-AVR3a^{EM} and YN-R3a do not. Two days after infiltration of mixtures of *Agrobacterium* cultures designed to express YC-AVR3a^{KI}, YC-AVR3^{EM}, YN-R3a or YN fusions to the R3a* variants, infiltrated *N. benthamiana* leaf tissue was examined under a confocal laser scanning microscope. Representative images from two experiments. Scale bar = 50 μm.

transient and stable expression systems were utilised. *Agrobacterium tumefaciens* transient assays (ATTAs) in *N. benthamiana* have successfully been used to demonstrate function for late blight resistance gene products such as R2, Rpi_STO1 [31] and R3b [32]. Selected R3a* clones Rd2-1, Rd3-1 and Rd4-1 were transiently expressed in *N. benthamiana* using the *R3a* promoter in ATTAs alongside wild-type R3a and an empty vector control. Infiltrated leaf areas were challenged two days after infiltration with AVR3a^{KI} or AVR3a^{EM} homozygous *P. infestans* isolates via drop inoculation. Disease progression was monitored by measuring visible lesion diameters in multiple independent experiments and analysis of variance was carried out using GenStat on the data from individual experiments to test for significant differences. Multiple comparisons were performed using Bonferroni correction with a significance level of 5% and a comparison-wise error rate of 0.005.

In three experiments ATTA sites were inoculated with the AVR3a^{KI} homozygous *P. infestans* isolate 7804.b. In all three

experiments the wild-type R3a and the R3a* variants significantly reduced spread of *P. infestans* relative to the empty vector control and there were no significant differences between the different R3a forms (Fig. 5a). This result indicates that the selected mutations in the LRR do not impair the resistance induced by AVR3a^{KI}. Likewise ATTA sites were inoculated with the AVR3a^{EM} homozygous isolate 88069 [9] in five experiments. In four of the five experiments there were no significant differences in *P. infestans* spread between any of the R3a forms and the empty vector control (Fig. 5b). In the fifth experiment there was significantly increased spread with the empty vector control, but the R3a* variants showed no significant differences from the wild-type gene which does not provide resistance against AVR3a^{EM} homozygous isolates. Co-infiltrations of R3a, R3a*, AVR3a^{KI} and AVR3a^{EM} constructs were carried out contemporaneously in all experiments to confirm that the conditions were conducive to HR development (data not shown).

(a)

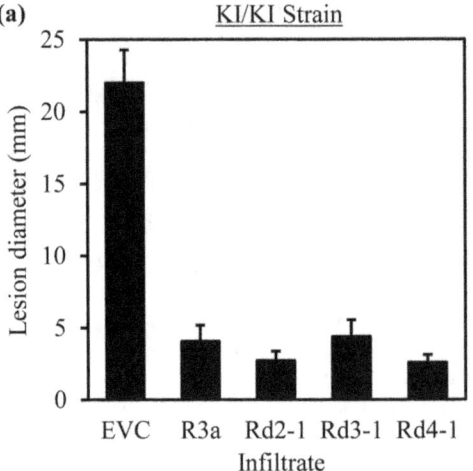

(b)

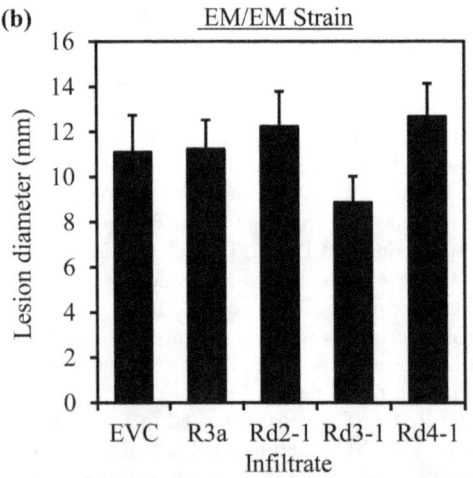

Figure 5. R3a and R3a* variants expressed from *Agrobacterium* reduce the spread of a *P. infestans* strain expressing AVR3a^{KI}, but not the spread of a strain expressing only AVR3a^{EM}. (a) Means of lesion diameters measured 12 days after drop inoculation of agro-infiltrated areas with strain 7804.b (KI/KI). (b) Means of lesion diameters measured 8 days after drop inoculation of agro-infiltrated areas with strain 88069 (EM/EM). (a) and (b) show representative experiments from sets of three and five repeated experiments, respectively. For both (a) and (b), error bars show +/− standard errors, N = 30. EVC indicates empty vector control.

To confirm these results and to rule out potential adverse effects and limitations of the ATTA system in *N. benthamiana*, transgenic potato plants were generated. The wild-type *R3a* gene and the *R3a** genes for Rd2-1, Rd3-1 and Rd4-1 were transformed into the potato cultivar Ranger Russet using the *R3a* promoter and terminator to regulate gene expression and stability. Transgenic Ranger Russet lines expressing R3a and the three R3a* variants were compared to Ranger Russet lines containing the *Rpi_vnt1* gene [33] and non-transgenic Ranger Russet plants as positive and negative controls, respectively. Transgenic lines were challenged with the Mexican isolate P6752, which is heterozygous for AVR3a^{KI} and AVR3a^{EM}, and the US isolate US-8 BF-6, which is homozygous for AVR3a^{EM}. The transgenic potato plants expressing R3a, Rpi_vnt1 and the R3a* variants, but not the non-transgenic Ranger Russet, demonstrated high levels of resistance towards the heterozygous isolate P6752 (Fig. 6). Thus, the transient ATTA data and the transgenic plants corroborate the conclusion that the selected mutations in the LRR do not negatively impact on resistance towards *P. infestans* isolates expressing Avr3a^{KI}. The transgenic Ranger Russet lines expressing Rpi_vnt1 provided resistance towards the AVR3a^{EM} homozygous isolate US-8 BF-6. However, none of the transgenic lines expressing the *R3a** or *R3a* genes, which had been shown to be resistant to isolate P6752, were able to control disease development of isolate US-8 BF-6 (Fig. 6).

Figure 6. R3a and R3a* variants expressed via the *R3a* promoter in transgenic plants protect the susceptible cultivar Ranger Russet from *P. infestans* strain P6752, which is heterozygous for AVR3a^{KI} and AVR3a^{EM}, but not from strain US-8 BF-6, which is homozygous for AVR3a^{EM}. Non-transgenic plants were used as a control for susceptibility. Transgenic plants expressing R3a or Rpi-vnt1 were used as positive controls for resistance to P6752 or US-8 BF-6, respectively. Representative plants were photographed at 11dpi.

Discussion

The relatively narrow genetic basis of clonal potato cultivars in agriculture provides pathogens such as *P. infestans* with sufficient opportunity to adapt and overcome inducible host resistant responses and to thus cause disease on a global scale. Resistance responses rely on the direct or indirect recognition of modified-self or pathogen-derived molecules [1]. For example, in the first layer of inducible resistance, also referred to as PAMP triggered immunity (PTI), conserved pathogen-associated molecular patterns (PAMPs) and/or damage-associated molecular patterns (DAMPs) are recognised [1,34,35]. Successful pathogens circumvent this recognition with the help of effectors that perturb host resistance responses and promote effector-triggered susceptibility (ETS). However, by being in close proximity to the host, pathogen effectors provide the innate plant immune system with another opportunity for detection that is dependent on the presence of cognate plant R proteins that subsequently yield ETI [1,36]. In nature, this closely entwined co-evolution between hosts and pathogens is evident in the diversification observed for effectors and *R* genes [4].

Indeed, *P. infestans* is known as a pathogen with 'high evolutionary potential' [37] and more than 560 RXLR-type candidate effectors have been described within the late blight pathogen genome [38]. The genomic organisation of RXLR effector genes, which are often in gene-poor and repeat-rich regions, is thought to facilitate their enhanced diversification by enabling non-homologous recombination. Oomycete RXLR-type effectors have been shown to evade detection by *R* gene products via transcriptional regulation [39], utilising functionally equivalent effectors that allow loss of recognised effectors [40], suppressor activity of unrelated effectors [41] and/or sequence diversity [11,42]. Sequence diversity underpins virulence or avirulence behaviour for the essential *P. infestans* effector AVR3a. AVR3a^{KI} determines avirulence on plants carrying *R3a* whereas AVR3a^{EM} promotes virulence. It is thought that only AVR3a^{KI} was present in the *P. infestans* strain responsible for the outbreak of late blight disease leading to the Irish Potato Famine in the 1840 s [43]. The AVR3a^{EM} allele may have come to dominate in *P. infestans* populations once the resistance gene *R3a* was deployed in potato crops, quickly usurping the AVR3a^{KI} allele. By using clonal potato varieties in current agriculture, the diversification of *R* genes is undermined and novel, naturally occurring resistances can only slowly be introgressed via breeding.

Functional *R* genes are often found in clusters that show evidence of duplication followed by diversification [44]. The phylogenetic NB-LRR gene grouping that contains homologs of the functional *R3a* gene in the potato genome has previously been described as CNL-8 [5]. This group is, after the *R2* cluster (CNL-5), the second largest *R* gene cluster in potato where more than 750 NB-LRR-like genes have been identified [5–6]. Functional *R3a*, a homolog of the tomato *I2* resistance gene that controls races of the fungus *Fusarium oxysporum* [45], was cloned alongside three paralogous sequences that provided insight into amino acid positions under diversification [13]. With one exception, positions under diversification reside in the LRR and cluster around two regions spanning LRRs 1 to 4 and 14 to 23 that were also identified as being important in this study.

Here we used, in addition to random mutagenesis and screening, DNA shuffling and targeted mutagenesis to enhance AVR3a^{EM} recognition by R3a*. DNA shuffling, which emulates the natural evolutionary processes of mutation, recombination and selection, has proven a highly effective method for evolving new specificities/properties for a wide range of proteins that cannot be rationally designed and is of particular use in identifying mutations that are beneficial in combination [18]. Artificial evolution of a resistance gene to broaden its specificity has previously been performed on the gene *Rx* that protects potato plants from strains of PVX. In an initial *Rx* study, mutagenesis of the LRR region was performed on the basis that this region is the primary determinant of recognition specificity. Screening identified four mutations in the LRR region that affected elicitor recognition and activation functions [16]. Introduction of the mutated *Rx* genes in to the model host *N. benthamiana* as transgenes extended resistance to a normally resistance-breaking strain of PVX, HB, and a distantly related carlavirus, PopMV [16].

Our primary screen identified eleven mutants with enhanced AVR3a^{EM} recognition containing in total 23 amino acid substitutions. However, only three of these (R618Q, K920E, Q931R), found in the clones with single amino acid substitutions, are known to be causative to the improved phenotype. The previous study performed by Segretin *et al.* [25] identified six amino acid substitutions in the LRR domain that enhanced AVR3a^{EM} recognition: two of these were also found in our primary screen (L668P, K920E). The fact that we did not identify all the LRR mutations identified by Segretin *et al.* [25] and that they did not identify more mutations in the LRR region, demonstrates that neither screen was exhaustive.

The initial screen identified mutants with enhanced recognition of AVR3a^{EM} when expressed from the strong 35S promoter. Amino acid changes in LRRs #3 and #15, and thus within regions known to be under diversifying selection [13], were prevalent in the clones from the first round of screening, suggesting these regions might be of particular importance. This suggestion was supported by the fact that all the clones from the second round of shuffling contained an amino acid change, E620D, in LRR #3 and sometimes additional changes in LRR #15, while all of the clones from the third round contained the E620D change and one or two amino acid changes in LRR #15. Using DNA shuffling it is more difficult to bring together combinatorially beneficial mutations that are in close proximity. Furthermore, random mutagenesis as a source of diversity has limitations in that single nucleotide changes can only convert a codon for one amino acid to a limited set of codons for other amino acids rather than all twenty possible amino acids. To circumvent the former problem a more directed approach was adopted in which a library that contained all permutations of the amino acid changes thought important (one in LRR #2, two in LRR #3 and four in LRR #15) was produced for screening. The aptness of this approach was shown by the fact that clones were obtained with enhanced recognition responses to AVR3a^{EM} compared to the best performing clone from the third round of shuffling, Rd3-1, and that one of the possible 128 forms was prevalent. This form contained two amino acid changes in close proximity in each of LRRs #3 and #15; a combination that would have been difficult to obtain through DNA shuffling. Interestingly, an amino acid change at one of these positions, Q931, was also found in one of the clones recovered by Segretin *et al.* [25], though their "de-convolution" did not show their substitution at this position, proline instead of arginine, to improve AVR3a^{EM} recognition. As shown by Segretin *et al.* [25] for the R3a mutants with enhanced AVR3a^{EM} recognition, we did not note any reduction in the AVR3a^{KI} recognition responses of the clones we isolated.

Our studies show that the recognition of AVR3a^{EM} by the R3a* mutants we have isolated recapitulates the mechanistic processes of recognition of AVR3a^{KI} by the wild-type *R3a* gene. It has previously been reported [14] that the HR triggered by R3a recognition of AVR3a^{KI} is dependent on the ubiquitin ligase-associated protein SGT1 and HSP90. VIGS of *SGT1* and, to a lesser degree, of *HSP90* in our experiments inhibited the cell death responses induced by recognition of AVR3a^{EM} by our R3a* mutants. Similarly, it has been shown that wild-type R3a re-localises from the cytoplasm to late endosomal compartments when co-expressed with AVR3a^{KI}, but not when co-expressed with AVR3a^{EM} [15]. We found that the mutants from the three later rounds still re-localised to endosomal compartments when co-expressed with AVR3a^{KI} and, importantly, also re-localised to the same vesicles when co-expressed with AVR3a^{EM}. The earlier study by Engelhardt *et al.* [15] also showed that the effector AVR3a^{KI} itself relocalises from the cytoplasm to endosomes when co-expressed with wild-type R3a and is in close physical proximity to R3a, whereas AVR3a^{EM} remains distributed through the cytoplasm. Our BiFC experiments show that AVR3a^{KI} and the normally unrecognized form, AVR3a^{EM} both traffic from the cytoplasm to vesicles when co-expressed with the R3a* forms. This re-localisation of R3a and AVR3^{KI} was shown to be a prerequisite for the development of the HR [15]. Thus, the gain of AVR3a^{EM} recognition R3a* variants have gained many aspects of the mechanism of the wild-type R3a response to AVR3a^{KI}.

Although the R3a* variants produced in this study responded to AVR3a^{EM} and produced HR responses when the elicitor was transiently expressed via *Agrobacterium*, critically they only provided resistance to *P. infestans* isolates that express AVR3a^{KI} and not to isolates that express only AVR3a^{EM}. Both transient expression via *Agrobacterium* in *N. benthamiana* and stable transgenic expression in potato corroborated this finding. Failure to protect from the pathogen itself was also reported for the R3a mutants identified by Segretin *et al.* [25]. That this was the case for mutants with single amino acid changes is perhaps not surprising given the large differences from the wild-type R3a/AVR3a^{KI} response we observed when first and second round clones were expressed from the *R3a* promoter with AVR3a^{EM}. For some pathogen/*R* gene combinations, e.g. PVX and *Rx*, the resistance responses can be separated from the HR [46] though the induction of necrotic responses by transient expression of the elicitor protein has been used to identify Rx mutations that, when expressed transgenically in the model host *N. benthamiana*, provide resistance to the pathogen itself. However, for *P. infestans* it has been suggested that the strength of the HR correlates with resistance levels [47].

A recent, secondary mutation study of *Rx* provides some evidence that stepwise artificial evolution could be required to obtain an optimum combination between effector recognition and subsequent *R* gene activation and signal transduction [17]. In the *Rx* study, the resistance provided by one of the mutations in the LRR domain to PopMV was improved by random mutagenesis of the CC-NB-ARC1-ARC2 domains [17]. In addition to constitutively active mutants that by themselves gave necrotic responses, four mutants with enhanced responses to PopMV were isolated. For three of these mutants the improved phenotype was conferred by a single amino acid change, while for the fourth a pair of amino acid substitutions was required. The mutations, which affect activation sensitivity, were found to be located around the nucleotide-binding pocket of Rx. As mentioned previously, we have evidence that neither this screening nor the efforts from Segretin *et al.* [25] were exhaustive as both approaches yielded novel beneficial mutations. Whereas our study of *R3a* focused solely on the LRRs, in Segretin *et al.* [25] the entire *R3a* gene was subjected to mutagenesis. The latter study identified eight single amino acid changes that enhanced responses to AVR3a^{EM}. Out of these, six occurred in the LRR domain and one in the CC domain. This substitution enhanced the response to AVR3a^{EM} but also showed some auto-activation. The final substitution, found in the NB-ARC domain, sensitised the AVR3a^{EM} response and broadened specificity to include an elicitor from another *Phytophthora* species [25]. Interestingly, this change occurred in the nucleotide-binding pocket and is adjacent to one of the sensitizing mutations found in *Rx* [17]. Broadening resistance gene specificity merely by introducing sensitizing mutations without improving recognition may have detrimental consequences in the field. However, a natural precedent for this has been found in PM3 resistance protein alleles in which the substitution of two amino acids in the NB domain enhances the HR and broadens the spectrum of resistance to wheat powdery mildew isolates [48]. Thus, additional efforts to combine novel mutations in R3a domains responsible for AVR3a^{EM} recognition (LRR) and response (CC-NB-ARC1/ARC2) could further improve R3a* variants that already display gain of recognition and mechanistic re-localisation to ultimately yield genes that provide effective resistance in potato against isolates expressing AVR3a^{EM}. Considering the importance of AVR3a to *P. infestans*, such a resistance, combined with other, mechanistically distinct *R* genes could provide a step towards more durable late blight control.

Supporting Information

Figure S1 R3a* variants are not auto-activators. *N. benthamiana* leaves were infiltrated with *Agrobacterium* cultures designed to express R3a or the R3a* variants from the strong 35S promoter. Mixtures of cultures designed to co-express AVR3a^{KI} (KI) were used as positive controls for the induction of cell death. Leaves were examined under white-light and UV-B illumination. Photograph of representative leaf was taken five days after infiltration. In the absence of elicitor the R3a* variants, like R3a, do not produce visible cell death.

Figure S2 HR responses resulting from R3a* recognition of AVR3a^{EM} and AVR3a^{KI}, like those caused by wild-type R3a recognition of AVR3a^{KI}, are dependent on SGT1 and HSP90. SGT1- and HSP90-silenced plants were produced using TRV-based vectors. These plants and control plants inoculated with TRV:*eGFP* were infiltrated with different combinations of *Agrobacterium* cultures designed to express R3a, R3a* variants, AVR3a^{KI} (KI) or AVR3a^{EM} (EM). Photographs show representative HR responses induced by each of the different mixtures on control TRV:eGFP inoculated plants, SGT1-silenced plants and HSP90-silenced plants. The non-host bacterial pathogen *Erwinia amylovora* was used as a control for an SGT1- and HSP90-independnet HR response.

Figure S3 Western blot analysis showing integrity of YFP fusion proteins. Soluble protein extracts were prepared from *N. benthamiana* leaf tissue two days after infiltration with cultures designed to express YFP fusions to R3a, Rd2-1, Rd3-1 or Rd4-1. The blot was probed with anti-GFP antibodies as described by Engelhardt *et al.* (2012). Protein sizes are indicated

in kilodaltons (kD) and protein loading is shown by Ponceau S (PS) staining.

Figure S4 YFP fusions to R3a and the R3a* variants localize to the cytoplasm in the absence of AVR3a. *N. benthamiana* leaves were infiltrated with cultures designed to express YFP fusions to R3a, Rd2-1, Rd3-1 or Rd4-1. Leaf tissue was examined two days after infiltration under a confocal laser scanning microscope. Representative images are from five independent experiments. Scale bar = 20 μm.

Figure S5 In the presence of AVR3a^{EM} YFP fusions to R3a* variants, but not YFP-R3a, re-localize to vesicles labelled by the prevacuolar compartment marker PS1-CFP. *N. benthamiana* leaves were infiltrated with mixtures of cultures designed to express PS1-CFP, AVR3a^{EM} and YFP fusions to R3a, Rd2-1, Rd3-1 or Rd4-1. Tissue was examined two days after infiltration under a confocal laser scanning microscope. The left-hand panel shows YFP signal, the right-hand panel CFP signal and the central panel displays the merged signals. Representative images are from three independent experiments. Scale bar = 10 μm.

Figure S6 YC-AVR3a^{EM} reconstitutes YFP fluorescence with YN fusions to the R3a* variants at vesicles labelled by the prevacuolar compartment marker PS1-CFP. Generation of the YFP signal indicates that AVR3aEM and the R3a* variants are in close proximity at the vesicles. *N. benthamiana* leaves were infiltrated with mixtures of cultures designed to express PS1-CFP, YC-AVR3a^{EM} and YN fusions to Rd2-1, Rd3-1 or Rd4-1. Tissue was examined 2 d after infiltration under a confocal laser scanning microscope. Left-hand panel, YFP signal; right-hand panel, CFP signal; central panel, merged signals. Representative images from three experiments. Scale bar = 20 μm.

Table S1 Primers used in this study.

Table S2 Nucleotide and amino acid changes found in R3a* clones.

Table S3 Mean symptom scores from four nine-day experiments for upper and lower leaves.

Acknowledgments

The authors acknowledge Dr Sanwen Huang for kindly providing sequences of wild type R3a and the native regulatory elements.

Author Contributions

Conceived and designed the experiments: SC PCB PRJB IH. Performed the experiments: SC LS PCB SE BH NC KM PSMVW XC. Analyzed the data: SC LS PCB CA NC PRJB IH. Wrote the paper: IH SC LS CA PRJB.

References

1. Jones JDG, Dangl JL (2006) The plant immune system. Nature 444: 323–329.
2. van der Biezen EA, Jones JDG (1998) Plant disease-resistance proteins and the gene-for-gene concept. Trends Biochem Sci 23: 454–456.
3. Heath MC (2000) Hypersensitive response-related death. Plant Mol Biol 44: 321–334.
4. Hein I, Gilroy EM, Armstrong MR, Birch PRJ (2009) The zig-zag-zig in oomycete–plant interactions. Mol Plant Pathol 10: 547–562.

5. Jupe F, Pritchard L, Etherington GJ, MacKenzie K, Cock PJA, et al. (2012) Identification and localisation of the NB-LRR gene family within the potato genome. BMC Genomics 13: 75.

6. Jupe F, Witek K, Verweij W, Sliwka J, Pritchard L, et al. (2013) Resistance gene enrichment sequencing (RenSeq) enables reannotation of the NB-LRR gene family from sequenced plant genomes and rapid mapping of resistance loci in segregating populations. Plant J 76: 530–544.

7. Birch PRJ, Boevink PC, Gilroy EM, Hein I, Pritchard L, et al. (2008) Oomycete RXLR effectors: delivery, functional redundancy and durable disease resistance. Curr Opin Plant Biol 11: 373–379.

8. Vleeshouwers VG, Raffaele S, Vossen JH, Champouret N, Oliva R, et al. (2011) Understanding and exploiting late blight resistance in the age of effectors. Annu Rev Phytopathol. 49:507–531.

9. Bos JIB, Armstrong MR, Gilroy EM, Boevink PC, Hein I, et al. (2010) *Phytophthora infestans* effector AVR3a is essential for virulence and manipulates plant immunity by stabilizing host E3 ligase CMPG1. Proc Natl Acad Sci USA 107: 9909–9914.

10. Vetukuri RR, Tian Z, Avrova AO, Savenkov EI, Dixelius C, et al. (2011) Silencing of the PiAvr3a effector-encoding gene from *Phytophthora infestans* by transcriptional fusion to a short interspersed element. Fungal Biol 115: 1225–1233.

11. Armstrong MR, Whisson SC, Pritchard L, Bos JIB, Venter E, et al. (2005) An ancestral oomycete locus contains late blight avirulence gene Avr3a, encoding a protein that is recognized in the host cytoplasm. Proc Natl Acad Sci USA 102: 7766–7771.

12. Cárdenas M, Grajales A, Sierra R, Rojas A, González-Almario A, et al. (2011) Genetic diversity of Phytophthora infestans in the Northern Andean region. BMC Genet 12: 23.

13. Huang S, van der Vossen EAG, Kuang H, Vleeshouwers VGAA, Zhang N, et al. (2005) Comparative genomics enabled the isolation of the R3a late blight resistance gene in potato. Plant J 42: 251–261.

14. Bos JIB, Kanneganti T-D, Young C, Cakir C, Huitema E, et al. (2006) The C-terminal half of *Phytophthora infestans* RXLR effector AVR3a is sufficient to trigger R3a-mediated hypersensitivity and suppress INF1-induced cell death in *Nicotiana benthamiana*. Plant J 48: 165–176.

15. Engelhardt S, Boevink PC, Armstrong MR, Ramos MB, Hein I, et al. (2012) Relocalization of late blight resistance protein R3a to endosomal compartments is associated with effector recognition and required for the immune response. Plant Cell 24: 5142–5158.

16. Farnham G, Baulcombe DC (2006) Artificial evolution extends the spectrum of viruses that are targeted by a disease-resistance gene from potato. Proc Natl Acad Sci USA 103: 18828–18833.

17. Harris CJ, Slootweg EJ, Goverse A, Baulcombe DC (2013) Stepwise artificial evolution of a plant disease resistance gene. Proc Natl Acad Sci USA, 110: 21189–21194.

18. Stemmer WP (1994) DNA shuffling by random fragmentation and reassembly: in vitro recombination for molecular evolution. Proc Natl Acad Sci USA 91: 10747–10751.

19. Bernal A, Pan Q, Pollack J, Rose L, Willets N, et al. (2005) Functional dissection of the Pto resistance gene using DNA shuffling. J Biol Chem 280: 23073–23083

20. Simpson CG, Lewandowska D, Liney M, Davidson DDavidson D, Chapman S, et al. (2014) Arabidopsis PTB1 and PTB2 proteins negatively regulate splicing of a mini-exon splicing reporter and affect alternative splicing of endogenous genes differentially. New Phytol 203: 424–436.

21. Leung DW, Chen E, Goeddel DV (1989) A method for random mutagenesis of a defined DNA segment using a modified polymerase chain reaction. Technique, 1: 11–15.

22. Lazo GR, Stein PA, Ludwig RA (1991) A DNA transformation-competent Arabidopsis genomic library in Agrobacterium. Biotechnology 9: 963–967.

23. van der Fits L, Deakin EA, Hoge JH, Memelink J (2000) The ternary transformation system: constitutive virG on a compatible plasmid dramatically increases Agrobacterium-mediated plant transformation. Plant Mol Biol 43: 495–502.

24. Duan H, Richael C, Rommens CM (2012) Overexpression of the wild potato eIF4E-1 variant Eva1 elicits Potato virus Y resistance in plants silenced for native eIF4E-1. Transgenic Res 21: 929–938.

25. Segretin ME, Pais M, Franceschetti M, Chaparro-Garcia A, Bos JIB, et al. (2014) Single amino acid mutations in the potato immune receptor R3a expand response to *Phytophthora* effectors. Mol Plant Microbe Interact 27: 624–637.

26. Liu Y, Burch-Smith T, Schiff M, Feng S, Dinesh-Kumar SP (2004) Molecular chaperone Hsp90 associates with resistance protein N and its signaling proteins

SGT1 and Rar1 to modulate an innate immune response in plants. J Biol Chem 279: 2101–2108.

27. Azevedo C, Betsuyaku S, Peart J, Takahashi A, Noel L, et al. (2006) Role of SGT1 in resistance protein accumulation in plant immunity. EMBO J, 25, 2007–2016.

28. Gilroy EM, Hein I, van der Hoorn R, Boevink PC, Venter E, et al. (2007) Involvement of cathepsin B in the plant disease resistance hypersensitive response. Plant J 52: 1–13.

29. Saint-Jean B, Seveno-Carpentier E, Alcon C, Neuhaus JM, Paris N (2010) The cytosolic tail dipeptide Ile-Met of the pea receptor BP80 is required for recycling from the prevacuole and for endocytosis. Plant Cell, 22: 2825–2837.

30. Walter M, Chaban C, Schütze K, Batistic O, Weckermann K, et al. (2004) Visualization of protein interactions in living plant cells using bimolecular fluorescence complementation. Plant J 40: 428–438.

31. Saunders DG, Breen S, Win J, Schornack S, Hein I, et al. (2012) Host protein BSL1 associates with Phytophthora infestans RXLR effector AVR2 and the Solanum demissum Immune receptor R2 to mediate disease resistance. Plant Cell 24: 3420–3434.

32. Li G, Huang S, Guo X, Li Y, Yang Y, et al. (2011) Cloning and characterization of R3b; members of the R3 superfamily of late blight resistance genes show sequence and functional divergence. Mol Plant Microbe Interact 24: 1132–1142.

33. Foster SJ, Park TH, Pel M, Brigneti G, Sliwka J, et al. (2009) Rpi-vnt1.1, a Tm-2(2) homolog from Solanum venturii, confers resistance to potato late blight. Mol Plant Microbe Interact 22: 589–600.

34. Maffei ME, Arimura G, Mithöfer A (2012) Natural elicitors, effectors and modulators of plant responses. Nat Prod Rep 29: 1288–1303.

35. Newman MA, Sundelin T, Nielsen JT, Erbs GI (2013). MAMP (microbe-associated molecular pattern) triggered immunity in plants. Front Plant Sci 16;4: 139

36. Deslandes L, Rivas S (2012) Catch me if you can: bacterial effectors and plant targets. Trends Plant Sci. 17: 644–655

37. Raffaele S, Win J, Cano LM, Kamoun S (2010) Analyses of genome architecture and gene expression reveal novel candidate virulence factors in the secretome of Phytophthora infestans. BMC Genomics. 16: 637.

38. Haas BJ, Kamoun S, Zody MC, Jiang RHY, Handsaker RE, et al. (2009) Genome sequence and analysis of the Irish potato famine pathogen *Phytophthora infestans*. Nature 461: 393–398.

39. Rietman H, Bijsterbosch G, Cano LM, Lee HR, Vossen JH, et al. (2012) Qualitative and quantitative late blight resistance in the potato cultivar Sarpo Mira is determined by the perception of five distinct RXLR effectors. Mol Plant Microbe Interact 25: 910–919.

40. Van Poppel PM, Guo J, van de Vondervoort PJ, Jung MW, Birch PR, et al. (2008) The Phytophthora infestans avirulence gene Avr4 encodes an RXLR-dEER effector. Mol Plant Microbe Interact 21: 1460–1470.

41. Wang Q, Han C, Ferreira AO, Yu X, Ye W, et al. (2011) Transcriptional programming and functional interactions within the Phytophthora sojae RXLR effector repertoire. Plant Cell. 23: 2064–2086

42. Gilroy EM, Tayor RM, Hein I, Boevink P, Sadanandom A, et al. (2011) CMPG1-dependent cell death follows perception of diverse pathogen elicitors at the host plasma membrane and is suppressed by *Phytophthora infestans* RXLR effector AVR3a. New Phytol 190: 653–666.

43. Yoshida K, Schuenemann VJ, Cano LM, Pais M, Mishra B, et al. (2013) The rise and fall of the *Phytophthora infestans* lineage that triggered the Irish potato famine. eLife, 2: e00731.

44. McDowell JM, Simon SA (2006) Recent insights into R gene evolution. Mol Plant Pathol 7, 437–448.

45. Ori N1, Eshed Y, Paran I, Presting G, Aviv D, et al. (1997) The I2C family from the wilt disease resistance locus I2 belongs to the nucleotide binding, leucine-rich repeat superfamily of plant resistance genes. Plant Cell. 9: 521–532.

46. Bendahmane A, Kanyuka K, Baulcombe DC (1999) The Rx gene from potato controls separate virus resistance and cell death responses. Plant Cell, 11: 781–792.

47. Vleeshouwers VGAA, van Dooijeweert W, Govers F, Kamoun S, Colon LT (2000) The hypersensitive response is associated with host and nonhost resistance to *Phytophthora infestans*. Planta 210: 853–864.

48. Stirnweiss D, Milani SD, Jordan T, Keller B, Brunner S (2014) Substitutions of two amino acids in the nucleotide-binding site domain of a resistance protein enhance the hypersensitive response and enlarge the PM3F resistance spectrum in wheat. Mol Plant Microbe Interact 27: 265–276.

Profound Impact of Hfq on Nutrient Acquisition, Metabolism and Motility in the Plant Pathogen *Agrobacterium tumefaciens*

Philip Möller[1], Aaron Overlöper[1], Konrad U. Förstner[2], Tuan-Nan Wen[3], Cynthia M. Sharma[2], Erh-Min Lai[3], Franz Narberhaus[1]*

1 Microbial Biology, Ruhr University Bochum, Bochum, Germany, 2 Research Center for Infectious Diseases (ZINF), Julius-Maximilian's University of Würzburg, Würzburg, Germany, 3 Institute of Plant and Microbial Biology, Academia Sinica, Taipei, Taiwan

Abstract

As matchmaker between mRNA and sRNA interactions, the RNA chaperone Hfq plays a key role in riboregulation of many bacteria. Often, the global influence of Hfq on the transcriptome is reflected by substantially altered proteomes and pleiotropic phenotypes in *hfq* mutants. Using quantitative proteomics and co-immunoprecipitation combined with RNA-sequencing (RIP-seq) of Hfq-bound RNAs, we demonstrate the pervasive role of Hfq in nutrient acquisition, metabolism and motility of the plant pathogen *Agrobacterium tumefaciens*. 136 of 2544 proteins identified by iTRAQ (isobaric tags for relative and absolute quantitation) were affected in the absence of Hfq. Most of them were associated with ABC transporters, general metabolism and motility. RIP-seq of chromosomally encoded Hfq[3xFlag] revealed 1697 mRNAs and 209 non-coding RNAs (ncRNAs) associated with Hfq. 56 ncRNAs were previously undescribed. Interestingly, 55% of the Hfq-bound ncRNAs were encoded antisense (as) to a protein-coding sequence suggesting that *A. tumefaciens* Hfq plays an important role in asRNA-target interactions. The exclusive enrichment of 296 mRNAs and 31 ncRNAs under virulence conditions further indicates a role for post-transcriptional regulation in *A. tumefaciens*-mediated plant infection. On the basis of the iTRAQ and RIP-seq data, we assembled a comprehensive model of the Hfq core regulon in *A. tumefaciens*.

Editor: Sander Granneman, Univ. of Edinburgh, United Kingdom

Funding: This study was supported by a grant from the German Research Foundation (DFG priority program SPP 1258; NA 240/8-1) to FN, a grant from Academia Sinica to EML, and a joint grant from the German Academic Exchange Service (DAAD) and the Taiwan National Science Council (100-2911-I-001-053) to FN and EML. The funders had no role in study design, data collection and analysis, decision to publish, or preparation of the manuscript.

Competing Interests: The authors have declared that no competing interests exist.

* Email: franz.narberhaus@rub.de

Introduction

Post-transcriptional gene regulation is a key strategy in the dynamic adaption to changing environmental conditions. In bacteria, small regulatory RNAs (sRNAs, non-coding RNAs, ncRNAs) rapidly adjust gene expression to the physiological needs and have also been implicated in virulence control [1,2]. Bacterial ncRNAs usually influence translation or stability of their cognate mRNA target. Translational regulation occurs via blockage or release of the Shine-Dalgarno (SD) sequence on the mRNA [3,4]. Further, the double-stranded ncRNA-mRNA duplex is often recognized and degraded by RNase III [5,6,7,8]. Rapid and dynamic adjustment of the cellular RNA pool also involves other RNases (e.g. RNaseE, PNPase) that interact with single-stranded RNAs and modulate processing, degradation and general quality control of ncRNAs and mRNAs.

ncRNAs are distinguished according to their location on the genome in anti-sense (asRNA) or trans encoded small RNAs (sRNAs). asRNAs partly or completely overlap with their target mRNAs encoded on the complementary strand. Due to the perfect sequence complementarity, asRNA-mRNA interactions are believed to be rather Hfq-independent [9,10,11,12,13]. In contrast,

sRNAs are encoded in intergenic regions and usually share imperfect sequence complementarity with their target mRNA. The association of sRNAs with their target mRNAs is often promoted by the RNA-chaperone Hfq [14].

Hfq is a homohexameric (L)Sm-like protein and was first identified as a host factor for phage Qβ in *Escherichia coli* [15]. The ring-shaped protein exposes a proximal, a distal and a lateral RNA-binding surface, allowing specific binding of RNA molecules [16,17]. ncRNAs bind primarily to the proximal face via U-rich internal regions and 3′ poly-U tails [18], while 5′ A-rich mRNA sequences, including ARN-motifs (A *adenosine*, R *purine*, N *any nucleotide*), bind to the distal surface [4,19]. The lateral surface supports sRNA-binding in *E. coli* [20]. Aside from mediating association of sRNA and mRNA, Hfq also directly influences its interaction partners [21]. On the one hand, Hfq-binding prohibits RNase-dependent decay, while on the other hand, ongoing interaction with Hfq promotes polyadenylation and subsequent RNA decay [22,23,24]. Deletion of *hfq* is usually accompanied by pleiotropic phenotypes and often results in reduced growth, motility and stress tolerance, as shown for numerous bacteria including the α-proteobacteria *Brucella abortus* [25], *Rhodobacter*

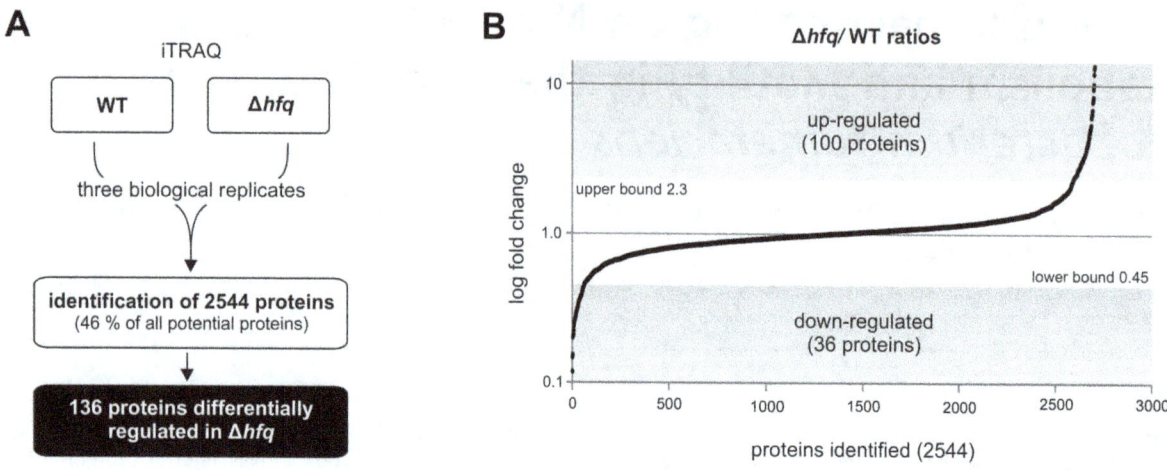

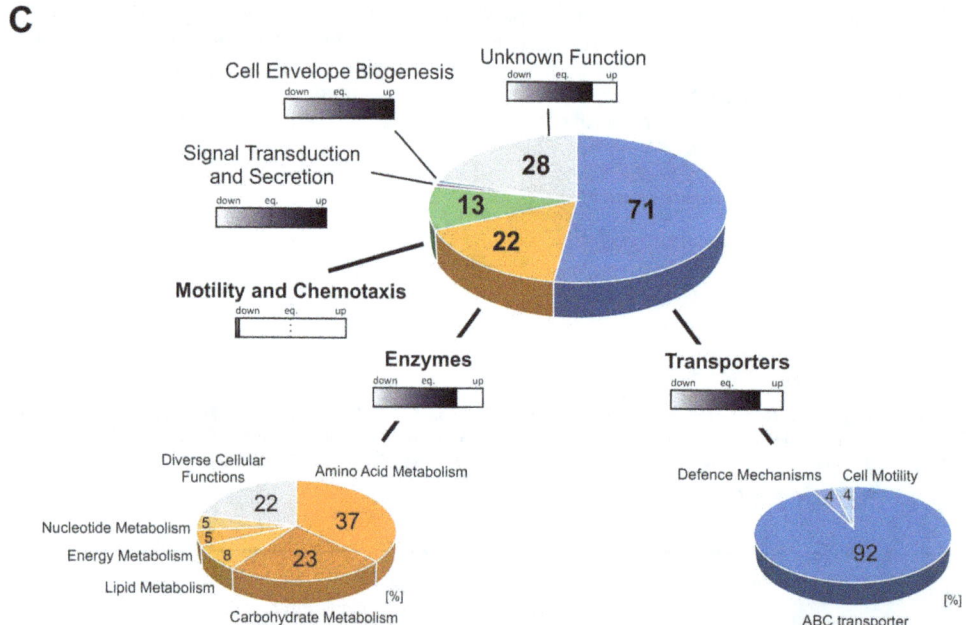

Figure 1. Quantitative proteomics of *A. tumefaciens* WT and Δ*hfq* mutant. A) iTRAQ experiments of 3 biological replicates from stationary phase cultures of WT and Δ*hfq* mutant revealed 136 proteins differentially expressed (2544 proteins identified). **B)** Distribution of all 2544 Δ*hfq*/WT logarithmic (log) fold-changes (FC). Calculating a confidence interval of 95% resulted in an upper bound of 2.3 and a lower bound of 0.45. 100 proteins were up-regulated (FC>2.3) and 38 down-regulated (FC<0.45) in absence of *hfq*. **C)** Classification of proteins into physiological relevant groups by KEGG ontology. Filled bars indicate up- or down-regulation of proteins within the different groups. eq., equilibrium.

sphaeroides [11], *Sinorhizobium meliloti* [26,27,28,29], *Rhizobium leguminosarum* [30] and *Agrobacterium tumefaciens* [31]. Most strikingly, Hfq is required for successful host-microbe interactions in several symbiotic and pathogenic bacteria [32].

Deletion of *hfq* in the phytopathogen *A. tumefaciens* leads to reduced viability and a severe reduction in plant infection efficiency but only eight direct targets are known so far [31]. The genome of *A. tumefaciens* consists of a circular chromosome, a linear chromosome, the At-plasmid and the Ti (*tumor-inducing*)-plasmid [33]. *A. tumefaciens* is capable of interkingdom DNA transfer, leading to tumor formation on infected plants [34]. Virulence is induced by exposure of the bacteria to plant wound

molecules (e.g. acetosyringone, low pH, monosaccharides). Signal perception leads to the activation of virulence (*vir*) gene expression and transport of a single-stranded DNA fragment (T-DNA) into the plant cell via the VirB/D4 type-4 secretion system (T4SS). Integration of the T-DNA into the plant chromosome results in phytohormone and opine biosynthesis. Two sequencing studies of the *A. tumefaciens* RNome identified 621 transcripts not dedicated to protein-coding thus constituting a large pool of potential regulatory RNAs [35,36]. These ncRNAs are distributed among all four replicons. Thirty-six (6%) of the identified ncRNAs were verified by Northern blot experiments, including the ABC-transporter regulator AbcR1, which turned out to be Hfq-dependent [35,36,37,38]. AbcR1 participates in bacteria-plant interactions by influencing susceptibility of *A. tumefaciens* to the plant defense molecule γ-aminobutyric acid (GABA) via negative regulation of *atu2422* encoding the periplasmic binding protein of the GABA uptake system. At least 15 additional AbcR1 targets suggest an extensive sRNA regulon [38]. Apart from that, it is still unclear how many additional ncRNAs and mRNAs are Hfq-dependent in *A. tumefaciens*. Despite the rapidly growing number of ncRNAs, identification of the corresponding mRNA target is still challenging. Determination of Hfq-bound mRNAs and ncRNAs has become an applicable technique to narrow down the involved transcripts and increase accuracy of subsequent analysis, e.g. biocomputational predictions.

To determine the underlying molecular mechanism of Hfq-dependent regulation in *A. tumefaciens* and to globally identify Hfq-dependent transcripts, we analyzed the Δ*hfq* global proteome by iTRAQ and the Hfq-bound RNA-interactome by RIP-seq. Our results demonstrate a major impact of Hfq on regulatory networks balancing nutrient acquisition, cellular metabolism and motility. Further, we revealed extensive binding of asRNAs to Hfq and validated the influence on asRNA-mediated regulation. Hfq also bound mRNAs of the major virulence genes thus indicating a distinct role in plant infection.

Materials and Methods

Bacterial strains, plasmids, media

All strains and plasmids used in this study are listed in Table S1. *Agrobacterium tumefaciens* C58 strains were cultivated at 30°C in YEB complex medium or AB minimal medium (pH 5.5, supplemented with 1% (w/v) glucose) [39]. For virulence induction overnight cultures of *A. tumefaciens* were inoculated in AB medium to an $OD_{600\ nm}$ of 0.1 and grown for 6 h at 30°C. Virulence gene expression was induced by addition of 0.1 mM acetosyringone (Sigma-Aldrich, Germany) and further cultivation at 23°C for 16 h. Non virulence induced cultures were treated with equal volumes of DMSO. For cultivation during mutagenesis, *A. tumefaciens* cells were grown in Luria-Bertani (LB) medium [40], supplemented with either 10% (w/v) sucrose or 50 μg ml^{-1} kanamycin (Km).

Chromosomal integration of *hfq*^{3xFlag}

The *hfq* (*atu1450*) gene of the *A. tumefaciens* C58 circular chromosome was tagged with a *3xFlag* at the 3′ end. For mutagenesis plasmid construction, a region upstream of *hfq* including its open reading frame without the TGA stop codon was amplified by PCR using primers *hfq*_up_*Pst*I_fw and *hfq*_up_*Sal*I_rv (Table S2). The 3xFlag tag was amplified from *E. coli* MG1655 *kdtA*::3XFLAG chromosomal DNA [41] with primers 3xFlag_*Sal*I_fw and 3xFlag_*Acc*65I inserting a TGA stop codon at the 3′ of the 3xFlag sequence. The *hfq* downstream region was amplified using primers *hfq*_down_*Acc*65I_fw and *hfq*_down_*Eco*RI_rv. The resulting PCR fragments *Pst*I_*hfq*_up_*Sal*I, *Sal*I_3xFlag_*Acc*65I and *Acc*65I_*hfq*_down_*Eco*RI were subsequently ligated into pK19*mobsacB* suicide vector [42], resulting in *hfq*_up_3xFlag_down mutagenesis plasmid. The plasmid was transformed into *A. tumefaciens* C58 wild-type cells by electroporation (800 Ω, 25 μF, 2 kV) and selected for homologous recombination by Km resistance on LB+Km agar plates. Single colonies were grown overnight in LB medium without antibiotics and plated on LB agar plates containing sucrose. Double cross over events resulted in sucrose tolerant and Km sensitive colonies. Putative mutants encoding *hfq*^{3xFlag} on the chromosome were verified by Southern blot analysis [40].

Total RNA preparation and Northern-blot analysis

RNA was isolated from *A. tumefaciens* strains as described in [31] by the hot acid phenol method [43]. Northern blot analyses were performed as previously described [37]. For RNA detection 8 μg (for mRNAs) to 10 μg (for sRNAs) total RNA were separated on agarose or polyacrylamide gels respectively, blotted on Hybond-N membranes (GE Healthcare, USA) and hybridized with specific DIG-labelled (Roche Applied Sciences, Germany) RNA probes. Oligonucleotides used for RNA probe synthesis are listed in Table S2. For signal detection a Hyperfilm ECL (GE Healthcare, USA) system was used.

Hfq^{3xFlag} co-immunoprecipitation

Co-immunoprecipitation (coIP) experiments of Hfq^{3xFlag} and bound RNAs were based on the procedure described in [9,44] with minor changes. 100 ml of *A. tumefaciens* wild-type and *hfq*^{3xFlag} cultures grown to $OD_{600\ nm}$ 0.5 and 1.0 in YEB medium or under non-induced (+DMSO) and virulence-induced (+ acetosyringone) conditions in AB medium were harvested and resuspended in 2 ml ice-cold lysis buffer (20 mM Tris (pH 7.5), 150 mM KCl, 1 mM MgCl$_2$, 1 mM DTT). Cells were disrupted by French Press (3 passes, 16,000 psi) and centrifuged at 10,000×g, 4°C, for 1 h. 10 ng monoclonal ANTI-3XFLAG M2 antibody (Sigma-Aldrich, Germany) were coupled to 50 μl Dynabeads Protein G (ThermoFisher Scientific, Life Technologies, USA) as described in the instruction manual, and incubated with the supernatant (3 h, 4°C). Dynabeads were separated on a magnet and washed 5x with PBS buffer. RNA was isolated using phenol/chloroform/isoamylalcohol and chloroform/isoamylalcohol followed by precipitation with ethanol and sodium acetate. After precipitation remaining DNA was digested by DNaseI (ThermoFisher Scientific, Life Technologies, USA) and RNA was precipitated as described before.

RNA-sequencing

Preparation of cDNA libraries was performed at Vertis Biotechnology AG (Germany). Equal amounts of RNA samples were poly(A)-tailed and 5′-PPP were removed. The RNA adapter was ligated to the RNA 5′-monophosphate and reverse transcription was performed with oligo(dT)-adapter primers resulting in first-strand cDNA. Higher yields of DNA (20–30 ng μl^{-1}) were gained by further PCR-based amplification using primers designed for TruSeq according to recommendations for Illumina (HiSeq). For multiplex sequencing a library specific barcode was part of the 3′- sequencing adapter. Purification was achieved using the Agencourt AMPure XP kit (Beckman Coulter Genomics, USA), followed by capillary electrophoresis. Final cDNAs were sequenced using a HiSeq 2500 machine in single-read mode and running 100 cycles. Raw (de-multiplexed) reads and normalized coverage files were deposited in the Gene Expression Omnibus (GEO) of the National Center for Biotechnology Information [45] and are accessible via the GEO accession GSE59123 (http://www.ncbi.nlm.nih.gov/geo/query/acc.cgi?acc=GSE59123).

RNA-sequencing data analysis

Illumina reads in FASTQ format were trimmed (cut-off phred score 20) by the fastq_quality_trimmer program from FASTX toolkit 0.0.13 (http://hannonlab.cshl.edu/fastx_toolkit/). Further processing was performed using "create", "align" and "coverage" subcommands of the READemption tool 0.2.6 with default parameters (Förstner et al., submitted). Poly(A)-tail sequences were removed and sequences shorter than 12 nt were eliminated. Collections of the remaining reads were mapped to the *A.*

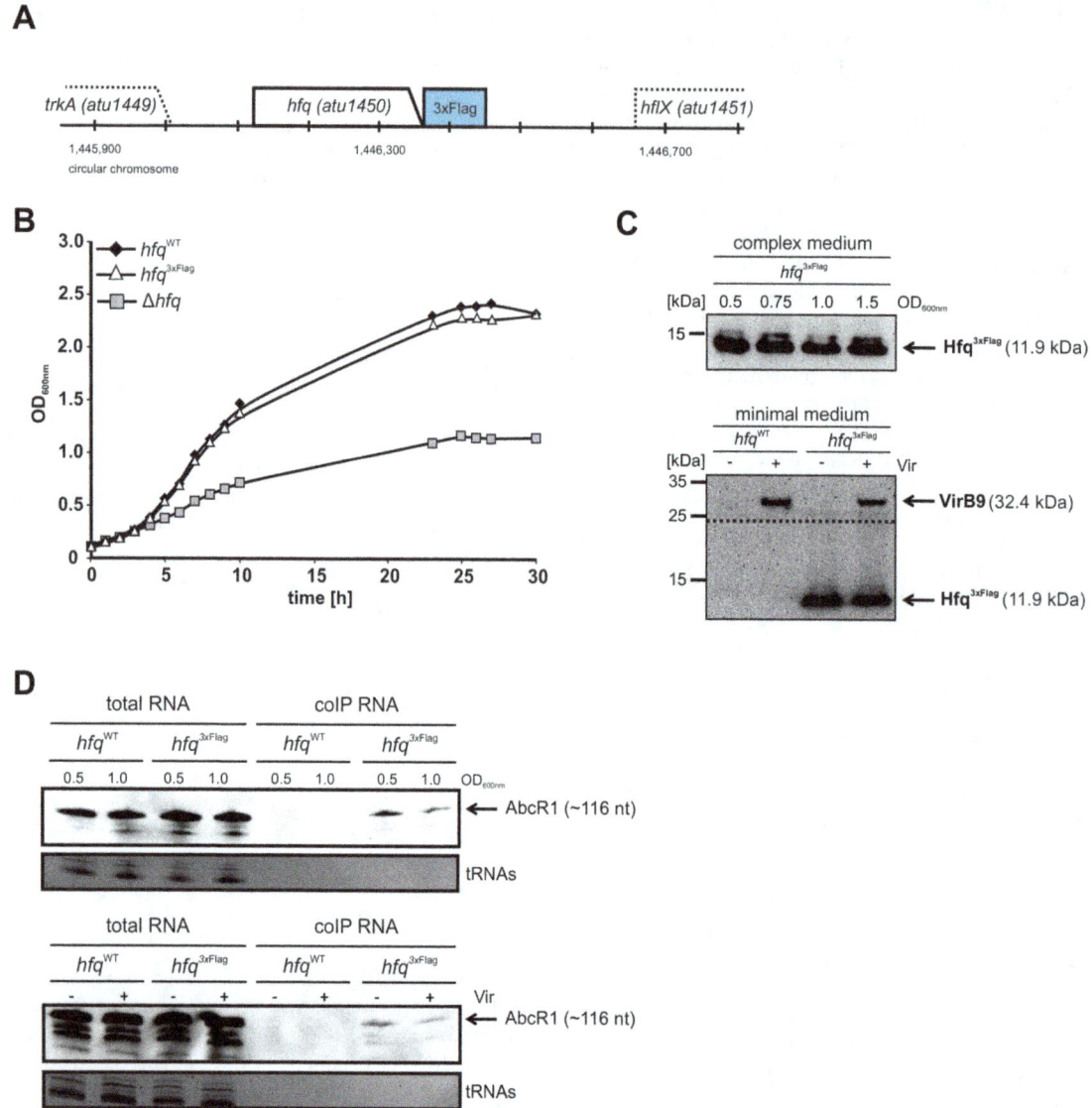

Figure 2. Construction of *hfq*^3xFlag for co-immunoprecipitation. A) Chromosomal integration of a *3xFlag* sequence at the 3′ end of the *hfq* sequence. **B)** Growth experiments in complex medium confirmed functionality of the Hfq^3xFlag fusion. Growth of *hfq*^WT and *hfq*^3xFlag strains was comparable, while the Δ*hfq* mutant exhibited a severe growth defect. **C)** Western blot analysis of Hfq^3xFlag protein (~11.9 kDa) from protein extracts isolated from *hfq*^WT and *hfq*^3xFlag strains. Proteins were isolated from exponential (OD 0.5) to stationary growth phase (OD 1.5). VirB9 was detected to confirm successful induction of virulence by addition of acetosyringone (+Vir). **D)** Purification of Hfq^3xFlag and isolation of co-purified RNA. Total RNA and co-immunoprecipitated RNA (coIP RNA) were analyzed by PAA gel electrophoresis and subsequent Northern blotting. AbcR1 was detected by a DIG-labeled RNA probe. Ethidium bromide stained tRNAs served as loading control. Equal amounts of coIP RNA were loaded, but no tRNAs were detectable in the corresponding lanes.

tumefaciens C58 reference genome (NC_003062.2, C_003063.2, NC_003064.2, NC_003065.3 - downloaded from the NCBI ftp server) using segemehl [46]. Mapping statistics (input, aligned, uniquely aligned reads etc.) are listed in Table S3. The numbers of aligned reads per nucleotide were represented by coverage plots (wiggle format) and visualized in the Integrated Genome Browser [47]. Normalization was performed based on the total number of reads aligned from the respective library. Multiplication of the corresponding graphs by the minimum number of mapped reads calculated from all libraries prevented rounding of small numbers to zero. For read quantification, annotation files in GFF3 format (accession numbers mentioned above) were obtained from the NCBI ftp server. Intergenic and antisense regions fulfilling the Hfq-dependency criteria (see below), were manually curated and

regions were adjusted according to IGB browser information from all libraries. The number of reads overlapping with annotation entries was calculated using the READemption "gene_quanti" subcommand. Reads overlapping in sense, anti-sense of all annotations were considered and counted separately.

Hfq dependent RNAs

5459 genes, 621 ncRNAs and 819 transcriptional start sites (TSS) identified in previous studies were included in the annotation [35,36]. Protein-coding sequences with so far undefined TSS were extended by virtual 54 nt at the 5′ UTRs, as described by [48]. All transcripts were additionally extended by virtual 20 nt at the 3′ UTR, applying the minimal transcriptional

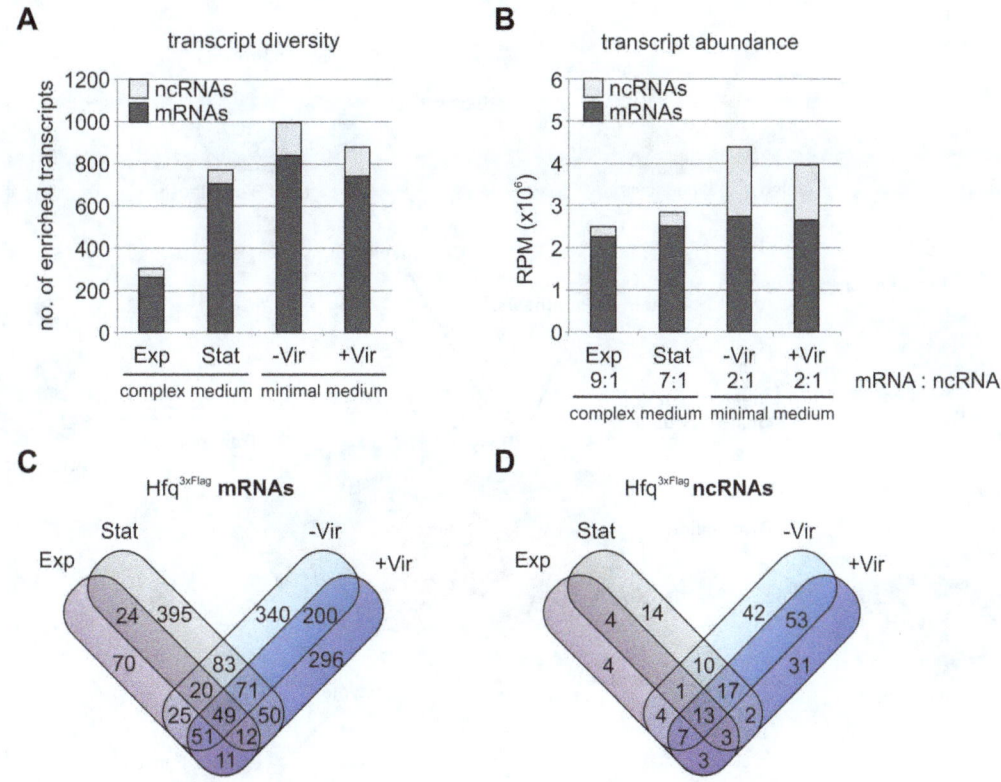

Figure 3. Hfq^{3xFlag} binds mRNAs and ncRNAs. A) Total numbers of mRNAs and ncRNAs enriched during Hfq^{3xFlag} coIP (transcript diversity) in Exp, Stat, -Vir and +Vir conditions. **B)** Abundance of mRNAs and ncRNAs enriched by Hfq^{3xFlag} (RPM) in the different growth phases. Ratio of mRNAs : ncRNA is indicated below the respective growth condition. Condition specific and overlapping enrichment of mRNAs **C)** and ncRNAs **D)** by Hfq^{3xFlag} at the different growth conditions. Exp, exponential; Stat, stationary; -Vir, non-induced; +Vir, virulence-induced; RPM, reads per million.

unit model (Fig. S1). Hfq enriched RNAs were further subjected to manual curation and reads overlapping the defined 5′ or 3′ features were merged with the cognate transcript to gain more accurate transcriptional units. RNAs with a minimal raw read count (RRC) of 50 and at least 2-fold enrichment (after a normalization by the total number of aligned reads of each library) in the Hfq^{3xFlag} library compared to the corresponding Hfq^{WT} library were considered enriched.

Identification of new ncRNAs

Transcripts fulfilling the Hfq-dependency criteria but not dedicated to any annotated feature were classified ncRNAs, when they reached a minimal length of 50 nt and a total number of aligned reads of 50 [48]. Transcripts not overlapping any feature were classified as trans sRNAs, while transcripts partly or completely overlapping a feature in anti-sense orientation were classified as anti-sense RNAs. Newly identified sRNAs were named (*Agrobacterium tumefaciens* Hfq associated ncRNA)

AhaR_X_Y, with "X" varying for the *A. tumefaciens* genomic replicons (C: *circular chromosome*, L: *linear chromosome*, At: *At-plasmid*, Ti: *Ti-plasmid*) and "Y" for ongoing numbering.

Bacterial growth, protein isolation and trypsin digestion for iTRAQ

A. tumefaciens C58 WT and Δhfq strains were grown to $OD_{600\,nm}$ 1.5 in YEB medium (30°C). Cells were harvested (4,000×g, 20 min, 4°C) and washed 2x in Tris-Cl (50 mM Tris-HCl, pH 7.5, 200 mM KCl) and suspended in Lysis buffer (50 mM Tris-HCl, pH 7.5, 200 mM KCl, 1 mM PMSF, 1x Protease inhibitor mix) to a final $OD_{600\,nm}$ of 10. Cells were disrupted by French Press (3 passes, 16,000 psi) and lysates were centrifuged 2x (10,000×g, 10 min, 4°C). Supernatants were precipitated overnight with 6 volumes pre-chilled TCA (10 w/v)/acetone. Precipitated proteins (10,000×g, 25 min, 4°C) were washed 3x with 85% cold acetone, dried and resuspended in urea buffer (8 M urea, 50 mM Tris-Cl, pH 8.5). Protein concentrations

Table 1. Total number of uniquely aligned reads.

		growth condition			
		exponential (Exp)	stationary (Stat)	non-induced (−Vir)	virulence-induced (+Vir)
total number of uniquely aligned reads	hfq^{WT}	5,490,606	4,429,841	265,804	1,489,476
	hfq^{3xFlag}	8,230,008	4,652,974	9,374,442	8,199,109

A

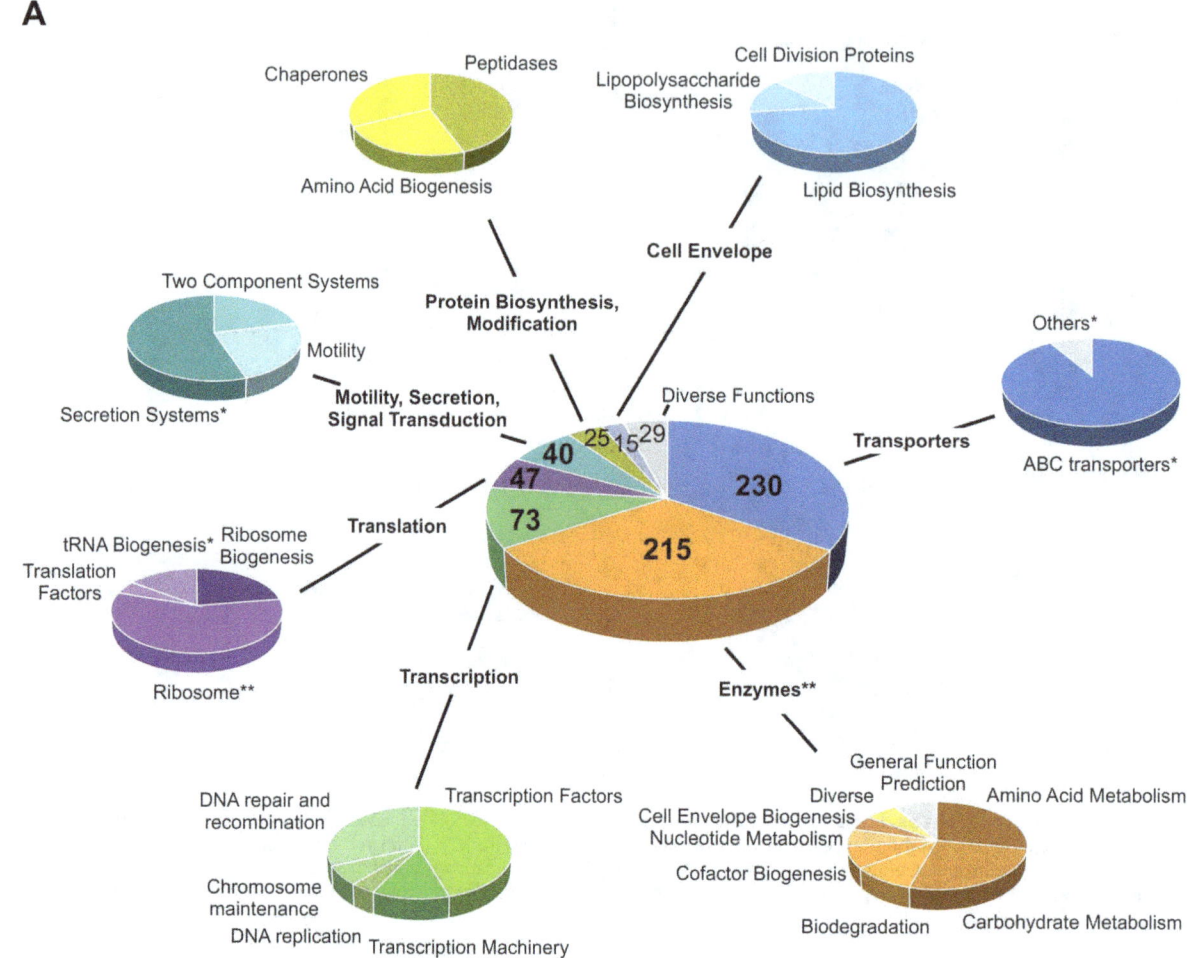

Figure 4. mRNAs enriched by Hfq³ˣᶠˡᵃᵍ cluster into physiological relevant groups. Clustering of Hfq³ˣᶠˡᵃᵍ enriched mRNAs into seven physiological coherent groups by KEGG ontology. Main groups are subdivided to specify the biological function of the respective proteins. Statistical relevance: * p-value <0.05, ** p-value <0.01.

were measured using Pierce 660 nm protein Assay kit (Thermo Scientific, Germany). Protein isolation, trypsin digestion and subsequent peptide treatment were performed as previously described in [49] with minor modifications. Total proteins (100 µg) were reduced by addition of DTT to a final concentration of 10 mM (1 h, 37°C). Further treatment with a final concentration of 50 mM iodoacetamide (30 min, RT, dark) was followed by consumption of any free iodoacetamide by 10 mM DTT (1 h, RT, dark). Proteins were diluted with 50 mM Tris-Cl, pH 8.5 (final urea concentration less than 4 M) and digested with 250 units/ml Benzonase (Sigma-Aldrich, Germany) (RT, 2 h), followed by Lys-C (Wako, Japan) digestion (1:200 (w/w), 4 h, RT). Proteins were diluted with 50 mM Tris-Cl pH 8.0 (final urea concentration less than 1 M) and digested with 2 µg of modified trypsin (Promega) (1:50 (w/w), 37°C, overnight). The peptide solution was acidified with 10% trifluoroacetic acid, desalted on an Oasis HLB cartridge (Waters, USA) and dried by SpeedVac.

Labeling of peptides with iTRAQ reagents and fractionation

Peptide pellets were dissolved in iTRAQ dissolution buffer and labeled with iTRAQ reagents according to the manufacturer's manual (Applied Biosystems). Wild-type samples were labeled with

reagent 114 while Δhfq-mutant proteins were labeled with reagent 115 (1 h, room temperature). iTRAQ labeled peptides were combined and further fractionated using a strong cation-exchange column (SCX, PolySulfoethyl A, 4.6×100 mm, 5 µm, 200 Å, PolyLC Inc.) on HPLC. The SCX chromatography was performed with an initial equilibrium buffer A (10 mM KH₂PO₄, 25% ACN pH 2.65), a 40 min linear gradient from 0% to 50% buffer B (1 M KCl in buffer B, pH 2.65), 5 min in 50% buffer B, 1 min in a linear gradient from 50% to 100% buffer B at a 1 ml min⁻¹ flow rate. According to the peak area (Abs 214 nm) the collected fractions were pooled into five final fractions. Samples were desalted using an Oasis HLB cartridge (Waters, USA) prior to LC-MS/MS.

LC-MS/MS analysis

SCX fraction samples were resuspended (0.1% formic acid) and analyzed using a nanoUPLC system (nanoAcquity, Waters, USA) coupled to an LTQ Orbitrap Elite hybrid mass spectrometer (Thermo Scientific, Germany). A C18 capillary column (1.7 µM particle size, 75 µM×250 mm, BEH130, Waters, USA) was used to separate peptides with a 120 min linear gradient from 5% to 40% ACN at a flow rate of 300 ml min⁻¹. The LTQ Orbitrap Elite MS was operated in the data-dependent mode with top 15

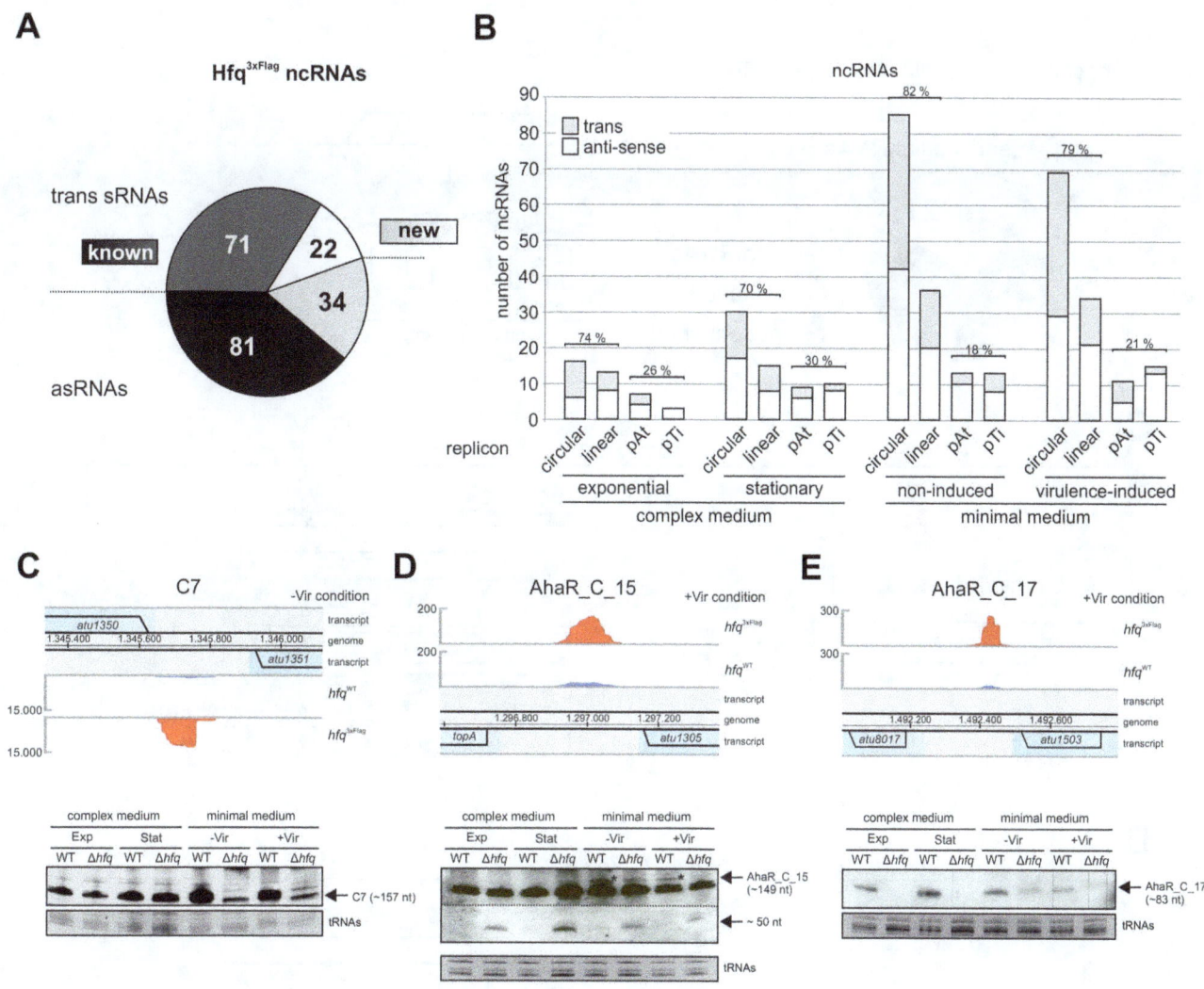

Figure 5. Identification of Hfq-dependent ncRNAs. A) Numbers of asRNAs and trans sRNAs enriched by Hfq[3xFlag]. 152 known and 56 new ncRNAs were enriched. **B)** Expression of ncRNAs during the different growth phases from the four *A. tumefaciens* replicons. Relative expression from chromosomes (circular, linear) and mega-plasmids (At, Ti) at the different conditions is indicated in %. Northern blots of RNA isolated from WT and Δ*hfq* strains validated Hfq-dependency of trans encoded sRNA **C)** C7 and the newly identified sRNAs **D)** AhaR_C_15 and **E)** AhaR_C_17 (lower panel). Expression locus and reads from the genome browser are indicated on top. Ethidium bromide stained tRNAs served as loading control. Exp, exponential; Stat, stationary; -Vir, non-induced; +Vir, virulence-induced.

ions (charge states ≥2) from the MS survey scan selected for subsequent HCD activation and MS/MS acquisitions in the Orbitrap cell. For MS and MS/MS the FT Orbitrap m/z range was set to 350–1600 with a resolving power of 120,000 and AGC of 500,000. HCD was set to MSn AGC target = 50,000 with a minimal signal of 5,000. Isolation width was 1.2 with an NCE of 35%, activation time of 0.1 ms and a resolving power of 15,000. Data dependent settings for dynamic exclusion were 1 for repeat count, 15 sec repeat duration and 90 sec for exclusion duration. Peptide identification was performed using the Proteome Discoverer software (v1.3, Thermo Fisher Scientific, USA) with SEQUEST and Mascot (v2.3, Matrix Sciences) search engines. MS data were searched against the *Agrobacterium tumefaciens* C58 protein sequence database (http://www.ncbi.nlm.nih.gov/). Peptides with 2 maximum missed cleavage sites after trypsin digestion were analyzed with a precursor mass tolerance of 10 ppm and a fragment mass tolerance of 50 mmu. Dynamic modifications were oxidation (M) while static modifications

included carbamidomethyl (C) and 4plex-iTRAQ tags (N-terminus and K). The identified peptides were validated using Percolator algorithm which automatically conducted a decoy database search and rescored peptide spectrum matches (PSM) using q-values and posterior error probabilities. All PSMs were filtered with a q-value threshold of 0.01% (1% false discovery rate) and proteins were filtered with a minimum of 2 distinct peptides identified per protein. For iTRAQ quantification, the ratios of iTRAQ reporter ion intensities in MS/MS spectra (m/z: 114, 115) from raw datasets were used to calculate the fold changes between control and treatment. Only unique peptides were used for peptide/protein quantification. All peptide ratios were normalized by the median protein ratio. The protein ratio was calculated from three biological replicates.

iTRAQ data analysis

Proteins with significantly changed abundance in the *hfq* deletion strain were selected as previously described in [50] with

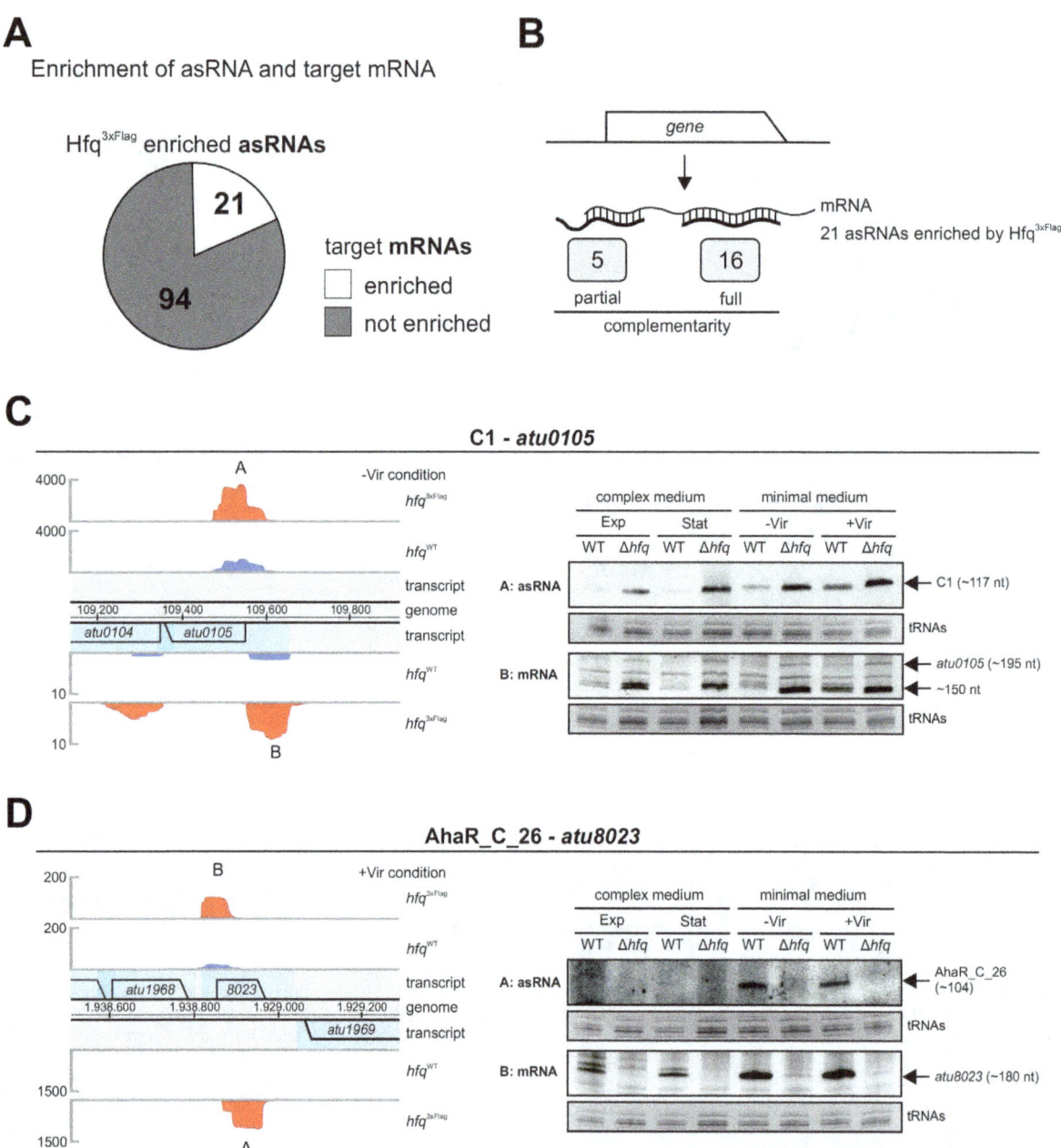

Figure 6. Hfq binds asRNAs and their cognate target mRNAs. A) 21 of the 115 asRNAs were enriched simultaneously with their target mRNA encoded on the complementary strand. **B)** Complementarity of the 21 asRNA-mRNA pairs. 16 asRNAs were fully complementary to their designated target mRNAs. **C), D)** Northern blot analysis of asRNAs and target mRNAs with full complementarity. Location and mapped reads of **C)** C1 and *atu0105* and **D)** AhaR_C_26 and *atu8023*, are indicated by the genome browser view (left). Northern blot analysis of RNA isolated from WT and Δ*hfq* strains grown under different conditions (right). Ethidium bromide stained tRNAs or 16S rRNAs served as loading control. Exp, exponential; Stat, stationary; -Vir, non-induced; +Vir, virulence-induced.

minor modifications. Mean and standard deviation (*SD*) from ln ratios of all 2544 identified proteins were calculated. A confidence interval of 95% (Z score = 1.96) was used to select proteins with a distribution outside the main distribution. For down-regulated proteins the confidence interval was 0.016 (*mean ratio*)–1.96×0.413 (*SD*), corresponding to a protein ratio of 0.4524. Proteins up-regulated were similarly calculated (*mean ratio* + 1.96×*SD*), corresponding to a protein ratio of 2.2823. Therefore, the cut-off value for down-regulated proteins was set 0.45-fold and for up-regulated proteins 2.3-fold. Proteins were considered

significantly regulated when reaching the cut-off value and a variability of less than 40% between the replicates. For protein ratios that failed the variance criterion, a combined ratio was calculated (*combined ratio = ratio +/− ratio×variance*). For up-regulated proteins variance was subtracted from the ratio, for down-regulated proteins variance was added to the ratio. Combined ratios that reached the cut-off criterion were also considered statistically significant. By this, proteins with higher variances were included when they explicitly reached the cut-off criteria for up- or down-regulation.

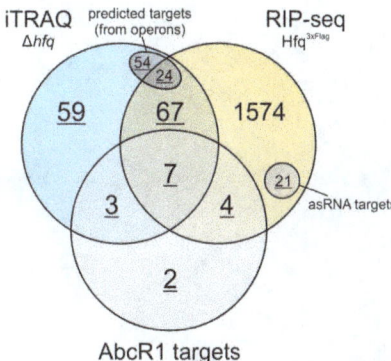

Figure 7. Hfq regulon in *A. tumefaciens*. Summary of mRNAs/proteins influenced by Hfq in *A. tumefaciens*. Results from iTRAQ and RIP-seq (this study) and AbcR1 targets [37,38] were combined. Operons encoding at least 2 equally regulated proteins (iTRAQ) were included and all encoded proteins were assumed to be Hfq-dependent. 21 target mRNAs simultaneously enriched with an asRNA were also included. Hfq-dependent proteins/mRNAs used for network predictions are underlined (Fig. 8).

Protein analyses by SDS PAGE and Western blot

Protein samples were separated by SDS gel electrophoresis on 12.5% polyacrylamide gels. Proteins were transferred to nitrocellulose membranes (Hybond-C, GE Healthcare, USA) by Western transfer using standard protocols [40]. 3xFlag protein fusions were detected using monoclonal ANIT-3XFLAG M2 antibody (Sigma-Aldrich) and the corresponding secondary goat anti-mouse IgG (H+L)-HRP conjugate (Bio-Rad, USA). Protein signals were visualized using Luminata Forte Western HRP substrate (Millipore). For signal detection the ChemiImager Ready system (Alpha Innotec) was used.

Network prediction, statistics and visualization

Operons were predicted using (http://meta.microbesonline.org/operons/gnc176299.html) [51]. Proteins were clustered based on KEGG Brite ontology for *A. tumefaciens* C58 (http://www.kegg.jp/kegg/tool/map_brite2.html) [52,53]. Statistical relevance of protein clustering was calculated using a two-tailed Fischer test [54]. Predictions of physical and functional protein interactions were performed using the String 9.1 webserver (http://string-db.org/) [55]. Results were visualized by Cytoscape 3.1.0 [56,57]. Additional ncRNA nodes and corresponding edges were added manually.

Results

Differential abundance of Hfq-dependent proteins

Deletion of *hfq* severely impacts *A. tumefaciens* fitness and virulence [31]. To reveal the molecular basis of this pleiotropic phenotype, we performed quantitative proteomics using isobaric tags for relative and absolute quantitation (iTRAQ). Almost half of all annotated *A. tumefaciens* proteins (2544 = 47.5%) were identified in three biological replicates of WT and Δ*hfq* cultures grown to stationary phase, indicating the high sensitivity of this method (Fig. 1A). Quantification of protein amounts was achieved by simultaneous LC-MS/MS analysis of the differently labeled peptides from WT and *hfq* mutant. Fold-changes (FC) from Δ*hfq*/WT ratios of all 2544 proteins were used to calculate a confidence interval of 95% (Fig. 1B). Given the calculated upper (FC 2.3) and lower (FC 0.45) bounds we found a total of 136 proteins, encoded from all four replicons, that were differentially abundant in the *hfq*

mutant compared to the WT (Fig. 1B). 100 proteins were up-regulated, whereas 36 proteins were down-regulated (Table S4). Hfq-affected proteins were clustered into six physiological groups based on KEGG ontology. The major group comprises 71 proteins involved in transport mechanisms. Sixty-five of these proteins belong to the ABC transporter class (Fig. 1C). Further, 22 enzymes participating in metabolism of amino acids, carbohydrates, lipids or nucleotides, and involved in energy production/conversion were identified. While proteins from the transporter or enzyme class were mainly up-regulated in the *hfq* deletion strain (Fig. 1C, filled bars), proteins related to motility and chemotaxis were consistently down-regulated. Abundance of 13 proteins assigned to this group was reduced in the *hfq* mutant (e.g. Fla, FlaB, FliF, McpA, CheW). One protein involved in signal transduction and secretion mechanisms (TraG, conjugal transfer protein) and one protein associated with cell envelope biogenesis (Atu4877, short-chain dehydrogenase) were up-regulated in the Δ*hfq* strain. Twenty-eight hypothetical proteins with unknown function followed the overall trend of up-regulation in the *hfq* mutant.

66 of the Hfq-dependent proteins were (at least) pairwise encoded in 27 putative polycistronic operons (Fig. S2). Proteins matching the same operon were consistently up- or down-regulated. The fact that 23 of the identified operons encode at least one protein associated with ABC transporter uptake systems further supports a principal role of Hfq in nutrient acquisition.

Establishment of a functional chromosomally encoded Hfq^3xFlag fusion

In a next step, we examined the Hfq regulon on the RNA level by using deep sequencing analysis of RNAs that were co-purified in a coIP with Hfq carrying a 3xFLAG epitope. The chromosomal *hfq* copy was replaced by a copy with a *3xFlag* sequence at the 3′ end (Fig. 2A). Whereas growth of the Δ*hfq* strain was severely reduced, growth of the *hfq*^3xFlag strain was indistinguishable from *hfq*^WT cultures, indicating that the C-terminal tagging did not interfere with its function (Fig. 2B). Hfq^3xFlag (~11 kDa) protein amounts were comparable in all growth phases as verified by Western blot analysis using a monoclonal anti-Flag antibody (Fig. 2C, upper panel). Similar Hfq^3xFlag amounts were present when cells were grown in minimal medium at non-induced (−Vir) or virulence-induced (+Vir) conditions (Fig. 2C, lower panel). The T4SS protein VirB9 served as control for successful virulence induction and was only detected at +Vir conditions. All these experiment suggest that Hfq^3xFlag is functionally equivalent to the WT protein.

In the next step, we validated the suitability of Hfq^3xFlag to enrich Hfq-interacting RNAs. RNA was isolated from Hfq^3xFlag proteins in exponential (Exp), early stationary (Stat), non-induced (−Vir) and virulence-induced (+Vir) conditions. The known Hfq-dependent sRNA AbcR1 [31] was successfully recovered by Hfq^3xFlag as shown by Northern blot analysis (Fig. 2D). While AbcR1 was present in total RNA samples from *hfq*^WT and *hfq*^3xFlag cells, it was only found in samples containing the Flag-tagged protein after co-IP.

Hfq-bound RNA pool varies in complexity and abundance

To identify Hfq-dependent RNAs, we performed RIP-seq experiments with *A. tumefaciens* *hfq*^WT and *hfq*^3xFlag strains grown to exponential (Exp) and stationary phase (Stat) in complex medium or at −Vir and +Vir conditions in minimal medium. Hfq^3xFlag protein was purified from cell extracts using DynaBeads

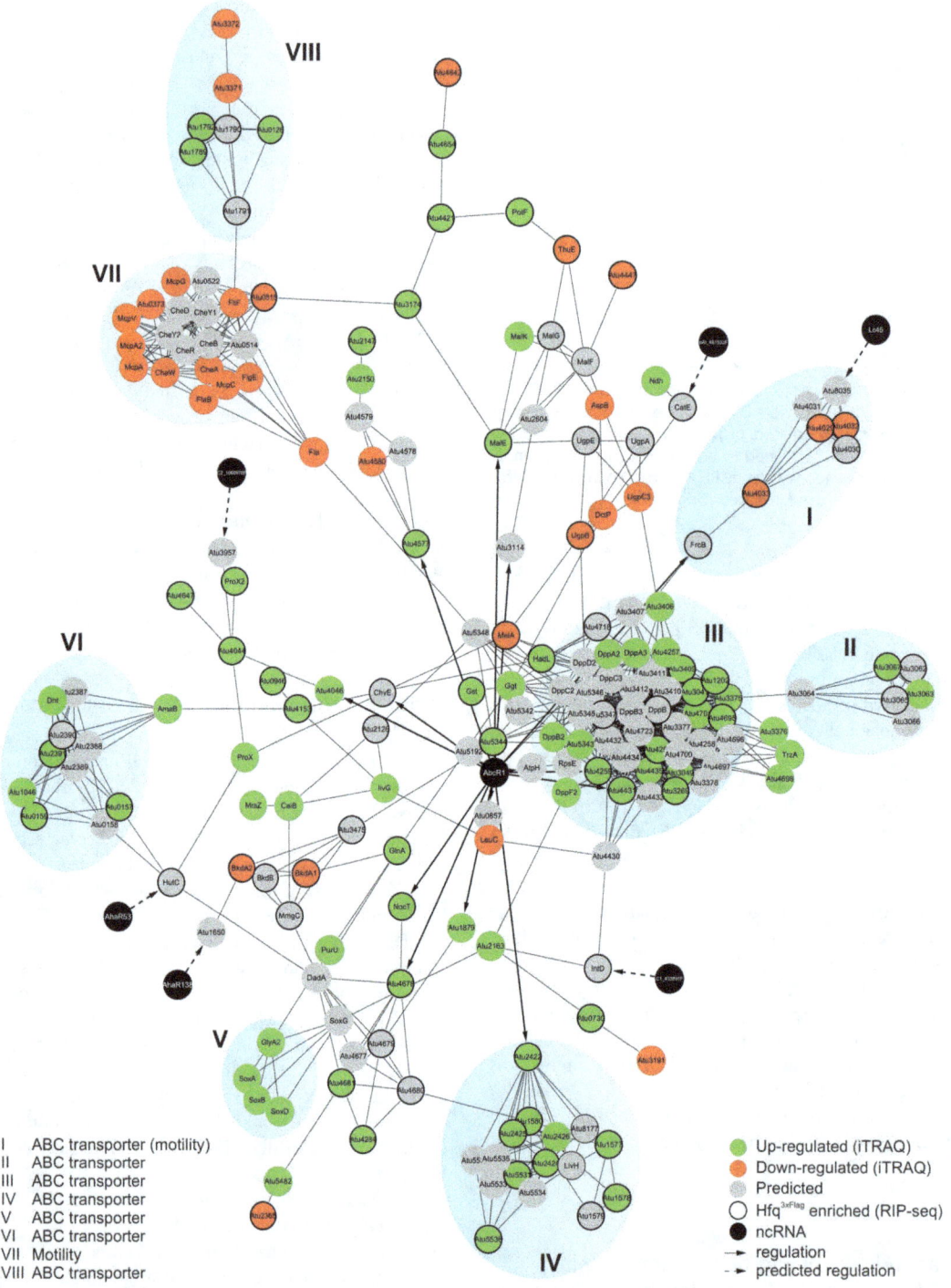

I   ABC transporter (motility)
II   ABC transporter
III ABC transporter
IV   ABC transporter
V   ABC transporter
VI   ABC transporter
VII Motility
VIII ABC transporter

● Up-regulated (iTRAQ)
● Down-regulated (iTRAQ)
● Predicted
○ Hfq^{3xFlag} enriched (RIP-seq)
● ncRNA
→ regulation
⇢ predicted regulation

Figure 8. Model of the Hfq core regulon. All 241 proteins associated with Hfq regulation (Fig. 7, underlined) were checked for putative interactions by String 9.1 software. Visualization of the resulting network was performed using Cytoscape 3.1.0. 207 proteins were interconnected and 7 ncRNAs (black) with predicted (dashed line) or validated (continuous line) target mRNAs inside the network were added. Proteins identified by iTRAQ were marked when up-regulated (green), down-regulated (red) or predicted to be influenced (grey). Corresponding mRNAs identified during Hfq^{3xFlag} coIP were indicated by bold circles. Hfq or AbcR1-dependent mRNAs validated by Northern blots in prior studies [37,38] were indicated by asterisks. Striking clustering of interacting nodes was marked by blue spheres and physiological functions were assigned according to protein functions of the involved proteins.

with monoclonal anti-Flag antibody prior to RNA isolation, reverse transcription and cDNA sequencing. As expected, the number of reads from the Hfq^{3xFlag} libraries was generally higher (Table 1). RNAs enriched at least 2-fold (Hfq^{3xFlag}/Hfq^{WT}) with a

minimal raw read count (RRC) of 50 in the Hfq^{3xFlag} libraries were considered Hfq bound. 1906 different RNAs were enriched, including 1697 mRNAs, 208 ncRNAs and one tRNA (tRNA-Gly, *atu0435*). Despite the rather large number of Hfq-bound mRNAs, many known targets were enriched about 2-fold (e.g. *atu2422*, 2.61-fold; *atu4678*, 2.2-fold; *malE*, 2.24-fold; *atu4113*, 2.3-fold) [31], indicating sufficient specificity of the observed interaction with Hfq. The transcript diversity of mRNAs and ncRNAs differed notably between Exp (262 mRNAs/39 ncRNAs), Stat (704 mRNAs/64 ncRNAs), −Vir (839 mRNAs/147 ncRNAs) and +Vir (740 mRNAs/129 ncRNAs) conditions (Fig. 3A). The highest number of different Hfq-associated RNAs was found during stress conditions (Stat growth, −/+ Vir).

Reads per million (RPM, raw read counting per gene divided by total number of aligned reads×10^6) from mRNAs and sRNAs enriched at the different conditions were summed up, to describe the actual load of RNAs on Hfq under different conditions (Fig. 3B). In complex medium, 70 to 90% of all Hfq-trapped RNAs were mRNAs. The relative number of Hfq-bound ncRNAs was highest in minimal medium suggesting that RNA-mediated regulation plays an important role when nutrients are scarce. A large number of transcripts (1101 mRNAs/91 ncRNAs) were specifically enriched at only one condition (Fig. 3C, 3D). The remaining RNAs were found in at least two conditions (547 mRNAs/104 ncRNAs) or in all conditions (49 mRNAs/13 ncRNAs). Although no Hfq-independent RNAs were described in *A. tumefaciens* so far, the well-known housekeeping RNAs 6S, tmRNA and SRP were not enriched by Hfq^{3xFlag} (Fig. S3). These observations are comparable to results from various other Hfq-RNA pull-down experiments [9,11,58,59], further supporting specificity of the observed RNA-binding to Hfq^{3xFlag} in our study.

Hfq primarily influences nutrient acquisition and cellular metabolism

The 1697 mRNAs enriched by Hfq^{3xFlag} represent 31% of the coding potential of *A. tumefaciens* (Table S5). Seventy-four mRNAs enriched by Hfq encode proteins that were differently abundant during iTRAQ analysis, strongly indicating post-transcriptional regulation. A large proportion (58%) of the Hfq-bound mRNAs encodes proteins of unknown function and was excluded from functional analysis. The remaining 721 mRNAs were clustered according to KEGG ontology into seven main groups ((1) transporters, (2) enzymes, (3) transcription, (4) translation, (5) motility, secretion and signal transduction, (6) protein biosynthesis and modification and (7) cell envelope biogenesis), each subdivided into several subgroups (Fig. 4). Due to the high number of enriched transcripts, statistical relevance (p-value <0.05) of the performed clustering was partly limited. Yet, the most prominent groups enriched during coIP correlate with the results obtained by mass spectrometry. Consistent with the observations from our proteome analysis, the largest group of 230 mRNAs enriched by Hfq was associated with transport processes. The vast majority (95%) encodes proteins from the ABC transporter class and was significantly overrepresented in the Hfq^{3xFlag} samples (p<0.05). 215 mRNAs encode enzymes participating in various cellular processes e.g. amino acid and carbohydrate metabolism, biodegradation of molecules, and cofactor and nucleotide biogenesis. Quite interestingly, Hfq bound 73 mRNAs encoding proteins involved in transcription. About one half of the corresponding proteins were transcription factors. Therefore, indirect influence of transcription by regulation of transcription factors might be an additional layer of Hfq-mediated regulation. The other half of proteins was involved in the transcription process itself, e.g. transcription machinery, DNA replication and DNA maintenance. Although ribosomal RNAs were not depleted from Hfq^{3xFlag} enriched RNA pools, neither rRNAs nor tRNAs (except for tRNA-Gly) were specifically enriched. Still, Hfq^{3xFlag} bound 23 mRNAs encoding ribosomal

proteins from both small and large ribosomal subunits as well as RNases (RNaseP, RNaseE) involved in ribosome biogenesis.

mRNAs enriched and assigned to motility, secretion and signal transduction partly encode the VirB/D4 type IV secretion system, including *virB1, B5, B6, B7, B8, B10, B11* and mRNAs directly associated with *A. tumefaciens*-specific DNA transfer and virulence like *virC1, C2, virD1, D3, D5* and *virE1-3*. The *virC2* and *virE3* mRNAs were enriched in −Vir and +Vir conditions, whereas the other *vir*-mRNAs were exclusively enriched at +Vir conditions. Additional mRNAs encoding motility (e.g. FlaF, FlgL) and chemotaxis-associated proteins (e.g. McpA, CheD) were enriched in accordance with Hfq-dependent changes in protein level already observed by iTRAQ analysis. Protein biosynthesis and modification was also associated with Hfq. Several mRNAs encoding chaperones (e.g. HslV), peptidases (e.g. protease ClpP) or proteins involved in amino acid biogenesis (e.g. AtrB, HemA) were Hfq-bound. Interestingly, the *hfq*^{3xFlag} mRNA was also enriched by Hfq^{3xFlag} in -Vir (3.8-fold) and +Vir (2.6-fold) conditions (Fig. S4). A rather small group of mRNAs encodes proteins participating in cell envelope biogenesis, lipid and lipopolysaccharide biosynthesis (e.g. KdtA, protease) as well as cell division (e.g. MinD).

Identification and validation of Hfq-dependent ncRNAs

Out of the 208 Hfq^{3xFlag} enriched regulatory ncRNAs (Table S6), 152 have been described in previous studies [35,36]. AbcR1 was strongly enriched under −Vir (33870 RPM, 5-fold) and +Vir (31705 RPM, 27-fold) conditions, confirming specific enrichment of Hfq-dependent RNAs. 56 ncRNAs were newly discovered in our study (Fig. 5A). The pool of ncRNAs enriched by Hfq^{3xFlag} was comprised of 93 trans-encoded sRNAs and 115 asRNAs with partial or complete complementarity to the target transcript encoded on the complementary strand. The ncRNAs were transcribed from all 4 replicons of the *A. tumefaciens* genome (Fig. 5B), and the distribution largely corresponded to the respective size of the replicons (Fig. S5). Both circular and linear chromosomes constitute 87% of the genome and harbor 70–80% of the ncRNAs, only slightly varying between the different conditions (Fig. 5B). In general, the number of ncRNAs was much higher in minimal medium than in complex medium.

To investigate Hfq-dependency of selected Hfq^{3xFlag} enriched ncRNAs, Northern blot experiments with total RNA isolated from WT and Δ*hfq* strains were performed. The three strongly enriched trans-encoded sRNAs C7, AhaR_C_15, AhaR_C_17 were chosen (Fig. 5C–E). C7 is encoded between *atu1350* and *atu1351* on the (−) strand and was 20-fold enriched in minimal medium (Fig. 5C, upper panel). The C7 RNA encoded on the circular chromosome has previously been identified in a screen for ncRNAs [35]. The amounts of C7 were comparable in complex medium. An influence of Hfq was apparent in minimal medium when C7 levels were reduced in the Δ*hfq* strain (Fig. 5C, lower panel). The two newly identified ncRNAs AhaR_C_15 and AhaR_C_17 were also affected by Hfq. AhaR_C_15 is transcribed from the (+) strand between *topA* and *atu1305* and was enriched 6.6-fold during coIP of the +Vir condition (Fig. 5D, upper panel). The sRNA full length signal was slightly larger than a non-identified signal present in all conditions. The RNA was barely detectable in Northern blots of total RNA from Exp and Stat growth phases (Fig. 5D, lower panel). In -Vir and +Vir conditions regulation was more evident, reflected by reduced amounts of the sRNA in the *hfq* mutant. In addition, a small fragment of about 50 nucleotides was detected in the Δ*hfq* deletion strain suggesting processing of AhaR_C_15. The AhaR_C_17 sRNA is encoded between *atu8017* and *atu1503* from the (+) strand and was enriched

10.3-fold in the coIP of the +Vir condition (Fig. 5E, upper panel). Northern blot analysis demonstrated a clear Hfq-dependence of the sRNA transcript under all tested conditions (Fig. 5E, lower panel).

Hfq influences asRNAs and target mRNAs

Out of the 115 asRNAs enriched by Hfq$^{3\times Flag}$, 21 asRNAs were enriched simultaneously with their cognate mRNA target encoded on the complementary strand (Fig. 6A). Sixteen of these Hfq-associated asRNA-target mRNA pairs are fully complementary (Fig. 6B, Table S7) and thus would normally be expected to not require the aid of Hfq. Northern blot analysis with two selected asRNAs and their putative target mRNAs, however, confirmed a clear influence of Hfq on both interaction partners.

The asRNAs C1 (3.6-fold) and AhaR_C_26 (24.8-fold) were enriched with their designated target mRNAs atu0105 (2.9-fold) and atu8023 (6.4-fold), respectively (Fig. 6C, D). C1 and atu0105 were enriched in -Vir conditions (Fig. 6C, left) and abundance was consistently higher in the Δhfq strain in all tested conditions compared to WT levels (Fig. 6C, right). The same was true for the sense RNA atu0105. AhaR_C_26 and atu8023 were enriched in +Vir conditions (Fig. 6D). Consistent with low read numbers AhaR_C_26 was barely detectable on Northern blots. Nonetheless, both AhaR_C_26 and its target mRNA atu8023 were notably less abundant in the Δhfq mutant strain at all conditions.

Discussion

Hfq, a global mediator of nutrient uptake, metabolism and motility

In this study, we combined two global approaches targeted at the identification of Hfq-affected proteins (iTRAQ) and Hfq-associated RNAs (RIP-seq) to further our understanding of the fundamental role of the RNA chaperone in Agrobacterium physiology. Our results reveal a huge Hfq regulon (Fig. 7) and extend it far beyond the eight previously identified Hfq-dependent genes and proteins in A. tumefaciens [31,38].

The abundance of a large number of ABC transporters (71), enzymes (22) and motility related proteins (13) was significantly altered in the hfq mutant. As one would expect, physiologically related proteins were consistently up or down-regulated, partly caused by coupled expression from polycistronic transcripts (Fig. S2). Most of the proteins (73%) were repressed depending on Hfq, including ABC transporters for oligopeptide (DppA), proline/glycine betaine (ProX), amino acid (EhuB), putrescine (PotF), maltose (MalE) and nopaline (NocT) import. 13 ABC transporter proteins identified in our combined study, among them MalE, NocT, Atu2422, were previously reported to be negatively regulated by AbcR1 (Fig. 7) and were therefore overrepresented in the hfq mutant [37,38] (Table S4). Whether more of the regulated ABC transporters are also controlled by AbcR1 or other sRNAs as demonstrated in various organisms [60,61,62,63,64] remains to be shown. A. tumefaciens encodes approx. 150 ABC transporters [33] and it is evident that precise regulation of these systems is necessary in the competitive rhizosphere. Unregulated permanent production of all 150 import systems would impose a costly metabolic burden [65]. Rapid translational control by Hfq-dependent sRNAs (e.g. AbcR1) helps to adjust the transporter repertoire in response to the metabolic demand and maintains bacterial fitness.

Apart from nutrient uptake, 13 proteins associated with motility were consistently down-regulated. FliF (MS-ring), FlgB (rod), FlgE (hook), Fla and FlaB (filament) are structural components of the flagellum, enabling rotation and functionality [66,67]. Misregula-

tion of motility-associated proteins upon hfq deletion in other bacteria resulted in reduced motility, independent of whether they were up-regulated (Serratia sp. ATCC 39006) or down-regulated (S. meliloti) [68,69]. Our results are in line with previously reported motility defects in the A. tumefaciens Δhfq strain [31]. Hfq-binding of the fliL mRNA as part of the flgB encoding polycistronic operon supports translational regulation of at least some motility-associated transcripts. Whether all of the affected proteins identified in our study are directly influenced by the RNA chaperone or alterations result from an indirect response to the hfq deletion, needs to be verified in further experiments.

Hfq bound mRNAs and ncRNAs – a layout of physiological state

RIP-seq of A. tumefaciens Hfq-bound RNAs from different growth phases revealed an extensive RNA binding potential of the RNA chaperone. 1697 mRNAs and 209 ncRNAs were enriched. The diversity of the mRNA and ncRNA pools varied notably between the tested conditions (Fig. 3). Under nutrient-limited conditions (minimal medium) the diversity of Hfq-enriched mRNAs and the relative amount of ncRNAs increased substantially. Condition specific binding to Hfq might give first hints about the physiological function of those ncRNAs. Hfq-mediated adaptation to stress is common in bacteria [11,12,25,59,70,71,72,73]. Riboregulation is less expensive than protein-mediated regulation since ncRNAs are shorter than most mRNAs and do not require translation [1]. This may explain the dramatic shift in the Hfq-associated mRNA:ncRNA population from 9:1 to 2:1 in response to nutrient limitation. Some regulatory sRNAs also encode small peptides, e.g. SR1 from Bacillus subtilis [74]. Determining potential protein-encoding ncRNAs identified in our study will be a challenging question for future studies.

Although E. coli Hfq was shown to bind tRNAs in vitro, our data on A. tumefaciens Hfq are consistent with reports on Salmonella enterica and S. meliloti Hfq showing specificity towards mRNAs and ncRNAs and reveal only tRNA-Gly associated with Hfq [10,59,75]. The regulatory mechanism of tRNA-Gly regulation and its physiological role are yet to be determined. It is conceivable, that tRNA-Gly enrichment is indirectly mediated by co-binding with another Hfq-interacting RNA. In Bacillus subtilis, T-Box riboswitches bind uncharged tRNAs and are also associated with Hfq [72]. So far, T-Box riboswitches were mostly known from Gram-positive bacteria, but have been recently found in Gram-negative bacteria as well [76,77]. A ternary complex of Hfq-riboswitch-tRNA could explain specific enrichment of a single tRNA species despite high homology in structure and sequence of this molecule class.

Although Hfq does not directly bind rRNAs, Hfq seems to play a role in ribosome biogenesis since we found 23 mRNAs encoding ribosomal proteins (13 L-proteins, 10 S-proteins) associated with Hfq. The rpsE mRNA was already shown to be Hfq-dependent [31], and binding of the co-transcribed rplN, rpmC, rplP, rplC and rpsC mRNAs further supports translational regulation of ribosome biogenesis by Hfq. Hfq-specific binding of mRNAs encoding ribosomal proteins seems to be widespread as it was also observed in E. coli, R. sphaeroides and S. meliloti [11,59,78]. Whether these transcripts are influenced by Hfq and/or underlie ncRNA-mediated regulation is a promising issue for future research.

Hfq binding to antisense RNAs

Among the 209 Hfq-associated ncRNAs identified by RIP-Seq, 56 ncRNAs were not found in previous studies [35,36]. They add to the growing list of A. tumefaciens ncRNAs and extend it to a total of 677. It seems that Hfq plays an important role in antisense regulation since 115 asRNAs were specifically enriched by the RNA chaperone. Strikingly, 21 asRNAs were enriched along with their cognate target mRNA derived from the complementary strand and we confirmed Hfq-dependent regulation of two asRNA-mRNA pairs (Fig. 6). asRNAs have rarely been found associated with Hfq [9,11,12,72]. This and the fact that asRNAs and target mRNAs are transcribed in close spacial proximity and by definition share perfect sequence complementarity, led to a plausible model concluding that Hfq is dispensable for cis-encoded antisense RNA regulation [13]. Although this may be true for many bacteria, asRNA-regulation was already described for the *E. coli* Tn*10*/IS*10* system (RNA-IN, RNA-OUT) [79], and genomic SELEX identified preferential binding of Hfq to AU-rich sequences antisense of protein coding genes [80]. Additional Hfq coIP experiments revealed binding of 67 asRNAs to the RNA chaperone [81]. Our study in concert with a recent Hfq-coIP study in *S. meliloti* [59], adds further evidence to the importance of Hfq-associated antisense transcripts in translational regulation.

Hfq contributes to *A. tumefaciens* virulence

An *A. tumefaciens hfq* mutant is severely impaired in tumor formation on plants [31]. Similar virulence defects have been reported for several other pathogens [32,73,82,83,84,85,86]. RIP-seq identified 296 mRNAs and 31 ncRNAs specifically enriched by Hfq under virulence-induced (+Vir) conditions in *A. tumefaciens*. Importantly, virulence-related mRNAs of the *virB*, *virC* and *virE* operons were associated with Hfq. Among those, the *virB1* (5,467 RPM, 3-fold) and *virE3* (13,770 RPM, 6.7-fold) mRNAs were enriched most explicitly. The *virB* and *virE* mRNAs are among the most highly induced transcripts under virulence conditions [87]. Yet, protein abundance of the *virB* operon encoded VirB2, B5, B8 and B9 proteins was not significantly affected in the *hfq* mutant [31] and Northern blot analysis with a *virE3* specific probe did not reveal obvious changes in RNA amounts in the Δ*hfq* strain as compared to WT levels. The identification of 31 ncRNAs specifically enriched under virulence conditions suggests a substantial regulatory potential. Further, regulation of the TraR anti-activator TraM (identified by iTRAQ) might also contribute to efficient infection. TraM (Atu6131) was down-regulated in absence of Hfq, while the conjugal transfer coupling protein TraG (Atu6124) was up-regulated. TraR is a transcriptional regulator and a key protein in replication and conjugation of the Ti-plasmid, directly linked to quorum-sensing and virulence [88]. TraR activates expression of the *tra*-operon including *traG*. In absence of Hfq, TraM amounts decrease, releasing TraR inhibition and resulting in an increase of TraG. Therefore, by influencing TraM, Hfq might contribute to Ti-plasmid replication and conjugation, thus modulating infection efficiency.

The *A. tumefaciens* Hfq core regulon

Our study places Hfq in the center of a complex posttranscriptional network that has a profound impact on the *A. tumefaciens* transcriptome and proteome. In order to visualize putative connections between Hfq-dependent proteins, we used the String 9.1 webserver [55] to predict physical and physiological interactions of the 241 proteins with known (or presumed) Hfq-dependent regulation highlighted in Fig. 7 (underlined). By this, we assembled a comprehensive network connecting 197 proteins (44 proteins did not connect to the main regulon) (Fig. 8). Since the network includes 6 asRNA targets, we manually included the corresponding asRNAs and the global regulator AbcR1 (black) with validated (continuous line) or predicted (dashed line) regulation of their

targets. Interconnection of most of the Hfq-dependent proteins highlights efficient regulation of whole physiological circuits (blue shaded). The impact of Hfq on the *A. tumefaciens* proteome already indicated uniform regulation of polycistronic operons. Strikingly, assembly of the Hfq-regulon demonstrates regulation of nutrient uptake and motility beyond influence on single proteins or transport systems. Mainly ABC transporters (II, III, IV, V, VI, VIII) but also motility and chemotaxis related proteins (I, VII) are integrated into a complex intertwined network. Multilayered regulation is further supported by two findings. First, 34 of the Hfq-enriched mRNAs encode transcriptional regulators, typically controlling transcription of multiple genes. Second, Hfq binds its own mRNA suggesting auto-regulatory control as in *E. coli* or *S. meliloti* (Fig. S3) [11,89,90].

In summary, our data show a fundamental role of Hfq in the genome-wide regulation of nutrient uptake, metabolism and motility in *A. tumefaciens*. Hfq-mediated riboregulation is not restricted to partially complementary sRNA-mRNA interactions but includes interactions between fully complementary antisense and messenger RNAs. More than 30 ncRNAs were associated with Hfq under virulence conditions suggesting that the RNA chaperone may be of prime importance in bacteria-plant interaction.

Supporting Information

Figure S1   Categorization of transcripts for RIP-seq quantification.

Figure S2   Operon predictions.

Figure S3   Housekeeping RNAs were not enriched by Hfq^{3xFlag}.

Figure S4   *hfq* mRNA bound by Hfq.

Figure S5   Relative size of the *A. tumefaciens* replicons.

Table S1   Bacterial strains and plasmids used in this study.

Table S2   Oligonucleotides used in this study.

Table S3   Mapping statistics from RNA sequencing data analysis.

Table S4   Proteins differentially abundant in *A. tumefaciens* Δ*hfq* (iTRAQ).

Table S5   mRNAs enriched by Hfq^{3xFlag}.

Table S6   ncRNAs enriched by Hfq^{3xFlag}.

Table S7   asRNA and mRNA target-enrichment by Hfq^{3xFlag}.

Acknowledgments

We are grateful to Christian Baron (Montreal) for VirB-specific antisera. We thank Jer-Sheng Lin, Chih-Feng Wu and the Proteomics Core Lab of the Institute of Plant and Microbial Biology for assistance in sample preparation for iTRAQ analysis and the Academia Sinica Common Mass Spectrometry Facilities for acquiring MS data on the LTQ-Orbitrap Elite. The authors acknowledge Sina Langklotz and Julia Bandow for insightful discussions on mass spectrometry data. Roman Moser and Johanna Roßmanith are acknowledged for critical reading of the manuscript.

Author Contributions

Conceived and designed the experiments: PM FN. Performed the experiments: PM AO. Analyzed the data: PM KUF FN. Contributed reagents/materials/analysis tools: KUF TNW CMS EML FN. Contributed to the writing of the manuscript: PM FN.

References

1. Waters LS, Storz G (2009) Regulatory RNAs in bacteria. Cell 136: 615–628.
2. Gottesman S, Storz G (2010) Bacterial small RNA regulators: versatile roles and rapidly evolving variations. Cold Spring Harb Perspect Biol.
3. Majdalani N, Cunning C, Sledjeski D, Elliott T, Gottesman S (1998) DsrA RNA regulates translation of RpoS message by an anti-antisense mechanism, independent of its action as an antisilencer of transcription. Proc Natl Acad Sci USA 95: 12462–12467.
4. Soper TJ, Woodson SA (2008) The rpoS mRNA leader recruits Hfq to facilitate annealing with DsrA sRNA. Rna 14: 1907–1917.
5. Saramago M, Barria C, Dos Santos RF, Silva IJ, Pobre V, et al. (2014) The role of RNases in the regulation of small RNAs. Curr Opin Microbiol 18C: 105–115.
6. Afonyushkin T, Vecerek B, Moll I, Bläsi U, Kaberdin VR (2005) Both RNase E and RNase III control the stability of sodB mRNA upon translational inhibition by the small regulatory RNA RyhB. Nucleic Acids Res 33: 1678–1689.
7. Bandyra KJ, Said N, Pfeiffer V, Gorna MW, Vogel J, et al. (2012) The seed region of a small RNA drives the controlled destruction of the target mRNA by the endoribonuclease RNase E. Mol Cell 47: 943–953.
8. Viegas SC, Silva IJ, Saramago M, Domingues S, Arraiano CM (2011) Regulation of the small regulatory RNA MicA by ribonuclease III: a target-dependent pathway. Nucleic Acids Res 39: 2918–2930.
9. Sittka A, Lucchini S, Papenfort K, Sharma CM, Rolle K, et al. (2008) Deep sequencing analysis of small noncoding RNA and mRNA targets of the global post-transcriptional regulator, Hfq. PLoS Genet 4: e1000163.
10. Sittka A, Sharma CM, Rolle K, Vogel J (2009) Deep sequencing of Salmonella RNA associated with heterologous Hfq proteins in vivo reveals small RNAs as a major target class and identifies RNA processing phenotypes. RNA Biol 6: 266–275.
11. Berghoff BA, Glaeser J, Sharma CM, Zobawa M, Lottspeich F, et al. (2011) Contribution of Hfq to photooxidative stress resistance and global regulation in Rhodobacter sphaeroides. Mol Microbiol 80: 1479–1495.
12. Chao Y, Papenfort K, Reinhardt R, Sharma CM, Vogel J (2012) An atlas of Hfq-bound transcripts reveals 3′ UTRs as a genomic reservoir of regulatory small RNAs. EMBO J 31: 4005–4019.
13. Georg J, Hess WR (2011) cis-antisense RNA, another level of gene regulation in bacteria. Microbiol Mol Biol Rev 75: 286–300.
14. Vogel J, Luisi BF (2011) Hfq and its constellation of RNA. Nat Rev Microbiol 9: 578–589.
15. Franze de Fernandez MT, Hayward WS, August JT (1972) Bacterial proteins required for replication of phage Q ribonucleic acid. Purification and properties of host factor I, a ribonucleic acid-binding protein. J Biol Chem 247: 824–831.
16. Wagner EG (2013) Cycling of RNAs on Hfq. RNA Biol 10: 619–626.
17. Sauer E (2013) Structure and RNA-binding properties of the bacterial LSm protein Hfq. RNA Biol 10: 610–618.
18. Olejniczak M (2011) Despite similar binding to the Hfq protein regulatory RNAs widely differ in their competition performance. Biochemistry 50: 4427–4440.
19. Link TM, Valentin-Hansen P, Brennan RG (2009) Structure of Escherichia coli Hfq bound to polyriboadenylate RNA. Proc Natl Acad Sci U S A 106: 19292–19297.
20. Sauer E, Schmidt S, Weichenrieder O (2012) Small RNA binding to the lateral surface of Hfq hexamers and structural rearrangements upon mRNA target recognition. Proc Natl Acad Sci U S A 109: 9396–9401.
21. Zhang A, Wassarman KM, Ortega J, Steven AC, Storz G (2002) The Sm-like Hfq protein increases OxyS RNA interaction with target mRNAs. Mol Cell 9: 11–22.
22. Mohanty BK, Maples VF, Kushner SR (2004) The Sm-like protein Hfq regulates polyadenylation dependent mRNA decay in Escherichia coli. Mol Microbiol 54: 905–920.
23. Folichon M, Allemand F, Regnier P, Hajnsdorf E (2005) Stimulation of poly(A) synthesis by Escherichia coli poly(A)polymerase I is correlated with Hfq binding to poly(A) tails. FEBS J 272: 454–463.
24. Worrall JA, Gorna M, Crump NT, Phillips LG, Tuck AC, et al. (2008) Reconstitution and analysis of the multienzyme Escherichia coli RNA degradosome. J Mol Biol 382: 870–883.
25. Robertson GT, Roop RM, Jr. (1999) The Brucella abortus host factor I (HF-I) protein contributes to stress resistance during stationary phase and is a major determinant of virulence in mice. Mol Microbiol 34: 690–700.
26. Barra-Bily L, Fontenelle C, Jan G, Flechard M, Trautwetter A, et al. (2010) Proteomic alterations explain phenotypic changes in Sinorhizobium meliloti lacking the RNA chaperone Hfq. J Bacteriol 192: 1719–1729.
27. Barra-Bily L, Pandey SP, Trautwetter A, Blanco C, Walker GC (2010) The Sinorhizobium meliloti RNA chaperone Hfq mediates symbiosis of S. meliloti and alfalfa. J Bacteriol 192: 1710–1718.
28. Gao M, Barnett MJ, Long SR, Teplitski M (2010) Role of the Sinorhizobium meliloti global regulator Hfq in gene regulation and symbiosis. Mol Plant Microbe Interact 23: 355–365.
29. Torres-Quesada O, Oruezabal RI, Peregrina A, Jofre E, Lloret J, et al. (2010) The Sinorhizobium meliloti RNA chaperone Hfq influences central carbon metabolism and the symbiotic interaction with alfalfa. BMC Microbiol 10: 71.
30. Mulley G, White JP, Karunakaran R, Prell J, Bourdes A, et al. (2011) Mutation of GOGAT prevents pea bacteroid formation and N2 fixation by globally downregulating transport of organic nitrogen sources. Mol Microbiol 80: 149–167.
31. Wilms I, Möller P, Stock AM, Gurski R, Lai EM, et al. (2012) Hfq influences multiple transport systems and virulence in the plant pathogen Agrobacterium tumefaciens. J Bacteriol 194: 5209–5217.
32. Chao Y, Vogel J (2010) The role of Hfq in bacterial pathogens. Curr Opin Microbiol 13: 24–33.
33. Wood DW, Setubal JC, Kaul R, Monks DE, Kitajima JP, et al. (2001) The genome of the natural genetic engineer Agrobacterium tumefaciens C58. Science 294: 2317–2323.
34. Pitzschke A, Hirt H (2010) New insights into an old story: Agrobacterium-induced tumour formation in plants by plant transformation. EMBO J 29: 1021–1032.
35. Wilms I, Overlöper A, Nowrousian M, Sharma CM, Narberhaus F (2012) Deep sequencing uncovers numerous small RNAs on all four replicons of the plant pathogen Agrobacterium tumefaciens. RNA Biol 9.
36. Lee K, Huang X, Yang C, Lee D, Ho V, et al. (2013) A genome-wide survey of highly expressed non-coding RNAs and biological validation of selected candidates in Agrobacterium tumefaciens. PLoS ONE 8: e70720.
37. Wilms I, Voss B, Hess WR, Leichert LI, Narberhaus F (2011) Small RNA-mediated control of the Agrobacterium tumefaciens GABA binding protein. Mol Microbiol 80: 492–506.
38. Overlöper A, Kraus A, Gurski R, Wright PR, Georg J, et al. (2014) Two separate modules of the conserved regulatory RNA AbcR1 address multiple target mRNAs in and outside of the translation initiation region. RNA Biol 11.
39. Schmidt-Eisenlohr H, Domke N, Angerer C, Wanner G, Zambryski PC, et al. (1999) Vir proteins stabilize VirB5 and mediate its association with the T pilus of Agrobacterium tumefaciens. J Bacteriol 181: 7485–7492.
40. Sambrook JF, Russel DW (2001) Molecular Cloning: A Laboratory Manual. Cold Spring Harbor Laboratory Press 3rd edn.
41. Katz S, Ron EZ (2008) Dual role of FtsH in regulating lipopolysaccharide biosynthesis in Escherichia coli. J Bacteriol 190: 7117–7122.
42. Schäfer A, Tauch A, Jager W, Kalinowski J, Thierbach G, et al. (1994) Small mobilizable multi-purpose cloning vectors derived from the Escherichia coli plasmids pK18 and pK19: selection of defined deletions in the chromosome of Corynebacterium glutamicum. Gene 145: 69–73.
43. Aiba H, Adhya S, de Crombrugghe B (1981) Evidence for two functional gal promoters in intact Escherichia coli cells. J Biol Chem 256: 11905–11910.
44. Pfeiffer V, Sittka A, Tomer R, Tedin K, Brinkmann V, et al. (2007) A small non-coding RNA of the invasion gene island (SPI-1) represses outer membrane protein synthesis from the Salmonella core genome. Mol Microbiol 66: 1174–1191.
45. Edgar R, Domrachev M, Lash AE (2002) Gene Expression Omnibus: NCBI gene expression and hybridization array data repository. Nucleic Acids Res 30: 207–210.
46. Hoffmann S, Otto C, Kurtz S, Sharma CM, Khaitovich P, et al. (2009) Fast mapping of short sequences with mismatches, insertions and deletions using index structures. PLoS Comput Biol 5: e1000502.
47. Nicol JW, Helt GA, Blanchard SG, Jr., Raja A, Loraine AE (2009) The Integrated Genome Browser: free software for distribution and exploration of genome-scale datasets. Bioinformatics 25: 2730–2731.
48. Schlüter JP, Reinkensmeier J, Barnett MJ, Lang C, Krol E, et al. (2013) Global mapping of transcription start sites and promoter motifs in the symbiotic alpha-proteobacterium Sinorhizobium meliloti 1021. BMC Genomics 14: 156.
49. Lan P, Li W, Wen TN, Shiau JY, Wu YC, et al. (2011) iTRAQ protein profile analysis of Arabidopsis roots reveals new aspects critical for iron homeostasis. Plant Physiol 155: 821–834.

50. Cox J, Mann M (2008) MaxQuant enables high peptide identification rates, individualized p.p.b.-range mass accuracies and proteome-wide protein quantification. Nat Biotechnol 26: 1367–1372.

51. Price MN, Huang KH, Alm EJ, Arkin AP (2005) A novel method for accurate operon predictions in all sequenced prokaryotes. Nucleic Acids Res 33: 880–892.

52. Kanehisa M, Goto S (2000) KEGG: kyoto encyclopedia of genes and genomes. Nucleic Acids Res 28: 27–30.

53. Kanehisa M, Goto S, Sato Y, Kawashima M, Furumichi M, et al. (2014) Data, information, knowledge and principle: back to metabolism in KEGG. Nucleic Acids Res 42: D199–205.

54. Agresti A (1992) A survey of exact inference for contingency tables. Statistical Science Vol. 7: 131–153.

55. Franceschini A, Szklarczyk D, Frankild S, Kuhn M, Simonovic M, et al. (2013) STRING v9.1: protein-protein interaction networks, with increased coverage and integration. Nucleic Acids Res 41: D808–815.

56. Cline MS, Smoot M, Cerami E, Kuchinsky A, Landys N, et al. (2007) Integration of biological networks and gene expression data using Cytoscape. Nat Protoc 2: 2366–2382.

57. Saito R, Smoot ME, Ono K, Ruscheinski J, Wang PL, et al. (2012) A travel guide to Cytoscape plugins. Nat Methods 9: 1069–1076.

58. Sittka A, Pfeiffer V, Tedin K, Vogel J (2007) The RNA chaperone Hfq is essential for the virulence of Salmonella typhimurium. Molecular Microbiology 63: 193–217.

59. Torres-Quesada O, Reinkensmeier J, Schlüter JP, Robledo M, Peregrina A, et al. (2014) Genome-wide profiling of Hfq-binding RNAs uncovers extensive post-transcriptional rewiring of major stress response and symbiotic regulons in Sinorhizobium meliloti. RNA Biol 11.

60. Sharma CM, Darfeuille F, Plantinga TH, Vogel J (2007) A small RNA regulates multiple ABC transporter mRNAs by targeting C/A-rich elements inside and upstream of ribosome-binding sites. Genes Dev 21: 2804–2817.

61. Sharma CM, Papenfort K, Pernitzsch SR, Mollenkopf HJ, Hinton JCD, et al. (2011) Pervasive post-transcriptional control of genes involved in amino acid metabolism by the Hfq-dependent GcvB small RNA. Molecular Microbiology 81: 1144–1165.

62. Urbanowski ML, Stauffer LT, Stauffer GV (2000) The gcvB gene encodes a small untranslated RNA involved in expression of the dipeptide and oligopeptide transport systems in Escherichia coli. Mol Microbiol 37: 856–868.

63. Antal M, Bordeau V, Douchin V, Felden B (2005) A small bacterial RNA regulates a putative ABC transporter. J Biol Chem 280: 7901–7908.

64. Bohn C, Rigoulay C, Chabelskaya S, Sharma CM, Marchais A, et al. (2010) Experimental discovery of small RNAs in Staphylococcus aureus reveals a riboregulator of central metabolism. Nucleic Acids Res 38: 6620–6636.

65. Akashi H, Gojobori T (2002) Metabolic efficiency and amino acid composition in the proteomes of Escherichia coli and Bacillus subtilis. Proc Natl Acad Sci U S A 99: 3695–3700.

66. Morimoto YV, Ito M, Hiraoka KD, Che YS, Bai F, et al. (2014) Assembly and stoichiometry of FliF and FlhA in Salmonella flagellar basal body. Mol Microbiol 91: 1214–1226.

67. Saijo-Hamano Y, Uchida N, Namba K, Oosawa K (2004) In vitro characterization of FlgB, FlgC, FlgF, FlgG, and FliE, flagellar basal body proteins of Salmonella. J Mol Biol 339: 423–435.

68. Wilf NM, Reid AJ, Ramsay JP, Williamson NR, Croucher NJ, et al. (2013) RNA-seq reveals the RNA binding proteins, Hfq and RsmA, play various roles in virulence, antibiotic production and genomic flux in Serratia sp. ATCC 39006. BMC Genomics 14: 822.

69. Sobrero P, Schlüter JP, Lanner U, Schlosser A, Becker A, et al. (2012) Quantitative proteomic analysis of the Hfq-regulon in Sinorhizobium meliloti 2011. PLoS ONE 7: e48494.

70. Guisbert E, Rhodius VA, Ahuja N, Witkin E, Gross CA (2007) Hfq modulates the sigmaE-mediated envelope stress response and the sigma32-mediated cytoplasmic stress response in Escherichia coli. J Bacteriol 189: 1963–1973.

71. Cui M, Wang T, Xu J, Ke Y, Du X, et al. (2013) Impact of Hfq on global gene expression and intracellular survival in Brucella melitensis. PLoS One 8: e71933.

72. Dambach M, Irnov I, Winkler WC (2013) Association of RNAs with Bacillus subtilis Hfq. PLoS One 8: e55156.

73. Kakoschke T, Kakoschke S, Magistro G, Schubert S, Borath M, et al. (2014) The RNA chaperone Hfq impacts growth, metabolism and production of virulence factors in Yersinia enterocolitica. PLoS One 9: e86113.

74. Gimpel M, Heidrich N, Mader U, Krugel H, Brantl S (2010) A dual-function sRNA from B-subtilis: SR1 acts as a peptide encoding mRNA on the gapA operon. Molecular Microbiology 76: 990–1009.

75. Lee T, Feig AL (2008) The RNA binding protein Hfq interacts specifically with tRNAs. Rna 14: 514–523.

76. Vitreschak AG, Mironov AA, Lyubetsky VA, Gelfand MS (2008) Comparative genomic analysis of T-box regulatory systems in bacteria. Rna-a Publication of the Rna Society 14: 717–735.

77. Saad NY, Schiel B, Braye M, Heap JT, Minton NP, et al. (2012) Riboswitch (T-box)-mediated control of tRNA-dependent amidation in Clostridium acetobutylicum rationalizes gene and pathway redundancy for asparagine and asparaginyl-tRNAasn synthesis. J Biol Chem 287: 20382–20394.

78. Zhang A, Wassarman KM, Rosenow C, Tjaden BC, Storz G, et al. (2003) Global analysis of small RNA and mRNA targets of Hfq. Mol Microbiol 50: 1111–1124.

79. Ross JA, Ellis MJ, Hossain S, Haniford DB (2013) Hfq restructures RNA-IN and RNA-OUT and facilitates antisense pairing in the Tn10/IS10 system. Rna 19: 670–684.

80. Lorenz C, Gesell T, Zimmermann B, Schoeberl U, Bilusic I, et al. (2010) Genomic SELEX for Hfq-binding RNAs identifies genomic aptamers predominantly in antisense transcripts. Nucleic Acids Research 38: 3794–3808.

81. Bilusic I, Popitsch N, Rescheneder P, Schroeder R, Lybecker M (2014) Revisiting the coding potential of the E. coli genome through Hfq co-immunoprecipitation. RNA Biol 11.

82. Michaux C, Hartke A, Martini C, Reiss S, Albrecht D, et al. (2014) Involvement of Enterococcus faecalis sRNAs in the stress response and virulence. Infect Immun.

83. Hu YH, Li YX, Sun L (2014) Edwardsiella tarda Hfq: impact on host infection and global protein expression. Vet Res 45: 23.

84. Zeng Q, McNally RR, Sundin GW (2013) Global small RNA chaperone Hfq and regulatory small RNAs are important virulence regulators in Erwinia amylovora. J Bacteriol 195: 1706–1717.

85. Zeng Q, Sundin GW (2014) Genome-wide identification of Hfq-regulated small RNAs in the fire blight pathogen Erwinia amylovora discovered small RNAs with virulence regulatory function. BMC Genomics 15: 414.

86. Schmidtke C, Abendroth U, Brock J, Serrania J, Becker A, et al. (2013) Small RNA sX13: a multifaceted regulator of virulence in the plant pathogen Xanthomonas. PLoS Pathog 9: e1003626.

87. Klüsener S, Hacker S, Tsai YL, Bandow JE, Gust R, et al. (2010) Proteomic and transcriptomic characterization of a virulence-deficient phosphatidylcholine-negative Agrobacterium tumefaciens mutant. Mol Genet Genomics 283: 575–589.

88. Costa ED, Chai Y, Winans SC (2012) The quorum-sensing protein TraR of Agrobacterium tumefaciens is susceptible to intrinsic and TraM-mediated proteolytic instability. Mol Microbiol 84: 807–815.

89. Sobrero P, Valverde C (2011) Evidences of autoregulation of hfq expression in Sinorhizobium meliloti strain 2011. Arch Microbiol 193: 629–639.

90. Vecerek B, Moll I, Bläsi U (2005) Translational autocontrol of the Escherichia coli hfq RNA chaperone gene. RNA 11: 976–984.

Cloning of Insertion Site Flanking Sequence and Construction of Transfer DNA Insert Mutant Library in *Stylosanthes Colletotrichum*

Helong Chen[1,⊚], **Caiping Hu**[1,⊚], **Kexian Yi**[1,2*,⊚], **Guixiu Huang**[2], **Jianming Gao**[1], **Shiqing Zhang**[1], **Jinlong Zheng**[2], **Qiaolian Liu**[1], **Jingen Xi**[2]

1 Institute of Tropical Bioscience and Biotechnology, Key Laboratory of Tropical Crop Biotechnology, Ministry of Agriculture, Chinese Academy of Tropical Agricultural Sciences, Haikou, China, 2 Environment and Plant Protection Institute, Key Laboratory of Integrated Pest Management on Tropical Crops, Ministry of Agriculture, Chinese Academy of Tropical Agricultural Sciences, Haikou, China

Abstract

Stylosanthes sp. is the most important forage legume in tropical areas worldwide. *Stylosanthes* anthracnose, which is mainly caused by *Colletotrichum gloeosporioides*, is a globally severe disease in stylo production. Little progress has been made in anthracnose molecular pathogenesis research. In this study, *Agrobacterium tumefaciens*-mediated transformation was used to transform *Stylosanthes colletotrichum* strain CH008. The major factors of the genetic transformation system of *S. colletotrichum* were optimized as follows: *A. tumefaciens*' AGL-1 concentration (OD_{600}), 0.8; concentration of *Colletotrichum* conidium, 1×10^6 conidia/mL; acetosyringone concentration, 100 mmol/L; induction time, 6 h; co-culture temperature, 25°C; and co-culture time, 3 d. Thus, the transformation efficiency was increased to 300–400 transformants per 106 conidia. Based on the optimized system, a mutant library containing 4616 mutants was constructed, from which some mutants were randomly selected for analysis. Results show that the mutants were single copies that could be stably inherited. The growth rate, spore amount, spore germination rate, and appressorium formation rate in some mutants were significantly different from those in the wild-type strain. We then selected the most appropriate method for the preliminary screening and re-screening of each mutant's pathogenic defects. We selected 1230 transformants, and obtained 23 strains with pathogenic defects, namely, 18 strains with reduced pathogenicity and five strains with lost pathogenicity. Thermal asymmetric interlaced PCR was used to identify the transfer DNA (T-DNA) integration site in the mutant that was coded 2430, and a sequence of 476 bp was obtained. The flanking sequence of T-DNA was compared with the *Colletotrichum* genome by BLAST, and a sequence of 401 bp was found in Contig464 of the *Colletotrichum* genome. By predicting the function of the flanking sequence, we discovered that T-DNA insertion in the promoter region of the putative gene had 79% homology with the aspartate aminotransferase gene in *Magnaporthe oryzae* (XP_003719674.1).

Editor: Brian Stevenson, University of Kentucky College of Medicine, United States of America

Funding: We thank the National Natural Science Foundation of China (NSFC, No. 31072076), Special Fund for Agro-scientific Research in the Public Interest (201303057) and The Major Technology Project of Hainan (ZDZX2013023-1) for funding this research. The funders had no role in study design, data collection and analysis, decision to publish, or preparation of the manuscript.

Competing Interests: The authors have declared that no competing interests exist.

* Email: yikexian@126.com

⊚ These authors contributed equally to this work.

Introduction

Stylosanthes guianensis, a diverse tropical and subtropical forage legume, is native to South America, Central America, and Africa. It is used for grazing cattle and raising livestock. Species of *Stylosanthes* are used for soil improvement through nitrogen fixation, reclaiming degraded wastelands, and water and soil conservation [1–3].

Introduction of *Stylosanthes* sp. to China from Australia, Africa, and South America began in the late 1960s and has continued to the present. *Stylosanthes* sp. is principally grown in Hainan and Guangdong Provinces as an annual crop for cut-and-carry forage, leaf meal, and hay [4].

Anthracnose of *Stylosanthes*, mainly caused by *Colletotrichum gloeosporioides* (Penz.) Penz. & Sacc., has been the most significant biotic factor limiting the production, persistence, and utilization of *Stylosanthes* in several countries [5]. The fungus initially infects leaves via an appressorium that develops from the germinating spore on the plant surface, followed by turgor-driven penetration of the cuticle. Fungal colonization on the leaf tissue follows and is associated with host cell necrosis, leading to a blight-like symptom and the formation of spore masses as acervuli [6,7]. Research in Australia, Colombia, Brazil, and China has identified two biotypes of *C. gloeosporioides* infecting *Stylosanthes* sp. [8–13]. Similarly, biotypes A and B and putative biotype C from Africa have been described [14]. The diversity among strains pathogenic on *Stylosanthes* and their relationship with other strains were analyzed

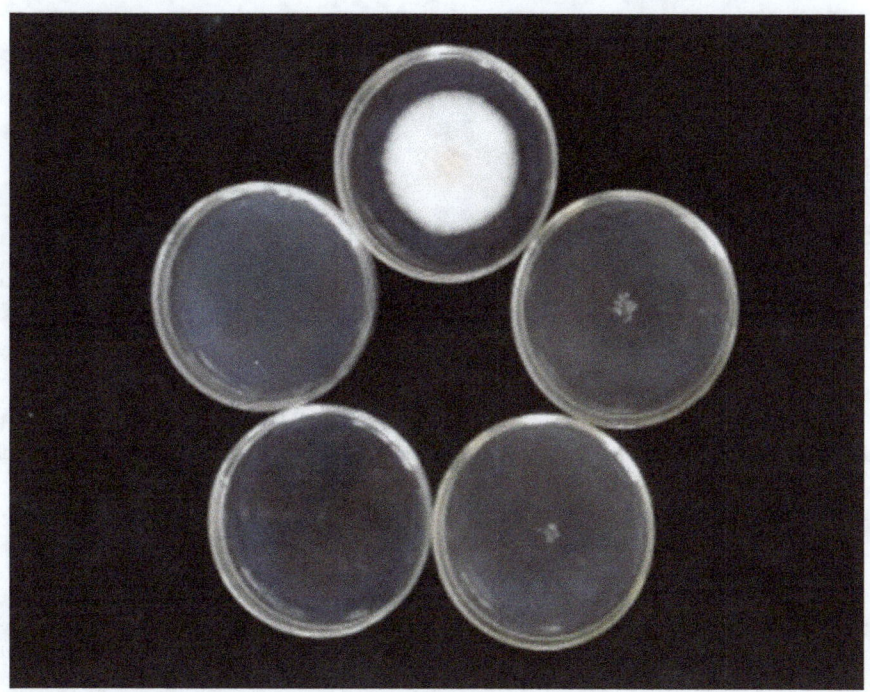

Figure 1. Tolerance test of a wild-type strain on medium containing concentration gradients of chlorimuronethyl. Note (from the top, clockwise): DCM plates (concentrations of 0, 5, 10, 20, and 30 μg/mL); cultivation was carried out at 28°C for 4 d.

at the molecular level using various markers, such as dsRNA [15], RFLP [16,17], RAPD [14,18–20], and ITS [21]. The diversity among the pathogen population from Brazil, Colombia, China, and India is extensive [19]. *Agrobacterium tumefaciens*-mediated transformation (ATMT) has been used to identify mutants of *C. gloeosporioides* impaired in pathogenicity to gain more insight into the molecular mechanisms of *C. gloeosporioides* pathogenesis [22].

ATMT is a suitable and efficient technique for insertion mutagenesis, genetic mapping, and related research in filamentous fungi [23]. ATMT has been used to transform over 50 different fungal species since it was first reported [24]. The advantages of ATMT are as follows: first, *A. tumefaciens* directly transforms fungal spores, hyphae, or tissues without protoplast preparation; second, the integration of transfer DNA (T-DNA) into the chromosome is random and generally involves a single copy,

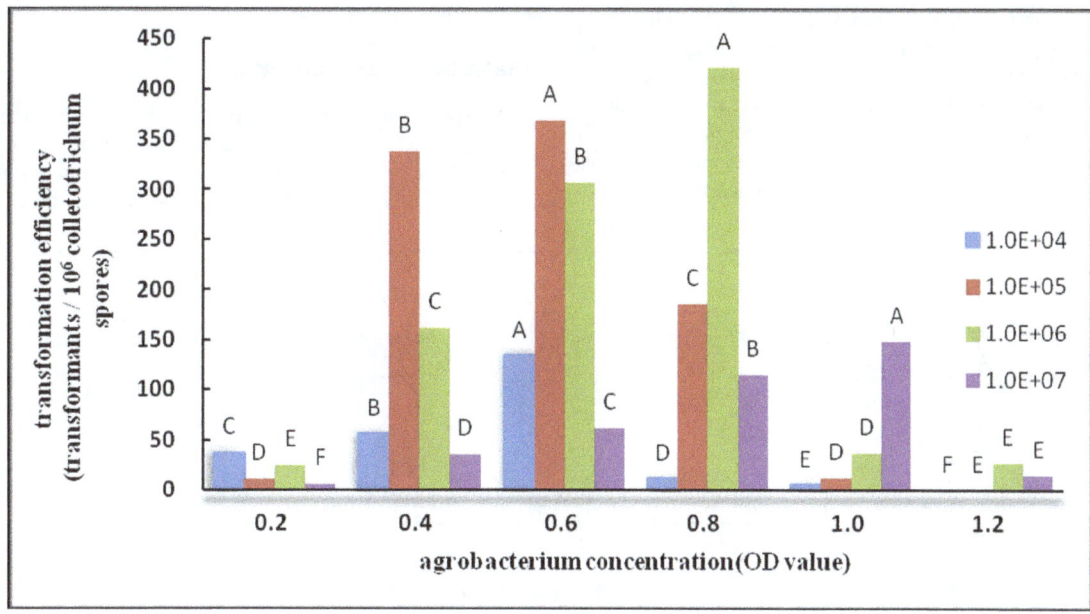

Figure 2. Effect of *Agrobacterium* concentration on transformation efficiency. Note: In the same series (four series), the same capital letters indicate no significant difference (p>0.01), whereas different capital letters indicate significant differences (p<0.01).

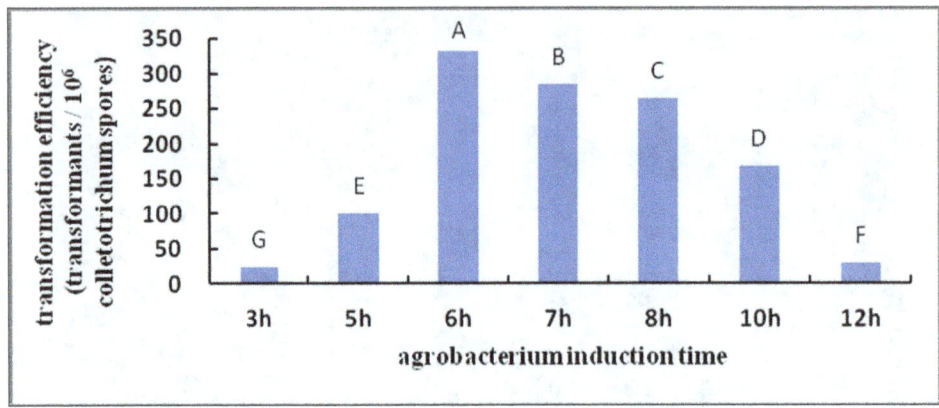

Figure 3. Effect of induction time on transformation efficiency. Note: Different capital letters indicate significant differences (p<0.01).

which can easily isolate and identify the insertion locus; third, ATMT is competent for the transformation of high-molecular-weight exogenous DNA [25,26]. *Agrobacterium*-mediated T-DNA tagging has been developed as a powerful tool for both random and targeted gene disruption; it is increasingly being regarded as the system of choice for many fungi [24]. *Agrobacterium*-mediated T-DNA tagging is a high-throughput system for identifying and analyzing novel genes [27–29], and the key for its success is the discovery of T-DNA-inserted mutants with altered phenotypes.

Traditional control measures for *Stylosanthes* anthracnose mainly involve chemical prevention and agricultural measure control. The use of pesticide produces the most direct effect, but it may easily cause a series of problems, such as pollution of the ecological environment. Theoretically, the most cost-effective and efficient control method is the cultivation of disease-resistant *S. guianensis* varieties. However, this method requires intensive manpower, material resources, and time. A single disease-resistant variety of *S. guianensis* cannot overcome the diversity and variability of *Stylosanthes Colletotrichum*. The resistance of cultivated disease-resistant varieties can only keep about 5a because *Colletotrichum* suffers from variation easily. Therefore, exploration on pathogenesis and causes for variation in *Stylosanthes colletotrichum* at the levels of molecular biology and functional genomics is of scientific significance and application value to effectively cultivate new varieties of *S. guianensis* with long-term disease resistance, and formulate permanent strategies for the reasonable control of *Stylosanthes* anthracnose.

Based on the constructed anthracnose genetic transformation system of *S. guianensis*, this study aimed to clone genes related to pathopoiesia of pathogenic bacteria. In this study, we utilized and inserted *A. tumefaciens*-mediated T-DNA, which contains an anti-chlorimuronethyl gene, into genes of *Stylosanthes colletotrichum gloeosporiodes* Penz strain CH008 with strong pathogenicity to generate insertion mutations, construct a library of mutants from *Stylosanthes colletotrichum gloeosporiodes* Penz strain CH008, and provide many mutation materials for future studies on functional genes analysis. Based on the selection of mutants related to pathogenicity in the mutant library, PCR and Southern blot were used for molecular verification. Flanking sequences of T-DNA insertion in virulence genes were obtained via thermal asymmetric interlaced PCR (TAIL-PCR). However, sequencing work on the whole genome sequence of *S. colletotrichum* remains unfinished. To predict information such as related gene functions of pathopoiesia, we compared the flanking sequences with the *Colletotrichum* gene libraries of known sequences, and used BLAST to analyze the homologous sequences. The results of this study could provide a basis for further investigations on functions of disrupted genes.

Materials and Methods

No specific permission was required for the sampling locations of this study. Moreover, ethical approval for this study was not required because we did not handle or collect animals involved in any animal welfare regulations, and no endangered or protected

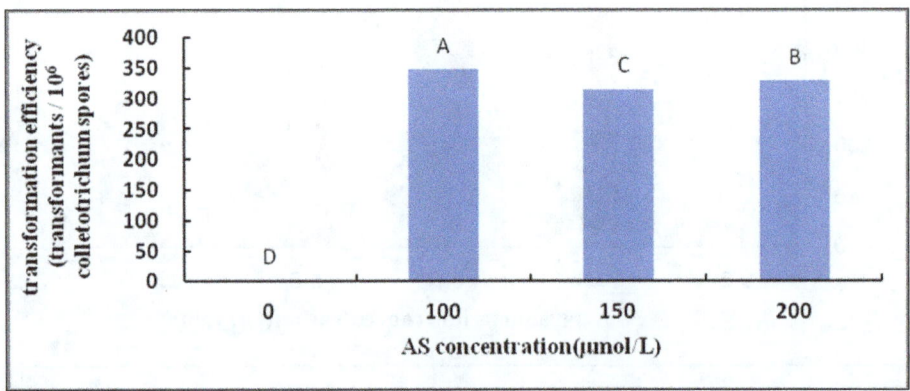

Figure 4. Effect of AS concentrations on transformation efficiency. Note: Different capital letters indicate significant differences (p<0.01).

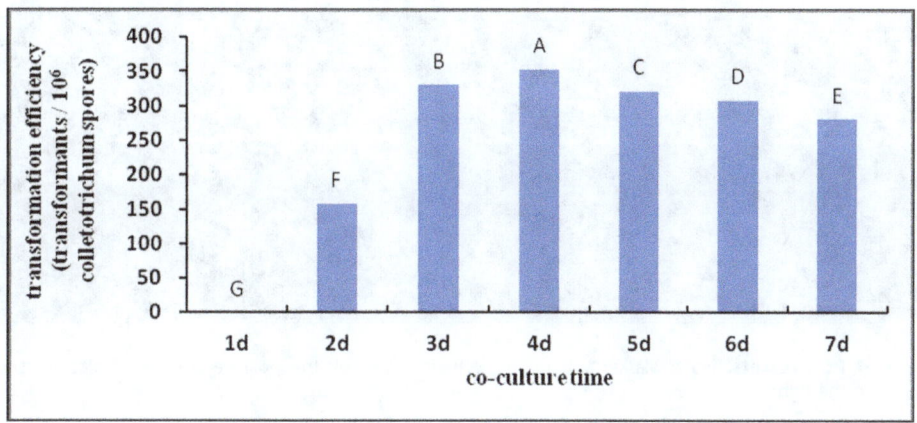

Figure 5. Effect of co-culture time on transformation efficiency. Note: Different capital letters indicate significant differences (p<0.01).

species were used as samples in the experiments. The study was conducted in Institute of Tropical Bioscience and Biotechnology, Chinese Academy of Tropical Agricultural Sciences (CATAS), Haikou, Hainan Province.

Materials

Bacterial strains and plasmid for tests. The bacterial strain used was *A. tumefaciens* AGL-1. The bacterium included a pSULF.gfp plasmid, which uses pCAMBIAl300 as a framework and contains the ILV1 gene of chlorimuronethyl resistance marker and binary carriers of reporter gene GFP. The strain was constructed by the Sainsbury Laboratory of John Innes Center in Norwich, Britain [30], and donated by Professor He Zhaozu of Hainan University. The recipient bacterium was collected from main flower and grass planting areas in China. *Colletotrichum* strain CH008 with strong pathogenicity was obtained via purification and single spore isolation [20].

Culture medium. PDA and LB were prepared by conventional methods. Details about the preparation of minimal medium (MM), induction medium (IM), and selective medium can be found in the literature [31,32].

S. guianensis for tests. Highly sensitive *S. guianensis* variant AFT3309 was collected from the Forage Germplasm

Garden of CATAS. The collected leaves were seven-day-old ternate compound leaves. Each compound leaf contained three small leaves. One small leaf was connected to a wild-type strain with strong pathogenicity, whereas the other two were connected to a transformant strain.

Methods

Construction of T-DNA insertion mutant library of *S. colletotrichum*. ATMT for *S. colletotrichum* was performed according to the methods of Hu XW [33] and Lin CH [34].

Determination of optimum working concentration of antibiotics (chlorimuronethyl). DCM plates with various concentrations of chlorimuronethyl (0, 5, 10, 15, 20, 30, 40, 50, and 60 μg/mL) were prepared. Approximately 1 μL of activated spore suspension of *S. colletotrichum* strain CH008 (concentration, 10^6 spores/mL) was dropped into the center of DCM containing various concentrations of chlorimuronethyl (each concentration gradient was investigated in triplicate), and cultivated at 28°C for 5 d to observe the growth of *C. gloeosporiodes* Penz and determine the working concentration.

Effects of acceptor materials on transformation. We used six different levels (0.2, 0.4, 0.6, 0.8, 1.0, and 1.2) for the OD_{600} value of *A. tumefaciens*, and four different levels (10^4, 10^5,

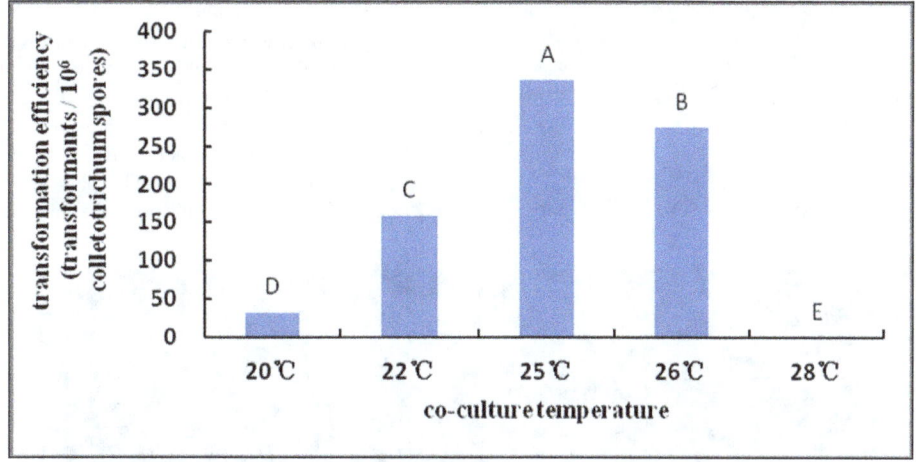

Figure 6. Effect of co-culture temperature on transformation efficiency. Note: Different capital letters indicate significant differences (p<0.01).

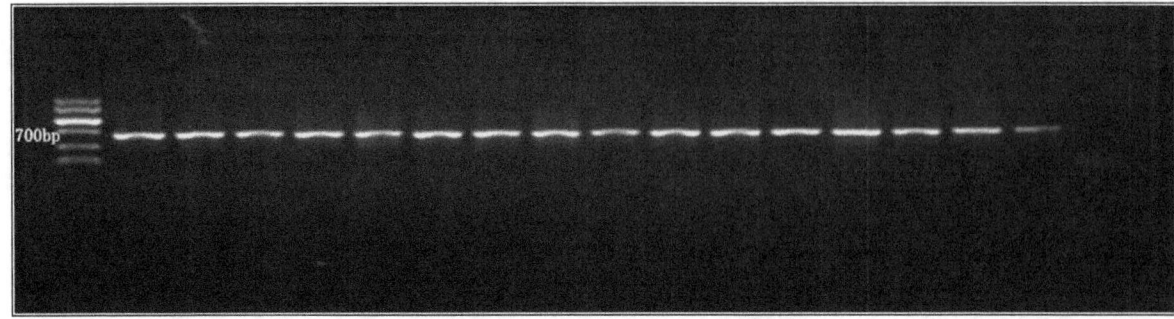

Figure 7. Verification of PCR results. Note: Marker v is on the left, followed by the positive control of plasmid and transformants, and untransformed CH 008 is on the right.

10^6, and 10^7 spores/mL) of *S. colletotrichum gloeosporiodes* spore fluid concentration. The effects of 24 different combinations on the transformation efficiency were examined. Experiments were performed in triplicate.

Effects of *Agrobacterium* induction time on transformation. The effects of seven different induction times of *Agrobacterium* (3, 5, 6, 7, 8, 10, and 12 h) on the transformation efficiency were determined. Experiments were performed in triplicate.

Effects of acetosyringone (AS) concentrations on transformation. The effects of four different levels of AS concentration (0, 100, 150, and 200 μmol/L) on the conversion efficiency were determined. Experiments were performed in triplicate.

Effects of co-culture time on transformation. The effects of seven different co-culture times (1, 2, 3, 4, 5, 6, and 7 d) on the conversion efficiency were determined. Experiments were performed in triplicate. Co-culture was conducted following the methods of Merer [35].

Effects of co-culture temperature on transformation during co-culture. The effects of five different temperatures (20°C, 22°C, 25°C, 26°C, and 28°C) on the conversion efficiency were determined. Experiments were performed in triplicate. Co-culture was performed according to the methods of Merer [35].

Verification of transformants
A. PCR

Total DNA of *S. colletotrichum gloeosporiodes* Penz transformants was extracted according to the conventional hexadecyltrimethylammonium bromide (CTAB) method. According to the GFP gene sequence in the plasmid pSuLF·GFP, primer pairs were designed. The primers were as follows:

GFP-F: 5'-TACTGCAGATGGTGAGCAAGGGCGAG-3'
GFP-R: 5'-CGGGATCCCTTGTACAGCTCGTCCATG-3'

For PCR reaction, a 20 μL system was used. The following reaction conditions were used: pre-degeneration at 94°C for 3 min, degeneration at 94°C for 45 s, annealing at 58°C for 45 s, extension at 72°C for 1 min, and final extension at 72°C for 10 min after 30 cycles. Samples were stored at 10°C.

B. Fluorescence microscopy test

Thirty transformants from the mutant library were randomly selected and cultivated on PDA with illumination at 28°C. Sterile water (1 mL) was dropped on the bacterial colony, and a pipette was used to carefully blow and mix the colony. Subsequently, 3 μL of the bacterial colony was placed on a clean glass slide. The slide was covered with a cover slip, and examined by confocal fluorescence microscopy. An excitation wavelength of 400–500 nm was used.

Analysis of T-DNA insertion mutant library of *S. colletotrichum*

Analysis of the copy number of T-DNA insertion mutants (Southern hybridization). Mycelia of *Colletotrichum* mutants were inoculated to PDA/CM fluid culture medium containing 200 μg/mL cephalosporin, 50 μg/mL tetracycline, and 10 μg/mL chlorimuronethyl. The medium was shaken at 28°C and 150 rpm for 7 d. Filter paper was used for filtration, and mycelial pellets were collected. Total DNA of *S. colletotrichum gloeosporiodes* transformants was extracted according to the conventional CTAB method. A detail protocol of Southern blot was performed following the specifications of DIG-High Prime DNA Labeling and Detection Starter Kit I.

Analysis of the growth rate of mutants. Purified untransformed CH008 was cultivated, and 30 transformants were randomly selected and cultivated on PDA plates for 5 d. The concentration of spore fluid was adjusted to 10^4 spores/mL. A pipette was used to extract 2 μL of spore fluid, which was inoculated onto a PDA plate. The plate was incubated at 28°C. After 4 d of cultivation, the colony diameter was measured once a

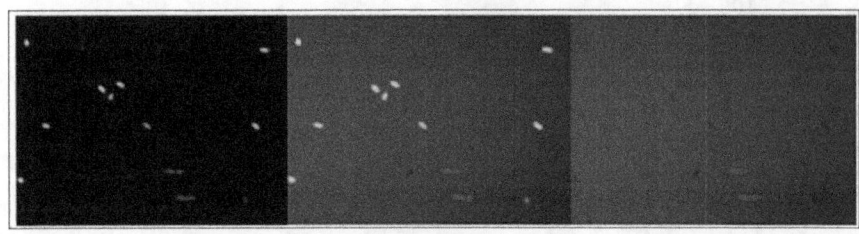

Figure 8. GFP fluorescence of *S. colletotrichum* T-DNA transformants.

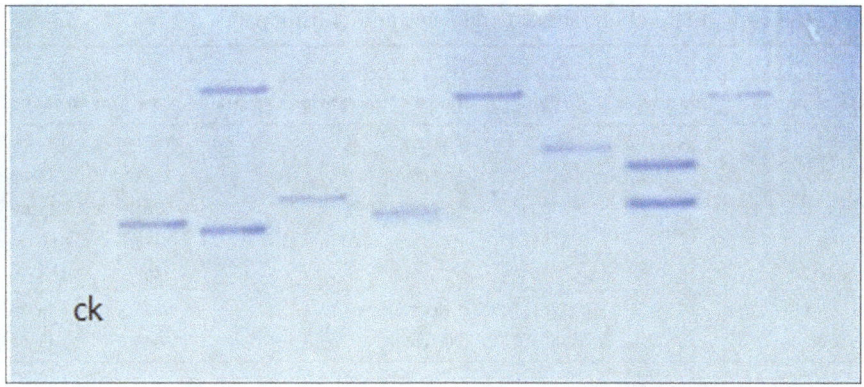

Figure 9. Southern blot analysis of genomic DNA from untransformed CH008 (CK) and transformed isolates.

day. The difference in colony diameter of two adjacent days was determined to represent the colony growth rate, and each strain was analyzed three times. The growth of colonies was observed for 9 d, and photos of their morphology were taken.

Sporulation quantity and spore morphology of mutants. Thirty transformants and untransformed CH008 were cultivated according to the methods specified above. After 6 d, sterile water was used to dilute spore fluid. A blood counting chamber was used for counting. Spore morphology was observed, and differences in spore morphology were recorded.

Conidial germination and appressorium formation of mutants. Fresh conidia suspension liquid was obtained using a water washing method. Its concentration was adjusted to 1.0×10^5 and 1.0×10^6 conidia/mL. Suspension liquid was dropped to a clean glass slide, blotted with a piece of bibulous filter paper, and incubated at 28°C. A total of 100 conidia were statistically analyzed, and the appressorium formation rate of germinated spores was recorded. Analyses were performed in triplicate. Samples were observed at 4, 6, 8, 10, and 12 h.

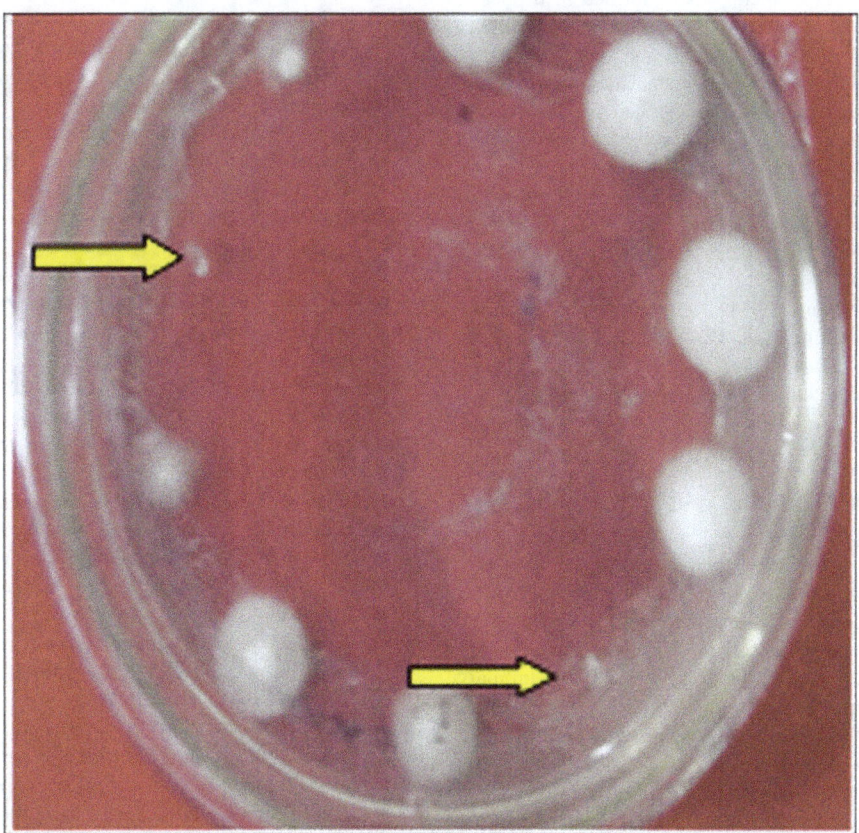

Figure 10. Determination of transformants' genetic stability (arrows indicate wild strains).

Table 1. Comparison of the growth rate of untransformed strains and transformants.

Approaching growth rate of untransformed strains		High growth rate of untransformed strains		Low growth rate of untransformed strains	
Ch008	0.80 cm/d	888	1.06 cm/d	1561	0.37 cm/d
1130	0.76 cm/d	328	1.40 cm/d	3532	0.53 cm/d
993	0.83 cm/d	2715	1.13 cm/d	678	0.40cm/d
1477	0.80 cm/d	3590	1.20cm/d	3425	0.60 cm/d
3200	0.87 cm/d	3416	1.23 cm/d	1801	0.56 cm/d
2181	0.73 cm/d	1761	1.16 cm/d	3393	0.46 cm/d
3616	0.86 cm/d	3605	1.00 cm/d	604	0.43 cm/d

Note: Data in the table are average values of three measurements.

Selection of pathogenicity-defective transformants of *S. colletotrichum* mutant library and analysis of flanking sequences at T-DNA insertion sites

Preliminary screening and re-screening of pathogenic defects of mutants. For preliminary screening, two inoculation methods, namely, spore fluid and mycelium cake, were used. In the spore fluid method, sterile water was used to wash *Colletotrichum* conidium cultivated on PDA for 5 d to prepare 5×10^5 CFU/mL suspension liquid. Conidium liquid (1 μL) was dropped on *S. guianensis* leaves with punctured parts and normal ones. Mycelium cakes at the edge of *Colletotrichum* bacterial colony were cultivated on PDA for 5 d (diameter, 5 mm). The surface of a mycelium cake, which contained hyphae, was attached to *S. guianensis* leaves with punctured parts and normal ones to observe morbidity. This experiment was performed in triplicate. The re-screening method was similar to preliminary screening, except that it was performed using potted and complete plants. According to the comparison of different inoculation methods, the most appropriate one was selected and applied for preliminary screening and re-screening of the pathogenic defects of each mutant.

Cloning of flanking sequences at T-DNA insertion sites of mutant strains with pathogenic defects. TAIL-PCR was used for amplification of the flanking sequences at T-DNA insertion sites. Random primers, composition of nested primers at left and right boundaries, and PCR procedures were based on the methods of Mullins [36]. After the PCR products accepted a 1.0% AGE test, cloning and sequencing were conducted. BLAST comparison was implemented for the obtained sequence.

Results

Construction of T-DNA insertion mutant library of *S. colletotrichum*

Optimum concentration of chlorimuronethyl. In accordance with Figure 1, *S. colletotrichum* CH008 was inhibited at 5 μg/mL chlorimuronethyl and this strain nearly showed no growth at 10 μg/mL chlorimuronethyl. Thus, the final concentration selected by the culture medium was 10 (preliminary screening) and 20 μg/mL (re-screening).

Effects of acceptor materials on transformation. Results of the differences in concentrations of *A. tumefaciens* and *Colletotrichum* spore liquid were analyzed by ANOVA (similar to Duncan's new multiple range method below). According to the related results, the difference in experimental results was significant at p<0.01. Based on Figure 2, each OD value had a corresponding and appropriate spore liquid concentration within

an appropriate range of OD_{600} (0.4–0.8). A high concentration of spore liquid resulted in a high transformation efficiency. In particular, when *Agrobacterium* OD_{600} was equal to 0.8 and the number of *Colletotrichum* CH008 spores was 10^6 spores/mL, the results were significantly higher than the other combinations, and the transformation efficiency peaked. This result may be related to the growth cycle of *Agrobacterium* and features of *Colletotrichum* CH008 spores. When the OD_{600} values of *Agrobacterium* were 1.0 and 1.2, the transformation efficiency showed a reducing trend with false-positive results.

Effects of induction time on transformation. By inducing the activation and expression of genes at the Vir region of *Agrobacterium*, AS promoted T-DNA processing and transfer so that *Agrobacterium* T-DNA could enter the target genome and integrate with it more easily. An appropriate induction time was directly related to the ability of AS to activate the Vir region sufficiently, and affected the efficiency of final recombination. The effect was most significant in this experiment, and the transformation efficiency peaked at 6 h of induction and transformation (Figure 3). However, extending the induction time did not improve the transformation efficiency. This finding may be due to the fact that the culture time of *Agrobacterium* was too long, so *Agrobacterium* gradually entered a decline phase and affected the transformation efficiency.

Effects of AS on transformation. When the AS concentration was 100 μmol/L (Figure 4), the results were significantly higher than those at other AS levels. However, the transformation efficiency did not improve as the concentration increased.

Effects of co-culture time on transformation. According to Figure 5, the transformation efficiency peaked when *S. colletotrichum* and *Agrobacterium* were co-cultured for 3–4 d, and the results were highly significant.

Effects of temperature on transformation during co-culture. In this experiment (Figure 6), the optimum transformation efficiency was obtained at a co-culture temperature of 25°C. Its effect was most significant at this temperature.

PCR. Among 20 selected converter strains, the positive control and all transformants demonstrated bright and clear strips at 750 bp. Thus, the T-DNA insertion rate was 100% (Figure 7).

Verification of fluorescence of transformant's conidia. Thirty transformants were randomly selected. After sporulation, conidia were examined under confocal fluorescence microscopy at an excitation wavelength of 400–500 nm (Figure 8). Spores of 13 transformants were bright under fluorescence, which indicates that GFP genes were carried to and integrated with *S. colletotrichum gloeosporioides* Penz genome, and exhibited good expression.

Table 2. Comparison of transformants' ability to produce spores.

Strain number	Ch008	2097	1447	1477	2181	961	2881	888	993
Spore concentration	1.05×10^6	4.00×10^4	2.15×10^5	1.00×10^6	2.15×10^7	6.03×10^4	1.67×10^7	1.50×10^5	0
Strain number	3590	3661	3605	2561	3443	844	1561	1761	604
Spore concentration	2.33×10^6	2.34×10^5	2.50×10^5	3.45×10^7	0	1.75×10^7	2.35×10^5	5.40×10^7	1.70×10^5

This study optimized the genetic transformation system conditions of *S. colletotrichum*. The optimized conditions were as follows: AGL-1 concentration of *A. tumefaciens* (OD_{600}), 0.8; concentration of *Colletotrichum* conidium, 1×10^6 conidia/mL; AS concentration, 100 mmol/L; induction time, 6 h; co-culture temperature, 25°C; and co-culture time, 3 d. The T-DNA insertion mutant library of *S. colletotrichum* was constructed successfully using the genetic transformation system. The transformation efficiency was determined to be 300–400 transformants/10^6 *Colletotrichum* spores.

Analysis of *S. colletotrichum* T-DNA insertion mutant libraries

The genetic transformation system was used to obtain 4616 *S. colletotrichum* genetic transformants. Some of these transformants were then selected for analysis.

Analysis of the copy number of T-DNA insertion mutants. The results of Southern blot are shown in Figure 9. Among eight randomly selected transformants, six demonstrated a single strip and the remaining two had two strips. Untransformed CH 008 and one transformant had no strip. These results indicate that most T-DNA insertions had a single locus, which could aid in amplification for sequences of insertion sites via Tail-PCR.

Genetic stability. After cultivating untransformed and mutant strains in PDA plates without chlorimuronethyl for 10 generations, they were inoculated to screening plates with chlorimuronethyl. The transformants grew normally (Figure 9), whereas the untransformed strains did not grow (indicated by the arrow in Figure 10). These findings show that resistance could be stably inherited in transformants.

Growth rate and colony morphology of transformants. Untransformed CH008 was cultivated, and 30 transformants were randomly selected. Thus, we observed an obvious difference in the growth rate of transformants. Specifically, mutant 1561 had the lowest growth rate, and its growth rate was much lower than that of untransformed CH008. Transformant 328 exhibited the highest growth rate (Table 1). Some strains demonstrated morphological variation, such as white hyphae that did not generate spores, greyish green hyphae, and yellow hyphae. However, the proportion of such strains was small.

Sporulation quantity of mutants. The sporulation quantities of wild strains and mutants were determined, and the results are shown in Table 2. The sporulation quantity of transformants 2181, 2881, 2561, 844, and 888 was much higher than that of untransformed CH008, whereas the sporulation quantity of transformants 1477, 1561, and 2097 was lower than that of untransformed CH008. Transformants 3443 and 993 showed nearly no sporulation quantity.

Conidial germination and appressorium formation in mutants. The results show that wild-type conidia and 10 randomly selected transformant spores could germinate after some time. The time that most of the transformants took to germinate was shorter than that of the wild-type strains, and the average length of germ tube growth in some transformants was longer than that of the wild-type strains. Three mutant strains demonstrated both low germination and appressorium formation rates, namely, t-960, t-604, and t-2327. By contrast, seven mutant spores (t-2393, t-2515, t-906, t-888, t-1130, t-2416, and t-3616) showed no significant difference from wild-type strains. In addition, wild-type strains and most transformants grew one to two germinal tubes from both ends of spores, and 80% of them demonstrated appressorium formation. Spore germination of mutant t-906 was abnormal; this mutant had three to four germinal tubes and its

Table 3. Comparison of conidial germination and germination rate.

Strain name	Initial germination time	Germination rate	Length and number of germinal tubes
Ch008	8 h	90–100	55–110 μm, 1
t-2515	6 h	80–90	110–300 μm, 2
t-2393	8 h	70–80	200–300 μm, 2
t-2327	6 h	40–50	300–400 μm, 1
t-888	6 h	70–80	200–250 μm, 2
t-906	4 h	95–100	300–400 μm, 3–4
t-960	8 h	60–70	30–45 μm, 1
t-1130	8 h	70–80	100–255 μm, 2
t-2416	6 h	90–100	110–350 μm, 2
t-3616	6 h	70–80	20–100 μm, 1
t-604	4 h	40–60	110–300 μm, 2

germination rate was high, but it germinated few appressoria (Table 3 and Figure 11).

Selection of transformants of the *S. colletotrichum* mutant library with pathogenic defects and analysis of flanking sequences at T-DNA insertion sites

Preliminary screening of mutants with pathogenic defects. Comparison of inoculation methods showed that the most appropriate preliminary screening method was the use of the wild-type strain CH008 and transformant spore liquid to infect punctured parts of leaves of *S. guianensis*, and select mutants with lost pathogenicity. Using this method, we selected 1230 transformants and obtained 23 strains with pathogenic defects (18 strains with reduced pathogenicity and five strains with lost pathogenicity). Figure 12 shows that brown scabs formed at the infected part of wild strain CH008, whereas transformant 2430 lost its infection ability.

Re-screening of pathogenicity determination. For re-screening, we selected defective mutant strains whose pathogenicity was lost when preliminary screening was performed. The potted *S. guianensis* was inoculated again. Comparison of different re-screening inoculation methods showed that the most appropriate re-screening method was spraying the wild strain CH008 and transformant spore liquid to punctured parts of leaves of *S. guianensis* for verification and selection of mutants with lost pathogenicity. This method was used to re-screen five mutants with completely lost pathogenicity. The results were similar to those of preliminary screening. For example, mutant 2430 was used to infect *S. guianensis* plants. Two weeks later, pathogenic symptoms appeared. Scabs were observed on some leaves and stems of plants, and slowly expanded to most of the stem leaves (Figure 13).

Cloning of T-DNA insertion flanking sequences of mutant strains with pathogenic defects. TAIL-PCR amplification was used in 15 mutants with reduced pathogenicity and defects, and three transformants with peculiar strips were obtained, i.e., *Anthrax* transformants 2430, 913, and 3521. Only transformant 2430 was successful in transformation and sequencing; and the flanking sequences of 476 bp were obtained (Figure 14).

BLAST was used to compare the sequence with the sequenced *Colletotrichum* whole genome database (unpublished), and the website 'The FGENESH Program' (Softberry Inc., Mount Kisco, NY, USA; http://linux1.Softberry.com/berry.phtml) was used to predict its functions. Therefore, a hypothetical gene in the regional code of the nucleotide sequence was noted. BLAST was subsequently adopted to compare the sequences in NCBI; 401 nt was completely consistent with the partial sequence of Contig464 of the database (Figure 15). T-DNA was a promoter subregion of the predicted gene. The full length of the predicted gene was 1251 bp. The code of the predicted gene was 416 aa, and the amino acid homology between the predicted gene and *Magnaporthe oryzae* gene (XP_003719674.1) was 79%. This type of gene codes aspartate transaminase. This code may play an important role in the infection process of pathogeny. Insertion of

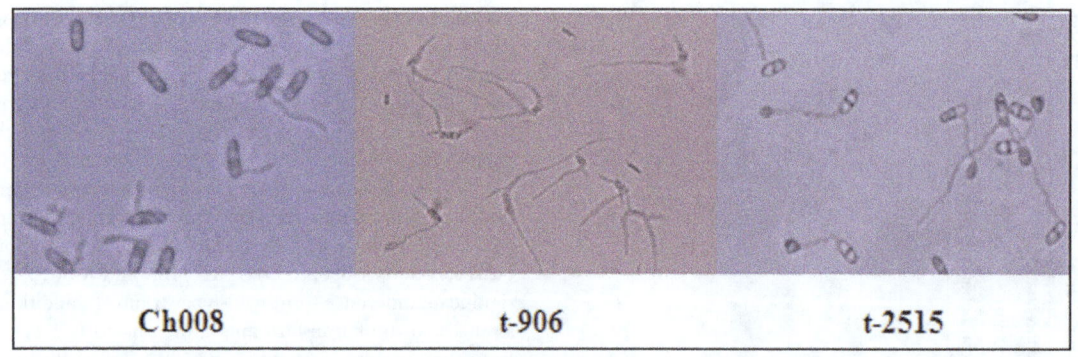

Figure 11. Conidial germination and appressorium formation after 8 h.

Figure 12. Determination of virulence during preliminary screening.

exogenous sections destroyed the gene's functions, so the mutants exhibited lost pathogenicity.

Discussion

In 2011, genetic transformation of over 60 fungi was realized successfully by ATMT [38]. When transformation is carried out, each fungus has optimum transformation conditions. The optimum transformation and screening conditions should be explored to construct a high-quality genetic mutant library with a large quantity in a short time period. Moreover, when high-quality and efficient transformation and optimization systems are used to construct necessary mutant libraries, manpower and material resources are not only saved but the library-establishment cycle is also shortened. Such systems can also reduce difficulty in screening and cloning related pathogenic genes [39].

The cultivation, growth status, and purity of *Agrobacterium* have significant effects on transformation, and they are important for preparing *Agrobacterium* infection liquid with high purity,

vigorous growth, and powerful infection capability [40]. *Agrobacterium* in the middle and late logarithmic phases is considered to be the optimum infectious bacteria. However, the time at which different species of *Agrobacterium* reach the logarithmic phase, as well as their concentrations, can differ. In the present study, after AGL-1 was cultured in MM for 36 h and cultivated in IM for 6 h, its OD_{600} value ranged from 0.6 to 1.0. By selecting *Agrobacterium* within this range and adjusting to an appropriate concentration, the obtained transformation efficiency was high.

AS is currently the most common Vir gene inducer. Numerous studies showed an optimum inductive effect at a concentration of 50–200 μmol/L. When the pH of the culture medium containing AS is 5.0–5.6, the induction of genes at the Vir region of *A. tumefaciens* peaks [41]. IM in the present study showed a final selected concentration of 100 μmol/L and pH of 5.2. This finding verifies the conclusion drawn by Holford et al., who reported optimum transformation effects at pH 5.2 [42]. When the AS concentration was 100 μmol/L, the results were significantly higher than those at other levels. However, the transformation

Figure 13. Incidence of different parts of plants after re-screening and inoculation.

```
TGTGAAGAAGCAGAGAGGATGAGGATGAGGAGAAGCAGAGAGAGAAAATGCAGAAGAAGAC

GACGAGGAGTCACCCGGCATTTTCAGTGGTGTCGGAGCTTTCGTGAGATGGAACGAGTCAGAG

GGAAGGGTCGCCTGGGGTCGTTGTTATTATTATGGCATCAGGAAGTGGTACGATTTAGACGGG

CGAATAGGAGGCTTTGCGATTACAGGCGCCAACCGGGGAAAAGGACGGGACATAGGGAGTGA

CCAAGTATACTAGAGAGAAAAGGGAGGTGGAAGCGGACATGCGCCTCCATGGCAGCGGGCAG

CTTGTTGCAGGCCGGGGGAAGCTACTGCACAATCCAAGGATTGGCTAGGGGAACCTGTGCAA

GTACCTGGGGACGAGGTGGGATGGTGAGGTAAGAGATGTACCACGGGGGCGGGGATGACGAC

GTAAGAGGACTTCCTCCCAGATAAAGGAGGTACTTTGCTGG
```

Figure 14. Flanking sequences of T-DNA at the insertion sites of mutant 2430.

```
Sequences producing significant alignments:                    Score      E
                                                               (bits)  Value

Contig464                                                        757    0.0

>Contig464
           Length = 80153

 Score =  757 bits (382), Expect = 0.0
 Identities = 401/406 (98%), Gaps = 1/406 (0%)
 Strand = Plus / Minus


Query: 71    tcacccggcattttcagtggtgtcggagctttcgtgagatggaacgagtcagagggaagg  130
             ||||||||||||||||||||||||||||||||||||||||||||||||||||||||||||
Sbjct: 2746  tcacccggcattttcagtggtgtcggagctttcgtgagatggaacgagtcagagggaagg  2687

Query: 131   gtcgcctggggtcgttgttattattatggcatcaggaagtggtacgatttagacgggcga  190
             ||||||||||||||||||||||||||||||||||||||||||||||||||||||||||||
Sbjct: 2686  gtcgcctggggtcgttgttattattatggcatcaggaagtggtacgatttagacgggcga  2627

Query: 191   ataggaggctttgcgattacaggcgccaaccggggaaaaggacgggacatagggagtgac  250
             || ||||||||||||||||||||||||||||||||||||||||||||||||||||||||||
Sbjct: 2626  atgggaggctttgcgattacaggcgccaaccggggaaaaggacgggacatagggagtgac  2567

Query: 251   caagtatactagagagaaaagggaggtggaagcggacatgcgcctccatggcagcgggca  310
             |||||||||||||| ||| |||||||||||||||||||||||||||||||||||||||||
Sbjct: 2566  caagtatactagaaagaaa-gggaggtggaagcggacatgcgcctccatggcagcgggca  2508

Query: 311   gcttgttgcaggccggggggaagctactgcacaatccaaggattggctaggggaacctgtg  370
             ||||||||||||||||||||||||||||||||||||||||||||||||||||||||||
Sbjct: 2507  gcttgttgcaggccggggggaagctactgcacaatccaaggattggctaggggaacctgag  2448

Query: 371   caagtacctggggacgaggtgggatggtgaggtaagagatgtaccacggggggggggatg  430
             |||||||||||||||||||||||||||||||||||||||||||||||||||||||||||
Sbjct: 2447  caagtacctggggacgaggtgggatggtgaggtaagagatgtaccacggggggggggatg  2388

Query: 431   acgacgtaagaggacttcctcccagataaaggaggtactttgctgg  476
             |||| ||||||||||||||||||||||||||||||||||||||||
Sbjct: 2387  acgatgtaagaggacttcctcccagataaaggaggtactttgctgg  2342
```

Figure 15. BLAST results of T-DNA flanking sequences in NCBI.

efficiency did not improve with increasing AS concentration. Excess AS concentrations may have toxic effects on explants, and influence further improvement in the transformation efficiency [41].

The co-culture time is one of the most important factors affecting the success of the transformation of *A. tumefaciens*. The process of transformation mediated by *A. tumefaciens* takes some time, and transformants are unable to form at a very short co-culture time [35,43–45]. In the present study, the number of transformants increased with increasing co-culture time. The number of transformants peaked when *S. colletotrichum* and *Agrobacterium* were co-cultured for 3–4 d, and the results were highly significant. Continuous co-culture can result in generation of false-positive clones [46], resulting in very large bacterial colonies that cannot be selected easily [47]. Campoy et al. revealed that the number of transformants peaks when the co-culture temperature is consistent with the optimum growth temperature of acceptors; reducing or increasing the co-culture temperature reduces the number of transformants [48]. The results of this study were consistent with those observed by Campoy et al.

Furthermore, selection of transformation acceptors is critical. Many forms, such as protoplast, mycelium, and conidium, can be used as acceptor materials [26]. Acceptor strains can be transformed by exogenous genes only when they stay in the phase of cell division. High requirements must be met in the preparation of protoplast, and this process is quite complicated. Materials such as mycelium and conidium can also achieve ideal transformation efficiency. Very high cell concentration of acceptor fungi can lead to excess fungal growth, so transformants cannot be selected [24]; very high *Agrobacterium* concentrations can result in serious *Agrobacterium* pollution [37].

The results of Southern blot show that the T-DNA insertion rate in transformants was 90%. The transformants, whose T-DNA insertion was at a single site, accounted for about 67%. The rate of a single copy was much higher than that of *Colletotrichum graminicola* (16%), slightly higher than that of other anthracnose species (65%) [49,50], but lower than that obtained by Jia Peisong [51] and Wang Haiyan [52] (100%). Single-site insertion in genome is usually expected because the derived phenotype is related to changes in single sites in the genome [24], which can help in finding marker genes from the fungal genome. Some studies implied that the insertion rate of T-DNA single copy numbers has an inverse relation with the co-culture time to some extent [53]. However, the co-culture time must not be shortened too much in single-copy insertion so that the transformation efficiency does not decrease. Moreover, Examination of a large number of mutants using PCR and Southern blot is time-consuming and laborious, and generally provides only a subjective measure of infectivity. Therefore it would be worthwhile to develop the high-throughput technologies for molecular verification of *Stylosanthes* anthracnose mutants, such as Luminex or transposon sequencing, Lin Tao et al. examined 434 signature-tagged mutagenesis mutants using Luminex-based multiplex PCR, which is an efficient and timesaving method [54–56].

This research showed that the growth rates and sporulation quantity of some transformants changed. T-DNA insertion causes inactivation for a certain gene of transformants, and this variation results in changes in other characters. However, most transformants did not demonstrate a significant difference from untransformed strains. Conidial generation is a premise for infection and pathopoiesia of many pathogenic bacteria; the quantity and germination rate of conidia, as well as appressorium formation, affect their pathogenicity to some extent [57,58]. In the future, relations among the pathogenicity of transformants, growth rate of mutants, sporulation quantity, conidial germination, appressorium formation, and pathopoiesis will be further examined.

The flanking sequences of T-DNA insertion sites were cloned. TAIL-PCR is commonly used to amplify unknown flanking sequences of the known T-DNA sequence. This technique can design reliable random degenerate primers that are appropriate for the background of genomes that need to be detected [59]. The present study identified the optimum random primers (i.e., AD4 and AD8), and four right flanking sequences and one left flanking sequence were obtained. The reason for these results might be the high probability at which the left border of T-DNA is cut off, which was consistent with several fungal results that have been reported [36,49,60]. The left border of T-DNA may be unnecessary for T-DNA transfer, whereas the right border is essential for T-DNA transfer; thus, transfer starts at the right border and continues toward the left [61].

TAIL-PCR was used in the present study to amplify the flanking sequences of five mutant genomes, and five flanking sequences were obtained. For one sequence, BLAST comparison showed that its pathogenicity was lost and the flank section amplified by t-2430 was approximately 0.5 kb. By analyzing and comparing the obtained sequences, these genes and genes of aspartate transaminase coded by the pathogenic bacterium *M. oryzae* demonstrated high homology. This enzyme mainly exists in the mitochondria of cells, and exerts important catalytic actions during nitrogen metabolism. The insertion of exogenous sections may alter the functions of genes, affect the coding of transferase, disturb some important metabolisms of the mutant, and result in the loss of pathogenicity in pathogenic infection processes.

Acknowledgments

We thank the National Natural Science Foundation of China (NSFC, No. 31072076), Special Fund for Agro-scientific Research in the Public Interest (201303057), and The Major Technology Project of Hainan (ZDZX2013023-1) for funding this research.

Author Contributions

Conceived and designed the experiments: KXY GXH. Performed the experiments: HLC CPH. Analyzed the data: CPH JLZ. Contributed reagents/materials/analysis tools: SQZ QLL JGX JMG GXH. Contributed to the writing of the manuscript: HLC.

References

1. Williams RJ, Reid R, Schhultze-Kraft R, Sousa Costa NM, Thomas BD (1984) Natural distribution of *Stylosanthes*. In: Stace HM, Edye LA, editors. The Biology and Agronomy of Stylosanthes. pp. 73–101.

2. Jiang CS, Jia HS, Ma XR, Zou DM, Zhang YZ (2004) AFLP analysis of genetic variability among *Stylosanthes guianensis* accessions resistant and susceptible to the Stylo Anthracnose. Acta Botanica Sinica 46: 480–488.

3. Segenet K, Jiang CS, H GX, Gustavo S (2005) Genetic transformation of the tropical forage legume *Stylosanthes guianensis* with a confers resistance to *Rhizoctonia* foliar blight disease rice-chitinase gene. Afr J Biotechnol 4: 1025–1033.

4. Liu GD, Phaikaew C, Stur WW (1997) Status of *Stylosanthes* development in other countries: II *Stylosanthes* development and utilization in China and southeast Asia. Trop Grassl 31: 460–467.

5. Lenné JM, Calderon MA (1984) Disease and pest problems of *Stylosanthes*. In: Stace HM,Edye LA, editors. The Biology and Agronomy of Stylosanthes. 279–293.

6. Manners JM, Masel AM, Braithwaite KS, Irwin JAG (1992) Molecular analysis of *Colletotrichum gloeosporioides* pathogenic on the tropical pasture legumes *Stylosanthes* spp. In: Bailey JA, Jeger MJ, editors. *Colletotrichum* Biology, Pathology, and Control. pp. 250–268.

7. Manners JM, Stephenson SA, He C, Maclean DJ (2000) Gene transfer and expression in *Colletotrichum gloeosporioides* causing anthracnose on *Stylosanthes*. In: Prusky D, Freeman S, Dickman M, editors. Host Specificity, Pathology and Host–Pathogen Interactions of Colletotrichum. pp. 180–194.

8. Chakraborty S, Fernandes CD, Charchar MJDA, Thomas MR (2002) Pathogenic variation in *Colletotrichum gloeosporioides* infecting *Stylosanthes* spp. in a center of diversity in Brazil. Phytopathology 92: 553–562.

9. Chakraborty S, Perott R, Charchar MJDA, Fernandes CD, Kelemu S (1997) Biodiversity epidemiology and virulence of *Colletotrichum gloeosporioides*. II Genetic and pathogenic diversity in *Colletotrichum gloeosporioides* isolates from eight species of *Stylosanthes*. Trop Grassl 31: 387–393.

10. Chakraborty S, Perott R, Ellis N, Thomas MR (1999) New aggressive *Colletotrichum gloeosporioides* strains on *Stylosanthes scabra* detected by virulence and analysis. Plant Dis 83: 333–339.

11. Iriwin JAG, Cameron DF (1978) Two diseases of *Stylosanthes* spp. caused by *Colletotrichum gloeosporioides* in Australia, and pathogenic specialization within one of the causal organisms. Aust J Agric Res 29: 305–317.

12. Kelemu S, Badel JL, Moreno CX, Miles JW (1996) Virulence spectrum of South American isolates of *Colletotrichum gloeosporioides* on selected *Stylosanthes guianensis* genotypes. Plant Dis. 80: 1355–1358.

13. Lenné JM, Burdon JJ (1990) Preliminary study of virulence and isozyme variation in natural populations of *Colletotrichum gloeosporioides* from *Stylosanthes guianensis*. Phytopathology 80: 728–731.

14. Munaut F, Hamaide N, Vander Stappen J, Maraite H (1998) Genetic relationship among strains of *ColLetotrichum gloeosporioides* isolated from *Stylosanthes* spp. in Africa and Australia using RAPD and ribosomal DNA markers. Plant Pathol 47: 641–648.

15. Dale JL, Manners JM, Irwin JAG (1988) *Colletotrichum gloeosporioides* isolates causing different anthracnose diseases on *Stylosanthes* in Australia carry distinct double-stranded RNAs. *Transactions of the British Mycological Society* 91: 671–676.

16. Braithwaite KS, Irwin JAG, Manners JM (1990a) Restriction fragment length polymorphism in *Colletotrichum gloeosporioides* infecting *Stylosanthes* spp. in Australia. Mycol Res 94: 1129–1137.

17. Braithwaite KS, Irwin JAG, Manners JM (1990b) Ribosomal DNA as a molecular taxonomic marker for the group species *Colletotrichum gloeosporioides*. Aust Syst Bot 3: 733–738.

18. Manners JM, Masel A, Irwin JAG (1993) Molecular genetics of *Colletotrichum gloeosporioides* infecting *Stylosanthes*. In: Isaac S, Frankland JC, Watling R, Whalley AJS, editors. Aspects of Tropical Mycology. pp. 233–51.

19. Weeds PL, Chakraborty S, Fernandes CD, Charchar MJdA, Amesh CRR, et al. 2003. Genetic diversity in *Colletotrichum gloeosporioides* from *Stylosanthes* spp. at centers of origin and utilization. Phytopathology 93: 176–185.

20. Yi KX, Huang JS, Liu GD, Weeds P, Charkraborty S (2003) Genetic diversity analysis of stylo anthracnose pathogens using random amplified polymorphic DNA. Wei Sheng Wu Xue Bao 43: 379–387.

21. Munaut F, Hamaide N, Maraite H (2002) Genomic and pathogenic diversity in *Colletotrichum gloeosporioides* from wild native Mexican *Stylosanthes* spp., and taxonomic implications. Mycol Res 106: 579–593.

22. Cai ZY, Li GH, Lin CH, Shi T, Zhai LG, et al. (2013) Identifying pathogenicity genes in the rubber tree anthracnose fungus *Colletotrichum gloeosporioides* through random insertional mutagenesis. Microbiol Res 168: 340–350.

23. Wu XY, Wang JY, Zhang Z, Jing JX, Du XF, et al. (2010) Isolation of gene mutation from a pathogenicity-enhanced mutant of *Magnaporthe oryzae*. Rice Sci 17: 129–134.

24. Michielse CB, Hooykaas PJ, Van den Hondel CAMJJ, Ram AFJ (2005) Agrobacterium-mediated transformation as a tool for functional genomics in fungi. Curr Genet 48: 1–17.

25. Mullins ED, Kang S (2001) Transformation: A tool for studying fungal pathogens of plants. Cell Mol Life Sci 58: 2043–2052.

26. De Groot MJ, Bundock P, Hooykaas PJ, Beijersbergen AG (1998) Agrobacterium tumefaciens-mediated transformation of filamentous fungi. Nat Biotechnol 16: 839–842.

27. Li AH, Zhang YF, Wu CY, Tang W, Wu R, et al. (2006) Screening for and Genetic Analysis on T-DNA-inserted Mutant Pool in Rice. Yi Chuan Xue Bao 33: 319–329.

28. Li M, Gong X, Zheng J, Jiang D, Fu Y, Hou M (2005) Transformation of Coniothyrium minitans, a parasite of Sclerotinia sclerotiorum, with Agrobacterium tumefaciens. FEMS Microbiol Lett 243: 323–329.

29. Walton FJ, Idnurm A, Heitman J (2005) Novel gene functions required for melanization of the human pathogen Cryptococcus neoformans. Mol Microbiol 57: 1381–1396.

30. Sesma A, Osbourn AE (2004) The rice leaf blast pathogen undergoes developmental processes typical of root-infecting fungi. Nature 431: 582–586.

31. Huang GX, Xue YC, Lu X (2008) Construction of a T-DNA insertional mutant library of *Colletotrichum gloeosporioides* of Mango and its molecular analysis. Re Dai Zuo Wu Xue Bao 29: 27–32.

32. Jennifer LF, Lisa JV (2005) Parameters affecting the efficiency of *Agrobacterium tumefaciens*-mediated transformation of *Colletotrichum graminicola*. Curr Genet 48: 380–388.

33. Hu XW, Jiang H, Huang HX, Guo JC (2005) RAPD analysis of mango anthracnose. Re Dai Zuo Wu Xue Bao 26: 38–42.

34. Lin CH, Liu XB, Cai JM, Li CP, Li JF, et al. (2009) *Colletotrichum acutatum* GFP tagged transformants generated from *Agrobacterium tumefaciens*-mediated insertional mutagenesis. Re Dai Zuo Wu Xue Bao 30: 1495–1500.

35. Meyer V, Mueller D, Strowig T, Stahl U (2003) Comparison of different transformation methods for *Aspergillus giganteus*. Curr Genet 43: 371–377.

36. Mullins ED, Chen PX, Romaine P, Raina R, Geiser DM, et al (2001) Agrobacterium- mediated transformation of Fusariium: An efficient tool for insertional mutagenesis and gene transfer. Phytopathology 91: 173–180.

37. Zeilinger S (2004) Gene disruption in Trichoderma atroviride via Agrobacterium mediated transformation. Curr Genet 45: 54–60.

38. Liu G, Yang CD, Kang K Xing M, Tian SL (2011) Agrobacterium tumefaciens-mediated transformation of Trichoderma reesei and optimization of the transforming conditions. Sheng Wu Ji Shu 21: 44–48.

39. Mishra NC, Tatum EL (1973) Non-Mendelian inheritance of DNA induced inositol independence in Neurospom. Proc Natl Acad Sci USA 70: 3875–3879.

40. Liu CP, Zeng BS, Qiu ZF, Li XY, Liu Y (2010) Analysis of the optimized factors on the efficiency of Agrobacterium mediated gene transformation. Anhui Nong Ye Ke Xue 38: 1141–1143.

41. Deng Y, Zeng BS, Zhao SD, Liu Y, Qiu ZF, et al. (2010). Mechanism and application of Acetosyringone in Agrobacterium mediated transformation. Anhui Nong Ye Ke Xue 2010, 38(5): 2229–2232.

42. Holford P, Hernandezn, Newbury HJ (1992) Factors influencing the efficiency of T-DNA transfer during co-cultivation of Antirrhinum majus with Agrobacterium tumefaciens [J]. Plant Cell Rep 11: 196–199.

43. Soichi T, Norio K, Shigeo N (2007) Isolation of pathogenicity- and gregatin-deficient mutants of Phialophora gregata f. sp. adzukicola through Agrobacterium tumefaciens-mediated transformation. J Gen Plant Pathol 73: 242–249.

44. Nyilasi I, Acs K, Papp T, Nagy E, Vágvölgyi C (2005) Agrobacterium tumefaciens-mediated transformation of Mucor circinelloides. Folia Microbiol, 50: 415–420.

45. Betts MF, Tucker SL, Galadima N, Meng Y, Patel G (2007) Development of a high-throughput transformation system for insertional mutagenesis in *Magnaporthe oryzae*. Fungal Genet Biol 44: 1035–1049.

46. Takahara H, Tsuji G, Kubo Y, Yamamoto M, Toyoda K, et al. (2004) *Agrobacterium tumefaciens*-mediated transformation as a tool for random mutagenesis of *Colletotrichum trifolii*. J Gen Plant Pathol 70: 93–96.

47. Rao ZM, Zhang JS, Wei S, Fang HY, ZhuGe J (2007) Genetic transformation of industrialized strain candida glycerinogenes by *Agrobacterium tumefaciens*. C Ying Yong Yu Huan Jing Sheng Wu Xue Bao 13: 868–871.

48. Campoy S, Perez F,Martín JF, Gutiérrez S, Liras P (2003) Stable transformants of the azaphilone pigment-producing Monascus purpureus obtained by protoplast transformation and Agrobacterium-mediated DNA transfer. Curr Genet 43: 447–452.

49. Maruthachalam K, Nair V, Rho HS, Choi J, Kim S, et al (2008) *Agrobacterium tumefaciens*-mediated transformation in *Colletotrichum falcatum* and C. *acutatum*. J Microbiol Biotechnol 18: 234–241.

50. Auyong ASM, Ford R, Taylor PWJ (2012) Genetic transformation of *Colletotrichum truncatum* associated with anthracnose disease of chili by random insertional mutagenesis. J Basic Microbiol 52: 372–382.

51. Jia PS, Ding LL, Zhou BJ, Gao HS, Gao F (2012) Construction of a T-DNA insertional mutant library for *Verticillium dahliae* Kleb. and analysis of a mutant. Mian Hua Xue Bao 24: 62–70.

52. Wang HY, Li BH, Zhang QM, Li GF, Dong XL, et al. (2013) Transformation of *Agrobacterium tumefaciens*-mediated *Colletotrichum gloeosporioides* and identification of transformants. Zhongguo Nong Ye Ke Xue 46: 1799–1807.

53. Maruthachalam K, Klosterman SJ, Kang S, Hayes RJ, Subbarao KV (2011) Identification of pathogenicity-related genes in the vascular wilt fungus *Verticillium dahliae* by *Agrobacterium tumefaciens*-mediated T-DNA insertional mutagenesis [J]. Mol Biotechnol 49: 209–221.

54. Lin T, Gao L, Zhang C, Odeh E, Jacobs MB, et al. (2012) Analysis of an ordered, comprehensive STM mutant library in infectious Borrelia burgdorferi: insights into the genes required for mouse infectivity. PLoS ONE 7: e47532 10.1371/journal.pone.0047532.

55. Troy EB, Lin T, Gao L, Lazinski DW, Camilli A, et al. (2013) Understanding barriers to Borrelia burgdorferidissemination during infection using massively parallel sequencing. Infect Immun 81: 2347–2357.

56. Van Opijnen T, Bodi KL, Camilli A (2009) Tn-seq: high-throughput parallel sequencing for fitness and genetic interaction studies in microorganisms. Nat Methods 6: 767–772.

57. Rogers LM, Kim YK, Guo W, González-Candelas L, Li D, et al. (2000) Requirement for either a host- or pectin-induced pectate lyase for infection of

Pisum sativum by *Nectria hematococca*. Proc Natl Acad Sci U S A 97: 9813–9818.

58. He CP, Lin CH, Liao QH, Li R, Zheng FC (2007) Phenotypes analysis for T-DNA insertional mutagenesis of Magnaporthe grisea. Nong Ye Sheng Wu Ji Shu Xue Bao 15: 877–883.

59. Sun BY, Piao HL, Park SH, Han CD (2004) Selection of optimal primers for TAIL-PCR in identifying Ds flanking sequences from Ac/Ds insertion rice lines. Chin J Biotechnol 20: 821–826.

60. Huser A, Takahara H, Schmalenbach W, O'Connell R (2009) Discovery of pathogenicity genes in the crucifer anthracnose fungus *Colletotrichum higginsianum* using random insertional mutagenesis. Mol Plant Microbe Interact 22: 143–156.

61. Miranda A, Janssen G, Hodges L, Peralta EG, Ream W (1992) *Agrobacterium tumefaciens* transfers extremely long T-DNAs by a unidirectional mechanism. J Bacteriol 174: 2288–2297.

Enhancement of Lipid Productivity in Oleaginous *Colletotrichum* Fungus through Genetic Transformation Using the Yeast *CtDGAT2b* Gene under Model-Optimized Growth Condition

Prabuddha Dey[1], Nikunj Mall[2], Atrayee Chattopadhyay[2], Monami Chakraborty[1], Mrinal K. Maiti[1,2]*

1 Adv. Lab. for Plant Genetic Engineering, Advanced Technology Development Center, Indian Institute of Technology Kharagpur, Kharagpur, India, 2 Department of Biotechnology, Indian Institute of Technology Kharagpur, Kharagpur, India

Abstract

Oleaginous fungi are of special interest among microorganisms for the production of lipid feedstocks as they can be cultured on a variety of substrates, particularly waste lingocellulosic materials, and few fungal strains are reported to accumulate inherently higher neutral lipid than bacteria or microalgae. Previously, we have characterized an endophytic filamentous fungus *Colletotrichum* sp. DM06 that can produce total lipid ranging from 34% to 49% of its dry cell weight (DCW) upon growing with various carbon sources and nutrient-stress conditions. In the present study, we report on the genetic transformation of this fungal strain with the *CtDGAT2b* gene, which encodes for a catalytically efficient isozyme of type-2 diacylglycerol acyltransferase (DGAT) from oleaginous yeast *Candida troplicalis* SY005. Besides the increase in size of lipid bodies, total lipid titer by the transformed *Colletotrichum* (lipid content ~73% DCW) was found to be ~1.7-fold more than the wild type (lipid content ~38% DCW) due to functional activity of the *CtDGAT2b* transgene when grown under standard condition of growth without imposition of any nutrient-stress. Analysis of lipid fractionation revealed that the neutral lipid titer in transformants increased up to 1.8-, 1.6- and 1.5-fold compared to the wild type when grown under standard, nitrogen stress and phosphorus stress conditions, respectively. Lipid titer of transformed cells was further increased to 1.7-fold following model-based optimization of culture conditions. Taken together, ~2.9-fold higher lipid titer was achieved in *Colletotrichum* fungus due to overexpression of a rate-limiting crucial enzyme of lipid biosynthesis coupled with prediction-based bioprocess optimization.

Editor: Kyoung-Heon Kim, Korea University, Republic of Korea

Funding: Grant support from Institute Scheme for Innovative Research and Development (ISIRD), Sponsored Research and Industrial Consultancy (SRIC), Indian Institute of Technology (IIT)-Kharagpur is acknowledged. The funders had no role in study design, data collection and analysis, decision to publish, or preparation of the manuscript.

Competing Interests: The authors have declared that no competing interests exist.

* Email: maitimk@hijli.iitkgp.ernet.in

Introduction

Lipid is an important raw material for the production of different essential compounds related to food and non-food industrial applications. Among the food applications, storage lipid has been considered since many years as source of important fatty acids particularly polyunsaturated fatty acids (PUFAs) and other nutraceuticals as dietary supplements for fishes, poultry birds, domestic animals and human [1,2]. Among non-food industrial applications, biodiesel production from lipid feedstock, specifically from plant seed oil, has massively increased recently. However, for sufficient production of oilseed crops fertile lands and several essential agro-inputs are required by plant. In addition to this, concerns about the global food security coupled with the increase in current food price and the competitive segregation of agricultural resources between the food industry and the energy sector have taken public awareness. Therefore, development of few sustainable and cost-effective alternatives to the traditional agricultural and forestry crops is of urgent need for biofuel production in the present scenario of escalating worldwide demand. Oil-rich microorganisms, specifically microalgae have been demonstrated to be a promising alternative source of lipids for biodiesel production [3,4,5]. There are several strains of bacteria, algae and fungi including yeasts and molds found in diverse natural ecosystems, which can accumulate storage lipid> 20% of their dry cell weight (DCW), and are considered as oleaginous microbes. The important criteria for a microorganism to be acceptable for production of lipid feedstock are its total amount of lipid and the types of fatty acids inherently present in it.

In last few years, production of microalgal lipid (also known as single cell oil) was primary research interest, but high amount of algal culture in industrial level either requires a large cultivation area or long incubation period in bioreactor system [6], which is also challenging due to necessity of appropriate light source [4]. Therefore, alternate oleaginous microbes, such as yeasts and filamentous fungi [6–8] are of special interest as they can be grown on a variety of starting materials (substrates), especially waste lignocellulosic materials; and biomass production can be scaled up in fermentation process to produce more total lipid. Filamentous fungi of Zygomycetes, like *Mortierella isabellina* and *Cunninghamella echinulata* have been reported with high amount of lipid content, i.e. 60–87% and 40–57% of DCW, respectively [9]. Our group has characterized two endophytic oleaginous fungi, *Colletotrichum* sp. DM06 and *Alternaria* sp. DM09, which can accumulate total lipid ~34–58% of DCW during standard and nutrient stress conditions of growth; and the fatty acid profiles of their storage lipids are suitable for biodiesel application [10]. It is worthy to mention here that another oleaginous fungi *Mucor circnelloides* (having 20% lipid of DCW) has been used for the first commercial production of microbial lipid [1].

From a long period of time, several strategies have been developed for overproduction of storage lipid in the selected microbial strain, besides bioprospecting of lipid hyper-accumulating novel isolate from natural resources. Among these, biochemical engineering approach that basically depends on creating a physiological stress such as nutrient-starvation (specifically nitrogen and phosphorous) to channel metabolic fluxes to lipid accumulation have been extensively carried out to enhance lipid accumulation in microorganism. The main requirement of fungi (i.e., molds and yeasts) for enhanced lipid production is the medium with excess carbon source and other limiting nutrients, mostly nitrogen [7], phosphorous [11], sulphur [12], aeration, inorganic salt, pH and temperature [13]. However, the biochemical engineering strategy (i.e., nutrient stress) limits the natural growth rate (cell division) of the organisms [1,14,9]. Since lipids are intracellular products, the overall lipid productivity is the product of cellular lipid content multiplied by total biomass titer. Hence, the overall lipid titer will be compromised due to the lower biomass productivity. Therefore to overcome this bottleneck, scientists have devised genetic or metabolic engineering approach.

In the past three decades, metabolic engineering strategies by overexpressing or down-regulating several important rate-limiting enzymes to create a channeling of metabolites towards lipid accumulation in different plant and microbial species have already been carried out. Many of the genes, like *ACC1*, *DGAT*, *LPAT*, *malE*, *ACL* and *PEPC* [15] involved in the lipid metabolism have been subjected to be overexpressed [16] and or silenced in oleaginous seeds of higher plant [17–20], bacteria [21], yeast [18], fungus [22] and algae [23]; and these genetic modifications have resulted in increased production of fatty acids and storage lipids or triacylglycerol (TAGs). In microbial systems, the most promising targets are diacylglycerol acyltransferase (DGAT) and malic enzyme (ME) for gene overexpression, and phosphoenolpyruvate carboxylase (PEPC) is for down-regulation in order to enhance storage lipid productivity [24]. The DGAT enzyme catalyzes the final step of TAG biosynthesis in most of the TAG-accumulating cells/organisms by transferring an acyl group from acyl-CoA to the sn-3 position of 1,2-diacylglycerol. The success with overexpression of DGAT could be explained by the fact that the substrate of DGAT i.e., diacylglycerol, naturally could be allocated to either phospholipid biosynthesis or TAG biosynthesis. Overexpression of DGAT would commit more diacylglycerol to TAG formation rather than phospholipid formation. The *DGAT* gene has been transgenically overexpressed in yeast and plant systems to enhance storage lipid production [18,20,25,26]. However, no report regarding the overexpression of this crucial enzyme in filamentous fungi was available in published literatures till the present study.

Model-based bioprocess optimization of media condition is one of the most important tools for obtaining increased yield of any fermentative product. The most effective method is the statistical approach to evaluate the relative significance of the variables in the medium for optimization of target metabolite production. Different statistical method like Plackett–Burman, response surface methodology (RSM) and factorial designs have been used in past few decades for optimization of medium. A mathematical approach such as artificial neural network (ANN) coupled to genetic algorithm (GA) is recently gaining popularity and found to be superior to other statistical approach like, RSM [27,28].

As already mentioned, the endophytic filamentous fungus *Colletotrichum* sp. DM06, which we have isolated, can accumulate>34% DCW and>49% DCW of total lipid without and with nutrient stress condition, respectively [10]. Moreover, our group has also recently characterized two structurally novel type-2 DGAT isozymes, CtDGAT2a and CtDGAT2b from an oleaginous rhizospheric yeast *Candida tropicalis* SY005 [29], and documented that the CtDGAT2b is catalytically 12.5% more efficient than CtDGAT2a for TAG production [30]. Therefore, in the present study we aimed to overexpress the more efficient CtDGAT2b isozyme of *C. tropicalis* in *Colletotrichum* fungus to increase the production of storage lipid. Another objective of this study was to establish the optimum culture conditions (with respect to temperature, pH, carbon: nitrogen and carbon: phosphorous ratio of growth medium) through model-based prediction so as to further increase the lipid productivity in the genetically modified *Colletotrichum* strain.

Materials and Methods

Strain, media and growth condition

Colletotrichum sp. DM06 strain, an endophytic wild type fungus isolated from the medicinal plant *Ocimum sanctum* [10], was cultured on potato dextrose agar (PDA) medium at 28°C. The composition of the basal liquid medium for lipid production has been described previously [10]. *Escherichia coli* strain DH10B was used for maintainance of binary plasmid and cultivated at 37°C on LB agar plates or liquid medium containing 50 μg/ml kanamycin. *Agrobacterium tumefaciens* LBA4404 was used as a transfer DNA (T-DNA) donor for fungal transformation. YEP medium (consisting of 10 g/l yeast extracts, 10 g/l peptone and 5 g/l NaCl) was used for cultivation of *A. tumefaciens* LBA4404 at 28°C.

Preparation of genetic constructs

The pCAM-GpdA-GusA-TrpC genetic construct was prepared using the backbone of pCAMBIA 1300 binary vector. Initially the pNOM102 vector carrying the intron-containing *gusA* (*uidA*) gene placed under the regulation of *gpdA* promoter of *Aspergillus nidulans* and *trpC* terminator of *A. nidulans* in pUC18 vector (kindly provided by Prof. Francisco JL Aragão, Embrapa Recursos Genéticos e Biotecnologia, Brazil) was doubly digested with *Eco*RI and *Hind*III to obtain the DNA fragment GpdA-GusA-TrpC, which was subcloned into the pCAMBIA 1300 using the same set of restriction enzymes. The resulting recombinant plasmid was named as pCAM-GpdA-GusA-TrpC (Figure 1A). Cloning, sequencing and characterization of the *CtDGAT2b* gene (GenBank accession number KJ437598) from *Candida tropicalis* SY005 have already been described previously [30]. The *CtDGAT2b* gene was obtained after digesting with *Bam*HI from pYES2/CtDGAT2b recombinant plasmid, which was used for the study of transgene expression in *Saccharomyces cerevisiae* [30]. The promoter GpdA was PCR amplified from pCAM-GpdA-GusA-TrpC recombinant plasmid (using Expand High Fidelity PCR mix, Roche) with a set of specific primers- GpdAFp and GpdARp (Table S1) to incorporate the restriction sites- *Eco*RI at 5' end and *Bam*HI at 3' end of the promoter. The pCAM-GpdA-GusA-TrpC construct was digested with *Eco*RI and *Bam*HI to remove the GpdA promoter and *gusA* gene. Thereafter, a tripartaite ligation was carried out using the *Bam*HI digested *CtDGAT2b* gene, *Eco*RI and *Bam*HI digested PCR amplified GpdA promoter and the above-mentioned *Eco*RI and *Bam*HI digested pCAM-GpdA-GusA-TrpC to generate the pCAM-GpdA-CtDGAT2b-TrpC recombinant plasmid (Figure 2A).

Agrobacterium tumefaciens-mediated transformation of Colletotrichum fungus

Prior to perform the *A. tumefaciens*-mediated transformation (ATMT), the antibiotic susceptibility test of wild type *Colletotrichum* sp. DM06 was carried out on hygromycin B (used as the selection marker for the transformants). Results showed that a dose of 200 μg/ml of hygromycin B completely inhibited the growth of wild type fungus (data not shown), therefore this concentration of antibiotic was used for the selection of fungal transformants.

ATMT was carried out following the method described earlier [31] with certain modification. An aliquot of 10% (v/v) of *Colletotrichum* sp. DM06 spore maintained in glycerol stock was spread on PDA plates and incubated at 30°C for 4 days for sporulation. Spores were collected, washed and suspended in induction media (IM) [32], and adjusted to a concentration of 10^3-10^6 conidia/ml. *A. tumefaciens* LBA4404 strain carrying the desired genetic construct was initially grown on YEP media supplemented with 20 μg/ml rifampicin, 75 μg/ml chloramphenicol and 50 μg/ml kanamycin at 28°C with shaking at 250 rpm for 24 h. Next day, 1.5 ml of *Agrobacterium* culture was pelleted down, washed and resuspended in IM either in absence or presence of 0.2 M acetosyrengone (AS), and allowed to grow until it reached $OD_{600} \sim 0.8$. One hundred microliter of the induced *Agrobacterium* cells and 100 μl of the fungal conidiospores were mixed. This 200 μl mixture was pipetted and spread evenly onto sterile nitrocellulose membrane (diameter 130 mm) placed over the solid IM with or without AS. Keeping the bacterial suspension constant, variable concentrations of the conidial suspension were tested (ranging from 10^2 to 10^5 conidia per plate). Different temperature (20°C to 25°C) and duration (2 to 5 days) of co-cultivation were tested to find out optimum condition of transformation for this fungal strain. After co-cultivation of the plates, the membranes containing the *Agrobacterium*–fungal mixture were removed and placed on PDA plates containing hygromycin B 200μg/ml. The plates were incubated at 28°C until transformed fungal colonies appeared. After obtaining some putative fungal transformants they were maintained on PDA media containing hygromycin. The stability of the randomly selected transformants was tested at least five times on fresh medium with or without hygromycin.

PCR-based screening and Southern hybridization of the transgene in fungal transformants

Screening of putative fungal transformants was carried out among several hygromycin positive clones by PCR using *gusA* or *CtDGAT2b* transgene-specific set of primers (Table S1). For Southern hybridization, 10 μg of genomic DNA from the wild type control fungus and from randomly selected few stable transformants were digested with restriction enzymes and size-fractionated on 0.8% agarose gel along with DNA molecular weight marker. Transfer of nucleic acids onto nylon membrane and hybridizations with $[\alpha-{}^{32}P]dCTP$ labeled transgene-specific probe were performed following standard techniques [33].

Histochemical GUS assay

Histochemical GUS assay of the stable fungal transformants was carried out following standard procedure [34]. Fresh mycelia were submerged in GUS assay buffer (0.1 M sodium phosphate buffer, pH 7.0, 0.1% Triton X-100 and 2 mM X-Gluc) and incubated at 37°C for ~8–10 h in order to develop the characteristic blue color. Following that, mycelia were washed with 70% ethanol and stored in 70% ethanol till photography.

Total RNA extraction and semi-quantitative RT-PCR

Total RNA was extracted from the wild type and fungal transformants using RNA extraction kit (RNeasy plant mini kit, Qiagen, India) according to the manufacturer's protocol. RNA quantity was measured by spectrophotometric reading at 260/280 nm, and the quality of RNA was determined by observing the integrity of rRNA bands in 1.2% agarose gel. For RT-PCR analysis, a set of specific primers was designed (Table S1). The 1st strand cDNA was synthesized by respective gene-specific reverse primers with the High-Capacity cDNA Reverse Transcription Kits (Applied Biosystems) using 3 μg total RNA as template according to the manufacturer's protocol. The products of RT-PCR were subjected to electrophoresis in 1.5% agarose gel and visualized under UV trans-illuminator following ethidium bromide staining.

Microscopy following Nile red staining

The Nile red is a lipid-specific dye. Confocal microscopic analyses of the Nile red-stained wild-type and transformed fungal cells were carried out to examine the presence of storage lipid bodies in mycelia as described previously [35] with certain modification [10]. The samples were observed under laser-scanning confocal microscope with 488 nm excitation and 585 nm emission filters (Fluo View FV1000 confocal microscope, Olympus).

Lipid extraction and fractionation

For extraction of total lipid from the wild-type and transformed fungal biomass, normal chloroform: methanol extraction procedure was followed, and estimation was carried out following Bligh and Dyer method [36] with slight modification as described earlier [10]. Total lipid was fractionated into neutral lipid and estimated using a silica cartridge (Sep-Pak Vac 6cc, Waters) following the method described earlier [10]. Both the quantitative estimations of total lipid and neutral lipid were conducted with three individual cultures for each set of experiments and with three replicates.

Preparation of fatty acid methyl esters and determination of fatty acid profiles

Fatty acid methyl ester (FAME) of the total lipid fraction from the wild type and transformed fungal cells was prepared following the method described earlier [10]. Finally, small volume (3-5μl) of the prepared FAME sample was analyzed by gas chromatography (GC) instrument (Clarus 500, PerkinElmer), fitted with a flame ionization detector (FID) and Omegawax-250 capillary column (30 m length, 0.25 mm internal diameter and 0.25 μm film thickness, Sigma). Identification and quantification of individual chromatographic peaks were carried out by means of external standard (Supelco 37-Component FAME Mix, Sigma) and their corresponding calibration curves. Samples were taken from three individual cultures for each set of experiments and with three replicates.

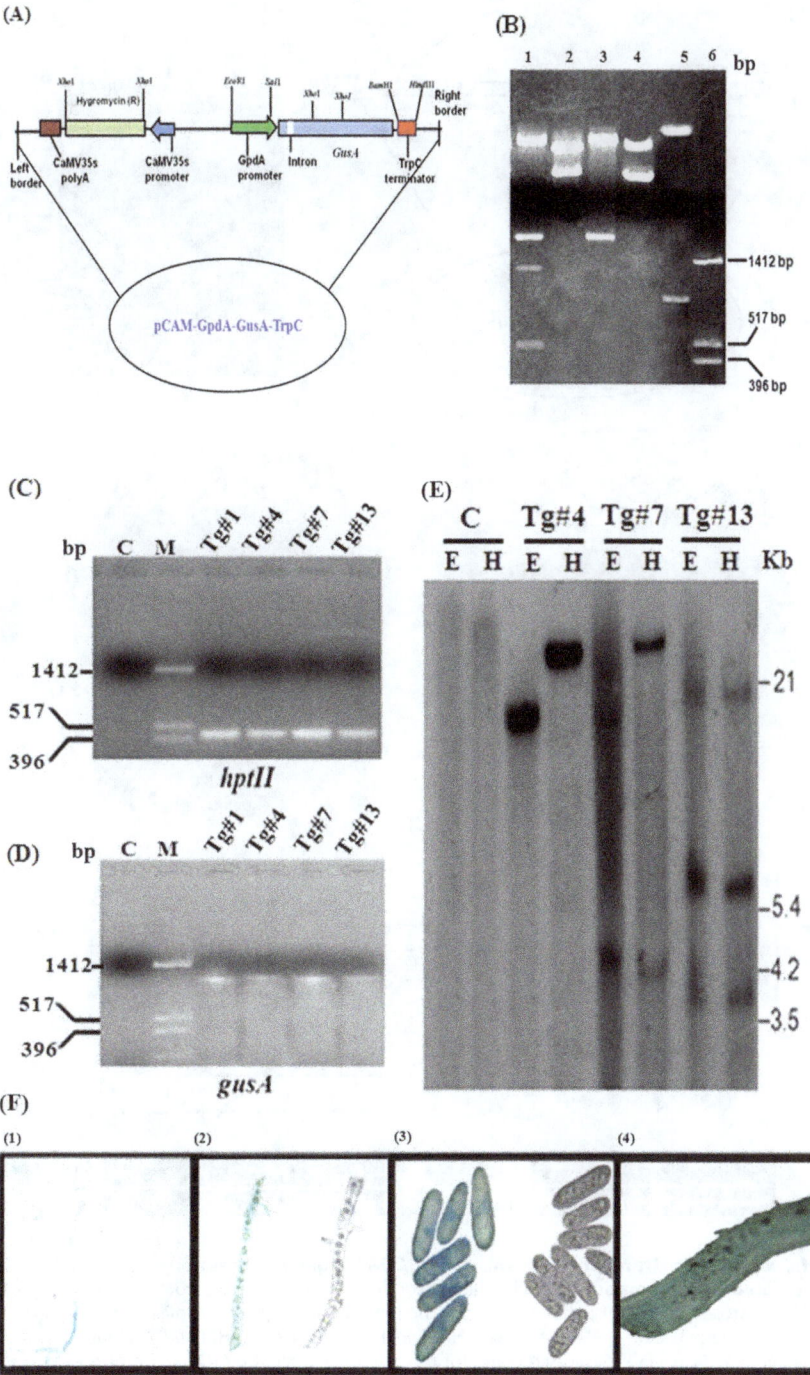

Figure 1. Standardization of *Agrobacterium*-mediated transformation of *Colletotrichum* fungus with the *gusA* reporter gene expression construct, pCAM-GpdA-GusA-TrpC. (A) Schematic diagram of the genetic construct prepared in pCAMBIA1300 binary vector. (B) Ethidium bromide-stained 1.2% agarose gel showing characteristic restriction enzyme digestion profile of the genetic construct with *Xho*HI (lane 1), *Eco*RI+*Hind*III (lane 2), *Eco*RI+*Sal*I (lane 3), *Eco*RI+*Bam*HI (lane 4) and *Bam*HI+*Hind*III (lane 5) along with the pUC18 DNA digested with *Hinf*1 as molecular weight marker (lane 6). (C) PCR-based screening for the presence of 450 bp *hptII* selection marker gene-specific amplicon in four (Tg#1, Tg#4, Tg#7 and Tg#13) chosen transformed lines. (D) PCR-based screening for the presence of 1100 bp *gusA* reporter gene-specific amplicon in the same four transformed lines. In (C) and (D), Lane M = *Hinf*I digested pUC18 DNA as molecular weight marker (E) Confirming genomic integration of the *gusA* transgene through Southern hybridization. Genomic DNA samples of three (Tg#4, Tg#7 and Tg#13) transformed lines were digested with *Eco*R1 (lane E) and *Hind*III (lane H) seperately, and hybridized with the *gusA* gene-specific probe. Lambda (λ) DNA digested with *Eco*RI+*Hind*III was used as molecular weight marker. In case of (C), (D) and (E), the wild type *Colletotrichum* sp. DM06 was taken as control (lane C). (F) Histochemical GUS assay in fungal transformant Tg#4 depicting functional activity of the *gusA* transgene along with the wild type strain as control. Microscopic visualization of GUS activity in different fungal tissues at variable magnifications: mycelium at 10X (1), hypha at 100X (2), spores at 1000X (3) and setae at 40X(4).

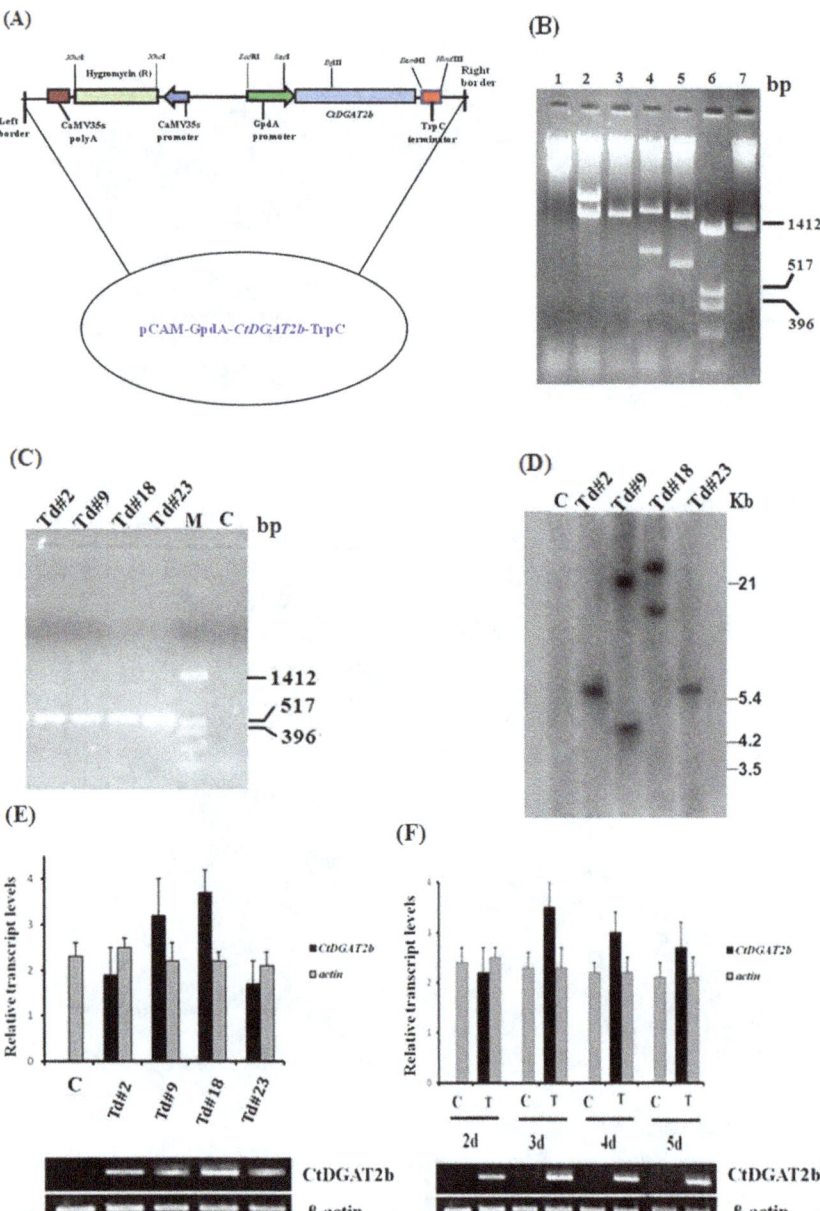

Figure 2. Development of *Colletotrichum* transformants with the *CtDGAT2b* gene expression construct, pCAM-GpdA-CtDGAT2b-TrpC through *Agrobacterium*-mediated transformation. (A) Schematic diagram of the genetic construct prepared in pCAMBIA 1300 binary vector. (B) Ethidium bromide-stained 1.2% agarose gel showing characteristic restriction enzyme digestion profile of the genetic construct with *Eco*RI+*Hind*III (lane1), *Eco*RI+*Bam*HI (lane2), *Bam*HI (lane3), *Sal*I+*Bgl*II (lane 4), *Bam*HI+*Hind*III (lane 5) and *Xho*HI (lane 7) along with the pUC18 DNA digested with *Hinf*1 as molecular weight marker (lane 6). (C) PCR-based screening for the presence of 550 bp *CtDGAT2b* transgene-specific amplicon in four (Td#2, Td#9, Td#18 and Td#23) chosen transformed lines. Lane M = *Hinf*I digested pUC18 DNA as molecular weight marker. (D) Confirming genomic integration of the *CtDGAT2b* transgene through Southern hybridization. Genomic DNA samples of four transformed lines were digested with *Eco*R1, and hybridized with the *CtDGAT2b* gene-specific probe. Lambda (λ) DNA digested with *Eco*RI+*Hind*III was used as molecular weight marker. In case of (C) and (D), the wild type *Colletotrichum* sp. DM06 was taken as control (lane C). (E) Transcriptional expression of the *CtDGAT2b* transgene in four fungal transformants as revealed by RT-PCR for 26 cycles. (F) Transcriptional expression of the *CtDGAT2b* in one of the fungal transformants, Td#18 (T) as revealed by RT-PCR for 26 cycles, showing upregulation of CtDGAT2b transcript upto 3rd day of growth and declined at 4th day onward. In case of (E) and (F), the wild type *Colletotrichum* sp. DM06 was taken as control (C), and the relative transcript level was measured based upon the densitometric scanning of the RT-PCR amplicons (550 bp for *CtDGAT2b* and 400 bp for endogenous *β-actin* as internal control) as shown in lower panel.

Results

Standardization of *Agrobacterium*-mediated transformation of *Colletotrichum* sp. DM06 strain using *gusA* reporter gene construct

Genetic transformation of *Colletotrichum* sp. DM06 was standardized using *Agrobacterium tumefaciens* LBA4404 strain harboring the *gusA* reporter gene construct pCAM-GpdA-GusA-TrpC (Figure 1A). Analysis of the restriction enzyme digestion profile of the recombinant plasmid confirmed the correct orientation of the genetic elements (Figure 1B). Prior to start of transformation the antibiotic susceptibility test of wild type *Colletotrichum* sp. DM06 was carried out on hygromycin B to be used as the selection marker for the transformants. Results showed that hygromycin concentration of 150 µg/ml inhibited the growth of wild type *Colletotrichum* sp. DM06 extremely, whereas the growth was completely inhibited at 200 µg/ml (data not shown). Therefore, a dose of 200 µg/ml of hygromycin B was used for the selection of transformants during *Agrobacterium*-mediated transformation of *Colletotrichum* sp.

After collecting the fungal spores (conidia), initial step of transformation was performed by spreading a mixed suspension of fungal spores and *Agrobacterium* cells on sterile nitrocellulose membranes placed on agar plates without or with acetosyringone (AS), which is necessary for the induction of *vir* genes, followed by successive incubation for cocultivation. Temperature, duration and concentration of *Agrobacterium* cells for cocultivation were optimized for this particular fungal strain. The optimum temperature for cocultivation was found to be 23°C for a period of 3 days at 0.6 cell density (OD_{600}) of *A. tumefaciens* cells. Further incubation for a prolonged period of 5 days increased excessive mycelial growth of *Colletotrichum*, which created difficulties in isolation of individual transformants in AS containing plates. Similar result was obtained with the increased number of *Agrobacterium* cells as well.

After cocultivation for 3 days at 23°C, the nitrocellulose membrane was transferred to hygromycin (200 µg/ml) containing plate, and incubated for 5 days at 28°C until putative transformants were visible. We could usually obtain 40 to 100 transformants per 5×10^5 numbers of conidia of *Colletotrichum* sp. DM06. Transformation with more number of conidia (1×10^6) produced a mycelial lawn, which made it complicated to isolate individual transformant (data not shown). Moreover, incubation of *Agrobacterium* cells with AS prior to cocultivation, and addition of AS to plates during cocultivation enhanced the number of putative transformants than the AS negative plates (data not shown).

Analysis of the fungal transformants for inheritance stability, genomic integration and functional expression of the transgene

The inheritance stability of the hygromycin resistant gene in putative transformants was examined by transferring mycelia from the edge of transformed colonies at least five times to new medium with and without hygromycin. Among the 40 transformants obtained, 15 were adjudged as stable, whereas other lost the hygromycin resistant phenotype after serial subculture on antibiotic selective medium. Among the 15 stable transformants, four (Tg#1, Tg#4, Tg#7 and Tg#13) were selected for further analyses because these four were significantly more stable, retaining their ability to grow rapidly on hygromycin selective medium after five serial subcultures on non-selective medium.

Verification of the transgene integration onto the fungal genome was carried out by PCR using two gene-specific primer pairs (Table S1) and subsequently confirmed by Southern hybridization. The presence of amplified fragments of 450 bp (Figure 1C) and 1100 bp (Figure 1D) in genomic DNA samples from transformed colonies revealed the existence of the *hptII* and *gusA* transgenes, respectively. The genomic DNA sample from the non-transformed colony did not show any PCR-amplified fragment specific for these two transgenes (Figure 1C, 1D). To confirm the integration and copy number of the transgene in the respective genomes of selected *Colletotrichum* transformants, Southern blot analysis was performed using genomic DNA samples isolated from the PCR-positive and non-transformed fungal colonies. Genomic DNA samples were digested with *Eco*R1 and *Hin*dIII seperately, and hybridized with the *gusA* gene-specific probe. All the three transformant lines tested (Tg#4, Tg#7 and Tg#13)) showed clear hybridization signal for transgene integration onto the corresponding genomes (Figure 1E). Genomic DNA from the non-transformed colony was used as a negative control and showed no hybridization signal (Figure 1E). The T-DNA region of the pCAM-GpdA-GusA-TrpC (Figure 1A) recombinant plasmid has only one *Hin*dIII and *Eco*RI site each. In transformed fungal colony Tg#7, two hybridization signals for *Eco*R1 and *Hin*dIII indicated two copies of T-DNA integration in this line (Figure 1E), while transformant Tg#13 showed three hybridization signals for both the enzymes indicating that three copies of T-DNA integrated onto the genome of this transformed line (Figure 1E). On the contrary, single hybridization signal for both the enzymes in transformed line Tg#4 specified one copy of transgene integration in this fungal genome (Figure 1E).

Functional expressions of the transgene in selected transformants were verified by histochemical GUS assay. A high percentage GUS positive cells were detected in the transformed mycelia, indicating successful expression and functional activity of the *gusA* gene. Characteristic blue staining was visible within the fungal hyphae, conidia and setae of Southern- positive fungal transformants through microscopic examination, as shown for Tg#4, one of the transformed lines (Figure 1F).

Generation of *Colletotrichum* transformants using *CtDGAT2b* gene construct

A. tumefaciens LBA4404 strain harbouring the recombinant plasmid pCAM-GpdA-CtDGAT2b-TrpC (Figure 2A) that contains the target gene *CtDGAT2b* was used to genetically transform the *Colletotricum* sp. DM06 following the transformation protocol standardized using the *gusA* reporter gene construct. Restriction enzyme digestion of the recombinant plasmid was carried out before transformation to confirm the correct orientation of different genetic elements including the *CtDGAT2b* gene (Figure 2B). Putative fungal transformants harbouring the *CtDGAT2b* gene were selected on hygromycin (200µg/ml) plate, as performed for *gusA* gene transformation. Few selected putative *Colletotrichum* transformants (Td#2, Td#9, Td#18 and Td#23) showing effective growth on both antibiotic selective and non-selective media after 5 cycles of subculture, were subjected to PCR screening using *CtDGAT2b* gene-specific primer pair (Table S1). An expected size 550 bp amplicon specific for *CtDGAT2b* gene was revealed in all four selected transformants, but no such amplified product was detected in untransformed control colony (Figure 2C). To confirm the integration of transgene in the fungal genome, Southern hybridization of the *Eco*RI digested genomic DNA isolated from each of the PCR-positive transformants along with untransformed fungal cells was carried out using the *CtDGAT2b* gene-specific probe (Figure 2D). Among the four PCR-positive transformants tested, Td#2 and Td#23 showed single copy genomic integration of the transgene, whereas two copies of transgene were found to be integrated into the genome of Td#9 and Td#18 transformsnts. On the contrary, no such Southern hybridization signal was observed for the untransformed control (Figure 2D). Further qualitative and quantitative analyses of storage lipid were carried out in these four stable transformants after examining the transgene expression.

To verify the expression of *CtDGAT2b* transgene in *Colletotrichum* transformants, semi-quantitative RT-PCR was carried out using the DNaseI-treated total RNA samples isolated from Southern-positive transformants along with the untransformed fungal cells. Under the conditions described in the methods, the exponential phase of RT-PCR using CtDGAT2b-specific primers was detected between 24 to 28 cycles (data not shown). Subsequently, the RT-PCR for 26 cycles yielded a 550 bp *CtDGAT2b*-specific amplicon in all the samples of fungal transformants, whereas no such RT-PCR amplicon was obtained in untransformed control (Figure 2E). No DNA fragment was PCR-amplified by transgene-specific primer set using the DNaseI-treated total RNA samples directly as the templates (without reverse transcription), indicating absence of genomic DNA contamination in the RNA samples used for RT-PCR (data not shown). When quantitative variations of the β-*actin* transcript were correlated with that of the *CtDGAT2b* transcript, a slightly enhanced expression level of *CtDGAT2b* transgene was observed in the fungal transformant till certain time periods of growth, as depicted for Td#18, one of the transformants (Figure 2F). A clear upregulation of the *CtDGAT2b* transcript was recorded at 3rd day of growth, and its expression declined at 4th day onward.

Enhanced accumulation of storage lipid with altered fatty acid profile in transformed fungal cells expressing *CtDGAT2b* transgene

Fungal mycelia from both the transformant and nontransformant were stained with lipid-specific Nile red dye, and visualized by confocal microscopy. Significant increase in fluorescence intensity was observed in the transformant with *CtDGAT2b* transgene compared to the control fungal cells at different magnification of microscopy (Figure 3A). Moreover, an average increase in size of lipid droplets was noticeable throughout the transformed mycelia on comparision to nontransformant at 60X magnification (Panel 4, Figure 3A). This result revealed that CtDGAT2b transformed fungal cells accumulated higher amount of storage lipid compared to the nontransformant. Quantitative details of the total lipid and neutral lipid production in four stable transformants are presented in the next subsection. Fatty acid profiles of the four stable transformants expressing the *CtDGAT2b* gene along with the untransformed fungal cells were examined through GC-FID of FAME samples prepared from total lipid. Analysis revealed that the contents of saturated fatty acids, particularly palmitic acid (C16:0) and stearic acid (C18:0), were found to be increased, but the amount of oleic acid (C18:1) decreased in all four transformants compared to the untransformed fungus (Figure 3B). The contents of C16:0 and C18:0 in transformants were approximately 30–58% and 40–58% more, respectively compared to the wild type; whereas the C18:1 content of the transformants was about 28–35% less with respect to wild type.

Functional expression of *CtDGAT2b* increases lipid titer in transformed *Colletotrichum* lines upon both standard and stress conditions of growth

Detailed estimation of the lipid productivity was carried out to determine the differences between the wild type and the transformed *Colletotrichum* lines, in order to analyze the effect of expressing yeast *CtDGAT2b* gene in fungal cells. Our previous study indicated that in high glucose-containing media, nitrogen and phosphorus are the limiting nutrient factors for lipid titer in this endophytic fungus [10]. Therefore, three culturing conditions were evaluated in the present study to find out *in vivo* effect of target gene overexpression: (i) standard condition (i.e., nutrient repleat condition), (ii) nitrogen stress condition and (iii) phosphorus stress condition.

Four stable transformants along with the wild type *Colletotrichum* sp. DM06 were cultured in the three above-mentioned conditions of growth, and total lipid was extracted and estimated from cell biomass of each sample. In standard condition (i.e., without nutrient stress) the wild type fungus produced about 4.8±0.1 g/l of total lipid that corresponds to 38.4±0.4% of dry cell weight (DCW), whereas in similar condition the four stable transformants produced on an average 8.1±0.2 g/l of total lipid (= 73.6±0.6% DCW) (Table 1, Figure 3C). Thus, the lipid titer of transformed cells was ~1.7-fold higher than the wild type strain at standard growth condition, indicating an increase of lipid content by ~92% DCW. Moreover, when tested in nitrogen stress (C:N 160:1) and phosphorus stress (C:P 86:1) conditions (previously standardized as optimum conditions for highest lipid titer of this fungus, [10]); the wild type fungus produced total lipid around 7.7±0.2 g/l (= 42.5±0.1% of DCW) and 7.9±0.4 g/l of total lipid (= 42.8±0.4% of DCW), respectively (Table 1, Figure 3C). In comparision to these data, the four stable transformants produced an average 12.2±0.7 g/l (= 75.2±0.4% DCW) and 12.1±0.1 g/l (= 75.2±0.4% DCW) of total lipid in nitrogen and phosphorus stress conditions, respectively (Table 1, Figure 3C). Therefore, the lipid titer of transformed cells was ~1.5-fold more than the wild type fungus in stress conditions of growth, implying an enhancement of lipid content by ~78% DCW.

To examine whether the heterologous expression of *CtDGAT2b* gene directly influences the neutral lipid production in fungal transformants, the total lipid of each sample was fractionated to recover neutral lipid component for its measurement. The wild type fungus produced neutral lipid ~2.2±0.7 g/l (= 17.6±0.8% DCW) in standard condition, ~3.6±0.7 g/l (= 19.7±0.6% DCW) in nitrogen stress condition, and ~4.3±0.4g/l (= 23.2±0.3% DCW) in phosphorus stress condition (Table 1, Figure 3D). On the contrary, the four transformants produced an average ~4.0±0.8 g/l (= 36.3±0.6% DCW), ~6.1±0.8 g/l (= 37.6±0.6% DCW) and 7.1±0.3 g/l (= 43.8±0.8% DCW) of neutral lipid in standard, nitrogen and phosphorus stress conditions, respectively (Table 1, Figure 3D). Taken together, it was observed that the neutral lipid titer in transformants increased upto 1.8-fold, 1.6-fold and 1.5-fold compared to the wild type when grown under standard, nitrogen stress and phosphorus stress conditions, respectively. From these findings it could be inferred that the *CtDGAT2b* gene has been successfully utilized for enhanced neutral lipid production in transformed *Colletotricum* lines.

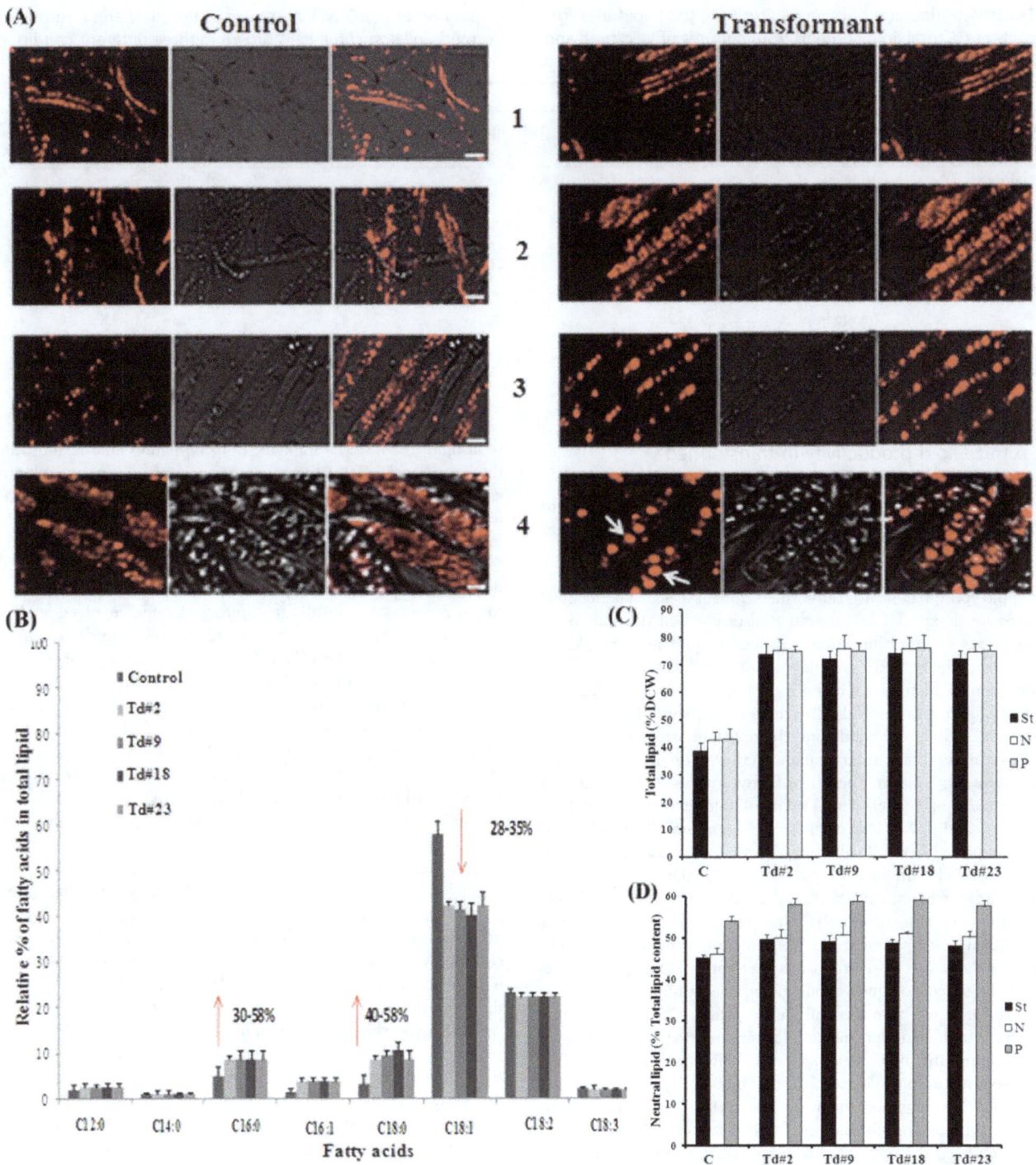

Figure 3. Alteration in the quality and quantity of storage lipid due to functional activity of CtDGAT2b in *Colletotricum* **transformants.** (A) Nile red stained confocal microscopic images of mycelia of the wild-type (control) and one of the transformed lines (Transformant). Panels 1 to 4 represent microscopic images of mycelia at different magnifications, and each panel has identical scale bar for both the control and transformant. In panels 1 and 2, scale bar = 10 μm showing same magnification but at different location of mycelia. In panels 3 and 4, scale bar = 5 μm and 2 μm, respectively. White arrow indicates increase in size of lipid droplets in mycelia of transformant. (B) Bar diagram representing fatty acid profile (relative % of fatty acids) in total lipid samples obtained from the wild-type and four transformants as revealed by GC-FID analysis. (C) Bar diagram showing total lipid content expressed as % of dry cell weight (% DCW) during standard growth condition (black bar), nitrogen starvation (white bar) and phosphorus starvation (grey bar) among the wild type control (C) and four (Td#2, Td#9, Td#18 and Td#23) fungal transformants. (D) Bar diagram depicting neutral lipid content expressed as % of total lipid content during above-mentioned three growth conditions among the wild type control (C) and the same four fungal transformants.

Table 1. Estimation of biomass titer (g/l), total lipid titer (g/l), total lipid content (%DCW), neutral lipid titer (g/l) and neutral lipid content (% total lipid content) from cultures of wild type and transformed *Colletotrichum* cells grown in three different conditions for 10 days.

Conditions of growth	Organism	Biomass titer (g/l)	Total lipid titer (g/l)	Total lipid content (%DCW)	Neutral lipid titer (g/l)	Neutral lipid content (%DCW)
Standard	Wild-type	12.5±0.4	4.8±0.1	38.4±0.4	2.2± 0.7	17.6 ± 0.8
	Transformant[a]	11±0.6	8.1±0.2	73.6±0.6	4.0± 0.8	36.3± 0.6
Nitrogen stress	Wild-type	18.2±0.8	7.7±0.2	42.5±0.1	3.6± 0.7	19.7 ± 0.6
	Transformant[a]	16.2±0.1	12.2±0.7	75.2±0.4	6.1 ± 0.8	37.6 ± 0.6
Phosphorus stress	Wild-type	18.5±0.5	7.9±0.4	42.8±0.4	4.3 ± 0.4	23.2 ± 0.3
	Transformant[a]	16.2±0.5	12.1±0.1	75.2±0.4	7.1 ± 0.3	43.8 ± 0.8

a= average value for four stable transformed lines- Td#2, Td#9, Td#18 and Td#23

Model-based optimization of culture conditions to maximize lipid productivity in transformed *Colletotrichum* cells

The four critical parameters influencing lipid productivity of *Colletotrichum* fungus i.e., temperature, pH, carbon: nitrogen ratio and carbon: phosphorus ratio as revealed by our previous study [10] were taken into account to find out the optimum conditions for lipid production in transformed fungal cells. The central composite design (CCD) ensured statistically well distributed 30 data points for modelling using artificial neutral network (ANN) based on these four growth parameters (Table S2). After estimation of biomass titer and total lipid titer of the wild type (Table S3, S5) and one (Td#2) of the transformants (Table S4, S6) grown in 30 different conditions, the optimum growth conditions for maximum lipid accumulation were set up (Table 2). The regression correlation coefficient between the ANN-simulated values and actual experimental values of lipid titer was 0.97724 for wild type and 0.98541 for transformed organism (Table S7). The Mean Squared Error (MSE) and Mean Absolute Error (MAE) of predicted lipid yield for wild type fungus were 0.0174 and 0.0118, respectively; whereas the MSE and MAE of predicted lipid titer for transformed fungus were 0.0161 and 0.0130, respectively (data not shown). Genetic algorithm was run for a maximum of 500 iterations to observe a gradual convergence of the best fitness values of successive generations giving the optima for both the strains. Once we have the optimum conditions of lipid accumulation for both wild type and transformant seperately, it was easier to compare the lipid accumulating ability among them. The experimental result showed that the highest lipid titer in best optimum condition for the wild type fungus was 9.1±0.8 g/l, whereas for the transformant it was 14.1±0.5 g/l (Table 3). We cross-validated the optimal conditions by switching the wild type strain under the transformant's optimal conditions, and the transformant under the wild type's optimal conditions. Analysis of the data revealed that switching the optimal condition for growth did not alter lipid content significantly (Table 3). Therefore, the lipid titer of transformed cells was ~1.5-fold greater than the wild type strain in model-optimized growth condition, indicating an increase of lipid content by ~73% DCW.

Discussion

Previously, we have characterized an endophytic fungus *Colletotrichum* sp. DM06, which is capable of accumulating significant amount of storage lipid having biodiesel properties [10]. This *Colletotrichum* strain can produce high amount of total lipid (~7.8±0.2 g/l) using simple sugar, and substantial quantity of total lipid (~84.3±3.9 mg/gds) utilizing lignocellulosic substrates that are degraded by the inherent secretory enzymes including cellulase. The results of both liquid culture and soild state fermentation for lipid production are promising and provided clue for further investigation towards strain improvement of this filamentous fungus to produce lipid feedstock in cost-effective way. Therefore, we have adopted the genetic engineering strategy to overexpress a critical rate-limiting enzyme involved in lipid biosynthesis along with the prediction-based optimization of culture conditions. The *CtDGAT2b* gene (encoding for a type-2 DGAT isoform from the oleaginous yeast *C. tropicalis*) has been transgenically introduced in *Colletotrichum* sp. DM06 to increase the metabolic flux towards the production of storage lipid in the transformed fungus. Additionally, ANN-based optimization of culture conditions (with respect to temperature, pH, carbon: nitrogen ratio and carbon: phosphorus ratio) has been established to further enhance the lipid productivity in the genetically improved *Colletotrichum* strain.

Agrabacterium tumefacience-mediated transformation (ATMT) in fungus is already in practice by scientists to genetically improve the strain for the production of desired metabolite, but the protocol varries among fungal genera. Generally, the most commonly used antibiotic selection agent for fungal transformants is hygromycin B, and the dose is 100 μg/ml [37,38]. However, there are previous reports on few species of *Colletotrichum* genus, where hygromycin sensitivity of 50 μg/ml to 300 μg/ml has been recorded [39]. Similar to different plant species, as various fungal strains of same genus have species-specific sensitivity on hygromycin; here firstly we have tested the antibiotic sensitivity dose on wild type isolate *Colletotrichum* sp. DM06, and finally 200 μg/ml concentration of hygromycin B was used for the selection of genetic transformants. Previous study showed that the effectiveness of transformation could be increased in presence of AS for different fungal species [40]. Moreover, few studies reported that the efficiency of fungal transformation is dependent primarily on the temperature, cocultivation duration, and concentration of the bacterial cell suspension [41,37,42,39]. In a variety of Ascomycetes fungi, generally 10^6 numbers of conidia have been used for ATMT experiment [41]. In the present study, we have standardized the ATMT of *Colletotrichum* sp. DM06 strain with *gusA* reporter gene, and it was found that cocultivation of $5×10^5$ numbers of fungal conidia with 0.8 cell density (OD_{600}) of *A. tumefaciense* (LBA4404) cells at 23°C for 3 days and incubation with AS were optimum. After confirmation of the inheritance stability and genomic integration of the transgene in few transformed lines,

Table 2. Predicted biomass titer, lipid titer and lipid content in the wild type and transformed fungal cells through model-based optimization of growth conditions

Organism	Temperature	pH	C:N	C:P	Predicted biomass titer (g/l)	Predicted lipid titer (g/l)	Predicted lipid content (%DCW)
Wild type	27.3	6.30	129.46	83.32	18.2	9.5	52.8
Transformant	29.6	6.61	152.07	108.08	17.5	14.74	86.7

Table 3. Experimentally determined values of biomass titer, lipid titer and lipid content in the wild type and transformed fungal cells cultured in model-optimized growth conditions.

Organism	Temperature	pH	C:N	C:P	Experimental biomass titer (g/l)	Experimental lipid titer (g/l)	Experimental lipid content (%DCW)
Wild type	27	6.3	129	83	18.5±0.7	9.1±0.8	49.2±0.6
Transformant	27	6.3	129	83	15.5±0.4	12.4±0.5	79.2±0.4
Wild type	29	6.6	152	108	17.6±0.5	8.4±0.6	47.5±0.6
Transformant	29	6.6	152	108	16.2±0.9	14.1±0.5	85.4±0.3

phenotypic expression of the transgene was validated in selected fungal transformants through histochemical GUS assay.

In order to improve the production of storage lipid (TAG) in *Colletotrichum* sp. DM06 strain, the *CtDGAT2b* gene was transgenically introduced in the filamentous fungus following the standardized ATMT protocol. Putative fungal transformants were selected on hygromycin (200μg/ml) containing plates, and few stable transformants with either single copy or two copies genomic integrations of the transgene were obtained. After successful expression of the *CtDGAT2b* transgene, qualitative improvement of storage lipid and increase in the size of lipid bodies were observed in *Colletotricum* transformants with respect to the wild type strain as revealed through Nile red fluorescence staining assay. Moreover, difference in fatty acid profile with significant increase in the content of saturated fatty acids (C16:0 and C18:0) was also recorded in transformants compared to the wild type fungus. This could be due to the fact that the overexpression of *CtDGAT2b* gene, which has been isolated from the high stearate-rich *C. tropicalis* yeast strain, produces TAG with more saturated fatty acids as the CtDGAT2b enzyme has a natural tendency to pull saturated fatty acyl-CoAs. In fact, this preferential substrate specificity of CtDGAT2b has also been documented in our earlier study through heterologous expression in non-oleaginous yeast *Saccharomyces cerevisiae* [30].

From the quantitative estimation data on total lipid and neutral lipid fractions of transformants and wild type strain grown in standard and nutrient-stress conditions, it is evident that the overexpression of *CtDGAT2b* gene has prominant effect on enhancing the accumulation of storage lipid in transformed fungal lines. However, the four different transformed lines (Td#2, Td#9, Td#18 and Td#23) did not display much variation in accumulation of either total lipid or neutral lipid amongst them under all three conditions of growth (standard, nitrogen stress and phosphorus stress). Another interesting observation was noticed during growth of the wild type and transformant cells cultured for quantitative analysis of lipid. Although phenotypically (colony morphology) no differences were observed between wild type and transformant in plate culture, slight variation in growth morphology was found in liquid culture. When cultivated during standard growth condition or nutrient stress condition, wild type *Colletotrichum* cells usually grow by forming small aggregates or pellet-like structures and the culture is viscous. On the contrary, when grow in similar condition, all four transformed fungal lines produce less pellet and form cake-like morphology (data not shown), suggesting that the CtDGAT2b overexpression might have slight influence in cell physiology. As a consequence of this differential behavior, growth rate and biomass formation of the wild type and transformant demonstrated discrepencies even when they were grown in similar conditions. Therefore, to further validate the comparison of lipid titers between the wild type and transformants of *Colletotrichum*, a model-based optimization of growth parameters was carried out to maximize lipid productivity in transformed fungal lines along with the wild type cells. A bioprocess optimization protocol has been established using ANN genetic algorithm, where 30 well distributed statistical data points were set with the help of CCD based on four critical parameters influencing lipid productivity in *Colletotricum* fungus such as temperature, pH, carbon: nitrogen (C:N) ratio and carbon: phosphorus (C:P) ratio according to our previous study [10].

We have found that the wild type fungus can produce about 7.8±0.6 g/l of total lipid under nutrient (either nitrogen or phosphorus) stress condition, which is ~1.6-fold more than the lipid titer (4.8±0.1 g/l) at standard growth condition. However, after model-based optimization of culture conditions, the wild type

strain produces 9.1±0.8 g/l of total lipid i.e., ~1.9-fold higher lipid titer without imposition of any nutrient stress (Figure 4). On the contrary, while the transformants at standard growth condition yield ~1.7-fold more total lipid compared to the wild type (4.8±0.1 g/l increases to 8.1±0.2 g/l), optimization of culture conditions further increases the lipid titer upto 14.1±0.5 g/l or ~1.7-fold higher in transformants (Figure 4). Taken together, ~2.9-fold greater lipid titer over the wild type has been achieved in transformed *Colletotrichum* fungus following model-based optimization of culture condition and without any nutrient stress (Figure 4). Therefore, it is evident from the present study that not only biochemical engineering (nutrient stress) but the metabolic engineering of fungus using suitable gene together with bioprocess optimization could be an effective combined strategy to increase the storage lipid productivity. Critical analysis of our experimental data (Table 3) reveals that the lipid content in the transformed *Colletotrichum* is ~1.9-fold higher under standard growth condition, and ~2.2-fold more after model-based optimization of culture conditions as compared to the wild type strain grown at standard condition. In a previous report, the fungal genes coding for maiC enzyme from *Mucor circinelloides* (malEMt) and *Mortierella alpine* (malEMc) were overexpressed in *M. circinelloides*, which resulted 2.5- and 2.4-fold higher lipid content in the transformed malEMt and malEMc strains, respectively [22]. A recent report has documented 4-fold increase in lipid content compared to wild type through overexpression of *DGA1* gene (encoding for a type-1 DGAT) in *Yerrowia lipolytica* oleaginous yeast [43]. In the same organism, increased lipid content of 2-fold over control was observed after overexpression of acetyl-CoA carboxylase (ACC1) gene. Moreover, when both the genes (*ACC1+DGA1*) were overexpressed a lipid content of 4.9-fold was recorded in the same study. This 'push-and-pull' strategy can achieve large flux towards lipid accumulation with less feedback regulation as it maintain a balance between upstream and downstream metabolite formation pathway.

Conclusions

We have genetically modified an oleaginous filamentous endophytic fungus *Colletotrichum* sp. DM06 with the introduction of heterologous *CtDGAT2b* gene, which encodes a highly efficient version of DGAT enzyme responsible for enhanced TAG accumulation in rhizospheris yeast *C. tropicalis* SY005. Our laboratory-standardized method of *A. tumefacience*-mediated transformation was carried out to develop several independent transformed lines of *Colletotrichum*. Four stable transformants were investigated for the expression of *CtDGAT2b* transgene, qualitative assay and quantitative estimation of total lipid as well as neutral lipid. One of the stable transformants having single copy CtDGAT2b transgene in the fungal genome was further examined for the fatty acid profile of the total lipid and subjected to model-based bioprocess optimization for enhancement of lipid titer. Functional expression of CtDGAT2b successfully increases storage lipid titer as well as lipid content in genetically modified *Colletotrichum* with fatty acid profile being altered towards more palmitate and stearate. Prediction-based bioprocess optimization has been carried out to find the appropriate culturing conditions for lipid accumulation in both the wild type and transformed *Colletotrichum*. The lipid accumulation ability has been tested at optimum state of growth, and a net increase to 2.9-fold lipid titer over the wild type has been achieved due to genetic transformation of *Colletotrichum* sp. DM06 with the *CtDGAT2b* gene, which was not possible only by applying biochemical stress. To our knowledge, this is the first report to document genetic engineering

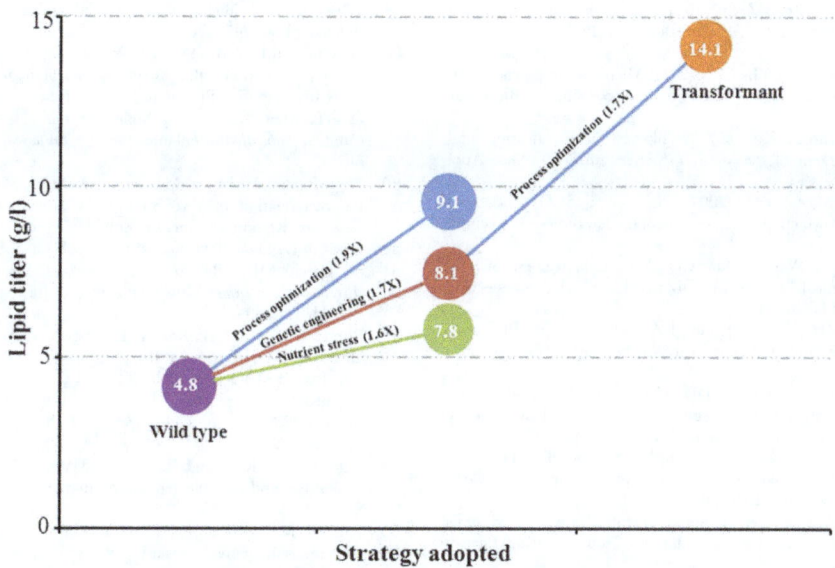

Figure 4. Schematic representation of changes in lipid titer in the wild type and transformed *Colletotrichum* fungus after adopting different strategies.

coupled with model-based bioprocess optimization in an endophytic filamentous fungus for the enhancement of lipid productivity. In future such kind of genetic or metabolic engineering approach with important rate-limiting enzymes or critical transcription factors of lipid biosynthesis pathway along with the optimization of culture condition could further increase the lipid accumulation ability of oleaginous fungi or other microorganisms.

Supporting Information

Table S1   Primers used in this study.

Table S2   Central composite design (CCD) of statistically distributed 30 different conditions based on four growth parameter.

Table S3   Experimentally determined biomass and lipid titers of the wild type *Colletotrichum* sp. grown in 30 different conditions.

Table S4   Experimentally determined biomass and lipid titers of transformed *Colletotrichum* sp. grown in 30 different conditions.

Table S5   Predicted and experimentally determined biomass and lipid titers of the wild type *Colletotrichum* sp. grown in 30 different conditions.

Table S6   Predicted and experimentally determined biomass and lipid titers of the transformed *Colletotrichum* sp. grown in 30 different conditions.

Table S7   Regression correlation coefficient (R^2) between ANN-simulated values and experimental values of lipid titer in the wild type and transformed *Colletotrichum* sp.

Acknowledgments

We thank Prof. S. K. Sen and Dr. A. Basu for their cooperation. Authors thank Gunna Seelan and Ankush Karemore for helping in bioprocess optimization, and Nitai Giri for technical help. PD acknowledges IIT-Kharagpur for providing research fellowship.

Author Contributions

Conceived and designed the experiments: MKM. Performed the experiments: PD NM AC MC. Analyzed the data: PD NM MKM. Wrote the paper: PD MKM.

References

1. Ratledge C (2002) Regulation of lipid accumulation in oleaginous microorganisms. Biochem Soc Trans 30: 1047–1050.
2. Gatenby CM, Orcutt DM, Kreeger DA, Parker BC, Jones VA, et al. (2003) Biochemical composition of three algal species proposed as food for captive freshwater mussels. J Appl Phycol 15: 1–11.
3. Walker TL, Purton S, Becker DK, Collet C (2005) Microalgae as bioreactors. Plant Cell Rep 24: 629–641.
4. Chisti Y (2007) Biodiesel from microalgae. Biotechnol Adv 25: 294–306.
5. Li Q, Du W, Liu D (2008) Perspectives of microbial oils for biodiesel production. Appl Microbiol Biotechnol 80: 749–756.
6. Meng X, Yang J, Xu X, Zhang L, Nie Q, et al. (2009) Biodiesel production from oleogenous microorganisms. Renew Energ 34: 1–5.
7. Beopoulos A, Cescut J, Haddouche R, Uribelarrea JL, Jouve CM, et al. (2009) *Yarrowia lipolytica* as a model for biooil production. Prog Lipid Res 48: 375–387.
8. Ageitos MJ, Vallejo AJ, Veiga-Crespo P, Villa GT (2011) Oily yeasts as oleaginous cell factories. Appl Microbiol Biotechnol 90: 1219–1227.
9. Fakas S, Papanikolaou S, Batsos A, Galiotou-Panayotou M, Mallouchos A, et al. (2009) Evaluating renewable carbon sources as substrates for single cell oil production by *Cunninghamella echinulata* and *Mortierella isabellina*. Biomass Bioenerg 33: 573–580.
10. Dey P, Banerjee J, Maiti MK (2011) Comparative lipid profiling of two endophytic fungal isolates – *Colletotrichum* sp. and *Alternaria* sp. having potential utilities as biodiesel feedstock. Bioresour Technol 102: 5815–5823.

11. Wu S, Hu C, Jin G, Zha X, Zhao ZK (2010) Phosphate-limitation mediated lipid production by *Rhodosporidium toruloides*. Bioresour Technol 101: 6124–6129.

12. Wu S, Zhao X, Shen H, Wang Q, Zhao ZK (2011) Microbial lipid production by *Rhodosporidium toruloides* under sulfate-limited conditions. Bioresour Technol 102: 1803–1807.

13. Naganuma T, Uzuka Y, Tanaka K (1985) Physiological factors affecting total cell number and lipid content of the yeast, *Lipomyces starkeyi*. J Gen Appl Microbiol 31: 29–37.

14. Papanikolaou S, Komaitis M, Aggelis G (2004) Single cell oil (SCO) production by *Mortierella isabellina* grown on high-sugar content media. Bioresour Technol 95: 287–291.

15. Courchesne NM, Parisien A, Wang B, Lan CQ (2009) Enhancement of lipid production using biochemical, genetic and transcription factor engineering approaches. J Biotechnol 141: 31–41.

16. Chen JQ, Lang CX, Hu ZH, Liu ZH, Huang RZ (1999) Antisense PEP gene regulates to ratio of protein and lipid content in *Brassica napus* seeds. J Agric Biotech 7: 316–320.

17. Zou J, Katavic V, Giblin EM, Barton DL, MacKenzie SL, et al. (1997) Modification of seed oil content and acyl composition in the Brassicaceae by expression of a yeast sn-2 acyltransferase gene. Plant Cell 9: 909–923.

18. Bouvier-Nave P, Benveniste P, Oelkers P, Sturley SL, Schaller H (2000) Expression in yeast and tobacco of plant cDNAs encoding acyl CoA: diacylglycerol acyltransferase. Eur J Biochem 267: 85–96.

19. Rangasamy D, Ratledge C (2000) Genetic enhancement of fatty acid synthesis by targeting rat liver ATP: citrate lyase into plastids of tobacco. Plant Physiol 122: 1231–1238.

20. Jako C, Kumar A, Wei Y, Zou J, Barton DL, et al. (2001) Seed-specific over-expression of an arabidopsis cDNA encoding a diacylglycerol acyltransferase enhances seed oil content and seed weight. Plant Physiol 126: 861–874.

21. Davis MS, Solbiati J, Cronan JE (2000) Overproduction of acetyl-CoA carboxylase activity increases the rate of fatty acid biosynthesis in *Escherichia coli*. J Biol Chem 275: 28593–28598.

22. Zhang Y, Adams IP, Ratledge C (2007) Malic enzyme: the controlling activity for lipid production? Overexpression of malic enzyme in *Mucor circinelloides* leads to a 2.5-fold increase in lipid accumulation. Microbiology 153: 2013–2025.

23. Dunahay TG, Jarvis EE, Dais SS, Roessler PG (1996) Manipulation of microalgal lipid production using genetic engineering. Appl Biochem 57–58: 223–231.

24. Liang MH, Jiang JG (2013) Advancing oleaginous microorganisms to produce lipid via metabolic engineering technology. Prog Lipid Res 52: 395–408.

25. Zheng P, Allen WB, Roesler K, Williams ME, Zhang S, et al. (2008) A phenylalanine in DGAT is a key determinant of oil content and composition in maize. Nat Genet 40: 367–372.

26. Oakes J, Brackenridge D, Colletti R, Daley M, Hawkins DJ, et al. (2011) Expression of fungal diacylglycerol acyltransferase2 genes to increase kernel oil in maize. Plant Physiol 155: 1146–1157.

27. Pal MP, Vaidya BK, Desai KM, Joshi RM, Nene SN, et al. (2009) Medium optimization for biosurfactant production by *Rhodococcus erythropolis* MTCC 2794: artificial intelligence verses a statistical approach. J Ind Microbiol Biotechnol 36: 747–756.

28. Guo Yl, Xu J, Zhang Y, Xu H, Yuan Z, et al. (2010) Medium optimization for ethanol production with *Clostridium autoethanogenum* with carbon monoxide as sole carbon source. Bioresour Technol 10: 8784–8789.

29. Dey P, Maiti MK (2013) Molecular characterization of a novel isolate of *Candida tropicalis* for enhanced lipid production. J Appl Microbiol 114: 1357–68.

30. Dey P, Chakraborty M, Kamdar MR, Maiti MK (2014) Functional Characterization of Two Structurally Novel Diacylglycerol Acyltransferase2 Isozymes Responsible for the Enhanced Production of Stearate-Rich Storage Lipid in *Candida tropicalis* SY005. PLoS ONE 9(4): e94472.

31. de Groot MJA, Bundock P, Hooykaas PJJ, Beijersbergen AGM (1998) *Agrobacterium tumefaciens*-mediated transformation of filamentous fungi. Nat Biotech 16: 839–842.

32. Bundock P, den Dulk-Ras A, Beijersbergen A, Hooykaas PJJ (1995) Trans-kingdom T-DNA transfer from *Agrobacterium tumefaciens* to *Saccharomyces cerevisiae*. EMBO J 14: 3206–3214.

33. Sambrook J, Fritsch EF, Maniatis T (1989) Molecular cloning: a laboratory manual, 2nd edition. Cold Spring Harbor Laboratory Press, Cold Spring Harbor, NY.

34. Jefferson RA, Kavanagh TA, Bevan MW (1987) GUS fusion: β-Glucuronidase as sensitive and versatile gene fusion marker in higher plants. EMBO J 6: 3901–3907.

35. Kimura K, Yamaoka M, Kamisakar Y (2004) Rapid estimation of lipids in oleaginous fungi and yeasts using Nile red fluorescence. J Microbiol Methods 56: 331–338.

36. Bligh EG, Dyer WJ (1959) A rapid method of total lipid extraction and purification, Can J Biochem Physiol 37: 911–917.

37. Rogers CW, Challen MP, Green JR, Whipps JM (2004) Use of REMI and *Agrobacterium*-mediated transformation to identify pathogenicity mutants of the biocontrol fungus, *Coniothyrium minitans*. FEMS Microbiol Lett 241: 207–214.

38. Weld RJ, Eady CC, Ridgway HJ (2006) *Agrobacterium*-mediated transformation of *Sclerotinia sclerotiorum*. J Microbiol Methods 65: 202–207.

39. Betts MF, Tucker SL, Galadima N, Meng Y, Patel G, et al. (2007) Development of a high throughput transformation system for insertional mutagenesis in *Magnaporthe oryzae*. Fungal Genet Biol 44: 1035–1049.

40. Rho HS, Kang S, Lee YH (2001) *Agrobacterium tumefaciens*-mediated transformation of the plant pathogenic fungus, *Magnaporthe grisea*. Mol Cell 12: 407–411.

41. Mullins ED, Kang S (2001) Transformation: A tool for studying fungal pathogens of plants. Cell Mol Life Sci 58: 2043–2052.

42. Takahara H, Tsuji G, Kubo Y, Yamamoto M, Toyoda K, et al. (2004) *Agrobacterium tumefaciens*-mediated transformation as a tool for random mutagenesis of *Colletotrichum trifolii*. J Gen Plant Pathol 70: 93–96.

43. Tai M, Stephanopoulos G (2013) Engineering the push and pull of lipid biosynthesis in oleaginous yeast *Yarrowia lipolytica* for biofuel production. Metab Eng 15: 1–9.

Permissions

List of Contributors

Daria Zdżalik
Department of Biosciences, University of Oslo, Oslo, Norway
Institute of Genetics and Biotechnology, University of Warsaw, Warsaw, Poland

Cathrine B. Vågbø and Hans E. Krokan
Department of Cancer Research and Molecular Medicine, Norwegian University of Science and Technology, Trondheim, Norway

Finn Kirpekar
Department of Biochemistry and Molecular Biology, University of Southern Denmark, Odense, Denmark

Erna Davydova, Erwin van den Born and Pål Ø. Falnes
Department of Biosciences, University of Oslo, Oslo, Norway

Alicja Puścian
Institute of Genetics and Biotechnology, University of Warsaw, Warsaw, Poland
Department of Neurophysiology, Nencki Institute of Experimental Biology, Polish Academy of Sciences, Warsaw, Poland

Agnieszka M. Maciejewska
Institute of Biochemistry and Biophysics, Polish Academy of Sciences, Warsaw, Poland

Arne Klungland
Clinic for Diagnostics and Intervention and Institute of Medical Microbiology,
Oslo University Hospital, Rikshospitalet, Oslo, Norway
Institute of Basic Medical Sciences, University of Oslo, Oslo, Norway

Barbara Tudek
Institute of Genetics and Biotechnology, University of Warsaw, Warsaw, Poland
Institute of Biochemistry and Biophysics, Polish Academy of Sciences, Warsaw, Poland

Mat Yunus Abdul Masani, Ghulam Kadir Ahmad Parveez and Ravigadevi Sambanthamurthi
Advanced Biotechnology and Breeding Centre, Malaysian Palm Oil Board (MPOB), Kuala Lumpur, Malaysia

Gundula A. Noll
Westfälische Wilhelms-Universität Münster, Institut für Biologie und Biotechnologie der Pflanzen, Münster, Germany

Dirk Prüfer
Westfälische Wilhelms-Universität Münster, Institut für Biologie und Biotechnologie der Pflanzen, Münster, Germany
Fraunhofer Institut für Molekularbiologie und Angewandte Ökologie, Münster, Germany

David P. Janos and Catalina Aristizábal
Department of Biology, University of Miami, Coral Gables, Florida, United States of America

John Scott
Research Institute for the Environment and Livelihoods, Charles Darwin University, Darwin, Northern Territory, Australia

David M. J. S. Bowman
School of Plant Science, The University of Tasmania, Hobart, Tasmania, Australia

Kedong Xu, Kun Liu, Ju Zhang, Yi Zhang, Fuli Zhang, Guangxuan Tan and Chengwei Li
Key Laboratory of Plant Genetics and Molecular Breeding, Zhoukou Normal University, Zhoukou, People's Republic of China

Xiaohui Huang, Manman Wu, Yunxia Chang, Liming Yi, Tingting Li and Ruiyue Wang
Department of Life Science, Zhoukou Normal University, Zhoukou, People's Republic of China

Yan Wang
Key Laboratory of Plant Genetics and Molecular Breeding, Zhoukou Normal University, Zhoukou, People's Republic of China
College of Life Science, Henan Agricultural University, Zhengzhou, People's Republic of China

Claire M. Smith, Alexandra J. F. Patel, Sarah Glenn and Peter W. Andrew
Department of Infection, Immunity and Inflammation, University of Leicester, Leicester, Leicestershire, United Kingdom

Stephen C. Fry
The Edinburgh Cell Wall Group, Institute of Molecular Plant Sciences, School of Biological Sciences, University of Edinburgh, Edinburgh, United Kingdom

Kevin C. Gough
School of Veterinary Medicine and Science, University of Nottingham, Sutton Bonington Campus, Leicestershire, United Kingdom

Marie Goldrick and Ian S. Roberts
Faculty of Life Sciences, University of Manchester, Manchester, United Kingdom

Garry C. Whitelam
Department of Biology, University of Leicester, Leicester, Leicestershire, United Kingdom

Pu Yan
South China Botanical Garden, Chinese Academy of Sciences, Guangzhou, China
Institute of Tropical Bioscience and Biotechnology, Chinese Academy of Tropical Agricultural Science, Haikou, China
Graduate University of Chinese Academy of Sciences, Beijing, China

Wentao Shen, Xiaoying Li and Peng Zhou
Institute of Tropical Bioscience and Biotechnology, Chinese Academy of Tropical Agricultural Science, Haikou, China

Xin Zheng Gao
Department of Basic Medical Science, Hainan Medical College, Haikou, China

Jun Duan
South China Botanical Garden, Chinese Academy of Sciences, Guangzhou, China

Huili Li, Kebo Xie and Shuning Wang
State Key Laboratory of Microbial Technology, Shandong University, Jinan, PR China

Haiyan Huang
Institute of Basic Medicine, Shandong Academy of Medical Science, Jinan, PR China

Yachun Su, Jinlong Guo, Hui Ling, Shanshan Chen, Shanshan Wang, Liping Xu and Youxiong Que
Key Laboratory of Sugarcane Biology and Genetic Breeding, Ministry of Agriculture, Fujian Agriculture and Forestry University, Fuzhou, Fujian, China

Andrew C. Allan
The New Zealand Institute for Plant and Food Research, Sandringham, Auckland, New Zealand

Matías D. Asención Diez
Department of Chemistry, Loyola University Chicago, Chicago, Illinois, United States of America
Laboratorio de Enzimología Molecular, Instituto de Agrobiotecnología del Litoral (UNL-CONICET), FBCB Ciudad Universitaria, Santa Fe, Argentina

Mabel C. Aleanzi and Alberto A. Iglesias
Laboratorio de Enzimología Molecular, Instituto de Agrobiotecnología del Litoral (UNL-CONICET), FBCB Ciudad Universitaria, Santa Fe, Argentina

Miguel A. Ballicora
Department of Chemistry, Loyola University Chicago, Chicago, Illinois, United States of America

François Le Tacon
INRA, UMR 1136, Interactions Arbres/ Microorganismes (IAM), Centre INRA de Nancy, Champenoux, France
Université de Lorraine, UMR 1136, Interactions Arbres/ Microorganismes (IAM), Faculté des Sciences, Vandoeuvre les Nancy, France

Bernd Zeller
INRA, UR 1138, Biogéochimie des Ecosystèmes Forestiers (BEF), Centre INRA de Nancy, Champenoux, France

Caroline Plain, Christian Hossann and Claude Bréchet
INRA, UMR 1137, Ecologie et Ecophysiologie Forestières (EEF), Centre INRA de Nancy, Champenoux, France
Université de Lorraine, UMR 1137, Ecologie et Ecophysiologie Forestières (EEF), Faculté des Sciences, Vandoeuvre les Nancy, France

Christophe Robin
Université de Lorraine, UMR 1121 « Agronomie & Environnement » Nancy-Colmar, Vandoeuvre les Nancy, France
INRA, UMR 1121 « Agronomie & Environnement » Nancy-Colmar, Centre INRA de Nancy, Vandoeuvre les Nancy, France

Dipak Kumar Sahoo and Indu Bhushan Maiti
KTRDC, College of Agriculture, Food and Environment, University of Kentucky, Lexington, Kentucky, United States of America

Nrisingha Dey
Department of Gene Function and Regulation, Institute of Life Sciences, Bhubaneswar, Odisha, India

Ryan Kessens, Tom Ashfield, Sang Hee Kim and Roger W. Innes
Department of Biology, Indiana University, Bloomington, Indiana, United States of America

Anna Pietraszewska-Bogiel, Maria A. Koini and Theodorus W.J. Gadella
Section of Molecular Cytology, Swammerdam Institute for Life Sciences, University of Amsterdam, Amsterdam, The Netherlands

Benoit Lefebvre, Dörte Klaus-Heisen and Julie V. Cullimore
INRA, Laboratoire des Interactions Plantes-Microorganismes (LIPM), UMR441, F-31326 Castanet-Tolosan, France
CNRS, Laboratoire des Interactions Plantes-Microorganismes (LIPM), UMR2594, F-31326 Castanet-Tolosan, France

Frank L. W. Takken
Section of Plant Pathology, Swammerdam Institute for Life Sciences, University of Amsterdam, Amsterdam, The Netherlands

René Geurts
Department of Plant Science, Laboratory of Molecular Biology, Wageningen University, Wageningen, The Netherlands

Jennifer N. Bragg and Jiajie Wu
United States Department of Agriculture- Agriculture Research Service (USDA-ARS), Western Regional Research Center, Albany, California, United States of America
University of California Davis, Davis, California, United States of America

Sean P. Gordon, Mara E. Guttman, Roger Thilmony, Gerard R. Lazo, Yong Q. Gu and John P. Vogel
United States Department of Agriculture- Agriculture Research Service (USDA-ARS), Western Regional Research Center, Albany, California, United States of America

Hiroaki Mano and Naomi Sumikawa
Division of Evolutionary Biology, National Institute for Basic Biology, Okazaki, Japan

Yuji Hiwatashi and Mitsuyasu Hasebe
Division of Evolutionary Biology, National Institute for Basic Biology, Okazaki, Japan
School of Life Science, Graduate University for Advanced Studies, Okazaki, Japan

Tomomi Fujii
School of Life Science, Graduate University for Advanced Studies, Okazaki, Japan

M. Ashraful Islam, Sissel Haugslien, Dag-Ragnar Blystad and Jihong Liu Clarke
Bioforsk - Norwegian Institute for Agricultural and Environmental Research, Ås, Norway

Henrik Lütken and Søren K Rasmussen
Department of Plant and Environmental Sciences, Faculty of Science, University of Copenhagen, Frederiksberg, Denmark

Sissel Torre and Jorunn E Olsen
Department of Plant and Environmental Sciences, Norwegian University of Life Sciences, Ås, Norway

Jakub Rolcik
Palacky University, Olomouc, Czech Republic

Ibrahim Njimona, Rui Yang and Tilman Lamparter
Karlsruhe Institute of Technology KIT, Botanical Institute, Karlsruhe, Germany

Sean Chapman, Brian Harrower and Kara McGeachy
Cell and Molecular Sciences, James Hutton Institute, Invergowrie-Dundee, United Kingdom

Laura J. Stevens, Pauline S. M. Van Weymers and Paul R. J. Birch
Cell and Molecular Sciences, James Hutton Institute, Invergowrie-Dundee, United Kingdom
Division of Plant Sciences, University of Dundee at James Hutton Institute, Invergowrie-Dundee, United Kingdom
Dundee Effector Consortium, Invergowrie-Dundee, United Kingdom

Petra C. Boevink, Xinwei Chen and Ingo Hein
Cell and Molecular Sciences, James Hutton Institute, Invergowrie-Dundee, United Kingdom
Dundee Effector Consortium, Invergowrie-Dundee, United Kingdom

Stefan Engelhardt
Division of Plant Sciences, University of Dundee at James Hutton Institute, Invergowrie-Dundee, United Kingdom
Dundee Effector Consortium, Invergowrie-Dundee, United Kingdom

Colin J. Alexander
Biomathematics and Statistics Scotland, Invergowrie-Dundee, United Kingdom

Nicolas Champouret
J.R. Simplot Company, Simplot Plant Sciences, Boise, Idaho, United States of America

Philip Möller, Aaron Overlöper and Franz Narberhaus
Microbial Biology, Ruhr University Bochum, Bochum, Germany

Konrad U. Förstner and Cynthia M. Sharma
Research Center for Infectious Diseases (ZINF), Julius-Maximilian's University of Würzburg, Würzburg, Germany

Tuan-Nan Wen and Erh-Min Lai
Institute of Plant and Microbial Biology, Academia Sinica, Taipei, Taiwan

Helong Chen, Caiping Hu, Jianming Gao, Shiqing Zhang and Qiaolian Liu
Institute of Tropical Bioscience and Biotechnology, Key Laboratory of Tropical Crop Biotechnology, Ministry of Agriculture, Chinese Academy of Tropical Agricultural Sciences, Haikou, China

Kexian Yi
Institute of Tropical Bioscience and Biotechnology, Key Laboratory of Tropical Crop Biotechnology, Ministry of Agriculture, Chinese Academy of Tropical Agricultural Sciences, Haikou, China
Environment and Plant Protection Institute, Key Laboratory of Integrated Pest Management on Tropical Crops, Ministry of Agriculture, Chinese Academy of Tropical Agricultural Sciences, Haikou, China

Jinlong Zheng, Jingen Xi and Guixiu Huang
Environment and Plant Protection Institute, Key Laboratory of Integrated Pest Management on Tropical Crops, Ministry of Agriculture, Chinese Academy of Tropical Agricultural Sciences, Haikou, China

Prabuddha Dey and Monami Chakraborty
Adv. Lab. for Plant Genetic Engineering, Advanced Technology Development Center, Indian Institute of Technology Kharagpur, Kharagpur, India

Nikunj Mall and Atrayee Chattopadhyay
Department of Biotechnology, Indian Institute of Technology Kharagpur, Kharagpur, India

Mrinal K. Maiti
Adv. Lab. for Plant Genetic Engineering, Advanced Technology Development Center, Indian Institute of Technology Kharagpur, Kharagpur, India
Department of Biotechnology, Indian Institute of Technology Kharagpur, Kharagpur, India

Index

A

Acetosyringone, 37, 57, 69-70, 117-118, 148-149, 151, 153, 155, 188-190, 202, 206, 214, 221

Acyltransferase, 216-217, 228

Adenosine, 85, 187

Agricultural, 28, 36, 39, 41, 51, 111, 133, 158, 202, 204-205, 216

Agrobacterium, 1, 3-4, 8-9, 14, 17, 20, 23, 36-44, 47-49, 52-60, 66, 68-72, 74-75, 85, 95-99, 101-107, 111-115, 118-119, 121-128, 131, 133-134, 136-137, 139, 145-149, 151, 153-159, 161, 166-168, 173-174, 177-178, 180-182, 184-189, 191, 193, 198, 200-204, 206, 208, 211, 213-215, 217-221, 228

Agroinfiltrated, 40, 70, 72, 104, 108, 110-111

Ampicillin, 4, 48, 62, 70, 95-96, 98-99, 108, 169

Amylovora, 180, 185, 201

Annealing, 47, 51, 138, 200, 206

Anthracnose, 202, 204, 213-215

Appressorium, 202, 207, 209-210, 213

Arabidopsis, 3, 5, 12, 36-37, 40-42, 49, 68, 77, 95, 97, 99, 102-107, 109-114, 116, 119-122, 128, 131-134, 139, 142, 145-146, 156-159, 162-168, 170, 173-175, 186, 200, 228

Autophosphorylation, 126-127, 129-130, 132, 168-171, 173-174

B

Bacteriophytochrome, 173-174

Bacterium, 4-5, 8-9, 11, 59, 64, 85, 205, 213

Baulcombe, 58, 111, 157, 186

Benzylaminopurine, 37, 150

Bimolecular, 41, 58, 177, 186

Biodiesel, 216-217, 224, 227

Biogenesis, 12, 145, 195, 197-198

Biosynthesis, 9, 14, 41, 47, 49, 65-66, 85, 111, 145, 158-159, 163-164, 166-167, 188, 197, 200, 216-217, 224, 227-228

Brachypodium, 69, 133-134, 136-146

C

Carbohydrates, 86, 94, 120, 195

Carboxylase, 98, 112, 217, 226, 228

Catabolism, 50, 60, 66

Cefotaxime, 48, 102, 148

Cerevisiae, 2, 4, 9-10, 217, 226, 228

Chaperone, 132, 186-187, 198-201

Chemotaxis, 174, 195, 197, 199

Chlorimuronethyl, 203-206, 208-209

Chloroacetaldehyde, 3, 6, 12-13

Chromosome, 46, 77, 95, 112, 133, 141-143, 145-146, 188-189, 191, 197, 200, 203

Colletotrichum, 121, 202, 204-206, 208-210, 213-222, 224, 226-228

Cotyledonary, 147-151, 153-157

Cytoplasm, 16, 20, 23, 67, 71-72, 74, 76, 176-177, 181, 184-186

D

Decarboxylation, 1, 3, 59, 66

Degradation, 33, 50, 59-60, 62, 64-66, 77, 116, 187

Dehydrogenase, 49, 59-60, 62, 64-66, 195

Demethylation, 12, 59

Denaturation, 47, 173

Dephosphorylation, 147, 168

Diacylglycerol, 131, 216-217, 228

Distachyon, 69, 71, 133, 141-142, 145-146

Dna Microinjection, 14, 16, 19-20, 22, 24

Dna Repair, 1-4, 8, 10, 12-13

E

Ectomycorrhizas, 25, 29, 32, 34, 86, 92

Electrophoresis, 23, 39, 43-45, 48-50, 52, 61, 69, 79, 160, 170, 189-190, 195, 218

Electroporation, 4, 95, 97, 99, 109, 148

Embryogenic, 14, 16, 20, 23, 42, 134, 136-138, 159

Endophytic, 216-217, 222, 224, 226-227

Escherichia, 1, 12, 39, 49-50, 66-68, 70, 76-78, 85, 95, 98-99, 108, 111, 119, 169, 187, 200-201, 217, 228

Ethenoadenine, 4, 6, 12-13

Eucalyptus, 25-27, 29-35

G

Gibberellin, 158-159, 163, 166-167

Gloeosporiodes, 204-206, 208

Glucuronidase, 58, 95, 97, 108, 111, 114, 118-119, 140, 148, 156, 228

Glutinosa, 25, 27-32, 34, 113-119

H

Heterologous, 43, 47, 49, 62, 105, 111, 120-122, 126, 130, 200, 222, 226

Holoprotein, 168-171, 173-174

Homeostasis, 42, 110, 134, 146, 166, 200

Homozygous, 133, 154-155, 176, 178, 181-183

Hybridization, 41, 68, 112, 145, 150, 160-161, 200, 206, 218-222

Hydrolysis, 12, 44, 48, 79, 92-93

Hypersensitive, 72, 77, 113, 119, 122, 131, 176, 185-186

Hypersensitivity, 1, 186

I

Intracellular, 67, 82-83, 85, 119-121, 126, 130-131, 147, 156, 201, 217

Irradiance, 160-161, 164

Isolation, 4, 16, 19-20, 23-24, 42, 48, 57, 62, 67, 69, 77, 85, 97, 99, 112, 146, 156, 160, 167, 174, 186, 190-193, 196, 205, 214, 221

L

Lipopolysaccharide, 197, 200

M

Magnaporthe, 202, 210, 214-215, 228

Mammalian, 1-3, 9-13, 58, 166

Maturation, 87, 89-90, 132

Membranes, 76-77, 160, 189, 195, 218, 221

Mesophyll, 24, 36-38, 40

Metabolite, 13, 78, 81, 83, 85, 131, 217, 224, 226

Methylated, 3, 6, 12-13

Methylumbelliferone, 104, 108, 118

Microsystems, 22-23, 97, 178

Mirabilis, 95-96, 98, 105, 111

Monoclonal, 189, 195-196

Monooxygenase, 61, 65-66, 167

Morphology, 114, 140, 165, 207, 209, 226

Mycorrhizal, 25-35, 89, 92-94, 131

N

Naphthaleneacetic, 148, 150, 156

Nicotiana, 24, 37, 39, 44, 48, 55, 57, 68, 72, 75, 77, 97, 111, 113-131, 176, 186

Nitrocellulose, 218, 221

Nucleotide, 5, 8, 11, 44, 48, 64, 76-77, 112-113, 119, 176, 178, 184-187, 190, 197, 210

Nyctinastic, 147, 153

O

Oil Palm Protoplasts, 14, 16-20, 23

Oligonucleotides, 4, 6, 11, 179, 189, 199

Oxoglutarate, 1, 3-6, 12-13

P

Pathogenesis, 49, 77, 129, 131, 202-204

Peronospora, 109, 111-112

Phaseolus, 119, 155, 157, 166

Phosphorus, 26, 31, 33-35, 87, 216, 222-224, 226

Phosphorylation, 113, 119-120, 128-131, 156, 168-171

Photosynthesis, 91, 94, 163

Phylogenetic, 69, 71-72, 117, 146, 175, 183

Phytohormone, 149-150, 188

Planta Transient Transformation, 36

Pneumococcal, 43-49

Polyadenylation, 108, 187, 200

Polysaccharide, 43-49, 78, 85

Prokaryotes, 3, 70, 74, 174, 201

Protoplast, 16, 19-24, 36-37, 41-42, 95, 97, 100-101, 104-105, 109, 203, 213-214

Protoplast Transfection, 36-37

Protozoan, 1, 3-4, 9-11

Pseudomonas, 59-60, 65-66, 113, 119, 131, 174-175

R

Recalcitrant, 20, 147, 155

Rhizobium, 3, 5, 111, 120, 130-131, 188

Rhizosphere, 34, 60, 87, 93-94, 198

Ribosomal, 88, 142, 197-198, 214

S

Saprotrophic, 27, 87, 93

Scitamineum, 67-69, 71, 74, 76-77

Segregation, 97, 112, 153, 155-156, 216

Spectrometry, 4, 8, 12, 64, 197, 200

Stylosanthes, 202, 204, 213-214

Succinoylpyridine, 59-60, 66

Symbiotic, 93, 111, 120, 122, 126, 128-131, 188, 200-201

T

Thermophila, 3-4, 7, 9, 12

Thiogalactoside, 70, 73, 79

Transcriptome, 130-131, 187, 199

Trna Modification, 1, 4, 9-11

www.ingramcontent.com/pod-product-compliance
Lightning Source LLC
Chambersburg PA
CBHW080412190526
45161CB00003B/210